Ultrasonic Instruments and Devices I

Reference for Modern Instrumentation, Techniques, and Technology

PHYSICAL ACOUSTICS: Principles and Methods

Volume XXIII

CONTRIBUTORS TO VOLUME XXIII

ARTHUR BALLATO
AARON J. GELLMAN
NEIL J. GOLDFINE
ALBERT GOLDSTEIN
ROBERT S. HARRIS
WILLIAM LORD
LAWRENCE C. LYNNWORTH
VALENTIN MAGORI
EMMANUEL P. PAPADAKIS
RAYMOND L. POWIS
STEPHEN R. RINGLEE
RICHARD STERN
SATISH UDPA

Ultrasonic Instruments and Devices I

Reference for Modern Instrumentation, Techniques, and Technology

Edited by R. N. THURSTON

BELL COMMUNICATIONS RESEARCH, INC.
RED BANK, NEW JERSEY

ALLAN D. PIERCE

PENNSYLVANIA STATE UNIVERSITY
UNIVERSITY PARK, PENNSYLVANIA

Volume Editor EMMANUEL P. PAPADAKIS

QUALITY SYSTEMS CONCEPTS, INC,
NEW HOLLAND, PENNSYLVANIA

PHYSICAL ACOUSTICS

Volume XXIII

ACADEMIC PRESS

San Diego London Boston
New York Sydney Tokyo Toronto

This book is printed on acid-free paper ∞

ACADEMIC PRESS
A Division of Harcourt Brace and Company
525 B Street, Suite 1900, San Diego, CA 92101-4495, USA
http://www.apnet.com

ACADEMIC PRESS
24–28 Oval Road, London NW1 7DX
http://www.hbuk/co.uk/ap/

International Standard Series Number: 0893-388X

International Standard Book Number: 0-12-477923-9

PRINTED IN THE UNITED STATES OF AMERICA
98 99 00 01 02 DS 9 8 7 6 5 4 3 2 1

Contents

3

Nondestructive Testing

Emmanuel P. Papadakis

4

Industrial Process Control Sensors and Systems

Lawrence C. Lynnworth and Valentin Mágori

Contributors

Numbers in parentheses indicate the pages on which the authors' contributions begin.

ARTHUR BALLATO (33)
U.S. Army CECOM
Fort Monmouth, NJ 07703-5201

AARON J. GELLMAN (1)
Northwestern University
Evanston, IL 60208

NEIL J. GOLDFINE (15)
JENTEK Sensors, Inc.
Watertown, MA 02172

ALBERT GOLDSTEIN (43)
Detroit Receiving Hospital
Detroit, MI 48201

R.S. HARRIS (24)
Iowa State University
Ames, IA 50011

W. LORD (24)
Iowa State University
Ames, IA 50011

LAWRENCE C. LYNNWORTH (275)
Panametrics, Inc.
Waltham, MA 02154

VALENTIN MAGORI (275)
Siemens AG, Munchen
Germany

EMMANUEL P. PAPADAKIS (7, 193)
Quality Systems Concepts, Inc.
New Holland, PA 17557

RAYMOND L. POWIS (43)
Redmont, WA 98053

STEPHEN R. RINGLEE (20)
E-Markets, Inc.
Ames, IA 50010

RICHARD A. STERN (33)
U.S. Army CECOM
Fort Monmouth, NJ 07703-5201

S. UDPA (24)
Iowa State University
Ames, IA 50011

Preface

The purpose of this book is to show examples of the successful commercialization of devices and instruments arising from research in ultrasonics carried out over previous years. Much of the research has been reported (in the research stage and in the mode of research reports) in earlier volumes of this series, *Physical Acoustics: Principles and Methods*. Basically, there is progression from idea through research, development, technology transfer, and commercialization to application or use by a set of customers. The "Water Slide" diagram in Figure 1 illustrates this progression (Papadakis,

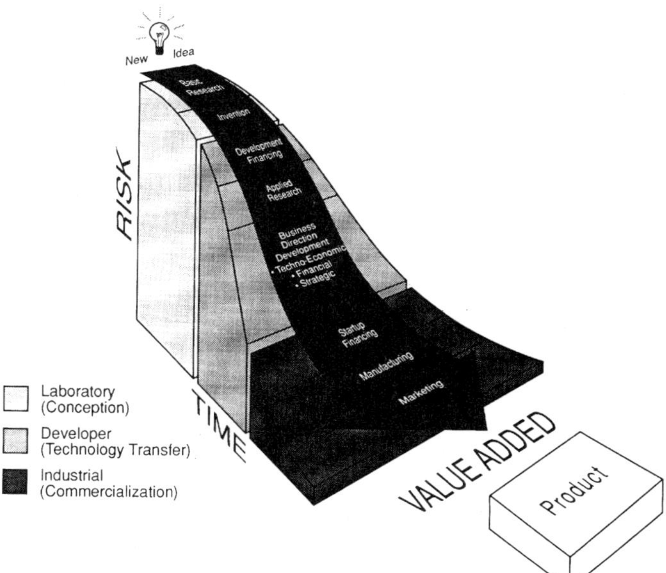

Figure 1 "Water Slide" diagram showing risk as a function of time as more value is added step by step to an idea to bring it to the status of a product. Because an idea may be rejected at any step of the process or may fail in the field with customers, the number of products is much smaller than the number of ideas initially. (*Materials Evaluation*. Used by permission.)

1992). The purpose here is to demonstrate that research in physical acoustics has led to successful commercialization of devices and systems useful to the public in the broad sense. The basic idea is that research, development, technology transfer, commercialization, and sales are part of a "food chain," so to speak, in which each step is necessary and all steps are interdependent.

Research can lead to development if the results of the research seem potentially useful. A conscious decision must be made by someone with resources to enter the development stage. Some developments with real utility go on to technology transfer, which means that the developed item is turned over to a user organization and actually employed. However, technology transfer has been defined as successful only when the transfer is a financial transaction between a buyer who is ready, willing, and able to buy and a seller who is ready, willing, and able to sell — just as in buying a house (See Professor Gellman's essay). When either the buyer or the seller is a "captive audience," technology transfer has not yet happened. For instance, a prototype handed over as the "deliverable" on a sole-source contract does not constitute technology transfer. Neither does the installation in a factory of a system developed by the R&D department of the same company—unless the factory's management has the power to refuse instead of just an obligation to accept. The idea of the captive audience and the consequence of incomplete technology transfer does not deprecate in any way the quality of the science and engineering that went into the development, nor does it mean that the item is not actually used.

In this book, the meaning of "commercialization" goes two steps further. Technology transfer is not commercialization even if money changes hands for the delivery of the first copy of an item. Even the sale of the second copy of the item is not defined as commercialization for the purpose of this book. This treatise does not deal in the delivery of only one or two copies of an item. Instead, commercialization is defined as the sale of three or more copies in arm's-length transactions. Subsequent numbers may be subject to modification, improvements, or customization, but the principle is there.

The idea of the captive audience is superseded when a vertically integrated organization uses many copies of an internally developed item. Then it is implicitly assumed that the organization would have carried out a "buy-make" decision vis-à-vis competitive items before it actually chooses its own. In a case of thorough-going vertical integration, a policy decision to buy only internally might have been made. Commercialization could still happen. The most well-known examples of a commercially successful firm with total vertical integration was the Bell Telephone System before the divestiture ruling. That firm as a regulated monopoly commercialized items (telephones) by renting without even selling them. It is still arguable that such regulated monopolies are the best type of economic organization. Every Sophomore economics book offers a proof that the opposite—namely, perfect competition—leads to zero economic profit, instability, bankruptcies, and a return to monopoly as only the strongest survives (see, for instance, Samuelson & Nordhaus, 1989).

However, economics per se is not the subject of this book. The concern here is with developed items that have been commercialized by selling three or more copies.

The seller may or may not be the inventor. Sometimes an inventor also has business acumen and starts a company to reap the benefits of the invention. Often the invention is transferred to another organization for sale. In the present era of downsizing and outsourcing, progressively fewer items are invented, developed, and utilized internally. Commercialization involving sales on the open market with competition, not sole-sourcing, is very relevant.

In Chapter 1, several authors address the processes of technology transfer and commercialization from the point of view of "how-to" and successful examples. This chapter introduces the concepts and points out difficulties. Following the lead chapter, there are chapters on various classes of ultrasonic devices and systems that have come to fruition. Included are medical ultrasonic diagnostics, nondestructive testing (NDT), process control, surface acoustic wave (SAW) devices, frequency control devices, research instruments, transducers, and ultrasonic microscopes. The exact title of any chapter may vary from this list a bit. The chapters are liberally illustrated with pictures of actual commercial objects that are or have been in use.

The list is not all-inclusive; this is a book and not an encyclopedia. One may object, for instance, that bulk wave delay lines are not given a chapter. One reason is that they were very well covered by John May in Volume 1A of this Physical Acoustics Series. A second reason is that they were subject to the captive audience phenomenon. Almost all were produced *ad hoc* for some government project in radar or sonar, although a few were utilized as volatile memories in early computers such as Univac. The third reason is that large-scale commercialization did not happen since the development of silicon technology for memories occurred just as the bulk wave delay line was poised to enter that commercial market.

Two chapters that were under consideration are not included. Their subjects are mentioned here for completeness. One is acoustic emission, which depends on the phenomenon of the generation of sonic and ultrasonic waves as a crack propagates. Other sources such as pressurized gas leaks also emit acoustic emission. The reader is probably familiar with audible acoustic emission from an ice cube right out of the freezer when it cracks as it is put into a beverage. Acoustic emission is often classed under nondestructive testing because materials and structures can be tested under stress (not destructive in extent) to determine by "listening" whether cracks propagate. Some monorails and amusement rides, among other things, are tested by acoustic emission when loaded with sandbags. Cracks that are acoustic emission sources can be located by triangulation with appropriate instrumentation. The second potential subject not covered is the uses of high-intensity ultrasound. This technology is used from medicine (to break up gallstones), to automobiles (to weld plastics into multicolored tail lights).

No single chapter is all-inclusive in its coverage of all inventions, all scientists, or all manufacturers in its domain. Inclusion of manufactured items as examples in a chapter should be taken as paradigms, not as recommendations for the items or as slights for other items not shown. The book is not a catalog of available merchandise.

We hope that this book will show the present success of much past research and will assist in the process of bringing research output into the marketplace, to the benefit of customers.

EMMANUEL P. PAPADAKIS
August 1998

References

Papadakis, E. P. (1992). Research and real world relationships. *Materials Evaluation* **50**(3), 352.
Samuelson, P. A., and Nordhaus, W. D. (1989). "Economics", 13th Edition. McGraw-Hill, New York.

— 1 —

The Process of Technology Transfer and Commercialization

Essay I

Achieving Successful Technology Transfer

AARON J. GELLMAN

Northwestern University, Evanston, IL 60208

Introduction

'Technology transfer' has become a popular phrase and a subject of great interest in myriad quarters. Not surprisingly, it has taken on various meanings. To consider it in any detail in the space available, the concept must be bound in several ways. First, technology transfer can be internal—that is, within the same enterprise (e.g., between the corporate R&D organization and an operating profit center). It can also be external (e.g., from one firm to another through, say, a licensing or joint venture arrangement). Second, it is assumed that external transfers follow strictly arm's-length negotiations. Third, presumably all transfers are undertaken with the expectation that the technology will

ISBN 0-12-477923-9 $30.00

be utilized in the market, again as the result of arm's-length bargaining. Also, successful transfer requires diffusion of the technology (or of what it can produce) and not just market introduction. It therefore follows that the interest of this paper is actually "technology transfer *and utilization*" (TTU).

Technology Transfer and the Process of Innovation

'Technology transfer' is but the outcome of a process called innovation. This process begins with an invention, an idea, or a concept and concludes with the introduction of a product or service in the marketplace on the basis of an arm's-length transaction. Joseph A. Schumpeter, the first to appreciate the importance of innovation for economies and societies, ultimately concluded that innovation can best be defined as "something newly tried." "Something" can be a product or a process — in modern vernacular, hardware or software, product or service. "Newly" refers to the market in which the "something" is to be advanced. Innovation is market-specific. The same product introduced in a different market represents the culmination of a separate process of innovation. "Tried" conveys that the innovative product or service is not a test article or prototype but rather the practical manifestation of a product or service, even if the underlying technology has many prior contemporaneous and subsequent deployments.

Much innovation takes place through the transfer of technology. As noted, such transfer can occur within an organization or involve different organizations (e.g., sellers and buyers). In a large enterprise, a technology or technique can be transferred between, say, its central R&D facility and a profit center unit of the firm. Or a firm (or government agency) can transfer technology by licensing it to another enterprise, public or private, in the same geographical area or in another. For the most part, this chapter will deal only with external transfers (although there are many attributes shared by the two types of transfer).

Supply-Push and Demand-Pull

For many elements of the process of innovation, from invention to market introduction, the driving force is either supply-push or demand-pull. Supply-push can be characterized as "I have, don't you need?" while demand-pull is reflected in "I need, don't you have?" Without doubt, the demand-pull force is the stronger of the two for moving technology into the marketplace through innovation. It is important to recognize that when considering technology transfer, the transferor represents supply-push and the transferee represents demand-pull. Thus it is far better when a potential transferee approaches an enterprise asking for help. (This is the case within firms as well as between enterprises.)

Often technology transfer is easier to achieve once the innovation has proved itself in some market. Finding a licensee or a joint-venture partner in another country or market becomes less difficult under such conditions. Put another way, achieving technology transfer is generally more challenging when supply-push is the primary force than when demand-pull is at work. But once success results from a supply-push effort, demand-pull becomes easier to galvanize for subsequent innovations. Depending on the nature of the market for a technology, it can be easier or more difficult to link supply-push and demand-pull in this way. For example, if there is a highly sophisticated, highly aggregated market to be served through transferring technology, supply-push has a higher probability of working than in markets at the other extreme of sophistication and aggregation. (This is one of the reasons that technology transfer has often been very successful when the transfer is to large government agencies or to very large firms producing a range of technology-intensive products and services.)

Notwithstanding the preceding, technology transfer is most readily accomplished through the exploitation of a demand-pull force. One of the more effective ways to generate such a force is to look for public or private enterprises that publish performance specifications for their inputs rather than design specifications. Performance specifications open the door to technology transfer and innovation in a way not otherwise possible. Indeed, if an enterprise lives or dies on the basis of the success it enjoys in transferring technology, that enterprise may well be wise to induce the target market to switch to performance specifications.

Promoting Technology Transfer

Achieving efficient, profitable technology transfer requires recognition of many of the fundamental "facts of life" regarding innovation and, therefore, of technology transfer. For example, one of the most effective ways to promote innovation and technology transfer and to achieve market diffusion beyond market introduction is to find a *champion* for the innovation or technology.

Innovation and technology transfer are people-processes; no matter how technology-dependent, no matter how technologically sophisticated, at base these processes must involve people who will be put in positions either to promote or to thwart them. Among the more powerful "people forces" available for advancing technology through its transfer to different settings is the champion for the technology or innovation. In fact, in most instances of technology transfer there is need for a champion from the originating enterprise as well as one in every transferee enterprise. Without these very special people, innovation is very difficult, if not impossible, to achieve based upon technology transfer.

But it is not the champions alone that matter. All along the path of technological innovation and transfer (and even diffusion) there are individuals who at one time or another can promote or obstruct, depending on how they are managed. It should always be recognized that it is easier to "prove" a cost associated with any given proposed transfer than it is to calculate the benefits that can be expected. This alone is sufficient to underscore the leverage individuals possess where innovation and technology transfer are concerned.

Mechanisms and Catalysts

The mechanisms and catalysts supporting the external transfer of technology fall into two categories: (1) overt and explicit mechanisms and (2) indirect or covert mechanisms.

Technology can be transferred to another entity through the outright sale of such technology or through a once-for-all payment that transfers title or the right to use such technology in all markets or in defined markets only. Then there is the licensing of technology through which the provider of the technology receives payment in one or a combination of forms, some of which are usually based on the market success of the transferee. Again, joint ventures can be a means for such transfer where the technology itself forms all or part of the equity of the transferor; similarly, a wholly-owned foreign subsidiary can be established explicitly to receive and exploit a technology.

Somewhat more complicated is transfer through the relationship that a prime bears to its suppliers of inputs. For example, the producer of a highly complex and technologically sophisticated product will often have developed designs and manufacturing techniques for components which are to be supplied by firms other than itself. Under such circumstances technology is transferred down the chain of supply rather than horizontally.

One of the more effective indirect or covert mechanisms for technology transfer has been patent documentation. Such documentation has proved to be highly catalytic for technology transfer in many cases. It is especially effective where the transfer is between different countries, given the great expense the unwilling (and probably unwitting) transferor must bear in order to pursue the matter in court. And, of course, there is industrial espionage, which everyone knows is quite ubiquitous but few are willing to discuss.

International setting of standards for products and processes often results in unintended transfers of technology. While the social benefits of such transfers may be substantial, individual generators of the technological possibilities that are plundered certainly suffer economic harm. Again, in such circumstances, it is usually very difficult for firms to pursue the matter given the cost and other constraints playing on the scene.

Reverse engineering is a time-honored if morally reprehensible mechanism for technology transfer. Over many years, even decades, some countries' economic performance has been substantially based on successful and unauthorized reverse engineering of products from other countries.

More of an indirect than a covert mechanism of technology transfer is the exchange of industrial personnel between firms in different countries. The exchange of academic faculty between universities can produce a similar result as can students pursuing studies abroad. There are a number of well-documented cases where exceptional graduate students from developing countries (and even those from developed ones) have taken home with them not only a diploma but also some commercially-valuable ideas based on scientific outcomes and technological possibilities they picked up while abroad.

Diplomats serving as commercial or scientific attachés have often been a source of intellectual capital leading to technology transfer back to their home countries. Rarely have such people been engaged in industrial espionage, but it has happened. Still, in most instances they operate legitimately but are nonetheless frequently invaluable conduits for the international transfer of technology.

Surprisingly, perhaps one of the most effective technology transfer mechanisms is the open literature, including scientific and engineering publications and trade journals. The power of technological intelligence derived strictly through such means to influence the course of technological innovation in a country or industry has been demonstrated time and time again.

Some Concluding Observations

Technology transfer, as technological innovation itself, faces many resistances. It is better to recognize and understand such resistances than pretend they do not exist. First among them is the general resistance to change that is found universally in both organizations and individuals. Innovation is always an uphill battle and so is technology transfer. And natural resistance to change is the primary reason why this must always be the case. Both the successful champion for a technology and the skilled technology transfer agent learn how to nullify or overcome such resistance.

It is more difficult to transfer technology where there is a need for system integration as contrasted with a technology or innovation that can be introduced on a stand-alone basis. Using the railroad industry as an example, it has been extremely difficult to introduce a new braking system for railway freightcars in North America because of the necessity to interchange freight equipment freely throughout the continent over many different railroads. Consequently, a system that was adopted more than half a century ago remains the basic standard today; the possibility that one can transition to a new and better form of braking without disrupting the freeflow of cars throughout the system is only now emerging. Had advanced braking systems been applicable to the fleet in a drawn-out manner, such technology transfer would have taken place many decades ago.

Another considerable barrier arises when a firm has to write off capital investment remaining on the books to exploit a transferred technology. Firms do not like to take capital losses; some even forbid it as a matter of policy. Thus they may not embrace new technology in many cases as quickly they should for their own benefit.

Insufficient data and information about the technology involved presents another barrier to transfer. Both technical and economic data and information must be adequate to support the case for the technology and to overcome the general resistance to change previously discussed.

An exceptionally powerful resistance comes from the strong propensity to avoid risk that is characteristic of many firms and some industries. Intelligent risk-bearing is too scarce in many firms and governments; this slows technological advance, which, of course, reduces opportunities for technology transfer. Perhaps only education can overcome this particular resistance and therefore incorporating material that stresses the value and importance of technological advance in appropriate academic curricula may be a matter for priority consideration.

Finally, market structure extremes present problems for technology transfer and innovation. The "pure" competitor does not have the resources to pursue technological advance even when the technology is available for transfer at no cost (which is usually not the case). At this end of the competitive spectrum, there simply are no excess funds. Such firms are operating at a subsistence level and cannot invest in innovation. Toward the monopoly end of the spectrum, firms generate increasing profits and could, if they so chose, deploy some of those profits to acquire or develop technology in the quest to make even more money. When the extreme is reached, however, the monopolist is so successful that it usually sees no reason to receive or develop technology with which to innovate. Although these remarks about the two extremes may be somewhat overdrawn, they do make the point that the most likely candidates to be transferees are firms in the middle of the spectrum between pure competition and monopoly.

In summary, it therefore can be observed that if a technology is to be transferred, whether internally or externally, the following must be true.

- The motives must be present in both the transferor and transferee.
- The technology must be available; that is, the people and resources necessary to accomplish the transfer must be present to ensure successful transfer and exploitation.
- The technology must be credible; that is, the data and information supporting the case for the technology to be transferred must be comprehensive and believable.
- The technology must be relevant to one or more of the markets the transferee seeks to serve.
- The price the transferee has to bear in receiving and exploiting the technology must be acceptable given the potential for profit generated by the endeavor.

Essay II

Difficulties in Technology Transfer: A Perspective

EMMANUEL P. PAPADAKIS

Quality Systems Concepts, Inc., 379 Diem Woods Drive, New Holland, Pennsylvania

This essay presents the author's opinions on a few of the difficulties experienced in technology transfer. In doing so, it addresses some factors that may be causes of these difficulties. As technology transfer is a prerequisite step to commercialization, it is valuable to see the process of technology transfer from the perspective of different people who have had experiences with the process. Other essays treat the subjects of technology transfer and commercialization from the didactic point of view to show ways and means to accomplish the goal of commercialization. Taking all the essays in this chapter together, the reader may be able to form an opinion on viable ways to bring products to the users through the marketplace.

The various types of organizations engaged in research and development experience varying degrees of difficulty in effecting technology transfer. For instance, the small company that decides to build a salable object can bring it to market relatively rapidly, provided it has the capital required. The large, vertically integrated company can do likewise. University professors seem to have the most difficulty, although they sometimes find an easier path if they happen upon an advantageous research topic.

So what is the secret of finding the optimum research subject? Observing a deeply felt need on the part of a genuine potential customer may be the key. A great deal of very interesting research has neither need nor customer in this sense. (Indeed, much pure research is not intended for market.) Even when the need is there, all books on marketing point out that of the multiplicity of ideas investigated at some early stage, only a few reach the stage of commercialization. Perhaps this book can show its readers ways to maximize potentials for successful commercialization, beginning with the observation of a market need.

As an example, consider Prof. Nicholas A. Milas (1896–1971) of MIT, the organic chemist who synthesized vitamin A and vitamin D (Johnson, 1996). Milas is a "Little Immigrant" success story. Prof. Milas (shortened from Miladakis) was an immigrant boy from Greece. He obtained a 4th-grade education in Greece, received some brief tutoring in math and German in Iowa, and began his college career at Coe College in Cedar Rapids, Iowa, before the United States entered World War I (Papadakis, 1977). He worked his way through, graduating Magna Cum Laude in 1922, and went on to Chicago for this Ph.D. in chemistry, after which he was given a National Research Fellowship at Princeton. Unfortunately, I do not have documentation on the method he used for technology transfer. However, his research undeniably bore the characteristic of having observed a real need with many potential customers: His synthesized molecules have found their way into almost every container of milk in the world as well as into vitamin pills.

Universities experience varying degrees of difficulty in getting their research into and through the process of technology transfer. Part of the difficulty must be ascribed to the sources of the ideas being brought to the attention of university researchers. Since this book concerns the commercialized products that originated as ideas *somewhere* and went through the research part of their life cycles in a university (or other) laboratory, any factor impeding the process or wasting resources calls for scrutiny.

"Where ideas originate" for university professors and graduate students is likely to be from Requests for Proposals issued by the funding agencies. To some degree, discussions in meetings of the industrial consortia in Research Centers may also contribute. Professional society meetings also play a role in disseminating ideas, but not nearly the primary role they once did.

Universities have been the sources of research in the scientific community ever since the Enlightenment. They have been joined more recently (in the 1850–1975 time frame) by a few great industrial and government laboratories. During and after World War II, universities were handed the responsibility of doing much of the scientific and engineering work of the United States government. Note, for instance, the Manhattan Project at the University of Chicago and the Radiation Laboratory at MIT. Now universities are also being asked to take on the responsibility of doing much of the industrial R&D as many industrial firms downsize and eliminate their research capability. This means that universities are under pressure to have more of their output realized as technology transfer leading to commercialization.

As companies deliberately eliminate their capabilities in certain technical areas, they must rely on supplier companies, universities, and/or the government to provide the technical know-how for their businesses to survive. The downsized companies reach a point where they are not doing the work but rather are choosing among suppliers and issuing contracts for the work to be done. The government has been doing just this for many years by closing facilities such as arsenals and shipyards while issuing more contracts to industry and universities for technology and its hardware. This leads to some degree of inadequacy in the knowledge base among funding agencies in the most knowledge-intensive field of all, R&D. Many businesses, meanwhile, try to gain economically through technology without complete internal capability by applying for benefits under the "dual-use" doctrine for military facilities, under CRADA (Cooperative Research and Development Agreement) arrangements, and under the concept of "leveraging of resources" at government-sponsored Research Centers in universities in which industrial consortia participate. Thus, university professors get a multiplicity of competing ideas. This might seem to be an advantageous new situation, but is it, the point of view of doing something relevant leading to technology transfer?

As a source of ideas, government has developed a degree of variability and fluctuation over time in the recent past. Government agencies have tended to find new subjects to emphasize on a rather frequent basis. For example, extending the lifetime of the infrastructure of the nation has come up as a research initiative; in previous years the repair of the infrastructure would have been treated by an infusion of tax money through an appropriation or an expenditure of corporate funds to improve profits. Many government agencies have found it necessary to present new ideas and to change the fundamental emphases of the agencies to satisfy needs that previously would have been handled by other means. For instance, the National Science Foundation (NSF), once only handling pure graduate-level and post-doctoral research, has developed a program to help junior college students transfer into degree programs at universities. Agencies have been caught up to a greater or lesser degree in the Areopagus Syndrome, in which it is necessary to address "some new thing" just to seem relevant. Much of the emphasis in some quarters is to provide "seed money" for a short period of time rather than to decide what is really important to do and to form a commitment to do it (fund it). Thus the nation has great difficulty generating the staying power to carry through a plan worthy of attention. The university professor as a consequence is bombarded with new ideas and funding possibilities that may not last long enough for the completion of relevant research to produce candidate inventions for technology transfer.

In this context the university professor must choose something relevant and important to work on, or at least something that will bring in money. Sometimes the requirements for bringing in funding are predominant. It is quite possible for a professor to propose, accept, and carry out research on a concept that his best judgment tells him will not work, although his supportive contract monitor has funding to pursue it. Untold millions of dollars are spent at universities and elsewhere on ideas that are infeasible or useless or too dangerous. A classic case of one too dangerous was the ANP (Aircraft Nuclear Propulsion) Project of the AEC (Atomic Energy Commission) in which a sodium-cooled breeder reactor was to supply the heat to power the jet engines of an airplane. (Although the work was done in-house at the AEC for security reasons circa 1955, professors and even students got special "ANP" clearances on top of "Q" clearances to carry on work at AEC facilities.) After a B-25 had crashed into the Empire State Building in 1945, many people believed that having a nuclear reactor flying over populated areas would be too dangerous to contemplate. Yet money was spent on the idea.

An example of an infeasible idea arose at about the time of Sputnik and the Mohole. It was reported that an RFP had been issued by a defense agency for the Orbiting Mole. A tunnel was to be dug secretly around the circumference of the earth in an essentially circular ellipse. When the tunnel had been evacuated, an earth satellite was to have been set into orbit within it for surveillance purposes, i.e., spying. It was further reported that big defense contractors privy to the secret RFP rushed to quote on this project to procure defense dollars. While it is very likely that the Orbiting Mole story is somewhere between apocryphal and fantasy, I take it as a fable (like Aesop's Fables) that shows the propensity for seeking and spending money from any spigot.

Much more recently, research on thick composites for submarine hulls was sponsored. Various academics worked on this research despite the fact that the builders and repairers of submarines knew that repairs to the internal mechanisms of submarines require the cutting of large sections out of the hulls. Whereas a metal hull can be welded after such repairs, there is no way (and none is envisioned at the present time) of patching such a gap in a composite hull. Hence the composite hull could not be built in the forseeable future, so the research was useless for its stated intent, namely, the hull of the next-generation manned nuclear submarine. The only possibility for technology transfer of the thick composites would be some serendipitous spin-off.

It can be argued that serendipity leads to great new things. NASA argued that the space program accomplished just this in the way of "spin-offs" for the general public. It even works in reverse. For example, the background radiation from the "Big Bang" was discovered by AT&T engineers who were looking for the source of static in microwave telephone transmissions. Having difficulty hearing Aunt Bertha on the telephone led to the confirmation of a scientific theory of fundamental importance—namely, that there is a point in time (the Big Bang) before which "before" has no meaning and "before which" scientists can measure nothing and hypothesize nothing physical. Thus serendipity is wonderful, but camouflaging the desire for it as nuclear airplanes, moon landings, or plastic submarines calls for scrutiny. The professor must look for the serendipity "in advance" if he or she is concerned with technology transfer opportunities.

University professors are put in a particularly difficult position by the mixed signals they receive about what is worth working on and why it is important. First, pure science for the sake of knowledge will always be worth working on. But the output of pure science may never be commercialized and hence may never be reported in a book like this. (The closest some pure research may come to popularization is in a book such as *A Brief History of Time* (Hawking, 1988).) Second, in the university there is always the "publish or perish" dictum. In the minds of some professors, however, worthy ideas that may require years of work prior to publication vie for importance with other ideas that are likely to lead to salable products. It is likely that a professor will yield to the pressures to conform to academic practices and eschew practical endeavors; the dean wants papers published in high-quality journals, not practical ones. Third, once projects have been started, the professors (just as all people) are likely to become fixated on their personal ideas and not see them as impractical even if such a condition were to be pointed out. Fourth, their funding agent is not interested in sales. Fifth, the professors rarely have the ongoing, vital feedback from a customer with a real need. The funding agent may perceive a real need, but is several layers of administration removed from the real customer.

By contrast, consider the millieu at a major corporation—one that I have experience with, the Ford Motor Company. The Vice President of Research was actively cognizant of the funder/customer relationship. The VP held an annual budget planning meeting to ascertain the value of research projects. He wanted to fund some and terminate others. Principal Investigators would present their work very succinctly (say, in $4\frac{1}{2}$ minutes) in the format of Problem/Value/Approach/Status/Plans. The Problem was the real company problem being addressed by the Investigator; the Value was the dollar figure that could be saved from costs or added to profit by solving the Problem; the Approach was Technical in a sentence or two; the Status included the percentage of completion of the R&D project; and the one-year (budget) Plans also defined the projected completion date and the probability of success. The Investigator is held to all these plans rigorously, just as the input to this little presentation had to be rigorously assembled and understood by the Investigator. (One year I had to fly home from a family vacation to give such a presentation. It was serious business.) The customer was not only the Vice President but also the Division of the Company that had the problem and had enlisted the support of Research; the investigator has plenteous feedback.

Even more direct was the feedback on one development project carried out at Panametrics, Inc., a small R&D company and builder of instruments. In 1970, I proposed building a commercial Pulse-Echo Overlap ultrasonic velocity metrology instrument of great versatility and accuracy. Edmund H. Carnevale, the President of Panametrics, agreed to underwrite the development expense *after* one copy had been sold. He put up the marketing money for this attempt. The NRC (National Research Council) in Canada ordered one sight-unseen; then design and production commenced. About 15 were sold in the first three years.

Other feedback is available in industry, also. In industry—a practical place—ideas that seem promising to researchers often are squelched by factory managers at the very outset so that little resource is expended upon them. One example is an X-ray system that would have been the ideal technical solution to an on-line quality monitoring problem in an automative plant. But the use of X rays was rejected because the labour union would have refused to have workers in the vicinity even with good shielding of the machine. Another example is a fluorescent additive that could have been added to oil in shock absorbers to make any leakage visible to an automatic UV camera. This idea was rejected because of the cost and delay that would have been needed to qualify the new oil mixture as noncorrosive to materials in the device. A final example is an ultrasonic device using attenuation to measure grain size that was proposed to a brass company. The company's response was that the small expense of having a person polish a small tab of the brass sheet and examine it under a microscope would render ultrasound not cost-effective. These reasons may seem crude and Philistine relative to the elegance of research, but that is the way researchers must learn to think in order to get research into production and not "spin their wheels" uselessly on never-to-be-wanted products.

At AT&T (the Bell system) before the divestiture ruling, technology transfer was frequently carried out by reassigning personnel to a more advanced division as the development progressed. A scientist in research might be moved to a device development department and then to a systems integration department, and so on, until finally finding himself in the long lines department making his device fit into a system to transmit telephone calls coast-to-coast. Although this personnel transfer did not happen often as a percentage of personnel, it happened often numerically in such a large establishment. By contrast, the university professor is not moved and generally does not want to move. He or she is held to pure research by the "publish-or-perish" philosophy of academic life. He or she must "toss the research over the wall" to the next stage of development, as Deming so aptly says. It is claimed (Deming, 1982) that even in industry this "walled enclave" mentality retards progress between research, development, engineering, manufacturing planning, production, and sales. How much more is the professor hemmed in! And feedback does not come back "over the wall," either.

In the context of the mixed signals coming as inputs, it is easy to see the difficulties that academics have in directing their output into the channels of technology transfer. Faced with a professor "wearing the hat" of a salesman trying to market some new invention, the industrialist occasionally makes the judgement that he or she is being presented with a "solution looking for a problem." This means that the professor was doing something interesting and publishable while being under the impression that it would also be useful. Had he been in industry, his market research department could have found out through surveys or by consulting factories whether the idea was useful. But the university system has no direct mechanism for such feedback. This is not criticism of universities — they were set up in the Middle Ages for an entirely different purpose, namely, saving and increasing knowledge while providing intellectual freedom for the professors. Few university founders (except the founders of the Colorado School of Mines in 1874 and the builders of the new MIT campus in 1916) ever envisioned government and industry being so dependent on the university nor, indeed, the university being so dependent on government and industry.

So, in this book, the items being reported have run the gauntlet from the idea phase through research, development, technology transfer, and commercialization into use. Many ideas have dropped by the wayside, being overwhelmed by negatives. This book reports on successes without any pejorative opinions about the ideas that have been rejected at any stage. And, as mentioned earlier, completeness of coverage is not claimed.

REFERENCES

Deming, W. Edwards (1982). "Quality, Productivity, and Competitive Position." Center for Advanced Engineering Studies, MIT, Cambridge, Massachusetts.

Hawking, Stephen W. (1988). "A Brief History of Time: From the Big Bang to Black Holes." Bantam Books, New York.

Johnson, Jean (1996). Private communication, Alumni Director, Coe College, Cedar Rapids, Iowa. Information in Coe tracking system for alumni. Dr. Nicholas A. Milas was awarded his B.S. there in 1922.

Papadakis, Philippos E. (1977). Private communication.

Essay III

Commercialization: From Basic Research to Sales to Profits

NEIL J. GOLDFINE
JENTEK Sensors, Inc., Watertown, Massachusetts

Initially, when Emmanuel Papadakis asked me to write an essay on commercialization for this book, I questioned whether readers of an ultrasound-oriented book would benefit from the perspectives and experiences of an eddy current and dielectric sensing company. Emmanuel, however, convinced me that describing my company's commercialization steps would provide insight for researchers and entrepreneurs hoping to commercialize products originating at universities.

I founded JENTEK Sensors, Inc. in January 1992 to develop and market a new sensor and measurement technology. This technology, including the Meandering Winding Magnetometer™ (MWM™) and the model-based measurement grid approach, was originally developed at the MIT Laboratory for Electromagnetic and Electronic Systems by Prof. James R. Melcher (now deceased) and me (U.S. Patent Numbers 5,015,951, and 5,453,689). The MWM is an advanced eddy current sensor that can either be scanned across a surface or surface mounted like a strain gage. In its array format, the MWM can build images of cracks, coating thickness variations, or early stage fatigue damage for metal components. The measurement grid approach uses models of the MWM to generate look-up tables for properties of interest such as coating thickness, conductivity as a function of depth from a part surface, or magnetic permeability.

The MWM modeling and sensor design research began in the early 1980s. This was proceeded by over 20 years of basic research by Prof. Melcher. After more than ten years of focused effort on the MWM, JENTEK is now successfully selling GridStation™ Measurement Systems with MWM probes and measurement grids for a wide range of applications in the aerospace, energy, and manufacturing sectors. (Note: A second, dielectric (capacitive) sensor has also been commercialized by JENTEK. This dielectrometer technology may be used to measure coating thickness; detect anomalies or disbonds; and measure moisture profiles, cure state, and other microstructure changes in relatively insulating materials such as polymers, ceramics, soil, paper, paint, plastic, concrete, electronic materials, and glass fiber composites.)

Through this essay, I hope I can provide some insight into the difficult path of building acceptance for a "theoretically complex" product by describing the path I chose for the commercialization of the MWM sensor technology.

When I first publicly presented the MWM technology at NIST in 1992, I understood little about the barriers to new products erected by the entrenched vendors in the nondestructive evaluation (NDE) industry. Of course, such barriers must be overcome to successfully market any new product. I was fortunate early on to meet industry experts who helped me navigate the painful process of converting the skeptics and building customer awareness of our new capabilities. For example, one mentor, with the U.S. Navy, provided an unusual understanding of the needs and problems of a large portion of the traditional NDE industry. She immediately told me that if I was going to sell to this community, I must remove the theoretical focus (i.e., "get rid of all those equations") and focus instead on the customer's needs. She also introduced me to several key players in the industry, opening doors that may otherwise have taken years to access.

The following is a brief description of the five steps along JENTEK's chosen commercialization path. A plot of the path is shown in Figure 1.

STEP 1: BASIC RESEARCH

The commercialization path begins with basic research and development. For fundamentally new technology, this step generally occurs over a period of ten to twenty years, *or more*. For JENTEK's technology, this research began at the

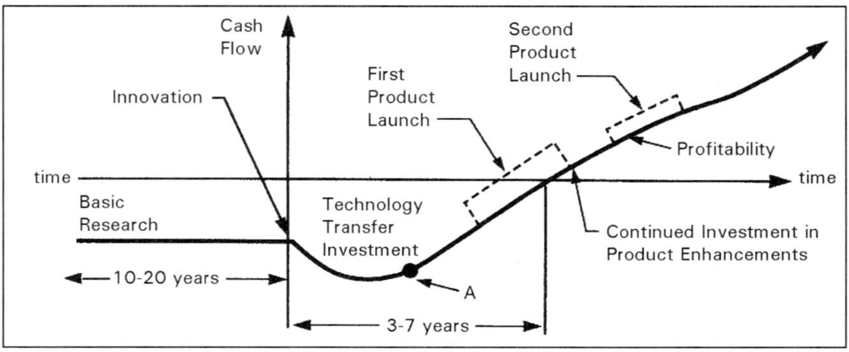

FIG. 1. Commercialization path.

MIT Laboratory for Electromagnetic and Electronic Systems and continued at JENTEK. *The purpose of basic research along the commercialization path is to provide the foundation for an innovation that can be "built out" into a product that provides a substantial technology lead over existing products in the market or provides solutions to a range of problems that cannot currently be solved.* JENTEK products are founded upon the capability to accurately model the interaction of magnetic and electric fields with multiple layered media, so that sensors can provide quantitative measurement of properties such as conductivity, permeability, dielectric constant, and layer thickness. These properties are then related to customer properties of interest such as coating thickness, cure state, crack size and depth, shotpeen quality, fatigue life, and other representations of quality or aging. Such a well-defined technology "asset" is critical to the sustainable success of a leading-edge company.

STEP 2: INNOVATION

The innovation phase is the most elusive of the commercialization path. While I was completing my Ph.D., I worked for several years for an investment banking firm. There I learned two things about innovation: (1) fundamental innovations that can sustain a profitable company by providing a maintainable competitive advantage are very rare and (2) innovations almost always require both substantial financial investment and a skilled champion to become a successful product. Unfortunately, you will not know if you have a true innovation that can result in a profitable product until you have moved far along the commercialization path. Many have tried (with limited success) to define and teach innovation. I will not attempt to elaborate on this most important step, beyond the following. Innovation begins during basic research and, at some difficult-to-define point, research "data" evolves into a technology that provides new value that *did not exist before*. Simply put, innovation, within the context of commercialization, results in development of a new technology "asset" that meets a clearly defined customer need.

STEP 3: TECHNOLOGY TRANSFER INVESTMENT

Many misguided entrepreneurs think that if they have a great idea for a product, investors or customers will come. They won't. *You must "build out" your technology with a substantial investment directed at meeting specific identified customer needs with substantial revenue potential in a market that can sustain profit margins while permitting continued product enhancement and supporting new product development.* For almost all technologies, this build out requires both time (3–10 years, depending on the nature of the innovation) and money (from a few hundred thousand dollars to several million dollars, depending on how capital-intensive the technology is and how large the technological lead). This is what I define as the *technology transfer investment*. In other words, technology transfer does not occur when a company licenses a technology from a university or national laboratory; it occurs during the several years that follow. This step results in the evolution of the technology asset produced by the innovation into a well-positioned and well-defined product or product line that meets specific customer needs and specifications. Cost and pricing issues must be dealt with in this step. For new technologies with fundamentally new capabilities, pricing strategies should consider objectives relating to market penetration, customer's return on investment, sustained profit margins, and initial cash flow requirements.

During the technology transfer investment step, companies must cover their cashflow requirements with outside sources (e.g., private investment, venture capital, strategic partners, loans). At point A indicated in Figure 1, these cash flow requirements can begin to be alleviated by customer-supported projects. Sources for this might be SBIRs, service revenues, or R&D funding from commercial customers. It is critical not to "sell" your future profits at this point by giving away excessive royalties or other similar (% of sales) payments.

STEP 4: PRODUCT LAUNCH

The product launch is the key to profits. Unfortunately, you never know if your product will meet your customers' needs until you are well along the commercialization path (i.e., steps 1–4 are completed). Only when you are selling and delivering product, and your customers are saving money or improving the quality of their products and services, will you know that you have met your customers' needs. Successful commercialization is not about luck, unless you subscribe to the philosophy that "luck is when opportunity meets preparation." Your product must be properly positioned in the market (*preparation*) to meet specific customer needs (*opportunity*). This is where you find out if your product is truly innovative.

JENTEK has been fortunate; the level of interest in our GridStation product with both of our sensor technologies has been overwhelming. For example, we anticipate that our MWM sensor will become a standard for in-service inspection and "health monitoring" of difficult-to-access locations for machinery, aircraft, and power plants. By mounting the MWM sensor like a strain gage, customers can monitor locations that currently require costly disassembly for inspection. Providing solutions to "hot" applications like health monitoring, combined with our core focus on manufacturing quality control systems, provides us with a balanced approach to building profitable product lines based on our new technologies.

STEP 5: THE SEARCH FOR PROFITS

Selling is only a small part of the mission of a business; the true goal is profitability. Achieving profitability requires as much innovation and tenacity as developing new technologies. There are numerous small companies in the NDE market that sell small numbers of systems each year, yet they never manage to grow and build profitability. Companies need both a leader with the vision and capability required to build a profitable enterprise and a truly innovative product that can generate substantial profits over a long period of time.

It is also important to remember that the limits of your market are only defined by the potential of your technology and your capabilities to meet customer needs. The traditional NDE market may have been somewhat confining, but, the emerging NDE market is expansive and continues to grow. Depending on which report you read and how you define its limits, the current worldwide NDE market ranges in size from $500 million to $1.5 billion *or more*. If you include fringe markets such as landmine and unexploded ordnance detection (estimated to have a latent market demand of over $33 billion), medical instruments, and electronic media inspection, then future "NDE" markets could clearly be placed among the largest in the land!

To those of you who plan to embark on the entrepreneurial path with a new technology you have developed, I can only hope that you have the support of a spouse, friends, and family as terrific as mine. You will need them every step of the way.

Essay IV

Perspectives on Technology Transfer and NDT Markets

STEPHEN R. RINGLEE
E-Markets, Inc., 125 South Third Street, Ames, Iowa

Overview

Major nondestructive test (NDT) research programs at institutions such as Iowa State and Johns Hopkins were established not merely to advance the state of the technologies in various NDT arts, but to transfer these advances to commercial markets. Despite the best efforts of these institutions, the NDT technology transfer record remains mixed at best. In fact, NDT markets may be too small, too specialized, and too conservative to absorb many new technical developments from university and institute laboratories. These institutions would be well advised to manage their technology transfer expectations down to more achievable goals.

NDT Markets

Few reliable market studies exist of the various industrial NDT markets. These markets are extremely diverse and include everything from equipment sales to inspection services, software, and engineering consulting. Available equipment includes disparate technologies such as ultrasound, acoustic emission, eddy current, particle and penetrants, x-ray and other radiography, and magnetic or particle emission. In addition, these technologies are used across a variety of end-use markets, such as aerospace, utility, chemical, electronics, energy, metalworking, and transportation. The firms selling equipment or services tend to be small (under $50 million in sales) and are frequently subsidiaries of larger technology companies. While the market has been growing consistently, this growth varies considerably by submarket and is in only a few cases above ten percent per year.

As a result, the NDT market can be understood not as a distinct market, but as a collection of solutions to various material and component testing problems. Each of these problems is unique, calling for an individual technical solution. Market participants must have a great deal of technical sophistication and flexibility to adapt various approaches to each problem. Even the largest NDT equipment marketers sell equipment for use in a number of end-use applications and must offer engineering assistance and customized software probes, or other devices to help the end user solve the particular technical problem at hand.

This assistance requires that NDT sales have margins high enough to justify the expense of customer service and customized engineering assistance. In many cases, these margins are achieved by the cost avoided by the customer from downtime, accidents or loss of use (as in the utility and aerospace industries). The price of inspection services and equipment can also be justified through lower life-cycle costs and increased equipment longevity (as demonstrated in military aircraft and ships). Many NDT firms have also achieved higher margins through the bundling of inspection services and equipment. Unfortunately, higher margins and prices have limited the size of the market for equipment and services. Customers will usually adopt the technique of least cost and effort — preferring lower cost (but less reliable) NDT technologies over higher cost, newer and more reliable solutions — unless they have a compelling reason to do otherwise.

Regulation paradoxically tends to lock NDT solutions to a "lowest-common-denominator." Technical gatekeepers such as aircraft manufacturers and government agencies limit the ability of end users to employ other than "approved" NDT equipment. These gatekeepers are resistant to new approaches unless they are of compelling utility or meet immediate needs of safety or reliability.

Limitations on Technical Innovation

The market forces that define the NDT markets — including the market variety, the intensity of customer service, the need for high margins, and the effects of regulation — all combine to limit the extent to which new technologies can be successfully introduced to end users. Technical developers, including those firms now selling NDT technologies, must justify the investments made to develop, perfect, manufacture, and market new designs. In a market characterized by small submarkets, limited growth, and frequent user resistance to new technology, an equipment or software developer must very carefully discriminate among NDT investment projects. The path that most have chosen, and that tends to yield the highest return on investment, is one of small, incremental improvements to existing technologies. In these cases, the investments required are limited to the engineering efforts needed to adapt equipment to a new problem and perhaps the development and manufacture of a new probe or transducer. The case histories of companies such as Zetec, Krautkramer-Branson, and Panametrics illustrate this path.

The Frustrations of University-to-Market Technology Transfer

Universities have created "centers of excellence" at institutions such as Iowa State to advance the state of the NDT art. These centers are supported through a mix of state and federal government, university, and industry funds. By design, the leading centers have sought active industrial sponsors, not merely for the incremental funding (which in most cases is modest) but also for the connection with industrial problems and existing solutions. The centers encourage a dialog among their supporting groups through regular conferences, publications, and informal conversations. They also solicit industry and government problems and propose research and development projects to solve these problems. Unfortunately, these centers have not had the success at technology transfer that was expected when they were first created. The author's experience with the Center for NDE at Iowa State illustrates many of the reasons for this disappointment.

Technologies developed at the university usually require substantial re-engineering or adaptation before they can be used in commercial applications. In many cases, they are not designed with the product economics in mind. In one instance, an ultrasonic instrument developed to measure the texture of rolled sheet steel was designed to a price point of $90,000 in 1989 economics assuming manufacture in batches of 10. Built to order, one instrument would have a price point of about $200,000 in 1995 economics. Although the instrument was of industrial quality and had technical merit as well as on-line potential, it neither materially improved the measurement desired by the customer nor saved the user any time compared with the existing off-line mechanical tester, which sold for approximately $12,000. Consequently, no users or potential NDT equipment makers showed any interest in buying, licensing, or developing this ultrasonic instrument.

In other cases, the university-developed technologies are embryonic and require substantially more refinement before they can be reliably used by less-sophisticated customers. Another development, a pulsed eddy-current flaw detector, had great technical merit but required almost six years of additional work in the laboratory before commercial users showed interest in licensing it. In many cases, the funding for the additional developmental work is difficult for a university lab to either obtain or justify, leading to many half-done projects that remain on a lab bench for years. This added investment is likewise hard for a commercial user to justify, and significantly limits the interest they show in licensing-developed ideas.

Further, many university-incubated projects are begun for motivations other than commercial success. Research papers, student projects, and personal interests are all very important in the academic culture. NDT developments occur during this research work, but are not often pursued through the difficult, expensive, and boring requirements of application development. This application development is a necessary precondition to commercial success. Although a researcher may have achieved significant results from a research project, the incremental work required to turn the results into a licensable technology or even a commercial product is not work at which the university culture excels, enjoys, or believes it can justify.

These cultural and economic imperatives create a gap between the university labs and the commercial NDT marketplace. Few technologies cross this gap. Those that do are introduced into a conservative NDT marketplace with growth rates low compared with markets such as communications, computing, and software. There are successes in NDT technology transfer, such as Krautkramer-Branson's license of eddy-current instrumentation technology developed by NASA for crack detection. However, the history of the last 15 years of university-based centers of NDT research illustrates the difficulties and infrequency of transfer of NDT technologies to industry. Those centers that advertise their success at NDT research need to keep this in mind when approaching sponsors. Center managers as well need to hold technology transfer expectations that are achievable, and should work with their funding sponsors to educate them to the realities of the NDT markets.

Essay V

Teaming — A Solution to the Problem of Integrating Soft Skills and Industrial Interaction into Engineering Curricula

WILLIAM LORD AND SATISH UDPA
Electrical and Computer Engineering Department, Materials Characterization Research Group

ROBERT S. HARRIS
Center for Advanced Technology Development, Iowa State University, Ames, Iowa

This essay describes two unusual R&D projects that have resulted in the development of a new paradigm for engineering education. The first project, funded by the Gas Research Institute, consists of a consortium of Battelle (Columbus), Southwest Research Institute, Iowa State University, and a number of gas pipeline inspection companies whose overall goal is to improve the state of the art in gas transmission pipeline inspection. The second project, funded by a Japanese company, Takano, Co., Ltd., involved the design and development to industrial prototype stage of an acoustic microscope. (This project with Takano has since been followed by two other projects for development of other NDE instruments.) Both the GRI and the Takano projects involved large teams of Ph.D., M.S., and undergraduate students that had to interact on a daily basis with faculty, visiting engineers, and postdoctoral researchers while balancing the hard deadlines imposed by the industrial partners and the academic concerns of working toward a degree. Issues relating to project reports, presentations, intellectual property, technology transfer, and industrial interaction were dealt with as a team, which has led to careful consideration of such teams as an integral part of the educational experience for all engineering students. Details of this new paradigm are presented together with suggestions for incorporating it into engineering curricula.

Introduction

Since its founding in 1858, Iowa State University (ISU) has developed a tradition of outreach excellence in fulfilling its role as the state's land grant institution. In keeping with this tradition, DOE's Ames Laboratory has been used as the core from which to spin off more specialized centers with targeted research, development, and outreach activities in the physical sciences and engineering (Snow, 1994). Eleven centers now form the university's Institute for Physical Research and Technology (IPRT), which has strong, interdisciplinary ties with the university's colleges and departments. Interactions with private industry are encouraged through IPRT's Center for Advanced Technology Development (CATD) and the Engineering College's Center for Advanced Technology Development (CATD) and the Engineering College's Center for Industrial Research and Service (CIRAS).

Although this structure is not vastly different from those existing at many other universities, it does provide a rich and varied R&D backdrop from which effective ties to industry can be developed and from which the ABET Engineering Criteria 2000 program outcomes and assessment for the university's Engineering College can be met.

There is no doubt that the pressures for change in graduate education (Holden, 1995; Brill and Larson, 1995), with their emphases on industrial relevance, flexibility, soft skills, etc., are equally present at the undergraduate level. This has been clearly recognized in the ABET Engineering Criteria 2000, where the necessary program outcomes and assessment delineates the expected abilities of graduating B.S. students. From work on two recent R&D projects involving both teams of research organizations and teams of researchers, it is the authors' contention that such team-based activities can themselves be the vehicle for a radical restructuring of the graduate/undergraduate experience that will meet the needs of a modern engineering education.

The Projects

Two large projects provided the basis for developing the new paradigm. The first project was supported by the Gas Research Institute (GRI) to develop new methodologies for analyzing data from nondestructive evaluation tools used for inspecting gas transmission pipelines. The second project was funded by a Japanese company, Takano Co., Ltd., to design and develop a state-of-the-art acoustic microscope. The projects provided a rich environment that called for extensive interaction, not only among a large group consisting of graduate students, post-doctoral fellows, undergraduate students, and faculty, but also with a number of external agencies. Although the needs of both projects were addressed using a team approach, differences in the nature of the sponsoring organizations, team structure, and deliverables contributed to subtle variances in the emphasis and character of the two project groups.

GRI PROJECT

The GRI project was charged with the task of improving the methods employed for detecting and characterizing flaws in underground gas transmission pipelines (Udpa *et al.*, 1996). Natural gas is a major source of energy that is normally transported from well sites to consumer locations through a vast network of pipelines. The pipelines are inspected periodically using devices called "pigs" that travel through them to gather information concerning their integrity. Most inspection devices currently use electromagnetic methods to test the pipeline. GRI funded a consortium consisting of Battelle, Iowa State University, Southwest Research Institute, and two pipeline inspection services vendors (Vetco Pipeline Services and Pipetronix) to investigate methods for improving inspection technology. The two inspection services vendors were deliberately included in the consortium to ensure that the work done is relevant to the industry needs and to allow for absorption of new technology. The consortium was required to meet several times a year to appraise each other and GRI of the progress made during the reporting period. In addition, GRI required monthly progress reports to be submitted.

A review of the project early on revealed the need for a systematic and comprehensive approach to solving the problem. First, it was felt that a full and thorough understanding of the physics underlying the inspection process was required to develop appropriate signal processing strategies for characterizing defects. This was accomplished through the use of complex numerical models adapted specifically for stimulating the physical process. The next important task involved the development of some new signal processing schemes for extracting information from the electromagnetic transducers in the pig. The sophisticated nature of the signal processing schemes gave rise to the third task, namely, the development of user-friendly software that the inspection services vendors could use without undergoing major retraining.

The ISU team was tailored to meet the needs of the project. At its peak, the group consisted of nine graduate students, two post-doctoral fellows, a visiting faculty member, three undergraduate students, and three faculty members. Each of the graduate students specialized in one of three areas; electromagnetics/numerical analysis, signal processing/pattern recognition, and software engineering. Students were also expected to be reasonably familiar with the two disciplines outside their area of specialization to facilitate free flow of ideas and develop an appreciation for the roles played by others in the team. This was accomplished by persuading students to take courses in all the subject areas.

The team met regularly every week and all students were expected to present brief summaries of their accomplishments, the problems they faced, and plans for the future. In addition, one of the students was given an opportunity to present his/her contributions in greater detail. The meetings provided the students an opportunity to develop a considerable level of confidence in giving presentations and face critical audiences. The graduate students were also recruited to submit monthly reports. These reports were integrated by the post-doctoral fellows and reviewed by the faculty members before being submitted to GRI every month.

Each undergraduate student was mentored by a graduate student and charged with specific responsibilities by the graduate student in consultation with a post-doctoral fellow and faculty members. The tasks assigned to the undergraduate student were usually a subset of the tasks required to be completed by the graduate students. The undergraduate students were rotated among the graduate students to provide them with a diverse set of experiences and capture the excitement that accompanies research. The undergraduate students gave presentations summarizing their work at the end of each semester. It must be mentioned that all the undergraduate students have since elected to pursue graduate studies and stay on at Iowa State University.

The interaction between the team and other members of the consortium provided an opportunity for the students to improve a number of soft skills. The consortium meeting locations were rotated through participant sites. Consequently, ISU had the opportunity of hosting the meeting about once a year. Each of the students presented a summary of their contributions to the project at these meetings. The presentations were rehearsed, and every effort was made to handle the event as professionally as possible. These meetings considerably improved the morale of the students and allowed them to take pride in their accomplishments. Sharing their results with the "outside world" gave them a proper perspective of their own contribution toward solving a real problem and prepared them for making presentations at technical conferences.

Students also visited several of the consortium member facilities. A visit to Battelle provided first-hand exposure to the extensive test facilities located in West Jefferson, Ohio. The students were also heavily involved in transferring technology to the inspection services vendors. Thus the students had to take several trips to industry sites and spend the time and effort to implement their systems in a real environment. The "leap" from an academic concept to a working software module is a long one, and the students had to overcome several challenges before seeing results. Apart from absorbing a number of soft skills, perhaps the most important lesson learned by the students was the need to work within the framework of a fixed time schedule.

TAKANO PROJECT

The Takano project received support to design and develop and acoustic microscope for inspecting small ceramic components and integrated circuits. The project called for the construction of a full prototype embodying both new hardware and software and required Iowa State to transfer all the technology and assist Takano in building a commercial product. Takano also signed an agreement with Battelle for evaluating ultrasonic transducers and requested mutual collaboration in finalizing the transducer specifications. This required periodic exchange of information and coordinating plans for testing and evaluating transducers.

The project team was structured in a manner similar to the GRI team. However, there was an additional group devoted to the development of hardware. A total of twelve graduate students, two post-doctoral fellows, one visiting faculty member, five undergraduate students, and three faculty members were involved in the project. The hardware group consisting of four students was responsible for building an advanced computer-controlled pulser-receiver with performance characteristics and features heretofore unrivalled in the industry. The students designed and assembled the circuits, and evaluated them before developing a printed circuit board layout. After the boards were fabricated by an outside agency, the prototypes were assembled and tested. The mechanical design of the system was also developed by the hardware group. The signal processing group was responsible for developing all the signal and image processing algorithms and pattern recognition routines. The software group was charged with the responsibility for developing the operating system software and integrating all the systems. The numerical analysis group was involved in simulating the test geometry both for estimating the parameters necessary for developing image restoration algorithms and for optimizing sensor location and characteristics. As in the case of the GRI group, students were expected to be familiar with disciplines outside their own area to facilitate the free flow of ideas.

The team met every week to discuss progress and plans for the future. Each student presented a brief summary, and students were encouraged to participate in discussions. The graduate students were required to submit a substantial quarterly report. The reports were integrated by the post-doctoral fellows and reviewed by all participants including faculty members before submission to Takano. The undergraduate students played a key role in the design and development activities, particularly in assembling and testing the hardware. This required them to gain familiarity with advanced circuit design and layout as well as simulation packages. The undergraduate students were required to submit a report at the end of each semester. As in the previous case, all the students eventually elected to pursue graduate studies at ISU.

Representatives from Takano paid quarterly visits to ISU to stay abreast of the progress being made by the team. All graduate students were required to make detailed presentations of their contributions. A Japanese translation of the presentation was provided by an interpreter. These presentations were particularly useful in developing an appreciation of the differences in the cultural backgrounds since the presentations had to be tailored to suit the Japanese corporate style. In the process, the students developed a remarkable level of sensitivity toward the Japanese culture and an ability to avoid committing social and cultural snafus. The awareness was heightened even further during the process of transferring technology when several students spent a few weeks in Japan to train Japanese engineers.

The project was an educational experience for both ISU participants and Takano. Both entities gained in the process. Takano personnel gained a healthy respect for the creative abilities of U.S. students and the university system that nurtures such talents. The ISU team was impressed by the industriousness of their counterparts and the environment in Japanese industry that brings out the best in their personnel. The direct benefits to both parties were truly manifold. As a consequence of their positive experience with this project, Takano decided to fund additional projects and pursue a long-term relationship with Iowa State University, endow a fellowship at ISU, and establish a manufacturing and marketing facility in Ames, Iowa.

Lessons Learned

Following is a list, in no particular order of importance, of the lessons learned through these two projects.

- Cooperation and teaming are not skills that are easily learned—perhaps more so for faculty than for students. A good example of this is the initial suspicion held by the consortium participants on the GRI project, who had previously had to complete for the same R&D funds.

- Undergraduate students who experience a good project involvement are more likely to stay on for graduate school.
- Graduate students who work on industrial projects are more likely to obtain job offers.
- It takes time and effort to adjust to industrial deadlines and the industrial mind-set. Faculty must be particularly alert to the need to protect the intellectual quality of graduate studies and to ensure that Ph.D. dissertations and M.S. theses are of appropriate academic standard.
- Academic administrators must learn how to evaluate and reward faculty participation in industrial project activity.
- Industrial-based projects are an excellent vehicle from which to develop student communication skills in that the need for both oral and written presentations is self-evident for the success of the project and not merely an academic exercise.
- Students underestimate the difficulty of dealing with the "bureaucracy" in terms of placing orders, getting parts made, etc.
- The mix of nationalities in most engineering graduate programs automatically ensures the development of social skills as well as sensitivity to cultural issues.
- Everyone underestimates the complexities of technology transfer and the degree of "hand-holding" that must take place in moving ideas from the lab to the field.
- Teaming and industrial project activity tends to be low cost in terms of university commitment in that R&D is a normal faculty function. Such interaction also fits in well with a land grant institution's mission and often fulfils the state's need for university/industry cooperation.
- Patent, intellectual property and ethical issues arise as a natural part of university/industry cooperation, allowing effective integration of special topical seminars and case studies.
- The synergistic effect of teaming is essential for project success. Three students in a team achieve much more than three students working individually.

The New Paradigm

An effective way to integrate soft skills into the engineering curriculum is via teaming (faculty, post-doctoral students, graduate students, and undergraduate students) on industrially relevant R&D projects. Such teams, depending on the degree of project complexity, could include all levels of graduate and undergraduate students. The exact logistics would vary from college to college, depending on contacts with industry, faculty involvement in R&D activities, numbers of graduate and undergraduate students, interdepartmental and center cooperation, etc. However, an appropriate hierarchy of involvement would be faculty and industry engineers, Ph.D. students, M.S. students, seniors, and undergraduates in the first three years of their degree programs. Student responsibilities at each level would be as follows:

- Freshmen, Sophomores, and Juniors would be the equivalent of "apprentices" on the projects and would be required to take "project" hours each semester. Their involvement with a project would grow with experience to include the building and conducting of experiments, data acquisition, analysis of data, and general "gopher duties" associated with all projects. They would be required to write short semester reports documenting their experiences and to give short oral presentations to their "host team."
- Seniors would be required to carry out a full two-semester "design project" that would include the identification, formulation, and solution of an engineering problem requiring the writing of formal technical reports and the presentation of formal seminars to their peers. Wherever possible, projects would be drawn from industry contacts or have relevance to faculty R&D interests.
- Graduate students, particularly at the M.S. level, would play an integrating and management role, organizing the team meetings, ensuring adequate progress toward project goals, and evaluating undergraduate work and development of team skills.
- Faculty and industry engineers would provide the motivation and overall organizational skills needed to ensure a quality educational experience. Each department would need to identify a "project czar" to organize the senior project seminar series and maintain uniform grading procedures. Coordination with class advisers would also be required to handle the complex logistics and to set up appropriate seminar series to cover some of the more general topics relating to professional and ethical responsibility, engineering and society, international issues, etc.

Implicit in this new paradigm is the concept that not all students need to spend the rest of their professional lives in industry. A high degree of adaptability and flexibility is required on the part of each college and department to match the project experiences at all levels to the students' own needs and expectations. A university, after all, is perhaps the only institution that can still provide a nurturing environment for the unique, eccentric, and unusually gifted individual.

Concluding Comments

Teaming and work on relevant industrial projects is essential in modern engineering curricula to provide all levels of students with an opportunity to experience first-hand the reduction of scientific principles to practice for the betterment of humankind. Such an educational framework encourages the development of those soft skills needed for survival in tomorrow's dynamic, fast-changing engineering profession.

Of paramount importance in the implementation of a project-oriented curriculum is both the commitment of the university and college administration to such an endeavor, and perhaps most importantly, the enthusiasm of each individual faculty member for the need to include this type of experience in the curriculum. The ABET Engineering Criteria 2000 goes a long way toward providing the necessary motivation for change.

References

Snow, Joel A. (1994). National research trends. *In* "Strategic Planning Position Papers." Iowa State University.

Holden, Constance *et al.* (1995). Careers 95: The future of the Ph.D. *Science* **270**, 121–130.

Brill, Arthur S. and Larson, Daniel J. (1995). Are we training our students for real jobs? *Academia*, **Nov–Dec**, 36–38.

Udpa, L. *et al.* (1996). "Developments in gas pipeline inspection technology. *Materials Evaluation*, **54** (4), 407–472.

Acknowledgments

The authors are indebted to the College of Engineering at Iowa State University for providing the environment in which to carry out university/industry projects. Financial support and encouragement from the Gas Research Institute (Harvey Haines, Project Manager) and Takano Co., Ltd., is also gratefully acknowledged.

Biography

Dr. William Lord is the first faculty member to hold the Palmer Chair in Electrical Engineering. His interests are in the area of nondestructive testing of materials and the application of numerical modeling techniques to the understanding of energy/defect interactions. Professor Lord was editor-in-chief of the *Institute of Electrical and Electronics Engineers (IEEE) Transactions on Magnetics* from 1991 to 1995. He is a Fellow of the IEEE, IEE (UK), and the British Institute of Nondestructive Testing, and served as a National Direction of the American Society for Nondestructive Testing from 1990 to 1994. Dr. Lord was Associate Dean for Research and Graduate Studies in the Engineering College from 1991 to 1995. In 1995, Dr. Lord was made Anson Marston Distinguished Professor of Engineering. He received the B.S. (1961) and Ph.D. (1964) from the University of Nottingham, England.

Dr. Satish Udpa's research interests lie in the broad areas of systems theory and numerical analysis. He is involved with the development of signal processing and pattern recognition techniques for solving inverse problems relating to nondestructive testing. He is also engaged in the application of numerical techniques for modeling a wide variety of physical processes underlying nondestructive evaluation methods. Dr. Udpa received the B.S. (1975) from J.N.T. University, India and the M.S. (1980) and Ph.D. (1983) from Colorado State University.

Robert Harris has over 30 years experience in the fields of applied research and development, technology assessment, licensing, and new venture start-up. His recent positions of responsibility include: Interim Director of the Center of Advanced Technology Development at Iowa State University (1996 to present), Manager of Industrial Outreach for the Ames Laboratory, (1995 to present), Director of the U.S. West/SBIR program for the state of Iowa, (which is administered by CATD), Associate Director of the Center for Advanced Technology Development (CATD) and Director, Office of Contract Research at Iowa State University.

Essay VI

Innovative Technology Transfer Initiatives

ARTHUR BALLATO and RICHARD STERN

US Army Communications—Electronics Command, Fort Monmouth, New Jersey

The Physical Sciences Directorate (PSD) of the Army Research Laboratory (ARL), Fort Monmouth, NJ, has achieved significant success in the technology transfer/strategic alliance arena by blending technology transfer statutes, regulations, and practices into its corporate culture. The essay describes the process and procedure PSD utilized in achieving that success, along with the lessons learned in introducing new and innovative methodologies, techniques, and approaches in transferring federal laboratory technology to the private sector.

Introduction

Physical Sciences Directorate (PSD) is one of eleven directorates of the Army Research Laboratory, with headquarters in Adelphi, MD. This directorate consists of 239 individuals, 177 of whom are scientists and engineers (S&Es), including 59 Ph.D.s. PSD is the Army's focal point for research and development in the physical sciences and generates more than 50 issued patents each year. It provides enabling technologies to solve critical barrier problems in photonics, optoelectronics, microelectromechanics, smart materials, solid-state and nanoscience, electrochemistry, energy science, bioscience, high-frequency electronics, rf acoustics, manufacturing science, and electrophysical modeling, advancing the physical sciences technology base consistent with combat needs of the Army.

PSD has developed a well-focused, broad-based technology transfer program that creates both strategic and tactical alliances with business, industry, and academia. The Directorate is committed to an "open laboratory" policy where entrepreneurial S&Es from the private sector may engage in on-site cooperative efforts with PSD's S&Es, utilizing the Directorate's intellectual property, unique and sophisticated facilities, and high level of expertise to develop or improve commercial products and processes of interest to both the private sector and the Army.

PSD has implemented more than 60 Cooperative Research and Development Agreements (CRDAs) and 18 Patent License Agreements (PLAs) since the Technology Transfer Act became law in 1986, with additional agreements in various stages of preparation. An income stream, established from these agreements, is being used to expand the PSD technology transfer program effort, cover CRDA operating expenses, and reward contributing S&Es.

The goal of PSD is to exploit the latest scientific advances to meet military needs while, at the same time, fostering the creation and improvement of commercial products and services within the civilian economy to foster both the economic and military success of the United States.

Changing the Culture

Early in the 1970s, PSD management recognized that the preponderance of R&D resources and technology needed by the Army existed outside the organization. Management therefore determined that it was essential for PSD to identify and interact with the best available outside research resources and that this could only be done successfully if the interaction was structured to be to the mutual benefit of all concerned. This approach required individual S&Es to develop an awareness of and familiarity with reseach and researchers throughout the world.

Toward that end, S&Es at PSD are required to complete a minimum of 80 hours of annual training in the areas of technology, technology transfer, intellectual property, and marketing. PSD provides a continual string of lecturers from business, industry, and academia who provide workshops, lectures, and seminars to supplement regular college and university curricula.

As a result of this approach, informal cooperative efforts were initially developed between PSD and technical contractors, colleges, and universities, and through technical working relationships borne of out personal interactions established at technology conferences, symposiums, seminars, workshops, etc. The aim of these teaming efforts was to address specific technical barrier problems using a larger and a more diverse R&D work force that approached a critical mass. The advent of the Technology Transfer Act of 1986 and the related Executive Orders of 1987 provided, at that point in time, a mechanism for formalizing, expanding, and strengthening informal cooperative efforts that already existed at PSD. The first CRDAs established at the Directorate were carried out with organizations with which PSD had informal or contractual relationships. This approach minimized difficulties and obstacles involved in establishing the initial formal collaborative agreements. Implementation of each of these new CRDAs, however, required continual nurturing by the PSD Technology Transfer Office; otherwise they would have languished.

Defining Technology Transfer

To define what is meant by "technology transfer," it is sometimes best to explain what technology transfer is not. Effective technology transfer is *not* merely the process of making known the availability of technology. It is *not* "throwing technology over the transom" for others to pick up and possibly utilize without further explanation or discussion. Little is accomplished by this hands-off approach, and disappointment in the process and its accomplishments will surely follow.

Some consider the transfer of federal funding to the private sector for the development of dual-use technology a form of technology transfer. Although useful technology is developed through contractual efforts, this should not be considered transfer of existing federal laboratory technology for the benefit of both the laboratory and the private sector. Likewise, establishing a CRDA for its own sake without specific, focused objectives (no matter what resources are made available for that collaboration) cannot be considered a potentially productive technology transfer effort. Without an objective and statement-of-work beneficial to all parties and to which all parties are committed, disappointment can be expected.

A fully functional technology transfer program is one where S&Es are motivated to transfer/transition federal laboratory technology (in-house technology, methodologies, expertise, capabilities, facilities use) to the private sector (business, industry, and academia) by means of a CRDA or PLA in a way that benefits all participating parties. The process should be reciprocal, with both federal and nonfederal partners contributing to the effort. A win–win situation should be a driving objective.

Technology Transfer Advancement

There are a number of positive steps that every federal laboratory can take to promote technology transfer. First, it is important that a laboratory intending to develop an effective technology transfer program have leadership that champions technology transfer. A laboratory whose management is unsympathetic to the cause is one where the technology transfer program will neither reach its potential nor satisfy any of those involved. Further, the lab must foster a strategic view of the technology transfer process, seeing technology transfer as a tool used by all elements of the lab to assist the lab in meeting its mission goals. An environment supportive of technology transfer needs to be created. Led by example, S&Es trained in the generation and use of CRDAs and PLAs by a technology transfer manager and a user-friendly legal staff will understand the value of intellectual property and appreciate the value of cooperative efforts. By encouraging S&Es to attend and network at technical meetings, conferences, workshops, and seminars and to generate working relationships with their peers outside of their laboratory, even marketing soon becomes a distributed activity. The use of database searches, publishing and presenting technical papers, and S&E exchanges all contribute to the open laboratory concept.

Certain ingredients are necessary for operating a successful technology transfer program. Among these are motivation, flexibility, trustworthiness, unique technical expertise, infrastructure equipment and easy access thereto, identification of technology ripe for commercialization, realistic expectations, clearly delineated responsibilities, fast response time, local negotiation authority, simple procedures, reasonable CRDA and PLA terms, and user-friendly attorneys.

Collaborative Research and Development Agreements

PSD has now established nearly all of its technology transfer and collaborative efforts under formalized CRDAs and PLAs. These businesslike contracts are useful because they define the technology transfer process — including allocation of intellectual property rights, termination procedures, dispute resolution, and publication rights — and they limit federal liability through indemnification.

Successful technology transfer efforts usually exhibit the following characteristics: participants have a strong vested interest in the outcome; all participants benefit in fair and equal portions; all parties are involved early on; the technology user/customer is involved in the early stages, with simultaneous development of technology and product; and the customer-oriented product is actively marketed. Conversely, barriers to technology transfer include a perceived lack of fairness, suspicion or lack of trust, NIH syndrome/arrogance/pride, a legalistic mind-set, and short-term focus. Each can hinder successful technology transfer.

The PSD Technology Transfer Office processes, on the average, one new CRDA a week. Although the content of these CRDAs is very similar, each CRDA has its own personality. Some CRDAs involve large corporations, such as Martin Marietta or Texas Instruments, whereas others establish partnerships with small start-up businesses. A CRDA may be a supplement to an ongoing contract, allowing the Army to assist in the advancement of the R&D effort, or may be initiated to aid a company that needs help in commercialization under an existing PSD PLA.

Some CRDA participants provide new and unique electronic materials to PSD in exchange for test and evaluation of those materials by PSD's sophisticated analysis equipment, facilities, technologies, and techniques.

Some CRDAs represent a "spin-on" to the Army, where a small business wants and needs technical assistance in the development of a new patentable idea that may be useful to the Army as well as the private sector. Several of the CRDAs with academia (Princeton, Rutgers, Stevens, and NJIT) involve placement of graduate students, post-doctoral resident associates, and professors into the PSD laboratory where they have the advantage of working together with PSD S&Es and using state-of-the-art facilities and equipment not readily available in academia. PSD has the advantage of enhancing its work force through the use of these individuals, thereby creating a win–win situation. CRDAs have also been used to formalize working relationships with industrial partners in the creation of collaborative team efforts in carrying out DARPA TRPs (Technology Reinvestment Programs) or other federal agency contract awards.

A PSD CRDA consists of approximately 12 pages and can be created in as little time as a few days, with Army formal approval received in fewer than 30 days. In the absence of federal acquisition regulations governing CRDAs, the CRDA contract terms anticipate conflict situations such as potential problems involving data rights and intellectual property.

When establishing a CRDA, PSD particularly looks for specific elements, including a mission-related project, a win–win situation, a leveraging opportunity, income, dual-use applicability, and the potential for a follow-on PLA. With these elements, the CRDA is one of the most effective and perfected tools in the PSD technology transfer repertoire.

Patent Licensing

PSD has found patent licensing to be a very efficient means of transferring technology and has granted licenses of PSD technology to both small and large companies. A number of valuable lessons have been learned from entering the patent licensing market.

The first and probably most important consideration in licensing is the breadth of the patent claims. The quality of patent claims is, in fact, that for which the licensee is paying. Claims that are not well crafted will not prevent others from "patenting around" those claims. Patents are too often rendered useless for licensing because their claims are too narrow. This situation arose from the old mind-set that considered patenting from a purely defensive standpoint: it was only used to prevent others from charging the government royalties for its own research. Sometimes, however, an older patent with typically narrow defensive claims can be rehabilitated by applying for new and related patents that expand around (picket fence) the claims and uses of the existing invention/patent. This is part of patent portfolio management.

Second, federal laboratories should be sensitive to the commercialization requirement (Code of Federal Regulations 37, Part 404) of any patent license they issue. This regulation requires that the licensee be capable of commercializing and intending to commercialize broadly the government patent in a reasonable time frame. This condition is not difficult for a large corporation to meet. Often, however, when dealing with a small business where this may be a problem, a federal laboratory should lean toward issuing nonexclusive licenses, at least until the licensee can show, either by internal growth or strategic partnerships, the capability of supplying the marketplace with the licensed technology. Further, milestones need to be a part of the patent licence to ensure that the licensee is bringing the technology to the market without delay. Failure to meet milestones can be a clear indication that effective commercialization will not be achieved. In such a case, the licence can either be terminated or converted to a less exclusive mode so that others may have the opportunity to license and advance commercialization. Options for greater exclusivity can be incorporated into nonexclusive licenses to allow a small business license to gain exclusivity at a later date when it is in a better position to prove its ability to commercialize the technology.

The PSD Technology Transfer Office usually requires an up-front fee when licensing its technology to business and industry. This fee is charged to partially cover the cost of applying for the patent and the related maintenance fees, and as a measure of good faith of the licensor. Up-front licensing fees charged to small business are usually paid in increments distributed over a period of years. Royalty rates vary with the technology field and can sometimes be partially written off against the initial up-front licensing fee. Knowing that the bottom line is the timely commercialization of the technology, PSD uses flexibility in licensing terms to bring PSD technology to the market as quickly as possible.

Each of the patent license issues raised here involves negotiations. Negotiation is a talent that must be developed through education, good legal advice, and experience. This seems to be one of the most difficult aspects of technology transfer with which federal technology transfer offices have to deal. Fortunately, a federal laboratory often has more maneuvering room in negotiating than private sector firms because its priority is technology commercialization.

Technology Marketing

There are many means by which PSD can and does market its technology. PSD showcases its technology at technical expos, advertises in trade journals, utilizes technology brokers, distributes booklets describing PSD patents available for licensing, provides inputs to major databases that are broadly available to the public, publishes in the Federal Register, examines citations to its patents that appear in other newly issued patents to expose potential areas of infringement, publishes and presents technical papers at conferences and symposiums, sponsors workshops and seminars, and has its technologists network with their peers in the private sector. By far, PSD's major marketing successes have been achieved through contacts made by its technologists networking with their peers in business, industry, and academia. Most PSD S&Es now operate in an entrepreneurial-like mode in dealing with potential technology users in the private sector. The results indicate that PSD S&Es are highly motivated to bring in new technology partners/users; act as consultants to the private sector; and deliver unique prototype designs, components, and devices to business and industry. These technologists, understanding the benefits of technology marketing, have changed the way PSD does business.

Examples of PSD CRDAs pertinent to the area of ceramics include:

- "Laser Ablation of Ferroelectric and High-Temperature Superconducting Thin Films," Rutgers University, New Brunswick, NJ.
- "Development of Smart Ceramic Materials," Rutgers University, New Brunswick, NJ.
- "Development of Hermetic Coatings for Optical Waveguides," Rutgers University, New Brunswick, NJ.
- "New Piezoelectric Materials with Application to Frequency Control," RF Monolithics, Dallas, TX.
- "Development of a Permanent Magnet System for a Microwave Tube," Martin Marietta, Rancho Bernardo, CA.

Because PSD has spent several years developing an effective, efficient, and productive technology transfer program, the major effort of the PSD Technology Transfer Office today is on managing the program rather than developing it. Areas such as marketing, licensing, negotiation, program tracking, exploring innovative approaches to technology commercialization, and managing the intellectual portfolio have now become the main tasks of the Technology Transfer Office at PSD.

The SBIR Program and Technology Transfer

The PSD Technology Transfer Office has developed a new and innovative method of promoting the commercialization of its technology: the Small Business Innovation Research (SBIR) program, which addresses the development, licensing, and commercialization of PSD patents. Proposals that are aimed at productizing specific PSD patents that have already been prototyped and demonstrated by Army technologists/inventors are solicited. These prototype designs still require commercially oriented development and optimization, with consideration given to manufacturing and transitioning into actual utilization for both military and private sector applications. SBIR contractors who successfully productize the subject patents are licensed by the Army to produce and market the technology.

The SBIR program appears to yield significant advantages. First, the small business is given funding to commercialize a relatively low-risk invention. This productization venture could in turn become the mainstay and contribute to the healthy growth and success of that particular company. Second, those inventions chosen to be commercialized under the SBIR program are handpicked, with strong consideration given to the need for that particular technology in both the military and private sector. Third, this methodology provides for the development of "niche" technologies where large corporations have either no capability or no desire to initiate development efforts and where small businesses have no funding base to pursue the technology development on their own. Lastly, using this approach, Army technology is fully developed, demonstrated, and effectively utilized for commercial applications, with patent licensing fees and royalties being paid to the inventors and the Army laboratory where the invention was conceived.

This new commercialization program effort is now in progress under two SBIR contracts and is advancing well. Even before completion of the SBIR contracts, the small business contractors are already producing results and expect to start delivering on orders from the private sector in the near future.

Follow-Up

One of the most important facets of a well-oiled technology transfer program is follow-up. Follow-up on personal contacts made at expos, conferences, seminars, etc. leads to new partnering arrangements under CRDAs and PLAs. Tracking milestones, royalty payments, commercialization plans, etc. with respect to existing patent licenses ensures that laboratory technology stays on the road to commercialization. Follow-up on progress and payments (if required) under existing CRDAs, and follow-up on investigative measures taken to identify possible infringements on patents are equally important. Continuous follow-up in all areas of the technology transfer cannot be overemphasized.

Conclusion

Even when a federal laboratory technology transfer program has been successfully developed and is in full operation, one cannot assume that technology and money will flow automatically. Marketing efforts must expand apace and technologists must be continually educated in the value of their technologies and their intellectual property. S&Es must be encouraged to become marketeers, while attorneys must be encouraged to become user-friendly. All participants — legal, technology-transfer, and marketing — must learn to be better communicators to nurture more effectively and harvest more fully the fruits of the strategic alliance that has become our new culture.

References

American Technology Preeminence Act of 1991 (PL 102-245)
Army Regulation 70-57, "RD&A Military–Civilian Technology Transfer," 7/91
Bayh-Dole Act of 1980 (PL 96-517)
Cooperative Research Act of 1984 (PL 98-462)
Defense Authorization Act for FY 1991 (PL 101-510)
Exec. Orders 12591 & 12618 (1987): Facilitating Access to Science & Tech
Federal Technology Transfer Act of 1986 (PL 99-382)
Intermodal Surface Transportation Efficiency Act of 1991 (PL 102-240)
Japanese Technical Literature Act of 1986 (PL 99-502)
Malcom Baldrige National Quality Improvement Act of 1987 (PL 100-107)
National Competitiveness Technology Transfer Act of 1989 (PL 101-189)
National Department of Defense Authorization Act for 1993 (PL 102-25)
National Department of Defense Authorization Act for FY 1993 (PL 102-484)
National Department of Defense Authorization Act for 1994 (PL 103-160)
NIST Authorization Act for 1989 (PL 100-519)
Omnibus Trade and Competitiveness Act of 1988 (PL 100-418)
Small Business Innovation Development Act of 1982 (PL 97-219)
Small Business Technology Transfer Act of 1992 (PL 102-564)
Stern, R. and Wittig, T. (in press). Technology transfer: Lessons learned — Preparation for the future. *In* "Proceedings of the Technology Transfer Society Annual Conference." Washington, D.C., July 1995
Stevenson-Wydler Technology Innovation Act of 1980 (PL 96-480)
Trademark Clarification Act of 1984 (PL 98-620)
Water Resources Development Act of 1988 (PL 100-676)

—2—

Medical Ultrasonic Diagnostics

ALBERT GOLDSTEIN
Wayne State University, Detroit, Michigan

RAYMOND L. POWIS
Consultant for Ultrasound Science, Education and Research, Redmond, Washington

PHYSICAL ACOUSTICS, VOL. XXIII
ISBN 0-12-477923-9 $30.00

I. Introduction

A. Triumph of Applied Science

In the roughly thirty years that medical ultrasonic imaging has been available clinically it has progressed rapidly from barely interpretable images to one of the premier methods of noninvasively imaging internal soft tissue structures and blood flow. However, even with all of its technological and image quality advances it is still an immature imaging modality. In a mature imaging modality, such as x-ray plain films, all instruments function essentially the same—a radiologist cannot identify the manufacturer from the radiograph alone. In ultrasonic imaging equipment, however, each manufacturer uses different transducer designs, system architecture, and signal processing for each clinical imaging task, and a trained radiologist *can* identify the equipment manufacturer from the ultrasonic image alone. Therefore, advances in medical ultrasonic diagnostics that will result in the best combination of system architecture, transducer beam patterns, and signal processing for each clinical imaging task can be expected as the modality matures.

B. Medical Imaging Concerns

Medical imaging utilizes various forms of energy transmitted into or through the body to obtain images depicting internal anatomy. Each image is used in an attempt to answer a specific medical question. The physical or cosmetic appearance of the image is secondary to its medical diagnostic information. While most present-day medical images are of such high quality that they appear to be easily understood, it is important to realize that only a highly trained medical professional can accurately interpret them.

Patient safety is also a concern. Many imaging modalities present a small, finite risk to the patient due to the energy used and a benefit/risk analysis must be performed prior to imaging. In today's medical environment, cost-effectiveness is also an important consideration.

C. Plain Film Radiography

The oldest (most mature) medical imaging modality is plain film radiography where a broad beam, short time duration, burst of x-rays passes through the body to produce a shadowgram of raw transmission data on medical x-ray film. This two-dimensional image actually represents three-dimensional patient anatomy with the real space dimension along the path of the x-rays collapsed in the image. Patient contrast on x-ray plain films is limited to discriminating between bone, soft tissue, and air. Soft tissue structures cannot be easily or routinely identified or studied using x-ray plain films.

D. High-Tech Imagers

The newest imaging modalities—ultrasound, computed tomography (CT), and magnetic resonance imaging (MRI)—permit soft tissue visualization, identification, and medical evaluation. These modalities, called high-tech imagers, share certain important characteristics. Each produces a tomogram, which is a two-dimensional image of a two-dimensional slice of patient anatomy (with a certain thickness). Tomographic images permit detailed searches for focal (well-localized) lesions and precise three-dimensional localization of internal structures. Each high-tech imager also utilizes digital signal processing to obtain the final image from the acquired raw image data, presenting a final image on a CRT.

There is one important difference between these high-tech imagers. CT and MRI present actual tissue data in their images (a tissue map) whereas ultrasonic images present raw echo amplitude data (a raw data map). In CT images, the image shades of gray represent CT numbers that are directly proportional to tissue x-ray attenuation coefficients. In MRI images, the image shades of gray represent a linear combination of the relaxation coefficients (T_1 and T_2) of the proton nuclear spins precessing about an externally applied magnetic field. The raw echo amplitude data presented by ultrasonic images are encoded as image shades of gray (a raw data map like x-ray plain films). Tissue ultrasonic attenuation and reflectivity combine to produce the raw transmission data, but at present it is not possible to separate their effects. If they could be separated, then each ultrasonic scan would produce two patient tissue maps—one of tissue attenuation coefficients and another of tissue reflection coefficients.

E. Overlapping of Disciplines

Medical ultrasonic imaging is a combination of several distinct disciplines: medicine, transducer development, ultrasonic physics, digital electronics, and display technology. It began when early pioneers (Kimmelman, 1988) developed the technology to produce crude but accurate images of internal soft tissue anatomy. As the medical interest rose in these fledgling images so did the projected medical need. Increased technology transfer satisfied the growing medical needs and raised possibilities for new imaging solutions to medical problems.

At present the synergy of these independent disciplines enables continuing improvements. Some evolutionary improvements come from satisfying stated medical needs. Some revolutionary improvements come from large, wholesale transfer of new technologies into the imaging equipment. The FDA, through the Medical Device Act of 1968, acts as a watchdog agency, limiting the acoustic output of ultrasonic imaging equipment in the name of patient safety.

Although the stochastic pace of progress comes from any and all of the separate disciplines, physicians—and they alone—provide the final judgments concerning the acceptance of any new development. This fact makes this melange of independent disciplines unique.

F. Advantages of Medical Ultrasonics

Ultrasonic imaging has several unique advantages over other medical imaging modalities that have aided its clinical acceptance. For instance, the image presentation is in real time, which permits not only the identification and study of moving internal structures but also rapid and complete survey scans of patients. Also, ultrasonic radiation is believed to be nontoxic to tissues at diagnostic levels (Barnett *et al.*, 1997), which permits obstetrical imaging without endangering the mother or fetus. Furthermore, ultrasonic imaging equipment is portable, so imaging exams can be performed at the bedside of critically ill patients. And last but not least, ultrasound is cost-effective because of its high patient throughput and low equipment cost. For example, a CT scanner costs about \$1M, an MRI scanner costs about \$2M, and an ultrasonic scanner costs \$0.15–0.25M. Ultrasonic scanners designed for single-purpose uses such as in-office ob/gyn scans or operating room scans are even less expensive.

II. Basic Imaging Principles

A. The Nature of Ultrasound

Sound waves are mechanical vibrations that propagate in a host medium. They are coupled modes between medium particles oscillating about equilibrium positions and a traveling ultrasonic wave. Solids support the propagation of both longitudinal waves (particles oscillating parallel to the wave propagation direction) and transverse waves (particles oscillating perpendicular to the wave propagation direction). Fluids (gases and liquids) only support longitudinal wave propagation. The lean body mass is approximately 72% water and the remainder is fat, which is fluidlike, so only longitudinal waves can be used to probe the human body. Transverse waves may be generated in bone due to mode conversion, but because of the bone's high attenuation, they do not contribute to ultrasonic image formation.

Propagating sound waves obey the standard relation

$$c = \lambda f, \tag{1}$$

where c is the acoustic velocity in the medium, f is the frequency of the wave, and λ is the acoustic wavelength. For single-frequency, continuous wave (CW) sound waves, at a single point in the medium f is the number of incident pressure (or any other wave parameter) cycles per second (Hz) and at a single instant of time λ is the basic spatial cyclic repetition distance of the single-frequency wave.

Sound waves with frequencies above 1 MHz (ultrasound) can be easily generated and focused and will propagate reasonable distances in soft tissue. Pulse-echo ultrasonic measurements, similar to those of sonar and radar, are now used routinely in medicine to provide detailed images of cross-sectional anatomy.

This section of the chapter will review the basic physical principles used in medical ultrasonic image formation. Only essential principles will be described; details are available in the references.

B. Transducer

1. *Piezoelectricity*

The generation and reception of ultrasonic waves is accomplished using piezoelectric crystals [whose properties have been described in detail elsewhere (Berlincourt *et al.*, 1964)]. Most medical ultrasonic transducers are fabricated using a polycrystaline ferroelectric ceramic material, lead zirconate titanate ($Pb(Zr,Ti)O_3$), called PZT. This material is produced by multiple cycles of ball-milling and sintering under pressure until the resultant ceramic grains are quite small and randomly oriented. A thin slab of PZT is then cut, shaped, and electroded on its flat surfaces with fired-on silver or sputtered chrome-gold. When heated to a temperature just above its Curie temperature and then cooled slowly with a strong electric field applied across the electroded surfaces, the PZT ferroelectric grains align. The resultant capacitive structure exhibits strong piezoelectric properties operating in a thickness-expander mode. When placed on the skin, the thickness-expander oscillations of the PZT crystal surfaces produce (and detect) longitudinal ultrasonic waves.

Figure 1 shows the essential components of a medical ultrasonic transducer. The thin PZT crystal is bonded to a rear backing material and a front plastic wear surface and the three components fit inside a plastic case. The purpose of the backing material is to dampen the crystal oscillations, producing a short output pulse. The purpose of the wear surface is twofold: It protects the thin front electrode from mechanical damage when scanning a patient, and it insulates the patient from potential electrical shock hazards. Even though the electrode facing the patient is kept at ground potential, an electrical short in the transducer could deliver a hazardous voltage on the front electrode, so this double insulation feature is essential for patient safety.

The piezoelectric crystal is energized into transmission by the application of an electrical voltage across its electroded surfaces that causes the crystal to deform slightly in thickness. When the voltage terminates, the deformed crystal surfaces attempt to recover to an undeformed state; this generates two longitudinal ultrasonic waves, one propagating into the crystal and the other propagating into the adjacent bonded layer.

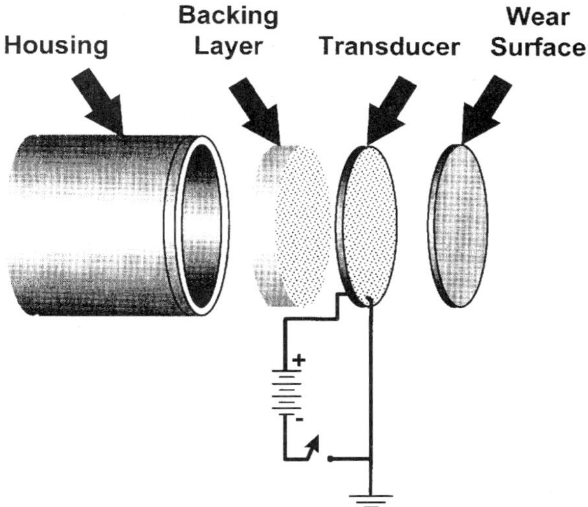

FIG. 1. Main components of a single element transducer. The piezoelectric crystal has electroded surfaces that are connected to both transmission and reception circuits. It is bonded to a backing layer and a wear surface. The backing layer acts as an absorber dampening the crystal vibrations. The wear surface functions both as an electrical insulator and as part of its focusing mechanism.

The two in-going longitudinal waves may reinforce each other as they reflect back and forth from the electroded surfaces, causing resonant thickness oscillations of the crystal. These resonance oscillations occur at a frequency at which the crystal thickness is half an acoustic wavelength or odd multiples of this frequency (Hunt *et al.*, 1983).

The outgoing wave that propagates into the wear surface will pass through it and into the patient. The outgoing wave that propagates into the backing material will be absorbed without reflection back into the crystal. The purpose of the backing layer is to absorb ultrasonic energy, causing the transmit time (or output pulse length) to be as short as possible.

2. *Frequency Bandwidth*

Medical ultrasonic transducers produce pulsed radiation with a bandwidth of frequencies. The relationship between the pulse length and the bandwidth of frequencies is given by the well-known uncertainty relation

$$\Delta t \cdot \Delta f \cong 1, \tag{2}$$

where Δt is the uncertainty in pulse arrival time (known as the pulse length τ) and Δf is the uncertainty or broadening of the pulse frequency, which is the frequency bandwidth, B. These two are inversely proportional, so short ultrasonic pulses will have a large frequency bandwidth. The ultrasonic pulses have a center frequency, f_c close to the half-wavelength resonance condition and a frequency bandwidth, B, determined by transducer construction and the degree of damping (AIUM, 1992). The transducer can be considered as a frequency passband filter both in transmission and reception.

Damped oscillations are generally specified by a quality factor, Q, equal to f_c/B. In medical ultrasonics it is preferred to specify the output pulse fractional bandwidth in percent:

$$FB = \frac{B}{f_c} 100 = \frac{100}{Q}. \tag{3}$$

Early single element static scan transducers had a typical fractional bandwidth of 40%.

The desire for short pulses leads to two trade-offs in system design. First, all system electronics must have a commensurate frequency bandwidth (or pulse stretching will occur). Also, due to the low Q caused by backing layer absorption, transducer efficiency in transmit and sensitivity in reception are reduced substantially.

3. *Reception*

It is instructive to consider in some detail the transducer in reception. The piezoelectric effect causes a surface charge density, σ_3, on the crystal electroded surfaces (Christensen, 1988),

$$\sigma_3 = -d_{33} P_3, \tag{4}$$

in the absence of an applied electric field. Here, d_{33} is the ceramic piezoelectric strain coefficient and P_3 is the echo wave pressure in the thickness direction. The instantaneous charge produced on one electroded crystal surface by the received echo is

$$Q_T = A\sigma_3 = -d_{33} P_3 A, \tag{5}$$

where A is the area of the electroded surface. The electroded crystal is a capacitive structure and the voltage across the crystal caused by echo reception is

$$V_T = \frac{Q_T}{C_T} = -\frac{d_{33}l}{\varepsilon'}P_3, \tag{6}$$

where $C_T = \varepsilon' A/l$, ε' is the crystal dielectric constant, and l is the crystal thickness.

The received echo voltage across the piezoelectric crystal is proportional to the average echo pressure incident on the crystal front surface independent of its area A. This pressure sensitivity explains the fact that the transducer is a coherent detector as well as a coherent transmitter of ultrasonic radiation. Due to the transducer pressure sensitivity, the following discussion on sound propagation will be presented only in terms of wave pressure and not intensity (power density) as in other ultrasonic texts.

Since the transducer is connected to the system electronics by a coaxial cable, the received system transducer voltage, V_S, is the crystal voltage, V_T, shunted by the cable capacitance, C_c,

$$V_S = \frac{Q_T}{C_c + C_T} = \frac{C_T}{C_c + C_T}V_T. \tag{7}$$

So, even though the received crystal voltage is independent of the transducer size, the system's received voltage is dependent on A. Small area transducers such as hydrophones or the individual crystals in multielement transducers (Section IV.A) produce weak signals unless connected to preamplifiers by short cables.

Medical ultrasonic transducers must occasionally be sterilized. Traditional autoclaving (boiling in water) cannot be used for two reasons: (1) the crystal Curie temperature may be exceeded, and (2) mechanical expansion stresses can cause the bonds between the crystal and the backing layer or wear surface to fail.

C. Transmit Beam Pattern

The transducer is the most critical component of an ultrasonic imaging system. It is the antenna that directs transmitted ultrasonic energy into the patient and receives the returning echoes. Its beam pattern determines image resolution both spatially and in contrast.

1. *Unfocused Flat Circular Piston*

Understanding of beam focusing begins with the beam pattern of an unfocused, flat, circular, single element transducer. Figure 2 shows a simplified representation of the radiated beam pattern for CW excitation at a single frequency. There are two regions: a cylindrical near-field (Fresnel Zone) and a conical far-field (Fraunhofer Zone). Actually, there is a natural focus of the beam at approximately three-quarters of the near-field length (Zemanek, 1971).

In the extreme far-field, the radiation pressure falls off inversely with distance according to the inverse-square law for point sources. In the near-field, interference effects caused by rays emanating from different portions of the front surface cause spatial variations of point pressure both axially and laterally. At the juncture of the two zones, the pressure has its last axial maximum (Kinsler *et al.*, 1982).

As shown in Figure 2, the length of the near-field and the conical half-angle of divergence in the far-field both depend on the transducer frequency (λ in the propagation medium) and aperture size. With a constant aperture, increasing the frequency increases the length of the near-field and decreases the angle θ, producing a more directional beam pattern suitable for imaging. Decreasing the aperture size at a constant frequency causes a shorter near-field and larger angle θ, producing the desired beam pattern for an omnidirectional hydrophone.

2. *Spherically Focused Piston*

A circular aperture, spherically concave single element transducer will produce a CW focused transmit beam pattern in an unattenuating medium

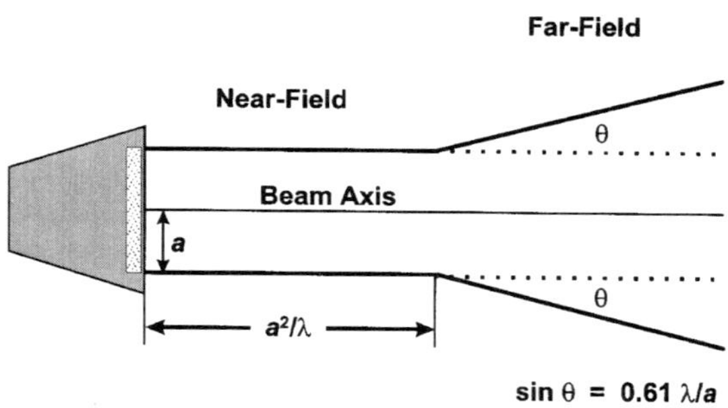

FIG. 2. Approximate beam pattern of an unfocused circular piston. It is composed of a cylindrical near-field and a conical far-field. The transition between the two is the position of the last axial pressure maximum. The length of the near-field and the sine of the half-angle of divergence in the far-field are given.

similar to that shown in Figure 3 (O'Neil, 1949) (Lucas and Muir, 1982) (Chen *et al.*, 1993). Due to the finite size of the ultrasonic wavelength, this diffraction-limited focus has several distinct characteristics. The peak axial pressure occurs before the geometric focal plane (located at the center of curvature of the spherically curved crystal) and the focused beam pattern is axially asymmetric. At the geometric focal plane and beyond, there are distinct low-amplitude side lobes along with the high-amplitude axial main lobe. Close to the crystal surface, the beam pattern is very complicated due to interference effects. The beam pattern has cylindrical symmetry due to the circular symmetry of the transducer aperture.

In the geometric focal plane, the lateral beam pressure (beam profile) is proportional to the jinc function (Goodman, 1968),

$$P(r) \propto \frac{2J_1\left(\dfrac{kar}{F}\right)}{\dfrac{kar}{F}}, \tag{8}$$

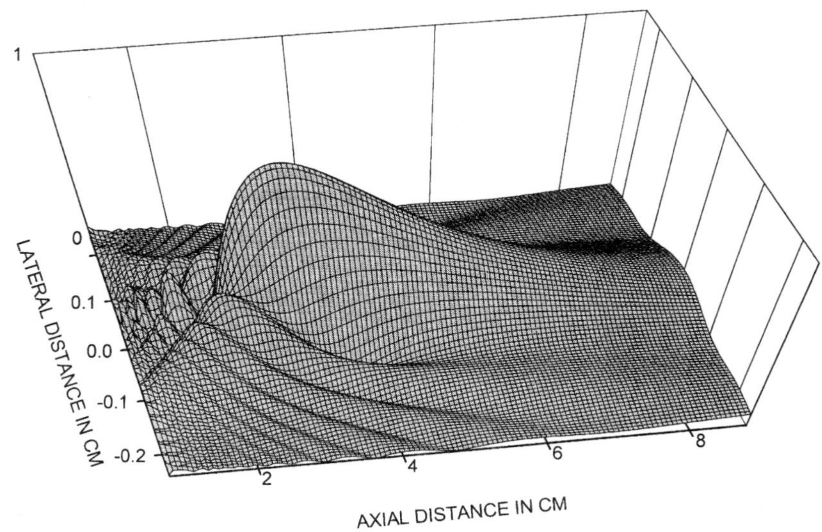

FIG. 3. Unattenuated transmitted beam pattern of a 0.5-cm-diameter, 19.7-MHz, spherically focused transducer with a 4-cm radius of curvature and a medium with $c = 1540$ m/s. The cross-sectional beam pattern shown has cylindrical symmetry. The vertical axis of this surface plot is the transmitted pressure magnitude. (Goldstein, A., unpublished calculation.)

where $k = 2\pi/\lambda$, F is the spherical radius of curvature, a is the circular aperture radius, J_1 is a Bessel function of the first kind, and r is the lateral radial distance coordinate.

If a spherically focused single element transducer has a rectangular aperture of dimensions l_1 and l_2, in its geometric focal plane the transmit beam profile is proportional to the product of two sinc functions (Goodman, 1968):

$$P(x, y) \propto \frac{\sin\left(\frac{kl_1x}{2F}\right)}{\frac{kl_1x}{2F}} \cdot \frac{\sin\left(\frac{kl_2y}{2F}\right)}{\frac{kl_2y}{2F}}, \tag{9}$$

where x and y are rectangular coordinate lateral distances parallel to the l_1 and l_2 dimensions. This beam pattern differs in the two orthogonal lateral directions due to the lower symmetry of the transducer aperture.

Figure 4 presents the geometric focal plane transmitted pressure for a circular aperture and one of the orthogonal directions of a rectangular aperture. The lateral distance has been normalized to half the main-lobe full width by expressing the jinc and sinc functions as

$$\text{jinc}(\varepsilon) = \frac{J_1(2\cdot 3.83171\cdot\varepsilon)}{3.83171\cdot\varepsilon} \qquad \text{and} \qquad \text{sinc}(\varepsilon) = \frac{\sin(2\cdot\pi\cdot\varepsilon)}{2\cdot\pi\cdot\varepsilon}, \tag{10}$$

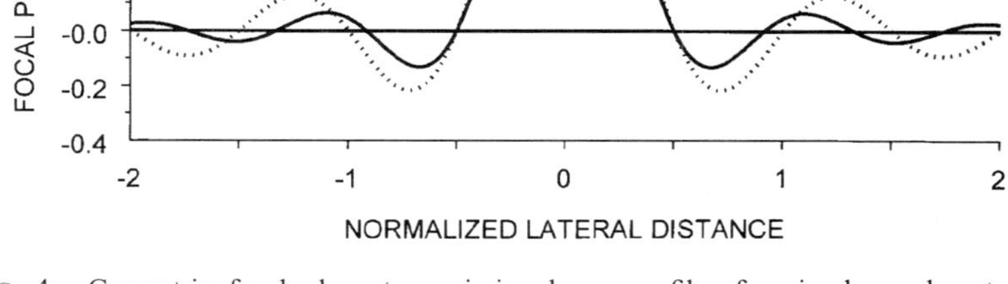

FIG. 4. Geometric focal plane transmission beam profiles for circular and rectangular aperture, spherically focused transducers. The lateral distance is normalized to half the main-lobe full width of the beam profile. The circular beam profile is proportional to a jinc function and the rectangular beam profile is proportional to a sinc function.

where ε is the normalized lateral distance. The main lobes are practically identical but the side lobes for the two apertures differ in spacing and magnitude.

Rules for the design of focused beams are derived from unfocused beam patterns, which are essentially beams focused at infinity. A circular aperture, spherically concave transducer will not form an adequate focus if its geometric focal length is in the far-field of the equivalent unfocused transducer (with the same aperture and frequency) (Kossoff, 1979). The shorter the geometric focal length with respect to the equivalent unfocused transducer near-field length, a^2/λ, the stronger the focus. So when a focused beam is desired at a certain depth in tissue, the proper combination of a and λ is chosen such that the equivalent unfocused transducer near-field length is larger by the appropriate amount.

D. Propagation in Soft Tissue

1. *Acoustic Velocity*

Soft tissue acoustic properties have been studied extensively (Goss *et al.*, 1978 and 1980). The acoustic velocity of soft tissues may be considered to be frequency independent. Typical values for air, water, various soft tissues, and bone are given in Table 1. Most soft tissue has an acoustic velocity within ±3% of an average value of 1540 m/sec. Fat is an exception at nearly 6% less. The various types of bones in the body have much higher acoustic velocities due to their higher density and bulk moduli. The 1540 m/sec

TABLE 1

ACOUSTIC VELOCITIES AND IMPEDANCES FOR SEVERAL MEDIA

	Velocity (m/s)	Impedance (MRayls) (Z values)
Air	330	0.00004
Water	1480	0.148
Fat	1450	0.138
Blood	1570	0.161
Kidney	1560	0.162
Soft tissue (average)	1450	0.163
Liver	1550	0.165
Muscle	1580	0.170
Bone	4080	0.780

(Goldstein, 1988) Reprinted by permission of John Wiley & Sons, Inc.

average acoustic velocity in soft tissue means a pulse-echo travel time of 13 μsec for each cm of range in the image.

2. *Attenuation*

Soft tissues have exponential attenuation coefficients that are proportional to frequency (Insana, 1995). Tissue attenuation is usually expressed in the convenient units of dB/cm-MHz. Table 2 lists the attenuation of the substances listed in Table 1. The attenuation of soft tissue is generally between 0.5 to 1 dB/cm-MHz.

The frequency dependence of tissue attenuation limits the tissue penetration of high-frequency ultrasonic pulses. To image 10 cm deep in tissue, for example, ultrasonic pulses must travel 20 cm round-trip and 5-MHz ultrasonic waves will be attenuated 50–100 dB. This frequency dependence of ultrasonic attenuation mandates that medical imaging equipment have a collection of transducers covering different frequency ranges: Lower-frequency transducers are used for deep imaging and higher-frequency transducers are used for shallow imaging.

The anisotropy of muscle ultrasonic attenuation poses no problem in medical ultrasonic imaging for two reasons: (1) most medical imaging is

TABLE 2

SOME TYPICAL VALUES FOR ULTRASONIC ATTENUATION COEFFICIENTS

	Attenuation (dB/cm at 1 MHz)	HVL[b] (at 1 MHz) (cm)
Air (STP)	12 (f^2)[a]	0.25
Water	0.002 (f^2)[a]	1500
Fat	0.63	4.76
Blood	0.18	16.67
Kidney	1.0	3.00
Soft tissue (average)	0.70	4.29
Liver	0.94	3.19
Muscle		
Along fibers	1.3	2.31
Across fibers	3.3	0.91
Bone	15	0.20

[a] (f^2) indicates a quadratic frequency dependence of these attenuation coefficients.

[b] The HVL (half-value layer) is the tissue thickness required to reduce the acoustic intensity (power density) by one-half.

(Goldstein, 1988) Reprinted by permission of John Wiley & Sons, Inc.

not done through thick muscle layers, and (2) the fibers of large muscles are generally parallel to the long body axis. So for both transverse images (cross section perpendicular to long body axis) and longitudinal images (cross section parallel to the long body axis), the ultrasonic beams are perpendicular to the large muscle fibers.

The attenuation frequency dependence does affect short ultrasonic pulses. As a pulse propagates through tissue, the higher frequencies in its bandwidth are attenuated more severely than the lower frequencies. This means that echoes from deep structures have longer pulse lengths and lower center frequencies than do echoes from shallow structures.

3. *Scattering*

Ultrasonic waves propagate through uniform media undisturbed. Three general types of scattering can occur, however, if there are any variations from uniformity in the medium: scattering (reflection) from large flat boundaries, known as *specular reflection*; scattering from point reflectors (or local variations in tissue structure or density), known as *diffuse scattering*; and scattering from structures whose dimensions are commensurate with λ, known as *resonance scattering* (Faran, 1951) (see also Section VI.B).

a. Specular Reflectors. Boundaries between two different tissues that are flat over several acoustic wavelengths, such as the kidney capsule, will specularly reflect ultrasonic waves like a mirror. Figure 5 depicts specular reflection with the tissue boundary seen on edge and an ultrasonic beam incident at an angle θ_i to the boundary normal. The law of reflection mandates that the angle of reflection, θ_r, be equal to the angle of incidence, θ_i. Since pulse-echo ultrasonic imaging is performed with a single transducer, specular reflections are only received by the transducer if the ultrasonic beam is normal to the tissue interface (or very close to normal).

The magnitude of the interface pressure reflection coefficient depends on differences of the acoustic impedance, $Z = \rho c$, of the tissues at the interface, where ρ is the tissue mass density. Some typical Z values are given in Table 1. For normal incidence from medium 1 to medium 2 (Wells, 1977),

$$\frac{P_r}{P_i} = R_p = \frac{(Z_2 - Z_1)}{(Z_2 + Z_1)} \tag{11}$$

and

$$\frac{P_t}{P_i} = T_p = \frac{2Z_2}{(Z_2 + Z_1)}, \tag{12}$$

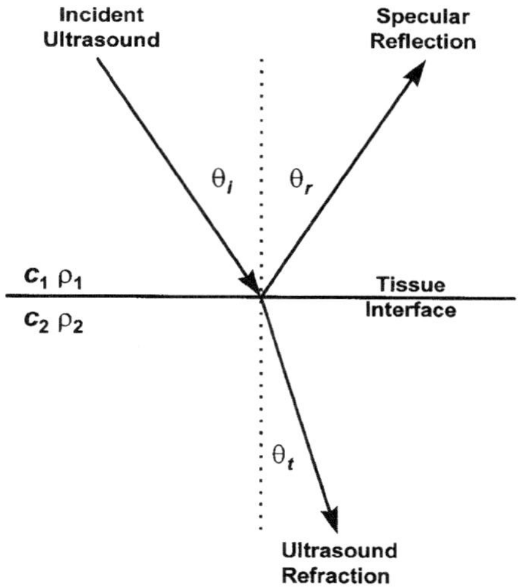

FIG. 5. A specular tissue interface is seen on edge with an ultrasonic beam incident at an angle θ_i to the interface normal. There is a specularly reflected echo at an angle θ_r to the normal. A transmitted beam is refracted by Snell's law to an angle θ_t to the normal.

where R_p is the pressure reflection coefficient and T_p is the pressure transmission coefficient.

If $Z_1 = Z_2$, the reflection coefficient is zero and only transmission occurs at the interface. This condition is called impedance matching. The larger the difference between Z_1 and Z_2, the larger the reflection coefficient. This condition is called impedance mismatching. If $Z_1 < Z_2$, the reflected wave has no phase change upon reflection and if $Z_1 > Z_2$, the reflected wave has a π radian phase change upon reflection.

Specular reflection produces high-amplitude, directionally dependent echoes that are only seen in medical ultrasonic images when the beam direction is perpendicular to tissue interfaces. Except for indicating organ size, they usually contain little medical diagnostic information.

b. Diffuse Reflectors. Point reflectors (tissue structures whose dimensions are small compared to λ) undergo Rayleigh scattering, which has a pressure scattering cross section proportional to frequency squared (McDicken, 1991) and produces omnidirectional scattering (left side of Figure 6). Some large tissue interfaces, such as the diaphragm, have surface roughness that produces a diffuse component of scattering along with specular reflection (right side of Figure 6).

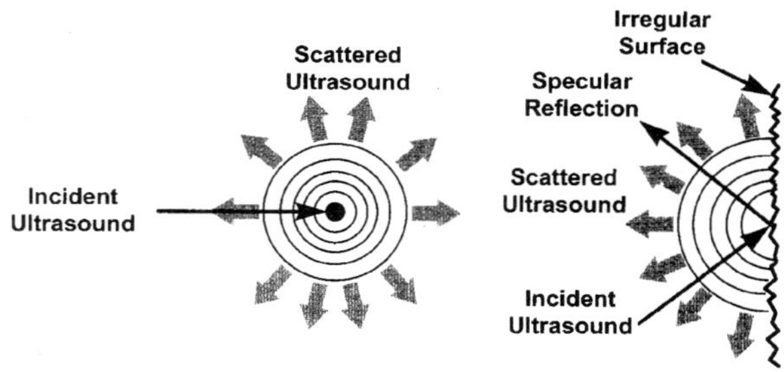

FIG. 6. Diffuse scattering. The point reflector on the left produces low amplitude, omnidirectional scattering. The rough flat interface on the right (seen on edge) produces diffuse scattering as well as specular reflection.

Diffuse scattering produces low-amplitude, omnidirectional echoes. Tissue parenchyma has local variations of density and structure that act like point reflectors in scattering ultrasound. Due to their omnidirectional scattering, diffuse echoes are always seen in medical ultrasonic images. It turns out that they contain a great deal of medical diagnostic information.

4. *Refraction*

Refraction at tissue interfaces changes the direction of the transmitted wave (see Figure 5). For a flat interface, Snell's law relates the angles of incidence, θ_i, and transmission, θ_t, as

$$\frac{\sin\theta_i}{\sin\theta_t} = \frac{c_1}{c_2}. \tag{13}$$

Refraction will not occur at a tissue interface if the two tissue acoustic velocities are equal or if there is normal incidence.

Substantial refraction occurs at a fat–soft tissue interface. Assuming a 30° angle of incidence and the acoustic velocities in Table 1, at 10 cm along the beam direction after the interface there will be a 3.6-mm misregistration of a reflector in the ultrasonic image (a difference between the actual location of the reflector and its imaged position). At shorter distances, the misregistration scales down proportionately. This amount of distortion (stretching) of a medical ultrasonic image is minor and changes none of the diagnostic information in the image. Because of these limits, refraction from flat interfaces usually plays no significant role in medical ultrasonic images.

E. Ultrasonic Image Formation

1. Pulse-Echo Measurement

Echo amplitude information is acquired using the pulse-echo principle. The central ray of the transducer-focused beam pattern is aimed along a specific direction in the patient and the transducer transmits a short pulse of ultrasonic energy that travels along the beam pattern. The transmitted pulse interacts with tissue along its path and generates a stream of echoes that travel back to the transducer. A great deal of information is contained in the returning echoes, but at present only the echo arrival time and amplitude are used for gray-scale ultrasonic imaging (Doppler measurements use the echo frequency or phase as well; see Section V). The range of the reflector, which is its depth from the transducer front surface, is determined from the echo arrival time by the range equation

$$\text{Range} = \frac{c}{2}t, \tag{14}$$

where t is the echo round-trip travel time. The factor 2 is present because the ultrasound travels the range twice: once as the transmitted pulse and once as the returning echo.

Ultrasonic imaging equipment must make three simplifying assumptions in using echo data to construct an image. First, all tissue has the same isotropic acoustic velocity. Since the acoustic velocity of most soft tissue is essentially independent of frequency and within $\pm 3\%$ of $1540\,\text{m/s}$, this assumption permits the formation of ultrasonic images that are reasonably accurate spatially. When misregistration of reflectors in the image occurs due to differing tissue acoustic velocities, the result is an image that is slightly distorted (stretched or contracted) along the beam axis direction.

Second, the ultrasonic pulse travels in a straight line so echoes returning at later times are interpreted by the range equation as being caused by proportionately distant reflectors. When there are large directional changes of the propagating ultrasonic pulse due to refraction, reflection from oblique highly reflecting interfaces, or reverberations between two highly reflecting interfaces, this assumption is no longer true and image artifacts will result (McDicken, 1991). When these image artifacts occur, they are immediately recognized by experienced operators.

Third, all detected reflectors are located on the beam central axis. This assumption is mostly untrue but gives reasonably accurate spatial information at image depths where the focused beam pattern is narrow. At image depths where the beam pattern is wide, this assumption leads to blurring of reflector position lateral to the beam central axis for high-contrast reflectors.

The ultrasonic image is composed of many individual pulse-echo transmissions and echo wave train receptions or transducer lines of sight. These image lines are restricted to a single, flat plane called the scan plane. It is important for tissue identification and image interpretation that the scan plane be flat. Medical images that represent flat planes in the patient are called tomographic images.

Even though a medical ultrasonic image is usually interpreted as representing a thin scan plane in the patient, the transducer beam pattern has a finite width perpendicular to the scan plane so the scan plane has a thickness varying with depth in the image. This finite slice thickness causes some image distortion and affects the identification of low-contrast objects in the image.

2. *Acoustic Velocity Limitation*

Rapid acquisition of ultrasonic tissue data results in a real-time display of tissue anatomy with frame rates varying from 5 to 30 Hz, but the finite acoustic velocity imposes severe limitations in real-time imaging. To avoid range ambiguity artifactual fill-in of echo-free image areas (Goldstein, 1981), the transducer cannot transmit the next line in the image until the deepest echoes from the last line have been received. This limits the number of transmitted image lines per second (equal to the transducer pulse repetition frequency, *PRF*). *LF* (lines per frame) lines are grouped into an image frame, and there are *FS* frames per second. So using Eq. (14), we have the relation

$$PRF = LF\left(\frac{\text{lines}}{\text{frame}}\right)\cdot FS\left(\frac{\text{frames}}{\text{sec}}\right) = \frac{c}{2\cdot\text{Depth}} = \frac{77{,}000}{\text{Depth}}, \tag{15}$$

where Depth is in cm. Here a fundamental trade-off in real-time ultrasonic imaging is seen; spatial and temporal resolution are in competition. Spatial resolution is dependent on *LF* while temporal resolution is dependent on *FS*, and they are inversely proportional.

As an example, consider imaging with a 20-cm-deep field of view. 3850 lines can be generated per second. The choice of *LF* and *FS* depends on the imaging task. In abdominal imaging with low-contrast, slow-moving tissue structures, spatial resolution is most important so a good choice is 15 frames/sec with 256 lines/frame. In cardiac imaging with high subject contrast (blood and myocardium), temporal resolution of valve and myocardium motion is most important so a good choice is 30 frames/sec with 128 lines/frame.

3. *Pulse-Echo Beam Pattern*

Thus far, only the CW-transmitted beam pattern of a single-element transducer has been considered. Taking a quasi-CW case (tone-burst transmission) and assuming that tissue reflectors are point reflectors, acoustic reciprocity (Morse and Ingard, 1968) ensures that the receive beam pattern is identical to the transmit beam pattern.

Figure 7 demonstrates the quasi-CW pulse-echo beam pattern of the circular aperture transducer in Figure 3. This pulse-echo beam pattern is the product of transmit and receive beam patterns and was obtained by taking the point-by-point square of the transmit beam pattern and then normalizing all lateral points at each axial depth to unity. The rationale for this normalization procedure is that at each image depth the beam profile determines lateral spatial resolution and absolute echo amplitude is irrelevant. Later it

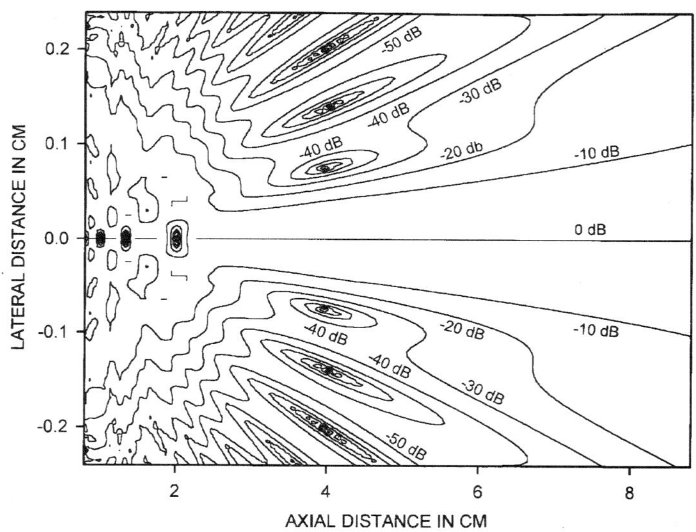

FIG. 7. Unattenuated pulse-echo beam pattern of a 0.5-cm-diameter, 19.7-MHz, spherically focused transducer with a 4-cm radius of curvature and a medium with $c = 1540$ m/s (see Figure 3). The 0-dB contour is along the beam axis by construction (see text for details). For clarity, only some of the dB contour lines are labeled. Only along the beam axis or in the geometric focal plane does the pulse-echo signal go to zero and the dB contour drop to $-\infty$. In the near-field lateral to the axial minima, the stronger side-lobe pulse-echo signals are normalized to 0 dB (at thin lines shown). (Goldstein, A., unpublished calculation.)

will be seen that amplifier swept gain in echo reception acts to roughly equalize echo amplitudes at all depths, further justifying this normalization (Section III.A.4.a).

In the geometric focal plane of a circular aperture, focused transducer the pulse-echo beam profile (pulse-echo signal due to a laterally positioned point reflector) is the square of the jinc function shown in Figure 4. Figure 8 presents this beam profile in dB down from the axial maximum using the same lateral distance (normalized to half of the main-lobe width) as in Figure 4. Only in the geometric focal plane does the beam profile go to zero ($-\infty$ dB down from the axial maximum) between its lobes.

The received echo signals are in phase for all lateral positions of a point reflector. In the first side lobe the jinc function is negative (see Figure 4), indicating that the transmitted pressure at those lateral positions is 180° out of phase with respect to the main-lobe pressure. But the reflected echo pressure also has a 180° phase shift, so the received pulse-echo signals from the first side lobe (and all other side lobes) are in phase with the main-lobe echo signals.

Two important characteristics of the beam profile are labeled in Figure 8. The beamwidth is the spatial spread of lateral echo signals when scanning a point reflector. The magnitude of the beamwidth depends on how many dB down it is defined from the main-lobe maximum. The dB difference between maximum echo signals from the main lobe and the first side lobe is called the beam echo amplitude dynamic range (von Ramm and Smith, 1978). To avoid

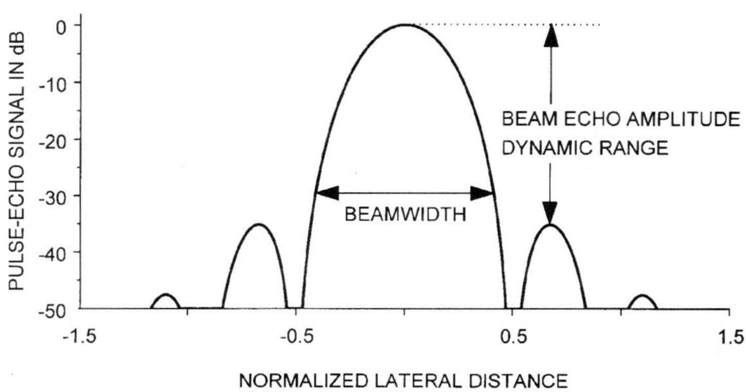

FIG. 8. Geometric focal plane dB pulse-echo beam profile of a circular, spherically focused transducer. The lateral distance is normalized to half the main-lobe width of the beam profile. The important beam parameters for high- and low-contrast resolution, beamwidth, and beam echo amplitude dynamic range are indicated. Since any 180° phase shifts that occur in the transmit beam pattern also exist in the receive beam pattern, all received lateral signals are in phase.

reception of side-lobe echoes, the equipment amplifiers should be limited to a signal dynamic range about each operating point (at each image depth) that is less than the beam echo amplitude dynamic range. For a circular aperture, the beam echo amplitude dynamic range is 35 dB. The geometric focal plane beam profile, in either orthogonal direction, for a spherically focused rectangular aperture has a beam echo amplitude dynamic range of only 26.5 dB due to its lower symmetry.

Due to the shape of the jinc function squared (see Figure 8), the beamwidth corresponding to the weakest echoes acquired at the geometric focal plane is approximately equal to the full width of the main lobe. Since the jinc function has a first lateral zero when its argument is 3.83, the pulse-echo lateral spot size diameter, *d*, in the geometric focal plane is usually given as (Christensen, 1988)

$$d = 2.44 \frac{F}{2a} \lambda = 2.44 fn\lambda, \tag{16}$$

where $fn = F/2a$ is the *f*-number of the focus. The spot size is seen to be proportional to λ, so higher frequencies give narrower beam patterns. For the spot size in either orthogonal direction of a rectangular aperture

$$d = 2 fn\lambda, \tag{17}$$

where $fn = F/l_r$ and l_r is the aperture length (see Eq. (9)).

The focal gain of a focused transducer is the relative increase in transmitted pressure (or intensity) at its geometric focal point compared to its equivalent unfocused transducer (Kossoff, 1979). Stronger focuses have smaller *f*-numbers, smaller spot sizes, and higher focal gains. So increasing the aperture of a focused transducer increases its focal gain and receive sensitivity (by acoustic reciprocity), which increases its pulse-echo tissue penetration.

In Figure 7, lateral beam contours are shown in −10-dB steps. For most antenna beam patterns, the −3-dB contour is taken as a measure of beamwidth. In medical ultrasonics, due to the large dynamic range of received echoes at each depth (over 50 dB), a larger dB down value must be used for beamwidth. If the focal length of a pulse-echo beam pattern is defined as the depth where the beamwidth is minimum, then, from Figure 7, this focal length depends on the dB down value chosen for the beamwidth. The larger this chosen dB down value, the larger the resultant focal length. Usually, for convenience, the focal length is specified as the element spherical radius of curvature.

In the focused near-field the beam pattern is wide and side-lobe magnitude can be greater than main-lobe magnitude, so reflector position misregistration is likely. Clearly, spurious information is present in images at these depths.

The range of depths (focal zone, FZ) over which the pulse-echo beam pattern will produce acceptable images has been defined in several ways: either as the range of depths over which the axial pressure signal is within $-6\,\mathrm{dB}$ of the peak signal (Steinberg, 1976) or as the range of depths over which the beamwidth is less than some multiple (2 or $\sqrt{2}$) of the minimum beamwidth (McDicken, 1991). The focal zone beamwidth definition is more appropriate for pulse-echo imaging. From Figure 7 it is seen that the beamwidth is larger and the focal zone shorter, the greater the number of dB down the beamwidth is defined.

The alternate definitions for the focal zone give roughly similar numerical results, so for the focal zone (Steinberg, 1976)

$$FZ = 7fn^2\lambda. \tag{18}$$

Equations (16), (17), and (18) for the pulse-echo spot size and the focal zone length indicate several trade-offs in focused beam pattern design. As the frequency is increased, the spot size gets smaller but so does the focal zone. At constant frequency if the focal strength is increased by reducing fn then the spot size gets smaller proportionately, but the focal zone reduces as fn^2. So as the beam is made narrower for better imaging, the depth range over which the image improves decreases.

The pulsed beam pattern in medical ultrasonic imaging may be synthesized from the CW beam patterns of the frequencies contained in the ultrasonic pulse bandwidth. However, for short pulse lengths the propagation distances involved must be carefully considered since the ultrasonic echo does not return to all parts of the transducer surface at the same time (this is only true for a point reflector at the geometric focal length). This phenomenon, known as transient diffraction (Lord, 1966), is included in so-called impulse computations of focused beam patterns (Arditi *et al.*, 1981).

Simply stated, the pulsed beam pattern is roughly similar to the CW beam pattern of the bandwidth central frequency, f_c, with near-field detail blurred due to the overlapping of different frequency interference patterns and the spot size (Eq. (16)) determined by the lower frequencies in the pulse bandwidth. So, CW computations can be used to get a good idea of focused beam properties.

An important concept associated with pulse-echo signal reception is the sample volume demonstrated in Figure 9 for a circular aperture, focused transducer. At any instant of time the ultrasonic echoes from a finite volume of tissue are received simultaneously. The lateral boundaries of the sample volume are determined by the pulse-echo beam pattern and the axial

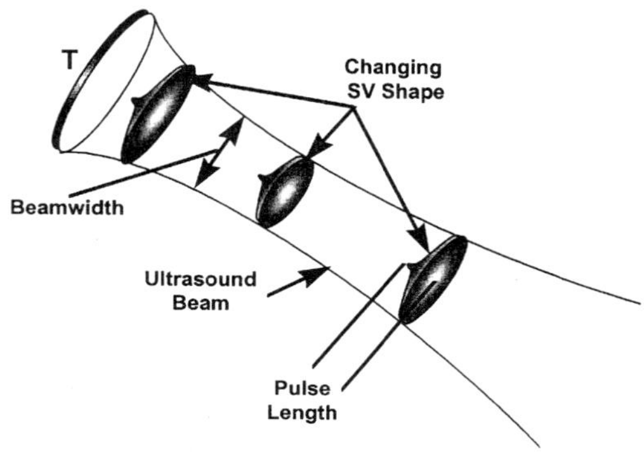

FIG. 9. Sample volume dimensions. The sample volume (SV) of a transducer (T) is shown at three successive positions along its focused beam pattern. For a circular aperture, spherically focused beam it resembles a thick round coin with axial dimples. Its thickness is governed by the ultrasonic pulse length and its diameter is governed by the lateral beam pattern.

boundaries are determined by the spatial pulse length. Axially, the proximal portion (close in) of the sample volume is determined by the point reflectors that have scattered ultrasound from the pulse leading edge, and the distal portion (further out) of the sample volume is determined by the point reflectors that have scattered ultrasound from the pulse trailing edge.

The echoes from the sample volume interfere coherently at the transducer front surface, producing a combined echo signal that will be interpreted by the equipment as a single reflector on the beam axis at the center of the sample volume. At depths where the beam pattern is wide (both in-plane and in the slice thickness direction), the sample volume can be quite large, resulting in substantial volume averaging (partial volume effects). Note that at a fixed depth in tissue the sample volume dimensions depend on the equipment output power and amplifier gain as well as the tissue attenuation — e.g., increasing the gain will increase the sample volume size.

Medical ultrasonics manufacturers have sophisticated computer simulation programs that mimic system front-end electronics and transducer characteristics so that representative images can be generated for a specific transducer design before it is fabricated. As a result, the present ambiguity in beam focal length and focal zone definitions is largely academic. Nevertheless, for these simulation programs to accurately predict the imaging performance of a transducer, several assumptions have to be valid:

1. Nonlinear propagation effects (Muir, 1980) should be negligible (or taken into account in the beam computations) (Section VI.C).

2. Tissue should scatter ultrasound as individual point reflectors and not as resonant scatterers.
3. Tissue acoustic velocity should be uniform, isotropic, and close to the assumed value of 1540 m/s.
4. Tissue should be uniform and homogeneous with respect to acoustic properties. Obese individuals have interspersed in their soft tissues tiny globules of fat that act like small defocusing lenses that scatter the ultrasound, widen the effective beam pattern, and reduce image detail.
5. Most of the attenuation of ultrasound in tissue should be due to absorption so that multiple scattering effects (Turner and Weaver, 1995) and their deleterious effects on the image are minimal.

4. Acoustic Coupling

Contact scanning with the transducer moving on the skin surface is the preferred manner of producing medical ultrasonic images. A thin layer of air between the transducer and dry skin will cause severe impedance mismatches at the wear surface–air and air–tissue interfaces (see Table 1). These mismatches prevent the ultrasound from entering the patient. Ample amounts of coupling gel must be applied to the skin surface. These gels have a suitable acoustic impedance to maximize the transmitted ultrasonic energy.

Another important concept is the "acoustic window." The acoustic window is the entrance area on and below the skin surface that permits the ultrasonic waves to reach the desired anatomical structures below. Sometimes the available acoustic window dictates the type and size of transducer to be used. For example, cardiac scans can only be performed with a transducer whose "footprint" fits into the space between ribs on the skin surface (an intercostal space) and obstetric fetal scans are best performed with a transducer with a long footprint to obtain an image with a large field of view at shallow depths. Thus a multiplicity of transducers is needed for medical ultrasonic imaging due to the different acoustic windows available.

F. Image Resolution

Ultrasonic pulse-echo measurements are performed in an uninterrupted medium (tissue) in order to produce a continuous (or quasi-continuous) image of tissue for medical diagnostic purposes. For identification and diagnosis it is crucial to be able to resolve the image tissue information. There are two types of image resolution: spatial and echo amplitude. Spatial resolution allows one to identify two adjacent structures in the image. It is usually associated with high-contrast objects in the image (such as two point reflectors). Echo amplitude resolution allows one to identify a region of echoes (reflectors) whose amplitude is slightly greater or less than the amplitude of the surrounding echoes (reflectors). Echo amplitude resolution is also called contrast resolution and is usually associated with low-contrast objects in the image (such as small cancerous focal lesions). While equipment-marketing literature usually touts spatial resolution, it is the contrast resolution of medical ultrasonic images that is of vital clinical importance.

There are two equipment parameters that contribute to image resolution: image line density and sample volume size. (Here it will be assumed for simplicity that the sample volume lateral size is its main-lobe full width (see Section II.F.1)). Both can vary over the image plane and can be axially asymmetric at each point in the image.

The number of lines in each image frame, *LF*, is an equipment design parameter that partially depends on the acoustic velocity, *c*, (Eq. (15)) and is limited by multielement transducer design and circuit complexity considerations. Image information is continuous along image lines and discontinuous across image lines (much like broadcast TV images). So the larger *LF*, the more potential information in the image.

Figure 10 demonstrates the relationship between image line density and sample volume size. A transducer is at the top aiming straight down. A parallel line image is being generated from left to right with a single point reflector in the magnified portion of the image. The projection of the sample volume on the scan plane is indicated by dotted lines. The sample volume dimension parallel to the lines is the spatial pulse length and the sample volume dimension across the lines is the beamwidth at that depth.

The top of Figure 10 demonstrates the instant when the sample volume of one of the lines first contacts the point reflector. The middle of the figure demonstrates a later instant when the sample volume loses contact with the point reflector. During the time the point reflector was inside the sample volume it was being registered in the image on the active line (beam axis). The result is an intensified short vertical section of the line. The bottom of the figure demonstrates a much later time when point reflector echoes are no longer being received from any lines in this image frame. Since medical ultrasonic displays interpolate between lines to fill in image voids, the point reflector creates an asymmetric image blur pattern whose thickness is equal to the pulse length and whose length is equal to the beamwidth (Goldstein and Clayman, 1983). Pulse length may increase slightly with image depth due to frequency-dependent attenuation. Beamwidth will vary with image depth according to transducer focus.

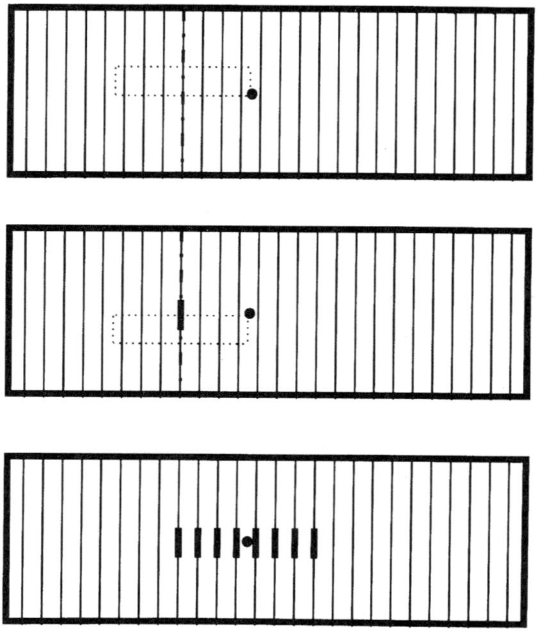

FIG. 10. Image of a point reflector. A magnified portion of an ultrasonic image frame is shown. The transducer is on top aiming down, and image lines are acquired from left to right. Top: The sample volume of one line is starting to receive point reflector echoes. Middle: The sample volume is ending reception of point reflector echoes. Bottom: All lines in the image frame have acquired point reflector echoes. The voids between image lines will be filled in to produce a blur pattern whose width is related to the pulse length and whose length is related to the beamwidth at the depth of the point reflector.

If the beamwidth is much larger than the line separation, it is the determining factor in lateral image blur. If the beamwidth is much smaller than the line separation, then low line density limits the lateral image blur. If the beamwidth and line separation are comparable, then the lateral image blur is due to both. (A more detailed discussion of the required image line density is given in Section IV.G.1.) The sample volume size is usually the determining factor in image resolution. The manner in which the sample volume affects spatial and contrast resolution will now be considered.

1. *Spatial Resolution*

Traditionally, high-contrast spatial resolution has been defined as the minimum separation permissible between two point reflectors that will produce two just-resolvable images (Figure 11). The image line density is assumed

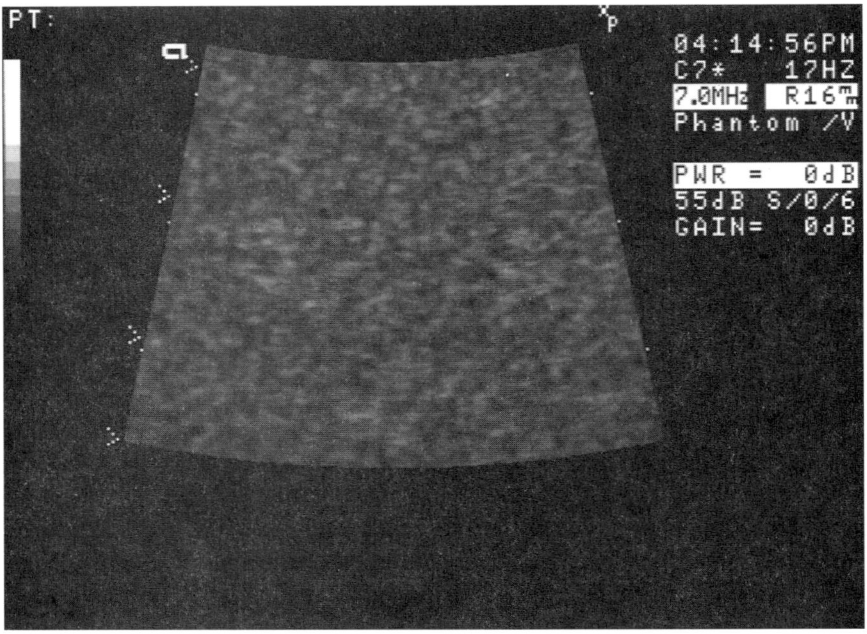
PT:
04:14:56PM
C7* 17HZ
7.0MHz R16mm
Phantom /V
PWR = 0dB
55dB S/0/6
GAIN= 0dB

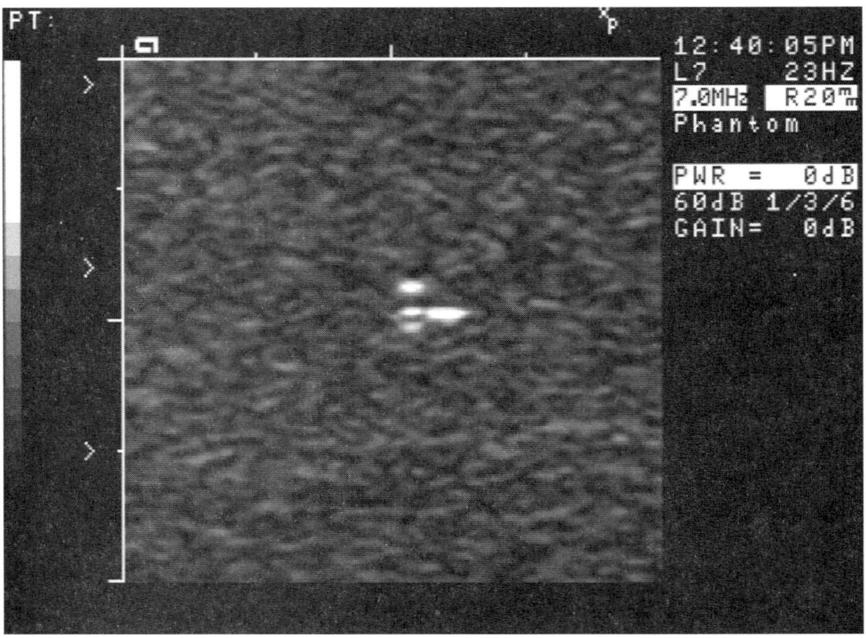
PT:
12:40:05PM
L7 23HZ
7.0MHz R20mm
Phantom
PWR = 0dB
60dB 1/3/6
GAIN= 0dB

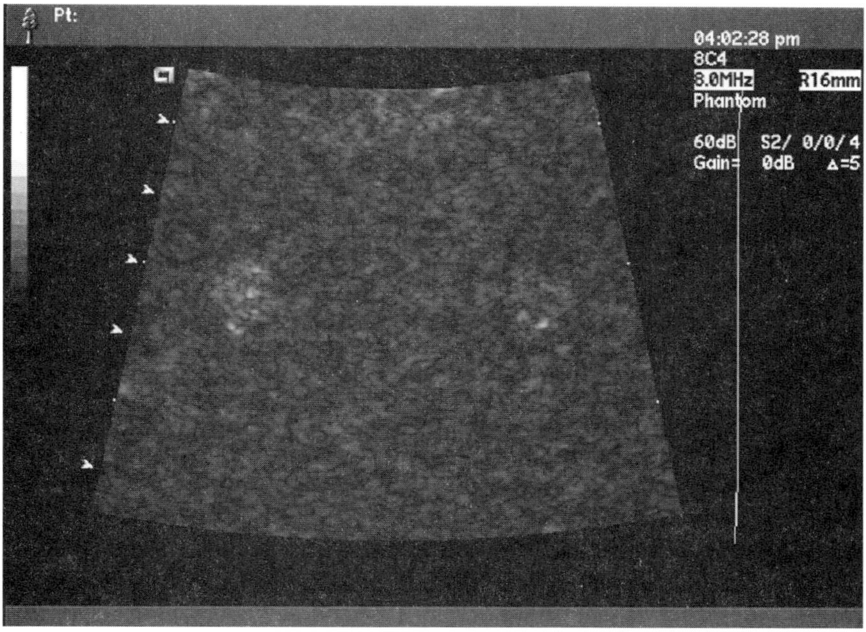
Pt:
04:02:28 pm
8C4
8.0MHz R16mm
Phantom
60dB S2/ 0/0/4
Gain= 0dB Δ=5

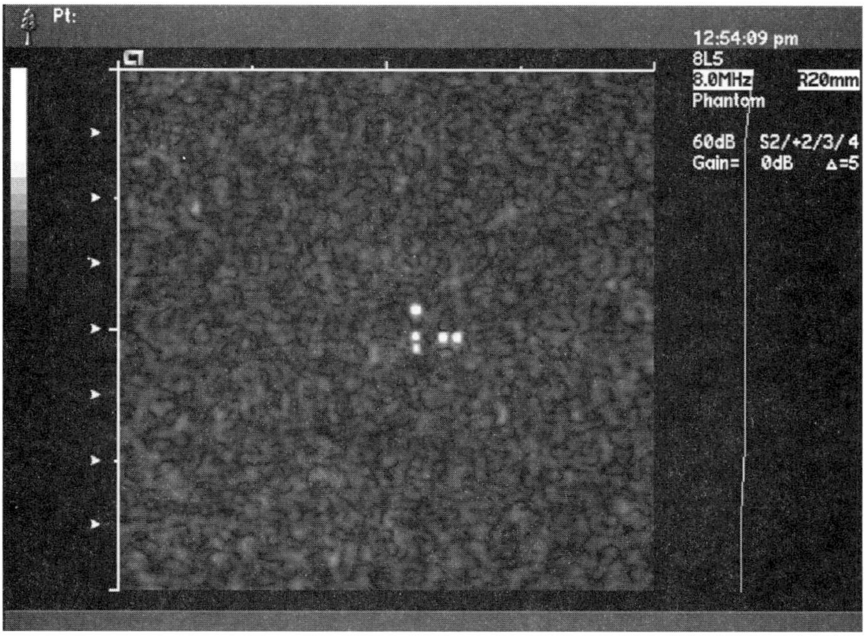
Pt:
12:54:09 pm
8L5
8.0MHz R20mm
Phantom
60dB S2/+2/3/4
Gain= 0dB Δ=5

FIG. 11. Image resolution. These are zoomed (magnified) images with cm marks on both lateral edges (upper convex array images) and on the top and left edges (lower linear stepped array images). The carets on the left edges indicate the focal lengths of the multiple transmit focuses in the composite transmit/dynamic focus utilized (Section IV.D.4). They were obtained with a liver-mimicking tissue-equivalent phantom containing (upper images) two simulated lesions whose echo amplitudes are 1–4 dB greater than the background echoes and (lower images) 0.1-mm diameter spatial resolution wires (seen on end) with varying spacing. The upper two images show the increase in image contrast resolution from an older $N = 128$ scanner (left) to a new $N = 512$ scanner (right). The two lesions are slightly visible in the $N = 128$ image after viewing the $N = 512$ image. The lower two images show the improved high-contrast spatial resolution of the $N = 512$ image (right) where the 500-micron (0.5-mm) laterally spaced wires have been resolved (and the wire image blur patterns are much smaller). These zoomed images clearly demonstrate the ever-present image speckle patterns. Note that the higher spatial resolution of the $N = 512$ images permits more accurate visualization of the speckle pattern image texture. (Images provided by Acuson Corporation.)

---

very high here so that only the sample volume dimensions determine the spatial resolution.

Lateral spatial resolution may be calculated using two identical point reflectors centered in the geometric focal plane. The calculation of the two-point-reflector beam profile is straightforward since the point reflectors produce separate, single-scattered, in-phase pulse-echo signals on the transducer front surface that add algebraically. For a circular aperture, spherically focused beam the two-reflector, pulse-echo lateral beam profile, normalized to half the main-lobe width, is proportional to

$$\left(\frac{J_1(2\cdot 3.83171\varepsilon)}{3.83171\varepsilon}\right)^2 + \left(\frac{2J_1(2\cdot 3.83171(\varepsilon - \xi))}{3.83171(\varepsilon - \xi)}\right)^2, \tag{19}$$

where ξ is the normalized separation between the two identical point reflectors. Figure 12 demonstrates this beam profile for point reflectors separated by (a) one-half the main-lobe width and (b) the main-lobe width.

Separation equal to one-half the main-lobe width is the Rayleigh criterion for image resolution in optics (Sears and Zemansky, 1957). The dip between main-lobe peaks is 2.67 dB and on the borderline of being detectable in a medical ultrasonic image. Separation equal to the main-lobe width produces two distinct images (or blur patterns) and will be the definition of lateral spatial resolution in this work. This definition is commensurate with an alternate definition of high-contrast spatial resolution as the two orthogonal widths of the image blur produced by well-separated, small, spherical latex reflectors in an anechoeic (echo-free) medium (Goldstein and Clayman, 1983).

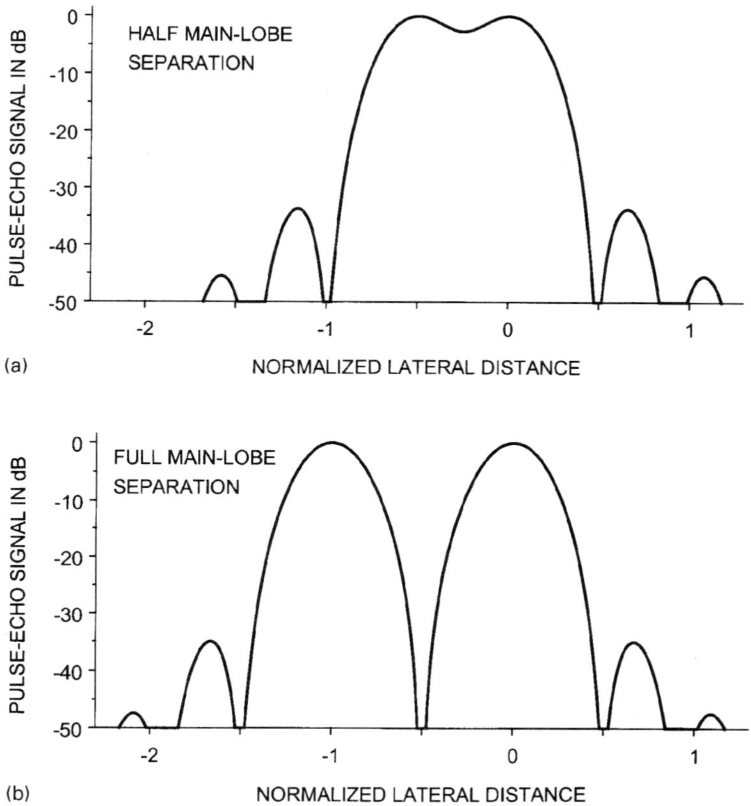

FIG. 12. Geometric focal plane two-point-reflector pulse-echo beam profile normalized to half the main-lobe width. (a) Point reflectors separated by half of main-lobe width (Rayleigh criterion in optics). (b) Point reflectors separated by main-lobe width.

Axial spatial resolution may be defined and calculated similar to lateral spatial resolution. The generally accepted definition of axial spatial resolution is the spatial pulse length. The need for highly damped, short ultrasonic pulses is now apparent.

2. *Contrast Resolution*

Most disease processes present in the ultrasonic image as low-contrast targets. Cancerous focal lesions are difficult to detect because the enclosed cancer tissue differs only slightly in ultrasonic properties from the surrounding normal tissue (Figure 11). The relationship between lesion tissue contrast and the resulting pulse-echo image contrast will be considered using a simple model.

A homogeneous spherical focal lesion composed of uniform tissue with an ultrasonic reflection coefficient, TR_l, is surrounded by homogeneous, uniform, background tissue with an ultrasonic reflection coefficient, TR_b. The subject contrast of the lesion, SC, is the dB difference between the two tissues' ultrasonic pressure reflection coefficients:

$$SC = 20\log\left(\frac{TR_l}{TR_b}\right). \tag{20}$$

The focal lesion is centered in the geometric focal plane of a spherically focused, circular aperture ultrasonic beam. It is assumed here that only the main lobe and the first side lobe of the ultrasonic beam determine the sample volume width. If the outer diameter of the first side lobe is smaller than the spherical lesion diameter, then both lobes are producing echoes from lesion tissue and the total received pressure signal at the front surface of the transducer is

$$Sig_l = TR_l \cdot PES_{\mathrm{ml}} + TR_l \cdot PES_{\mathrm{sl}} = TR_l(PES_{\mathrm{ml}} + PES_{\mathrm{sl}}), \tag{21}$$

where PES_{ml} and PES_{sl} are the average pressure signals received at the front surface of the transducer from the main and side lobes (respectively) from homogeneous, uniform tissue with a unity ultrasonic pressure reflection coefficient. Tissue attenuation has been neglected here since it will factor out of the calculated ratios below. The relative strength (or dB separation) of the first side lobe with respect to the main lobe is the dB ratio

$$\zeta = 20\log\left(\frac{PES_{\mathrm{sl}}}{PES_{\mathrm{ml}}}\right), \tag{22}$$

which is a negative quantity. If the ultrasonic beam is moved laterally to just intercept the background tissue, then the average received pressure signal at the front surface of the transducer is

$$Sig_b = TR_b(PES_{\mathrm{ml}} + PES_{\mathrm{sl}}). \tag{23}$$

The image contrast, IC, of the imaged lesion's center with respect to the imaged background tissue is the dB difference between the average received echo signals at these two image points:

$$IC = 20\log\left(\frac{Sig_l}{Sig_b}\right) = SC. \tag{24}$$

So when a lesion is much larger than the sample volume width, the image contrast at its center is equal to its subject contrast.

Smaller lesions (early stages of cancer) have an image contrast that is less than their subject contrast because when the beam is centered on the lesion, both lesion and background tissues are producing echoes. To see this explicitly, consider the special case where the lesion diameter is equal to the main-lobe diameter. Then, the average received pressure signal at the front surface of the transducer is

$$Sigsp_l = TR_l \cdot PES_{\text{ml}} + TR_b \cdot PES_{\text{sl}} \tag{25}$$

and the image contrast of the lesion's center to the tissue background is found to be

$$IC = 20\log\left(\frac{Sigsp_l}{Sig_b}\right) = 20\log\left(\frac{10^{SC/20} + 10^{\zeta/20}}{1 + 10^{\zeta/20}}\right). \tag{26}$$

For this special case Figure 13 demonstrates the variation of image contrast with side-lobe magnitude for focal lesions with subject contrasts of ± 20 dB, ± 40 dB, and ± 60 dB. Several important characteristics of contrast resolution are apparent from this figure: (1) High side lobes (smaller values of ζ) cause a reduction of image contrast (from the subject contrast); (2) hypoechoeic lesions ($SC < 0$) have a greater reduction in image contrast at the same side-lobe level than hyperechoeic lesions ($SC > 0$); and (3) the greater the subject contrast of hypoechoeic lesions the more their image contrast depends on side-lobe magnitude. When the side-lobe signal equals the main-lobe signal, the hyperechoeic lesions have an image contrast 6 dB lower than their subject contrast, whereas the hypoechoeic lesions have their image contrast reduced to about -6 dB. This side-lobe fill-in effect in hypoechoeic lesion image contrast make them more difficult to identify in ultrasonic images.

G. Image Noise

Any erroneous information included in the ultrasonic image can be considered image noise. There are many sources of image noise. For example, electrical signals picked up by the transducer structure (which acts like a

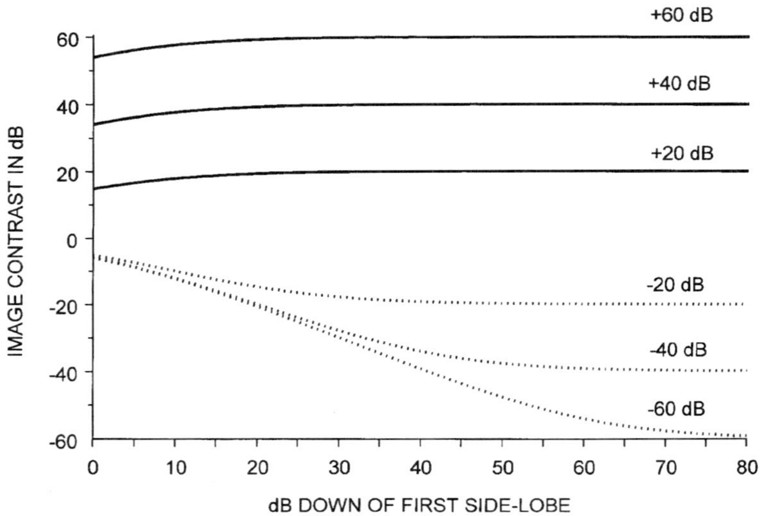

FIG. 13. Effect of the first side-lobe magnitude on the image contrast of the center of a focal lesion centered in the geometric focal plane of a circular aperture, focused transducer. The lesion diameter is equal to the main-lobe width. Three hyperechoeic lesions with subject contrasts of 20, 40, and 60 dB and three hypoechoeic lesions with subject contrasts of −20, −40, and −60 dB are presented. The lesion center image contrast is plotted vs the dB down of the first side lobe signal with respect to the main lobe signal.

miniature electrical antenna) or generated in the system electronics will mimic low-level echoes. Proper electrical shielding of the transducer structure and the above-mentioned limiting of amplifier signal dynamic range (by setting a low echo threshold level) will eliminate most of this noise.

A failure of any of the above-mentioned assumptions used in equipment image generation (Section II.E.1) or beam pattern design (Section II.E.3) will place false echo signals in the image. Usually this echo noise is minimal since the assumptions are quite reasonable. When an assumption fails dramatically, an image artifact may appear.

A very important source of image noise is the finite size of the sample volume and the coherent interference of the resulting echo signals at the transducer front surface. This interference produces an image speckle pattern that reduces the sharpness of high-contrast boundaries and limits the visualization of low-contrast lesions in uniform tissue areas (Figure 11).

In the uniform point reflector tissue model used to generate Figures 8 and 12, there is no variation in echo phase laterally in the sample volume. So any interference is due to echoes from different depths in the sample volume. A more realistic tissue model includes local variations in tissue impedance (tissue density and acoustic velocity) from an average value. These variations act like individual reflectors scattering ultrasonic waves (Jensen, 1991). The phase of the scattered waves depends on the local impedance variation (see Eq. (11)). Thus there is a two-dimensional random pattern of reflectors in uniform tissue that generates an arbitrary but deterministic speckle pattern in the ultrasonic image. The speckle pattern is deterministic because it is reproducible when the tissue is scanned repeatedly (with no change in scan parameters).

When not corrupted by nonlinear signal processing, if there are more than about eight random unresolved scatterers in the sample volume the speckle pattern is said to be "fully developed" with the following predictable statistical properties (Wagner *et al.*, 1983):

1. The individual speckle cells have a uniform echo amplitude.
2. The speckle cell echo amplitude probability density function is a Rayleigh distribution.
3. The mean speckle cell echo amplitude is 1.91 times its standard deviation (the ratio of mean to standard deviation is known as the speckle cell point signal-to-noise ratio, SNR_0).
4. The individual speckle cell dimensions are comparable to the sample volume dimensions.

For fully developed speckle patterns these statistical properties have three important consequences in ultrasonic imaging. The first is that the magnitude of speckle SNR_0 is independent of echo amplitude; increasing echo amplitude does not change the speckle pattern image texture. The second is that the speckle pattern image texture is characteristic of the ultrasonic imaging system and not the tissue being imaged. Only sample volume size or frequency bandwidth changes will affect the image texture. The third is that only the mean speckle cell echo amplitude is characteristic of the tissue being imaged. Different tissue types can only be identified in the ultrasonic image by differences in their mean speckle cell echo amplitudes (average shades of gray).

If there are less than eight random unresolved scatterers in the sample volume or the scatterers have some regular spatial distribution or the sample volume contains some resolved tissue structure, then the speckle pattern is not fully developed and its image texture may potentially be used to identify the scanned tissue (Wilhelmij and Denbigh, 1984).

Most medical ultrasonic image speckle patterns are fully developed, so speckle noise limits their fine structure detail. A great deal of research effort has been given to speckle reduction (e.g., Bamber, 1993). If many statistically independent speckle patterns of the same tissue area can be acquired, they can be averaged to reduce the speckle noise. Uncorrelated speckle patterns can be accomplished by either changing the sample volume (different beam focusing or frequency bandwidth) or scanning from a different angle or with a displaced beam pattern. Some amount of speckle reduction can be achieved in real-time imaging by temporally averaging over a number of frames to take advantage of the decorrelation due to small physiologic movements of either the tissue or the handheld transducer (Section IV.H.5).

If statistically independent speckle patterns can be obtained, spatial averaging will improve speckle SNR in proportion to the square root of their number. Mean echo amplitude still will be the only tissue characteristic in the image area. However, the increase in speckle SNR will improve viewer perception of low-contrast structures, as described next.

H. Image Feature Perception

The signal processing chain in medical imaging starts in patient anatomy and ends in the visual cortex of the sonologist's (image viewer's) brain. The eye–brain interface is sometimes the weakest link. For example, color coding of medical images is not effective because of the absolute sensitivity of the human eye to color. Most medical imaging modalities present relative tissue data and are subject to partial volume effects. If this data were presented with color-coded images, normal variations between patients or the exact location of the tomographic scan plane would change the colors of their imaged organs. However, due to the brightness constancy psychovisual property of the human eye (Cornsweet, 1970) normal variations between patients do not change the gray-scale appearance of their imaged organs (the eye adapts to the average image shade of gray).

High-contrast image features are perceived partially because of their edges. The image speckle pattern distorts their edges but not enough to prevent perception of the high-contrast feature. Low-contrast image features are perceived by spatial averaging over their image area (Rose, 1973). The ultrasonic speckle pattern affects the detectability of low-contrast structures in the image. Low-contrast lesions are detectable in the image if the difference between the lesion SNR_l and the background tissue SNR_b is greater than a threshold value (Smith *et al.*, 1983). This SNR difference is given by (modified from Insana and Hall, 1993)

$$\mathrm{SNR}_l - \mathrm{SNR}_b = \mathrm{SNR}_0 C_l \left(\frac{S_l}{S_c}\right)^{1/2}, \tag{27}$$

where C_l is the lesion image contrast, S_l is the lesion image area, and S_c is the image area of a single speckle cell. Since each speckle cell has a uniform echo amplitude, it is a single sample of the lesion echo amplitude. The number of samples of the lesion echo amplitude is the number of contained speckle cells, which is the area quotient in Eq. (27).

There are three ways to improve the detectability of low-contrast lesions in ultrasonic images. The first is to increase C_l. Decreasing sample volume size and/or slice thickness beamwidth will reduce partial volume effects and increase C_l. Also, increasing lesion subject contrast will increase the resulting lesion image contrast (see Figure 13). The image contrast of hypoechoeic cancerous lesions of the prostate has been shown to be frequency dependent (Goldstein, 1987) with lower frequencies producing greater image contrast.

The second way to improve detectability is to improve the statistics of the image averaging by increasing the number of speckle cells involved. This can be accomplished by narrower beamwidths and higher frequencies. The third way is to decrease image speckle noise by either suitable averaging of statistically independent speckle patterns or by image processing techniques that smooth the speckle image texture (Loupas *et al.*, 1994).

High frequencies always have been desired in ultrasonic imaging because they lead to smaller sample volumes and improved spatial resolution. However, the highest attainable frequencies may not always give the best contrast resolution of focal lesions. Each clinical imaging task must be evaluated separately.

I. Doppler Frequency Shift

Ultrasonic echoes contain a great deal of information concerning tissue reflectors. However, only echo amplitude and time delay are used in the formation of gray-scale medical images. Another piece of echo information used in medical ultrasonics, the echo frequency shift (compared to the transmitted frequency), is caused by the motion of reflectors toward or away from the transducer. While these Doppler-shifted echo frequencies can be used to measure the motion of any moving reflectors in the body, their main clinical application has been the study of blood flow in the heart and vascular system.

A simple pulse-echo phenomenological derivation demonstrates explicitly that the frequency shift is caused by an extra time dependence of echo phase due to target motion (Atkinson and Woodcock, 1982). The phase of a transmitted ultrasonic wave is

$$\phi = \omega_0 t + kz, \tag{28}$$

where ω_0 is the transmitted angular frequency and $k = 2\pi/\lambda$. The frequency (in cycles/sec) of the wave is obtained from the time rate of change of its phase at a fixed position z:

$$f_0 = \frac{1}{2\pi}\frac{d\phi}{dt} = \frac{\omega_0}{2\pi}. \tag{29}$$

The phase of an echo received from a target at a range z_R lags the phase of the transmitted wave by (Eq. (14))

$$\Delta\phi = -\omega_0 t_{\mathrm{tof}} = -\omega_0 \frac{2z_R}{c}, \tag{30}$$

where t_{tof} is the ultrasonic wave time of flight. If the target is stationary, the time rate of change of this phase difference is zero and there is no difference in frequency between the transmitted and received waves. If the target is moving with a constant velocity V, then there is a frequency difference given by

$$\Delta f = \frac{1}{2\pi}\frac{d\Delta\phi}{dt} = -f_0 \frac{2V}{c}. \tag{31}$$

The Doppler shift is positive (echo frequency shifted higher) when the target is moving toward the transducer and negative (echo frequency shifted lower) when the target is moving away from the transducer. In general, the target will be moving in an arbitrary direction, so the proper Doppler equation must be used (Evans *et al.*, 1989):

$$f_{DS} = \frac{2f_0 V}{c}\cos\theta_D, \tag{32}$$

where θ_D is the angle between the ultrasonic beam axis and the target velocity vector (Doppler angle). Ultrasonic equipment will detect the Doppler-shifted frequency and then solve a transposed Eq. (32) to obtain the target velocity,

$$V = \frac{f_{DS}c}{2f_0 \cos \theta_D}, \tag{33}$$

if the Doppler angle is known.

An important outcome of the medical ultrasound situation is that the Doppler-shifted frequency (DSF) falls into the human audio range. This is a consequence of the typical target velocities experienced within the cardiovascular system along with the typical Doppler carrier frequencies that range from 1 MHz to 25 MHz. An audio output is a standard feature found on all medical Doppler devices, and listening to that audio output in stereo is the first form of Doppler frequency analysis.

III. Analog Gray-Scale Imaging

The pioneers in ultrasonic imaging began their initial investigations in the 1940s (Kimmelman, 1988). In the early 1970s, the first crude medical ultrasonic scanners were available commercially. The early equipment was analog and went through many transitions throughout the years to produce today's digital imaging platforms. The early analog equipment will be described here. The latest digital equipment will be presented in Section IV.

A. Analog Static Scanners

The first medical ultrasonic scanners utilized handheld transducers and produced a static image. Figure 14 shows an early clinical examination room with a patient couch and the scanner (a modified Tektronix storage oscilloscope) with the scan arm and transducer.

The scan arm performed two functions. First, the scan arm monitored the position and angulation of the transducer front face. Sine/cosine potentiometers in the joints between the three segments of the arm sensed this spatial information. Second, the arm confined the transducer line of sight (beam axis) to a flat plane (scan plane), producing the required tomographic section. Transverse, longitudinal, and oblique patient images could be obtained by suitably orienting the scan-arm-defined scan plane.

The operator moved the transducer by hand over the patient's skin surface (which had a gel or oil covering) while monitoring the image display and adjusting various machine controls. The image line density was variable and

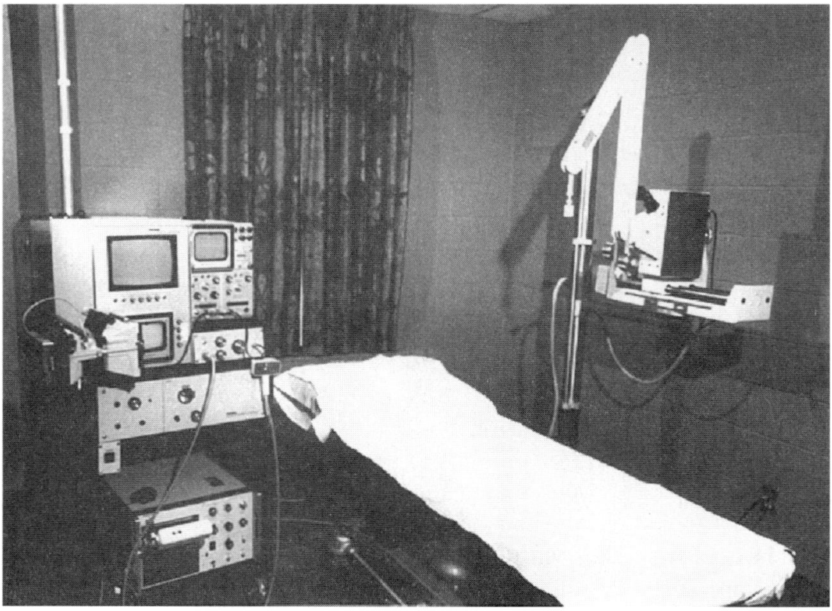

FIG. 14. Clinical ultrasound examination room circa 1972. The static image ultrasonic scanner is a modified Tektronix bistable tube storage oscilloscope. A three-segment scan arm is used to monitor the transducer position and angulation in space. The ultrasound operator manually moved the transducer over the patient's skin surface to obtain tomographic cross-sectional images.

under operator control. The image acquisition time varied between 5 and 20 seconds. After a suitable image was obtained and photographed, the scan plane was moved in increments of $\frac{1}{2}$ to 1 cm to obtain other images. The circular aperture, single element transducer produced a focused cylindrical beam pattern with a typical 5-mm spot size (Eq. (16) with $fn = 4$–6) in the geometric focal plane. If the distance increment was too large between imaged scan planes, the sonologist could have missed important anatomy.

Figure 15 presents an early clinical ultrasonic static image obtained using a bistable storage tube display. The receiving system gains had to be reduced until only the strongest specular reflection echoes were recorded. The operator had to work to find the correct beam angulation at all points in the image to capture these specular reflections. In these images only the boundaries of internal organs, bones, and blood vessel walls are recorded. Image quality depended greatly on operator skill and often only the operator could properly

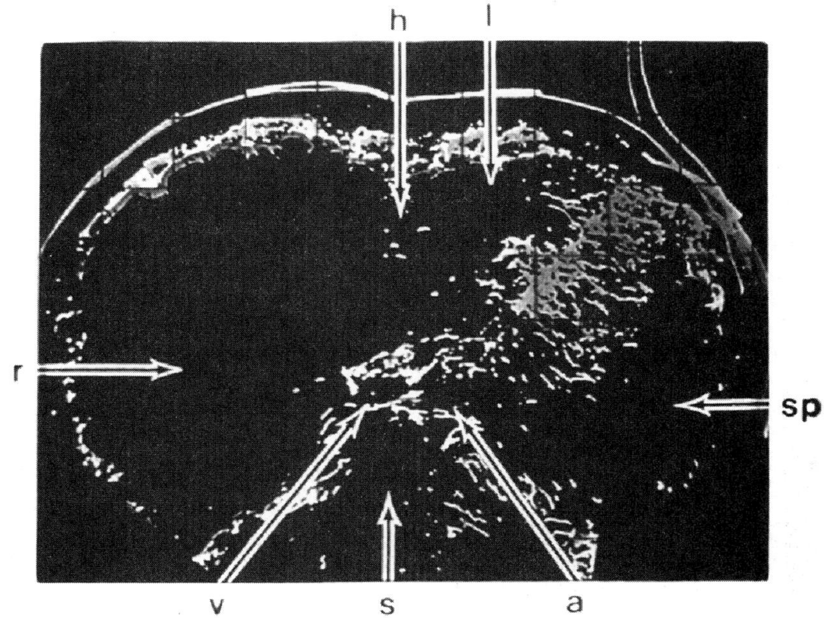

FIG. 15. Bistable tube, clinical ultrasound image circa 1972. The scan plane is transverse to the long body axis. The system gains have been set to only record the high-amplitude specular echoes from organ and bone surfaces and blood vessel walls. The transducer front surface echo maps out the patient skin surface with a layer of subcutaneous fat beneath. r: right lobe of liver, l: left lobe of liver, h: hepatic artery, sp: spleen, v: vena cava, s: spinal vertebral body, a: descending aorta. Note that in static images a very large field of view is possible with patient spatial orientation maintained (gravity is always pointing down).

interpret the obtained image. Nevertheless, these early ultrasonic images were an important advance over x-ray plain film images in many clinical soft tissue imaging situations.

Figure 16 presents a typical block diagram of an analog static imager. A master oscillator synchronizes the operation of all functions so that after transmission, the received echoes are processed according to their time of arrival, combined with the scan arm transducer spatial line-of-sight information, and stored in the scan converter (Section III.A.7.b) prior to display and hardcopy generation. The various components in Figure 16 will be discussed next.

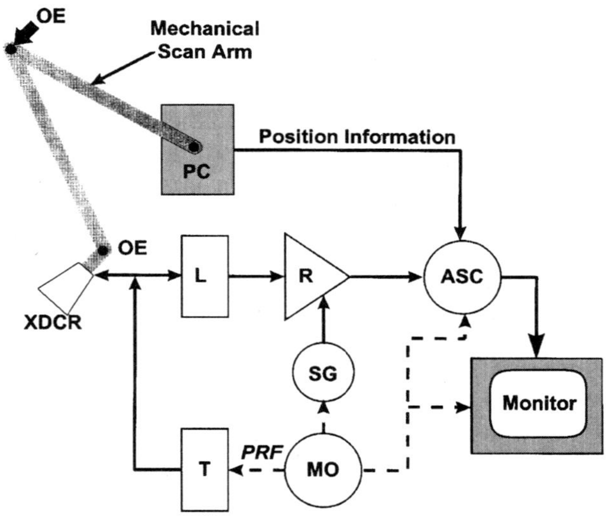

FIG. 16. Block diagram of an analog static scanner. The solid line indicates ultrasonic data flow. The dotted lines indicate control. The master oscillator determines the output pulse repetition frequency (*PRF*) in accordance with the selected image field-of-view depth and synchronizes all ultrasonic data processing in each of the functional blocks. XDCR: transducer, PC: position calculator, OE: optical encoder, L: limiter, T: transmitter, MO: master oscillator, SG: swept gain, R: receiver, ASC: analog scan converter.

1. Transducer

A cross-sectional diagram of a typical circular aperture, static imaging transducer is shown in Figure 17. The spherical shell, focused piezoelectric element is bonded to a plastic wear surface and a backing layer. A cylindrical acoustic insulating ring surrounds the acoustically active components to prevent spurious lateral acoustic modes from reverberating off the rf shield, causing artifactual echoes. The rf shield is a brass cylinder electrically connected to the front electroded surface of the piezoelectric element, forming a Faraday cage that eliminates EMF pickup in the electrically noisy clinical environment. The tuning element is used for electrical impedance matching. The plastic case provides electrical insulation for patient and operator protection in case of an electrical short inside. The connector permits rapid change of transducers while assuring that the acoustic axis of the transducer beam is parallel to the third segment of the scanning arm.

a. Matching. The transducer has maximum broadband sensitivity when properly matched electrically to its driving and reception circuits and

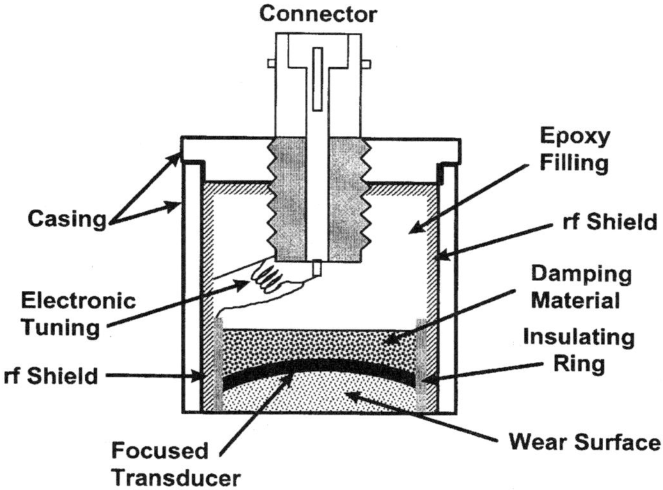

FIG. 17. Cross section of a circular aperture, static imaging transducer. A standard connector was used for rapid changing of transducers on the scan arm. The plastic case provides the required electrical insulation for patient safety. The components are described in the text. (Aero-Tech Reports, 1980.)

acoustically to the patient. A shunt inductance coil (see Figure 17) used to cancel the reactive component of the element impedance provides simple electrical matching (Thurston, 1960). When the coil and element reactive impedances cancel at the element center frequency, the impedance of the element is better matched to the typical $50\,\Omega$ circuit impedance than the untuned case (Hunt *et al.*, 1983). Complex broadband matching networks, which provide a wider frequency response with minimum distortion, have also been used (Augustine and Anderson, 1979).

Early static imaging transducers used a heavily damped backing layer to attain a wide transducer bandwidth to produce short echoes. However, because a great deal of acoustic energy was absorbed in the backing layer, these transducers were inefficient and had low sensitivity. Improvements in transducer efficiency and sensitivity came from better acoustic matching of the element to the patient. This acoustic matching was accomplished by using $\lambda/4$ layers of intermediate impedance materials (between the $Z_T = 34$ MRayls of the ceramic and the $Z_M = 1.5$ MRayls of the patient). Even though the $\lambda/4$ matching condition can only be attained at the crystal center frequency, the crystal loading caused by the matching layers produces wider bandwidth and more efficient transducers (Kossoff, 1966). Still more efficient transducers can be made from air-backed crystals with several front $\lambda/4$ layers by using the transmission line properties of thin crystals (Desilets *et al.*, 1978). For minimum transducer insertion loss, the matching layers should be made of low-loss materials and the glue bond thickness should preferably be less than one twentieth of a wavelength to have a negligible effect (Hunt *et al.*, 1983).

b. Focus. Single element, circular aperture transducers were spherically focused using the three different methods shown in Figure 18. Initially, internally focused transducers were used for general abdominal scanning. The spherical shell piezoelectric crystal was bonded to a wear surface with a flat front face with an intermediate acoustic impedance for impedance matching. Cardiac transducers were externally focused with a flat piezoelectric crystal and a wear surface with a spherical front face. Focusing occurred at the spherical front face–gel interface. Later, more sensitive $\lambda/4$ focused transducers had a spherical shell piezoelectric crystal and a $\lambda/4$ wear surface. Air

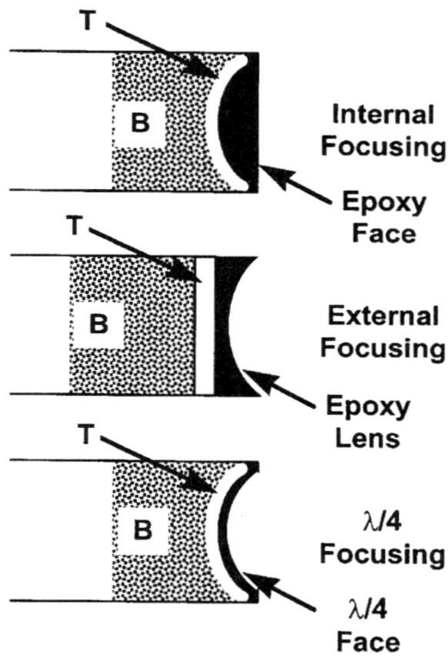

FIG. 18. Single element transducer focusing methods. The epoxy or plastic wear surface played an important role in the focus of the internal and external focusing methods. Its acoustic impedance was between that of the element and skin for impedance matching purposes. In the $\lambda/4$ focusing method with one or more different matching layers, the acoustic impedance match was better and the matching layers did not affect the focus. Since the coupling gel has an acoustic velocity close to 1540 m/s, there is no defocusing at the patient's skin surface. (Aero-Tech Reports, 1980.)

bubbles were sometimes trapped between the wear surface and skin in the latter two methods, causing image artifacts.

The internal focusing method was the best suited for the manual scanning used in static imaging because the wear surface flat front face could easily slide on the patient's gel-covered skin. Refraction at this front face, however, degraded the spherical focus of the element. Figure 19 demonstrates the refracted path of one ray from the crystal's spherically curved surface (radius of curvature R). This ray intersects the flat interface a distance L from the center of the aperture of radius a at an angle θ_p to its normal. The layer of coupling gel is neglected here because it has a negligible effect on the focus. According to Snell's law, since c_p was approximately 2950 m/s and c_m is taken as 1540 m/s, this ray is refracted to a smaller angle θ_m, creating a longer effective radius of curvature R_{eff}. h is a characteristic parameter of spherically focused transmitters (O'Neil, 1949)

$$h = R\left(1 - \sqrt{1 - \frac{a^2}{R^2}}\right). \tag{34}$$

The tangents of the two angles are

$$\tan\theta_p = \frac{L}{R-h} \qquad \text{and} \qquad \tan\theta_m = \frac{L}{R_{\text{eff}} - h}, \tag{35}$$

so that

$$R_{\text{eff}} = R\frac{h + \dfrac{L}{\tan\theta_m}}{h + \dfrac{L}{\tan\theta_p}}. \tag{36}$$

Expressing θ_p in terms of the dimensions in Figure 19 and using Snell's law (Eq. (13)) to give θ_m in terms of θ_p, Eq. (36) becomes

$$R_{\text{eff}}(L) = \frac{c_p}{c_m}\sqrt{L^2\left(1 - \frac{c_m^2}{c_p^2}\right) + (R-h)^2} + h. \tag{37}$$

So each ray exiting the spherically curved crystal has a different R_{eff} depending on its L value.

The larger the circular aperture (or the smaller the focus f-number), the more the departure from a pure spherical focus. A 3.5-MHz, 19-mm diameter transducer with $fn = 4$ was one of the standard long internal focus transducers

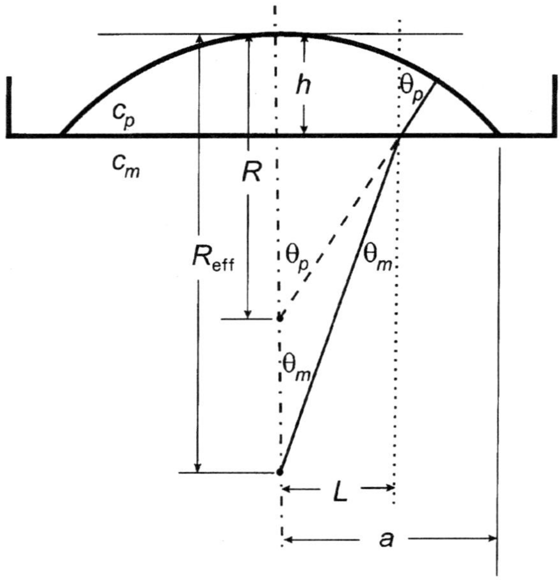

FIG. 19. Focus of internally focused transducer. An acoustic ray emanating from the front surface of the spherical shell piezoelectic element suffers refraction at the wear surface–skin interface. The effective radius of curvature of the spherical focus R_{eff} varies across the transducer aperture according to Eq. (37).

used routinely in static imaging (with a focal zone from 7 to 11 cm). R_{eff} only varies 1.7% across the aperture of this transducer. If this transducer were focused to $fn = 2$, Eq. (37) predicts an 8% variation of R_{eff}. The use of weak focuses for static imaging transducers was mandated by the desire for a long focal zone, so the defocusing caused by this front surface refraction was minimal. Any scars or other deformities on the skin surface would also affect the refraction at the wear surface–skin interface and further degrade the focused beam pattern.

2. *Transmitter*

In analog static imagers, a short high-voltage unipolar electrical pulse was used to shock-excite the heavily damped transducer to vibrate in its fundamental resonance thickness expander mode. Voltages of 400–1000 V were used to increase output energy and overcome transducer damping losses. These high-voltage pulses represented a source of electrical interference and potential shock hazard to both patient and operator. They could also gradually depolarize the piezoelectric ceramic material, reducing the transducer sensitivity with time (Schafer and Lewin, 1984).

These high-voltage pulses lasted for less than 100 nsec and had a very wide frequency spectrum so the transducer *FB* (Eq. (3)) determined the output pulse frequency spectrum. However, the wide frequency spectrum of the shock pulse could excite undesired modes of vibration in the crystal, which in turn would degrade transducer performance by distorting its frequency response (Schafer and Lewin, 1984).

3. *Limiter*

The pulse-echo transducer is connected to both transmitting and receiving circuits. Thus, the strong shock pulse also is applied directly to the receiver input (see Figure 16). This pulse must not be allowed to enter the high gain amplifiers; it would either blow out circuits or cause a "paralysis" period during which echoes from close reflectors cannot be distinguished (Maginness, 1979). Early analog equipment used diode clipping circuits to limit the transmitted amplitude of the shock pulse while passing low-level echoes. This clipped shock pulse defined the transducer/skin surface in the resulting static image.

4. *Receiver*

"Receiver" is the generic term applied to all circuit functions that amplify the weak echoes and then determine their magnitude (Wells, 1974). As such, an analog receiver would include a preamplifier, logarithmic amplifier, rectifier, and low-pass filter.

The preamplifier performs two functions. Small single element transducers or the individual elements in multielement arrays cannot drive long lengths of coaxial cable (c.f. Eq. (7)). Thus a preamplifier close to the piezoelectric crystal is used for some initial signal amplification to overcome cable capacitance. Also, the preamplifier output is electrically impedance matched to standard 50 Ω cable impedance.

A logarithmic amplifier processes the weak echo signals. The echo amplitude dynamic range of the received echoes is over 100 dB. The log amplifier performs needed dynamic range compression. Usually, the log amplifier is calibrated in dBs for echo signal comparisons.

The echo amplitude along with the spatial location of the reflectors form the ultrasonic image. This explains the two additional receiver stages shown in Figure 20. The amplified echo signals are full-wave rectified and then low-pass filtered to obtain the echo envelope. The peak of this "video" signal represents the echo amplitude. Two different operator controls for this amplifier are used to properly process the echo amplitude video signal for inclusion in the medical ultrasonic image: swept gain and overall gain.

a. Swept Gain. At each depth in the image the echoes have an amplitude dynamic range of at least 50 dB. (Single element, circular aperture transducers should only process a unique 35 dB dynamic range, as explained in Section II.E.3; newer multielement arrays can process a larger unique dynamic range, as explained in Section IV.C.) Tissue attenuation adds another 50+ dB of echo amplitude dynamic range over the full image depth. This echo amplitude dynamic range presents several problems.

One problem is the difficulty of image interpretation when similar tissue reflectors at different depths are represented with different amplitude echoes. The other problem is one of system design or dynamic range matching — the input echoes have over 100 dB of dynamic range but the display CRT has only a 20–25 dB dynamic range of available shades of gray (Schafer and Lewin, 1984).

Both of these problems can be resolved by noting that the average echo amplitude at each image depth decreases with depth due to tissue attenuation. Increasing the log amplifier gain in synchrony with the arrival time of echoes provides a simple means of compressing the echo amplitude dynamic range closer to the limited CRT display dynamic range by approximately compensating for tissue attenuation (Maginness, 1979). This swept gain function is known also as time gain compensation (TGC).

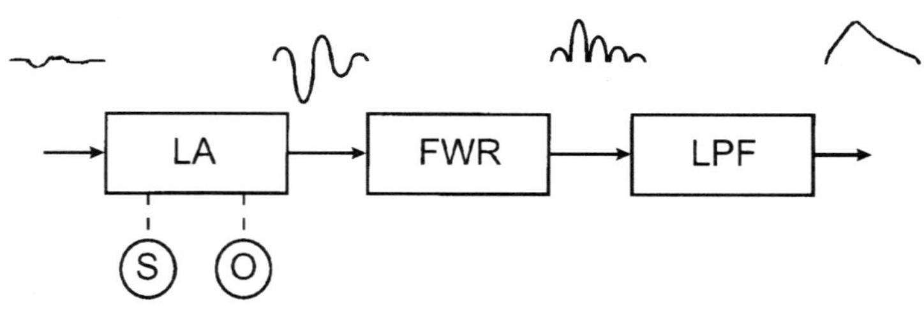

FIG. 20. Echo amplitude processing in the receiver. The low-level echo signals from the transducer pass through the limiter and are amplified by the log amplifier (LA). Full wave rectification (FWR) and low-pass filtering (LPF) produces the echo envelope signal whose peak is the desired echo amplitude. The operator sets the swept gain (S) and overall gain (O) controls to produce uniform image patterns of uniform tissue regions in the patient.

Early analog equipment utilized a constant slope swept gain signal (increasing the log amplifier gain by an operator-selectable number of dB per cm). However, this could not compensate for echo amplitude variations due to the focused beam pattern. Figure 21(a) demonstrates the focused beam echo signals obtained from identical reflectors in a uniform tissue type. In this case, the constant slope swept gain has been set to compensate exactly for the tissue attenuation. Each reflector scatters echoes with the same scattering cross section but the incident pressure varies along the focused beam pattern so the echoes from the focal zone are largest.

An alternative swept gain control is the use of a set of slide potentiometers to divide the image depth into equal range increments. Each slide pot determines the swept gain level at the center of a given range in the image. A smooth swept gain control voltage with depth is then automatically generated from the slide pot settings. Figure 21(b) demonstrates the use of the slide pots to compensate for both medium attenuation and the focused beam pattern.

b. Overall Gain. Along with swept gain the operator also has control of overall gain (a constant gain level independent of image depth). Using these

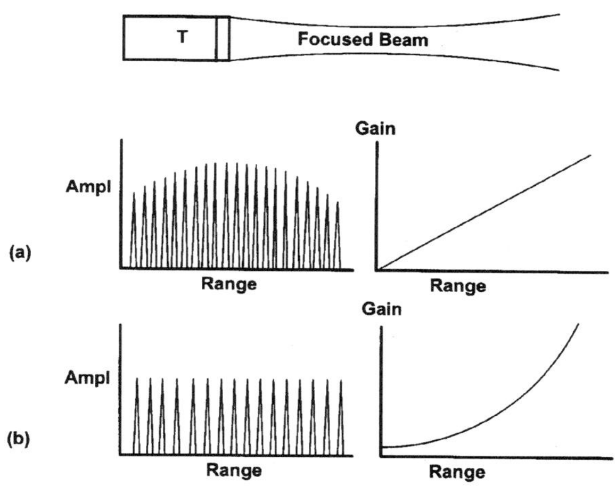

FIG. 21. Swept gain. The pulse-echo focused beam pattern receives echoes from a set of identical reflectors in a uniformly attenuating medium. On the left top are the echo signals obtained when the constant slope swept gain curve (top right) is set to exactly compensate for the medium attenuation. On the left bottom are the echo signals obtained when a range-selectable swept gain curve (bottom right) is used to compensate for all causes of echo amplitude variation: medium attenuation and focused beam pattern.

two controls, the operator can accurately process arriving echo signals to approximately compensate for tissue attenuation and beam focusing variations at all depths in the image. While viewing the ultrasonic image the operator simply sets the slide potentiometer at each depth to produce a uniform shade of gray of (assumed) uniform tissue anatomy. Then the overall gain control is used to set the overall brightness of the image.

c. Adaptive Gain Control. It is possible to automate the swept gain to produce approximately equalized echo amplitudes with image depth (DeClercq and Maginness, 1975). Several commercial scanners have offered this function. For the most part, it was rejected by many clinical users due to their reluctance to lose their sense of control in producing the ultrasonic images. Recently, adaptive gain control has been reintroduced in echocardiography.

5. Analog Signal Processing

Any procedure performed on the echo amplitude signal after transducer reception and before display on the image qualifies as signal processing. Many signal processing steps and their rationale, based on ultrasonic physics or circuit considerations, have already been considered. These include thresholding, logarithmic amplification, swept and overall gain, and echo amplitude detection.

Some early analog scanners differentiated the video signal to emphasize those echoes that have more rapid distance rates of amplitude change (Wells, 1974). This differentiation lost some information since the more slowly varying echo signals were suppressed. This was partially compensated for by the addition of an appropriate fraction of the undifferentiated video signal (Kossoff, 1972).

6. Display

There were several different modes of echo display in early analog equipment. The A-Mode (amplitude mode) was essentially an oscilloscope display of the line of received echoes with the time axis converted to distance using the range equation (Eq. (14)). Although it had limited spatial information, it presented better echo amplitude information than the bistable storage tube. Ophthalmologic A-Mode equipment continues to give useful clinical information within the relatively simple geometry of the eye.

B-Mode (brightness mode) was used for cross-sectional imaging. The two dimensions of the TV display face represented the miniaturized patient cross-sectional dimensions and the brightness of the display represented echo amplitude. A white on black echo amplitude encoding scheme is generally used, with the lighter shades of gray representing the higher-amplitude echoes. This display mode is now called gray-scale imaging.

Another display variant, M-Mode (motion mode) provides useful quantitative information in cardiac scanning. Figure 22 demonstrates an M-Mode echo display. The vertical axis of the TV display represents tissue target depth. The echo amplitude is encoded as CRT spot brightness (B-Mode). The vertical beam axis is swept horizontally across the image at a known constant velocity. The motions of the tissue targets toward and away from the transducer appear as oscillatory traces on the display. Calculating the slope of a target's trace along with the known display horizontal sweep velocity permits a calculation of the target velocity along the beam axis.

M-Mode is used to quantate cardiac valve motion. Because the valves are surrounded by blood and appear as high-contrast targets, their traces are easily identified in M-Mode images. Figure 23 presents an M-Mode display of a

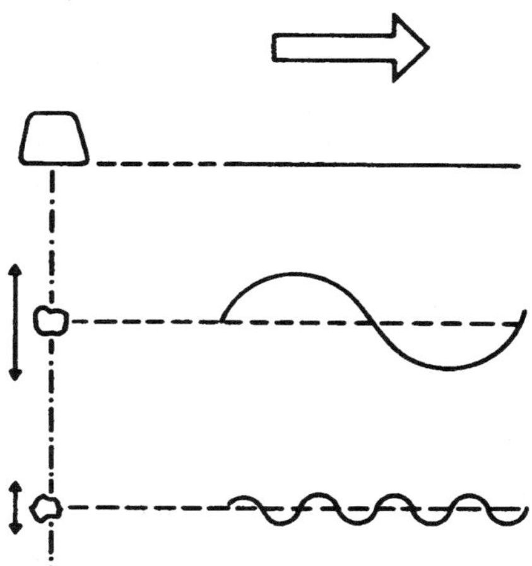

FIG. 22. M-Mode. The vertical axis represents depth along the transducer beam. The echoes are presented in B-Mode. The horizontal constant velocity sweep of the vertical beam axis provides a display of target velocity and extent of motion. The upper target moves a large distance but slowly. The lower target moves a smaller distance but much more rapidly. Goldstein (1988). Reprinted by permission of John Wiley & Sons, Inc.

fetal heart. Note the presence of a gray-scale image with the highlighted beam direction, which was used to generate the M-Mode display. This "road map" image is very important as it documents a piece of medical/legal information not present in the M-Mode image, the location of the M-Mode beam axis relative to cardiac anatomy. The advantage of M-Mode imaging is that it has much finer temporal resolution (on the order of 1000 lines per second) than gray-scale imaging, which can be important when analyzing detailed cardiac valve motion (Quistgaard, 1997).

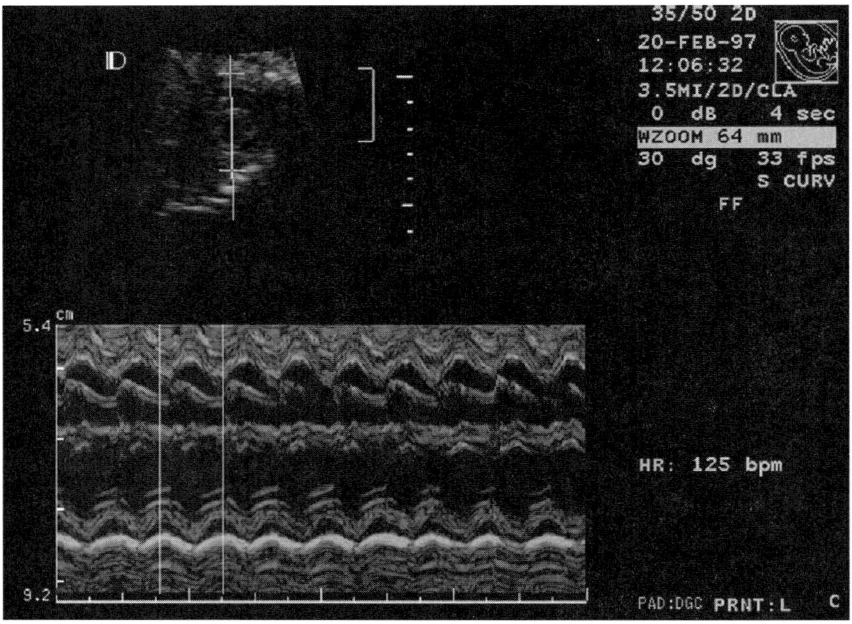

FIG. 23. Fetal M-mode examination. The small pictogram in the upper right of the display indicates optimization for a fetal scan. On top is a zoomed (magnified) gray-scale image of the fetal heart. The square bracket on the right of the zoomed image defines the range of depths over which the "confocal imaging" multizone focus is performed (Section IV.D.4). The vertical line superimposed on the fetal heart defines the transducer line of sight that was used to generate the M-mode display below it. The crosses on this line indicate the depth extent (5.4 to 9.2 cm) of the M-mode display. In the M-mode display the black regions are blood, the broad horizontal band of tissue just above center is the interventricular septum with the right ventrical above and the left ventrical below. The equipment has calculated a fetal heart rate of 125 bpm (beats per minute) from the placement of the two vertical lines in the M-mode display. (Courtesy of Diasonics Ultrasound.)

7. Memory

a. Bistable Storage Tube. Early ultrasonic images were displayed using bistable storage tubes that combined both image memory and display functions. The storage tube displays were binary, and only high-amplitude echoes were shown. Some early analog equipment used a zero dynamic range system in which a pulse of defined amplitude and duration was generated whenever the echo amplitude exceeded a threshold level (Wells, 1974). A dead time was employed to prevent multiple registrations from a single echo complex. These images presented little diagnostic information except organ size, location, and occasionally the presence of unusual intraorgan high-amplitude echoes.

An early attempt at gray-scale imaging used two B-Mode displays: a bistable tube and an ordinary CRT driven by the compressed dynamic range of the processed echo signals (Kossoff and Garrett, 1972). Photographic hardcopy of the CRT display was obtained with a opened shutter technique using low-contrast black-and-white film. The bistable display was used to help the operator achieve the smooth and steady beam sweep necessary for a well-exposed CRT film image. This laborious technique required extensive training and patience.

b. Analog Scan Converter. Medical ultrasonic image quality and diagnostic content improved dramatically when image memory and display functions were separated. Modified high-bandwidth TV monitors were used for display. The new image memory was called a scan converter because its input and output formats were different. The acquired echo data (amplitude and target position) were scanned in commensurate with the random, operator-guided motion of the transducer beam axis. The image data was scanned out commensurate with U.S. broadcast TV format—525 lines (only 480 displayed), 30 frames per second.

In 1973 the analog scan converter was introduced commercially (Ranalli, 1975) (Torrence and Ranalli, 1975). Figure 24 shows its essential features. A CRT tube was modified by replacing the output phosphor with a 1000×1000 matrix of silicon oxide capacitive elements whose plane represents the minified patient cross section. In write mode (during patient scanning), the deflection plates guide the electron beam to the anatomical location of the echoes being received. It deposited a charge on the silicon oxide capacitors proportional to the received echo amplitude so the charge distribution on the capacitor plane represents the echo amplitude patient cross-sectional image.

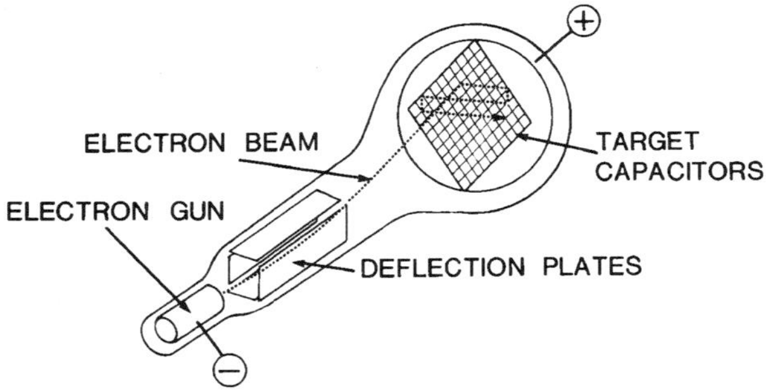

FIG. 24. Analog scan converter. A modified CRT tube was used to store the echo amplitude patient cross-sectional image as a charge distribution on the plane of the target capacitors. The 1000×1000 matrix of silicon oxide capacitors was tilted 45° to the horizontal to avoid Moiré patterns on the TV display.

In read mode, a weak electron beam scans the capacitor plane rectilinearly in broadcast TV format, sensing the stored charge and generating a driving signal for the display TV. Since writing and reading had to be done sequentially, during patient scanning they were multiplexed, producing a venetian blind pattern of dark bars (when the scan converter is writing) over the patient gray-scale image. This flickering pattern was distracting and obscured the details of the image during its acquisition.

The analog scan converter was large and expensive, required supplemental electronic circuitry, and needed frequent service calls for focusing and other adjustments. However, because the TV monitor produced a gray-scale image presenting a large dynamic range of received echo amplitudes, the new image was not only cosmetically pleasing but also contained vital diagnostic information, especially in the low-amplitude echoes from tissue parenchyma (Leopold, 1974).

c. Digital Scan Converter. The clinical success of the gray-scale TV display was frustrated by the cumbersome analog scan converter. Digital techniques had been used by several investigators to control image acquisition and display (Fry *et al.*, 1968) (Erikson and Brill, 1970) or to digitally process off-line the echo amplitude data (Hubelbank, 1970). Further advances were the use of digital dynamic shift registers to store the echo data acquired during a rectilinear immersion scan (using water as the acoustic coupling medium) (Yokoi and Ito, 1972) or a sector scan (Yokoi and Ito, 1973). In the latter, the added digital computer was used also for echo data processing.

A prototype digital scan converter for gray-scale imaging using computer memory to store the image data was reported in 1974 (Goldstein *et al.*, 1974). It was developed for feasibility studies (Goldstein, 1976) (Ophir and Goldstein, 1977) and only had a 128 × 128 pixel map with 16 display shades of gray. Compound scanning (multiple passes over the same anatomy) could be used in data acquisition. The echo amplitudes were digitized and stored in the image pixel map one pixel per pulse transmission. The data processing schemes included an average store, which stored the average of the acquired pixel samples, and a peak store, which stored the highest-amplitude pixel sample. The average processing scheme was designed for diffusely scattered echoes and the peak processing scheme was designed for specularly reflected echoes. Figure 25 demonstrates the improved image quality using the average processing scheme.

In 1976, the first commercial digital scan converter was available (Schorum and Fidel, 1977). A very useful feature in the commercial version was a last-value processing scheme in which newly acquired echo amplitude data overwrites the previously stored data. Now static imaging surveys could be performed where the image scan plane was slowly moved across the patient while the transducer was rapidly scanned on the patient's skin surface. The important clinical benefits of these survey scans were reduced patient examination time while assuring the interrogation of *all* patient anatomy (previously the scan plane was moved incrementally and could miss some important patient anatomy).

With careful circuit design, the writing of echo data in image memory could be synchronized with the TV retrace times so that no image blanking was necessary. Many other data processing schemes could also be implemented, such as using line buffers to temporarily store the acquired data so that signal processing could be performed on the image line data prior to insertion in the image map (Ophir and Maklad, 1979). The digital scan converter completely replaced the analog scan converter commercially after several years due to its low cost, stability, and flexible data processing capabilities (Maginness, 1979).

8. *Hardcopy*

Image hardcopy is clinically essential for medical and legal reasons. It was obtained initially using a Polaroid oscilloscope camera on a second dedicated hardcopy TV. Then multiformat cameras were developed that contained a dedicated hardcopy TV and the mechanical means to place 4, 6, or 9 images

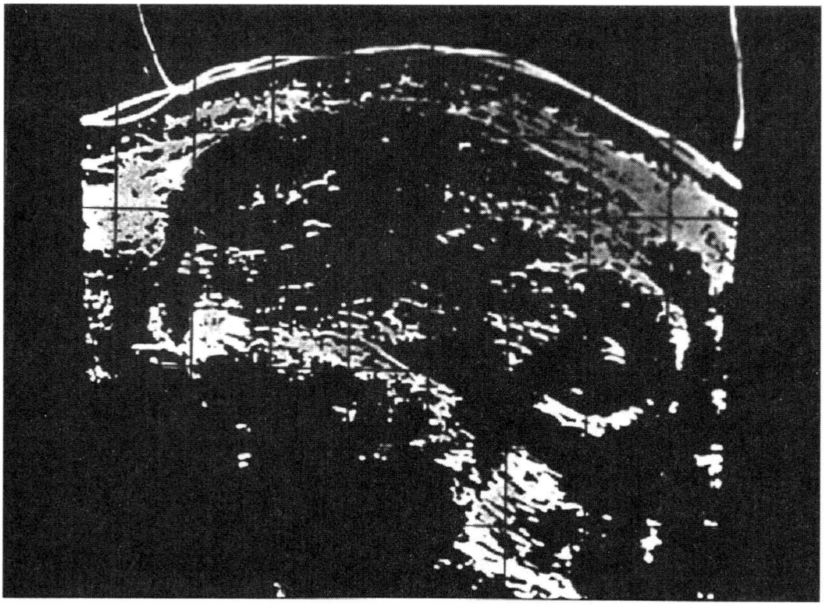

(a)

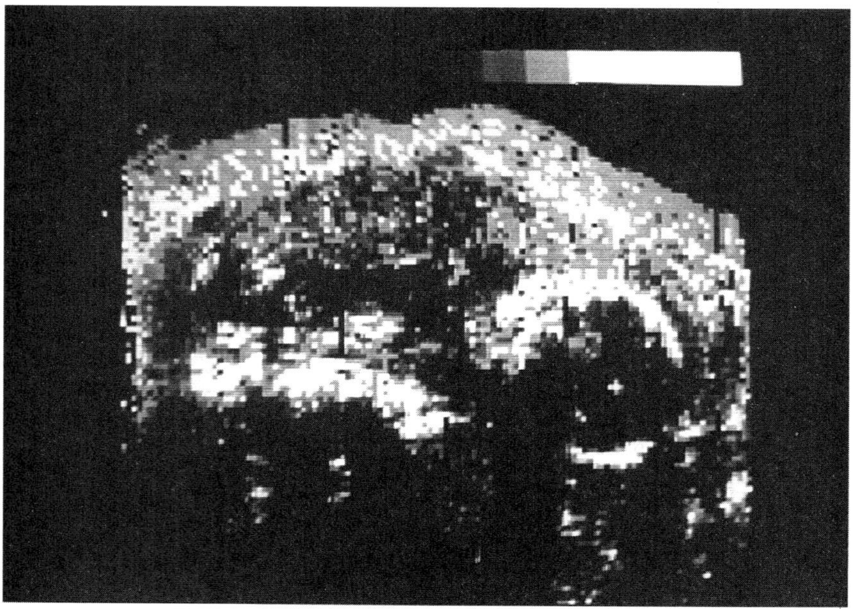

(b)

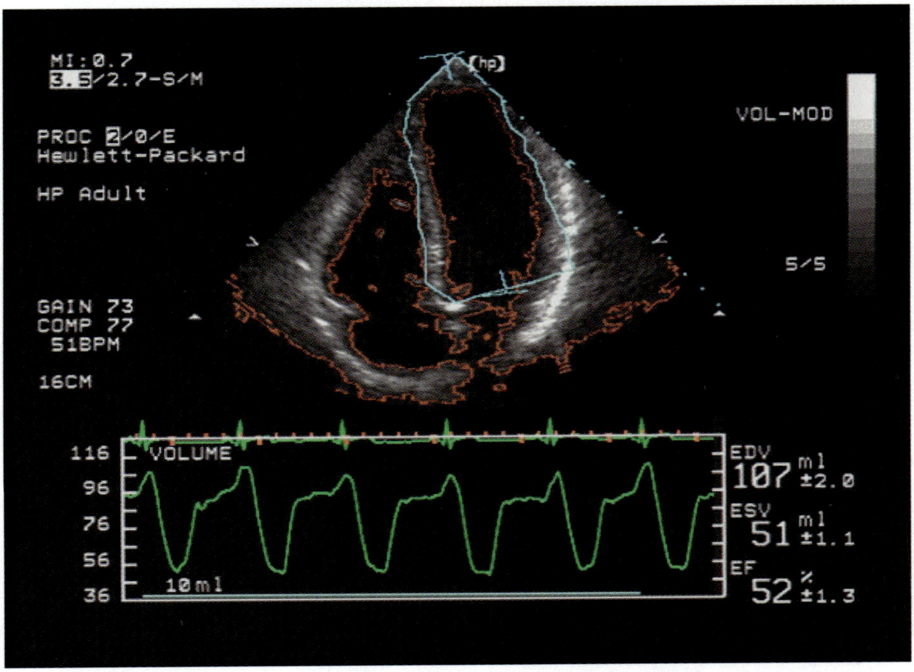

Fig. 40. Phased array cardiac scan. In this equipment the borders of the cardiac chambers are automatically located in real-time by an edge detection algorithm and a red line denoting the detected myocardium/blood boundary added to the image. The operator uses a freehand trace function (green line around rightmost chamber) to outline the left ventricle. The equipment, using an ellipsoidal model for the chamber volume, automatically calculates this chamber volume in real-time and displays it in the graph shown below the scan. The end diastolic volume (EDV) [high point of the curve], end systolic volume (ESV) [low point of the curve], and ejection fraction (EF) [difference between the two], averaged over five operator-selected cardiac cycles, are continually computed and displayed. The cardiac rate of 51 BPM (beats per minute) is automatically calculated from the measured EKG (displayed at the top border of the graph) and displayed on the left side above the graph. The mechanical index (MI) is 0.7 (AIUM, 1992). (Ultrasound image courtesy of Hewlett-Packard Company.)

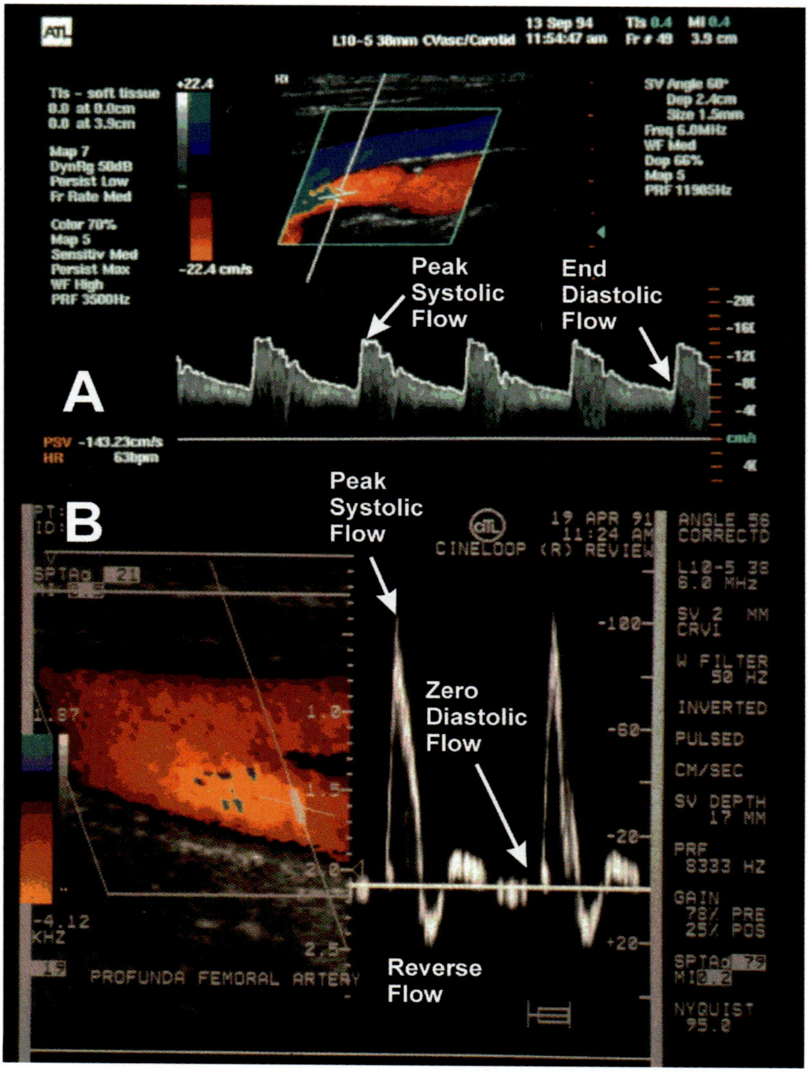

Fig. 43. Examples of flow for low- and high-resistance vessels. The upper waveform (A) shows a low-resistance, internal carotid artery with flow throughout the cardiac cycle. The CFI portion shows carotid flow from image right to left. The brighter portion of the spectral waveform tracks the maximum frequency. The lower image (B) shows high-resistance flow in a femoral artery. The high resistance causes reflected pressure waves that produce reversed flow in late systole, and a small forward flow rebound. This pattern is called triphasic flow. The CFI portion shows flow from image left to right. Note in both images that the Doppler spectrum vertical axis is calibrated in velocity due to the use of the flow angle indicator line at the range gate position. (Courtesy of ATL Corporation.)

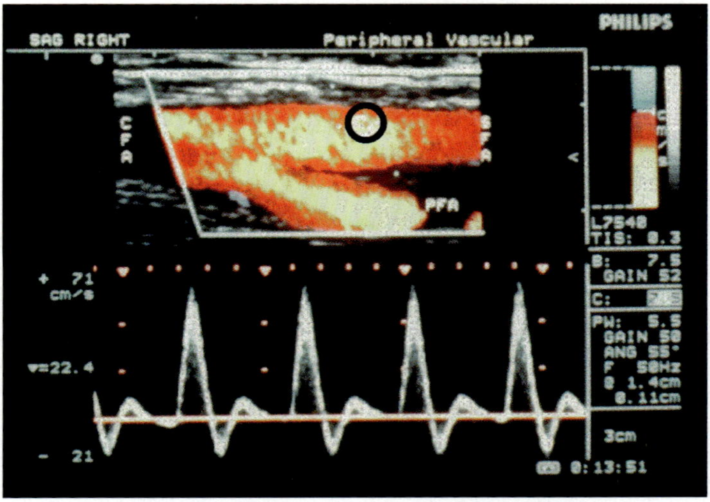

Fig. 55. CFI hue encoding. This example of color velocity Imaging of the femoral artery uses color hue (red to yellow) to encode flow away from the transducer. The color bar (image right) also shows that flow toward the transducer would be blue. The image incorporates both CVI (time domain analysis) for the CFI and duplex Doppler to produce the flow spectrum. The black circle surrounds the location of the Doppler range gate and angle corrector. (Courtesy of Philips Ultrasound International.)

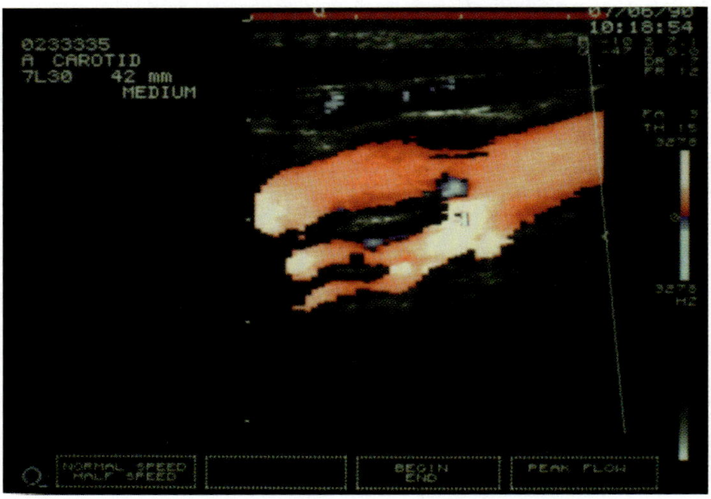

Fig. 56. CFI saturation encoding. The color bar on the image right shows the color code. Red is flow away from the transducer; blue is flow toward it. The image shows a common carotid artery bifurcating into the internal (upper) and external (lower) carotids, with a further branching off the external carotid (lowest vessel). (Used with permission of R. Knighton 11/5/97.)

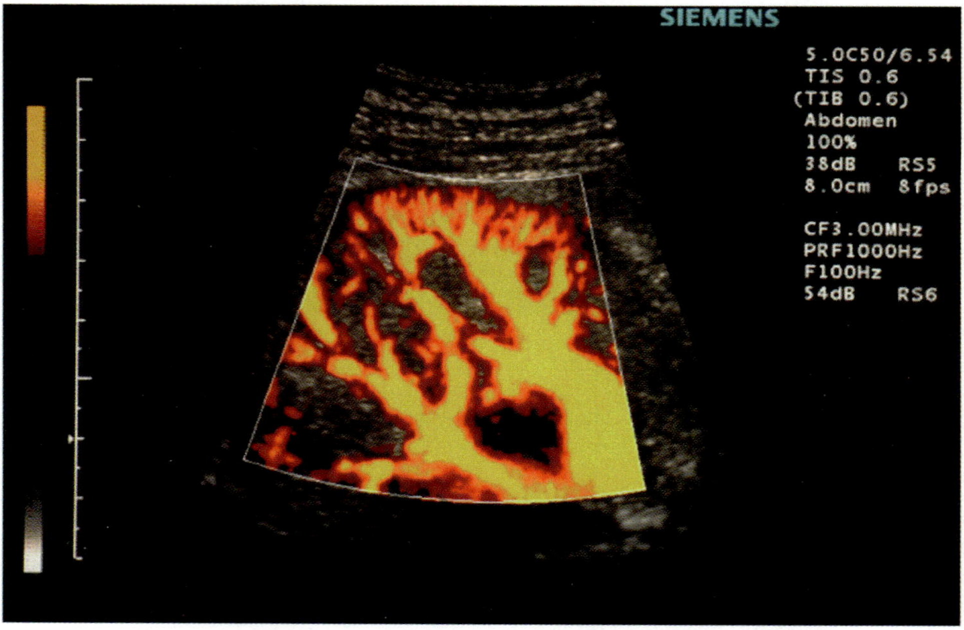

Fig. 57. Power Doppler color flow image. Power Doppler encodes the amplitude of the Doppler signals into color rather than frequency. The color bar on image left shows the amplitude color encoding. The image shows flow in the renal vessels, both arteries and veins. (Used with permission of R. Knighton 11/5/97.)

on a single sheet of 8×10 inch film, which was compatible with the automatic film processors in radiology departments. Each ultrasonic scanner required a dedicated multi-image camera. Since each acquired image was erased after hardcopy photography, many patient images were lost due to problems in film processing. These difficulties persisted until analog hardcopy was replaced by digital image archiving.

B. Analog Mechanical Sector Scanners

Mechanical sector scanning was a simple improvement to analog equipment, providing a real-time image. Single element transducers are mechanically scanned to provide a sector shaped image of patient anatomy. Figure 26 shows an oscillating, single-transducer, mechanical sector scanner. The element/backing layer combination oscillates about a mechanical pivot. An internal fluid acoustically couples the ultrasonic beam to the scanhead case, with coupling gel (not shown) between the scanhead and the patient. The *PRF* had to be varied commensurate with the sinusoidal beam motion to produce equally spaced image lines. The oscillating motion was mechanically unreliable.

These difficulties were reduced by replacing the single oscillating transducer by three continuously rotating transducers with slip-ring electrical contacts (Barber *et al.*, 1974). Each transducer transmitted and received echoes as long as its beam was inside the defined image sector. An added advantage of this design was the ability to use transducers with different focal lengths or frequencies. Then, only one transducer could be selected for targeted imaging or a composite image formed.

The sector angles could be varied from 30° to 90° with image depths up to 20 cm. Due to the image sector format and the small scanhead footprint, these scanners could be used on small-area acoustic windows such as the intercostal spaces used for cardiac scanning. Since the image lines are in a sector format, image line density and spatial resolution decreased with image depth.

FIG. 25. Longitudinal cross-sectional bistable and digital fetal images. (a) The bistable image shows the maternal skin surface, a layer of subcutaneous fat, and the proximal uterine wall along with the outline of the fetal skull. (b) The digital (128×128 pixel map, 16 shades of gray) image displaying the average of the acquired echo amplitudes also demonstrates the placenta just below the proximal uterine wall. Note the black unsampled pixels in this early digital image. (Goldstein *et al.*, 1975).

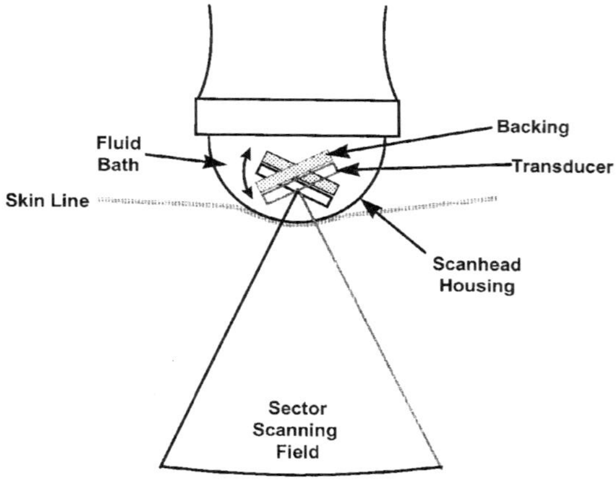

FIG. 26. Mechanical sector scanner. A single element transducer with backing is oscillated back and forth around a pivot to produce a sector-shaped image of patient anatomy. The acoustic coupling to the patient is through the enclosed coupling fluid and scanhead housing.

A difficulty with these scanners was beam integrity inside the scanhead. Reverberation in the enclosed coupling fluid, mode conversion in the scanhead case, and beam distortion due to refraction were all serious problems (Schafer and Lewin, 1984). Another difficulty was that mechanical inertia did not permit random access to selected transducer lines of sight, which is needed in M-Mode and Doppler studies (Thomenius, 1995). Perhaps the most important limitation of mechanical sector scanning (even with annular arrays — Section IV.B.5) is the difficulty in acquiring Doppler information from a moving transducer (Quistgaard, 1997).

IV. Digital Gray-Scale Imaging

With the development of multielement array transducers and the extensive use of digital circuitry and signal processing techniques, medical ultrasonics has evolved into a high-quality, cost-effective imaging modality. High-resolution digital ultrasonic imaging equipment first became commercially available in 1983 (Maslak, 1985). The improved image quality came at a price, however — an increase in scanner cost from \$60,000 to \$125,000. But soon the radiology community became convinced that these new high-quality images both improved patient care and permitted ultrasonic imaging to compete with CT imaging in many diagnostic situations. This improved image quality —

along with its low cost, real-time capabilities, and relative safety compared to CT and MRI—has established ultrasonic imaging as one of the leading medical imaging modalities (Brice, 1993).

All modern digital gray-scale equipment present real-time images with frame rates from several to 40 or more frames/sec. Motion of rapidly moving soft tissue structures, such as cardiac valves, can be resolved temporally at the higher frame rates. But even general ultrasonic imaging has benefited from real time. Slowly moving the real-time transducer over the patient's skin surface generates survey scans that present all internal patient anatomy with decreased total scan time.

High-resolution ultrasonic scanners have evolved from the initial hybrid (digital control of analog hardware and signal processing) to fully digital systems. At the present state of the art, the image quality of the real-time frames is much improved over the old static scans that contained many more image lines. This is mainly due to modern digital beamforming technology (Thomenius, 1996), which will be explained in Section IV.F. The operation of modern fully digital equipment will be presented in this section.

A. Multielement Transducers

Single element transducers have a fixed focus, pulse-echo beam pattern that permits high-quality imaging only in the focal zone. Dividing the transducer into individual elements that can be independently fired and utilized in reception provides great flexibility in the design of transmit and receive beam patterns.

1. Structure

Present multielement transducers are configured as a linear array of thin rectangular elements, as shown in Figures 27 and 28. The long axis of the rectangular elements are in the slice-thickness direction (perpendicular to the scan plane). The width, spacing, and number of elements depends on the type of transducer (Section IV.C). Computer-controlled diamond saws are used to fabricate the arrays to attain the required spatial uniformity.

2. Composite Piezoelectrics

Early linear arrays were fabricated from a thin piezoelectric slab bonded to a backing layer and diced into individual elements with the necessary dimensions both in and out of the scan plane. Due to their required small widths (as small as $\lambda/2$), the frequency characteristics of the individual array elements lead to several difficulties.

Single element piezoelectric transducers have a large ratio of width to thickness (W/T), so the frequency bandwidth of their thickness-expander mode vibrations is well separated from undesired modes of vibration. This permitted the use of shock-excitation voltage pulses with large frequency bandwidths. The small array elements have small W/T ratios with overlapping of the frequency bandwidths of desired (thickness-expander) and undesired (shear, lateral, and surface wave) modes of vibration. Excitation of these undesired modes of vibration: (1) wastes energy, which reduces element sensitivity; (2) degrades the single element directivity (Section IV.C) by distorting the shape of the radiating face; and (3) leads to acoustic cross talk (Section IV.C) between elements, which degrades the beam shape (Schafer and Lewin, 1984). The higher the desired operating frequency, the lower the W/T ratio.

Composite piezoelectric structures were designed to reduce these effects. They are fabricated using a dice-and-fill technique. According to Smith (1992), "Two sets of deep grooves are cut in a block of piezoceramic at right angles to each other, a polymer is cast into these grooves, and the solid ceramic base is ground off. After polishing the plate to final thickness, electrodes are applied to the faces and the ceramic is poled. For high-frequency operation, fine spatial scales are required. Kerfs of 25 microns and below are achievable, using diamond impregnated dicing wheels on OD saws developed for the semiconductor industry to dice chips from a processed silicon wafer. A fine-grained, high-density piezoceramic is essential if the pillars are to survive this machining."

The tall, thin piezoelectric pillars have a W/T ratio much less than unity, which separates their desired and undesired mode frequency bandwidths and reduces the deleterious effects of the undesired modes (Kino and DeSilets, 1979). Ceramic-air composites with potentially better engineering characteristics than ceramic polymers have also been developed (Oakley and Marsh, 1992).

There is an accepted trade-off associated with the use of composite piezoelectrics. Since the electroded area of piezoelectric material is reduced due to the saw kerfs and polymer (or air) fill, the element sensitivity is reduced proportionately.

3. *Acoustic Impedance Matching*

An important feature of composite piezoelectrics is the reduction of element acoustic impedance. If the composite substructure scale is small compared to the shortest wavelength in the frequency bandwidth, the acoustic impedance of the composite piezoelectric is the volume average of the piezoceramic and fill (or air) acoustic impedances (Smith, 1992). This lowers the transducer element acoustic impedance and reduces the required acoustic matching.

Acoustic matching technology has improved as well. New techniques such as acoustic matching by micromachined graded volume fraction silicon structures (Haller and Khuri-Yakub, 1992) and active piezoelectric layers (Hossack and Auld, 1992) provide improved transmission and reception characteristics. Besides increased element sensitivity, the improved acoustic impedance matching produces better near-field image quality (less acoustic ring-down after transmission), broader bandwidth operation, and less energy dissipated within the transducer. This latter factor is very important when the transducer is pulsed rapidly—otherwise the transducer structure would soon become too hot to handle.

4. *Broadband Operation*

The recent improvements in composite piezoelectrics and acoustic matching permit the design of transducers with *FB*s exceeding 80% of their center frequency (Thomenius, 1995). By using driven excitation (Section IV.E), all or part of this bandwidth can be selected for each output ultrasonic pulse. Manufacturers have utilized the large available bandwidth differently. For example, the complete available bandwidth can be used. Since echo bandwidth is truncated at higher frequencies due to frequency-dependent attenuation, bandpass filtering commensurate with target depth is used to reduce input noise. Or, the received echo signal bandwidth can be divided into several discrete frequency subbands that are combined together to reduce (average out) speckle noise.

The large bandwidth can also be divided into several smaller bandwidths so that the clinical user can select different frequency ranges from the same transducer at the touch of a button. This provides very rapid change in transducer frequency while the tissue of interest is visible on the gray-scale display. By using the same frequency bandwidth for all depths in the image, the speckle pattern is more constant with depth.

The large bandwidth can be divided into several smaller selectable bandwidths for more flexibility in designing strong transmit focuses. These strong transmit focuses can be used in a multizone focusing scheme (Section IV.D.4) in which the transmit focus as well as the receive focus is optimized for each image depth.

B. Types of Transducers

Several types of transducers are used in medical imaging to take advantage of the different acoustic windows available or to conform with the geometry of the soft tissue being examined. Most are multielement arrays, although some single element transducers are still used for special applications.

1. Phased Array

A phased array is a multielement linear array that uses all of its elements to form each image line (Thurstone and von Ramm, 1973). If the elements are fired simultaneously, an unfocused "plane wave" ultrasonic pulse will be transmitted parallel to the normal of the array surface. This is because the cylindrical waves from each small element combine at a distance from the array to form a plane wave parallel to the array surface and in a homogeneous medium the direction of wave propagation is normal to the surfaces of constant phase. The direction of wave propagation may be changed by firing the elements along the array with a constant time delay between element transmissions. Then the cylindrical waves will combine at a distance from the array to form a plane wave tilted at an angle commensurate with the array geometry and the constant time delay. This is called beam steering (von Ramm and Smith, 1983).

The transmitted ultrasonic pulse may be focused at any depth by timing the element firings such that the cylindrical waves all arrive at the chosen focal point (in the scan plane) at the same instant of time. When the focal point is along the linear array axis, there is no beam steering and the phase (time) delays needed to focus the beam vary quadratically from the array center (Christensen, 1988). When the focal point is at an arbitrary angle θ to the linear array axis, beam steering is necessary and the time delays needed for each element may be computed using the law of cosines.

In reception, the echo signals received from each element are time delayed to form a beam-steered and focused receive beam pattern. Assuming point targets that generate spherical wave echoes, the receive beam pattern will be identical to the transmit beam pattern (in the scan plane) if the receive time delays are identical to the transmit time delays. Figure 27 demonstrates a 12-

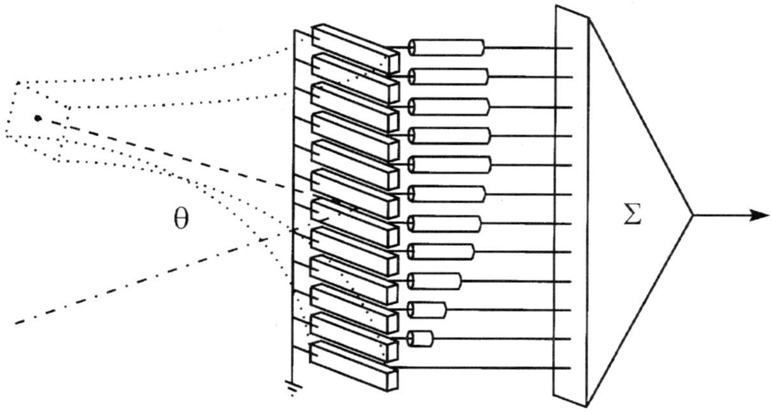

FIG. 27. Phased array in receive mode. The receive beam pattern is shown. All array elements are used for each image line. The time delays of the echo signals from each element are depicted as cylinders whose length is proportional to the time delay. The time delayed element signals are summed in a summing circuit to produce the echo wavetrain received from the beam-steered, focused beam pattern shown. Note that the front electroded face of each element is connected to electrical ground to aid in forming a Faraday cage shield against EMF pickup inside of the transducer structure.

element phased array with a receive beam pattern focused at the point shown at the beam-steered angle θ. The time delayed signals from each element are summed to form the array pulse-echo signal from the beam-steered image line.

The image format of a phased array is the sector format shown in the upper left of Figure 29. Most phased array transducers have between 64 and 128 elements (Thomenius, 1995) and typically steer through a 90° sector angle. When the available acoustic window is limited, a small-footprint phased array is useful. Initially, phased arrays were used for cardiac scanning where the acoustic window is an intercostal (between the ribs) space and the expanding field of view permits visualization of the cardiac chambers. Abdominal phased arrays, with larger apertures for deeper focusing, were developed later to get around pockets of bowel gas at shallow image depths. The large aperture prevents fine focusing in the near-field, but this trade-off is acceptable because of a limited field of view in the sector image format close to the sector apex.

Besides the required circuit complexity caused by the large number of individual elements and the need for different focus time delays at each beam-steered angle, there are several other disadvantages of phased arrays. Due to the sector image format, the image line density decreases with depth, causing a commensurate decrease in spatial resolution. At the same time, as the beam-steered angle θ changes, the effective transducer aperture is foreshortened by a factor $\cos\theta$, which diminishes the maximum possible focal length by a factor $\cos\theta^2$ (Section II.C.2) and increases the beam spot size by a factor $\cos\theta^{-1}$ (Eq. (16)). Consequently, phased array image quality falls off at large beam-steering angles.

2. *Linear Stepped Array*

A linear stepped array is a multielement linear array that uses subgroups of its elements to form individual image lines (Bom *et al.*, 1971). It is much longer than a phased array and has more elements up to a typical maximum of 512. It is useful with large acoustic windows such as obstetrical and pelvic examinations. Beam steering is usually not used, so its image format is rectangular (see Figure 29, upper right), and its image line density is constant with depth. It has better image quality in its large near-field than a phased array because smaller apertures (fewer elements) can be selected to form beams with better short-range focuses.

The operating principle of this array is demonstrated in Figure 28, where a 9-element aperture is used in receive. With no beam steering, the required element receive time delays vary quadratically from the aperture center. For each successive image line, the 9-element aperture is shifted one element further along the array. This is called full-stepping (Thomenius, 1995) and results in image lines separated by the array pitch (center-to-center distance between adjacent elements). Half-stepping is also possible. When the number of elements in the aperture is odd (see Figure 28), the beam central ray is at an element center. When the number of elements in the aperture is even (see Figure 35), the beam central ray is between elements. So if the aperture toggles appropriately between an even and odd number of elements for adjacent image lines, the distance between these lines will be half the array pitch.

For a linear stepped array with no beam steering, it is possible to cut in half the number of signal processing channels. Since the element time delays are symmetric about the aperture center, the element signal outputs may be summed in pairs then time delayed and finally summed with no change in the resultant pulse-echo signal. Although this scheme reduces circuit complexity and cost, the trade-off for its use is reduced signal processing flexibility.

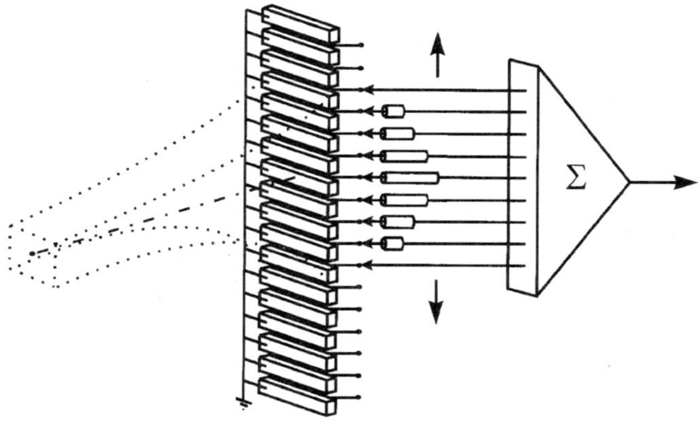

FIG. 28. Linear stepped array in receive mode. Only a 9-element subset of the 17-element array is used for each image line. The subset receive beam pattern is shown. The time delays of the echo signals from each element are depicted as cylinders whose length is proportional to the time delay. The time delayed element signals are summed in a summing circuit to produce the echo wavetrain received from the unsteered, focused beam pattern shown. The 9-element aperture is stepped sequentially along the array by electronic switching to produce the unsteered image lines.

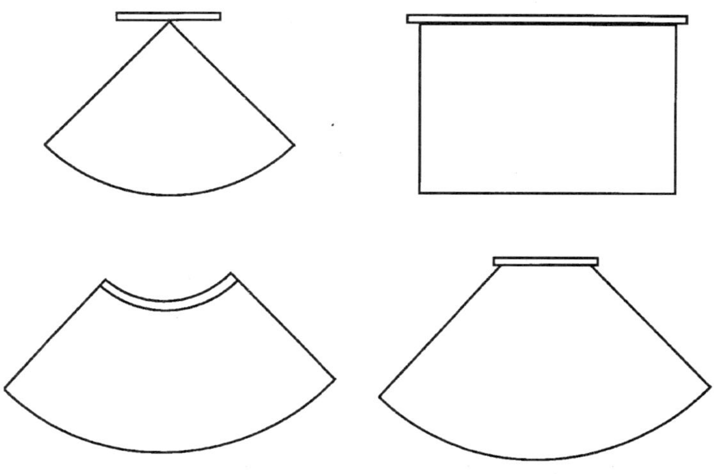

FIG. 29. Transducer image formats. The image formats of four multielement array transducers are presented. Upper left: phased array, Upper right: linear stepped array, Lower left: convex array, Lower right: vector array.

When Doppler color flow imaging was added to ultrasonic imaging equipment (Section V), beam steering was added to linear stepped arrays. This was necessary because many of the peripheral blood vessels to be studied (such as the common, external, and internal carotid arteries) flow parallel to the skin surface and unsteered ultrasonic beams would have a 90° Doppler angle with the flowing blood. Commensurate with the multielement array beam pattern (Section IV.C), a 20° beam-steering capability (operator-selectable — to the left or right) was implemented for all stepped linear array image lines in the Doppler mode of operation. Once the necessary capability was built into the equipment, it also became possible to use these steered image lines in gray-scale mode. This 20° gray-scale beam steering is known as microsteering and is utilized, in addition to unsteered image lines, in some high-resolution equipment.

The signal processing flexibility inherent in a multielement transducer can also be used to improve the focused beam pattern by reducing side-lobe magnitudes. One standard technique used is aperture apodization ('t Hoen, 1982). Here, the sensitivity of the outer elements in the aperture is gradually reduced, causing the beam side lobes to decrease in magnitude. In general, different apodization functions are used in transmit and receive (Maslak, 1985). In Figures 27 and 28, the apodization function would be added to the summing operation. By judicious choice of the aperture function (variation of element sensitivities), the side lobes can be almost completely eliminated for ultrasonic pulses.

The altered beam pattern does have two trade-offs: First, when the side lobes have been sufficiently reduced, the focused beam width (spot size) is increased by typically 50%. And second since lateral aperture elements contribute less to the pulse-echo signal than central elements, the aperture sensitivity is reduced by about 10 dB for optimized apodizations. These trade-offs are accepted because in the search for cancerous focal lesions image contrast resolution is more important than image spatial resolution or sensitivity. Modern ultrasound imaging equipment use different degrees of apodization for different clinical imaging tasks. This is part of "tissue-specific imaging" where the operator selects a tissue mode of operation before beginning the patient examination.

Another technique for reducing side-lobe magnitude is to use different aperture sizes for the transmit and receive beam patterns (Moshfeghi, 1987). Since the pulse-echo beam pattern is their product, placing one beam pattern side lobe at the position of the other beam's zero will cause a reduction of the resultant pulse-echo beam side-lobe magnitudes. The optimum reduction of

5.4 dB occurs when the transmit aperture is 72% of the receive aperture. Once again, the trade-off is an increase in focused pulse-echo beam width, this time by 15%.

It is evident from Figure 28 that the ultrasonic beams at the two lateral edges of the rectangular field of view will not have symmetric beam patterns if the image lines cover the full array length. For small stepped apertures, a slight reduction of the image field of view by omitting these asymmetric beam patterns is acceptable. However, for deep focusing with large stepped apertures, they must be utilized. They will not focus as deep or have as small a spot size as the symmetric beams in the center of the image field of view. As a consequence, some reduction in image quality will exist at the lateral edges of the rectangular field of view of linear stepped arrays (and convex and vector arrays), especially at depth.

3. *Convex Array*

A disadvantage of the linear stepped array is its limited field of view at depth. Manual scan, static imagers had much larger fields of view. To obtain larger fields of view for pelvic and abdominal scanning, the convex array was developed. It is essentially a linear stepped array with the array plane mechanically curved into an arc (see Figure 29, lower left).

If the radius of curvature of the convex array is made too small the aperture size and focusing at depth will be limited. This is because the outer elements in the aperture cannot propagate and receive at a steep angle to their acoustic axes — the steep angle reduces beam sensitivity (Section IV.G.1).

The gentle curvature of the convex array is actually an asset in imaging through many acoustic windows. Thus it is a popular transducer type in many clinical scanning situations. The obvious trade-off associated with the convex array is its decrease in image line density with depth.

4. *Vector Array*

The vector array was developed to increase the image field of view of the abdominal phased array (see Figure 29, lower right). It is basically a phased array transducer, driven using stepped array principles with image lines that are also steered. It has a larger field of view at shallow depths than the phased array. Due to the stepped array operation, the beam aperture can be made small to obtain a good focus at shallow image depths. It is an improvement over phased arrays in abdominal imaging. Compared to linear stepped arrays it has a larger field of view at depth, but the trade-offs are decreased image line density with depth and more complicated driving circuitry due to the different beam steering required for each image line.

5. *Annular Array*

Circular aperture transducers can be divided into multiple annular concentric rings, as shown in Figure 30. This structure is called an annular array. Due to the circular aperture and resulting cylindrical symmetry of its beam, beam steering is not possible. Consequently, annular arrays must be mounted in mechanical sector scanner heads with all of their attendent disadvantages (Section II.B). The number of elements in an annular array is usually less than 12, typically 6 or 8 (Thomenius, 1995).

Time delays are used to focus the beam at selected image depths. An advantage of the annular array is that due to the cylindrical beam symmetry it focuses electronically out of plane (slice thickness direction) as well as in plane. This is important because the thinner the scan plane the greater the image contrast for cancerous focal lesions. The difficulties caused by the mechanical sector scanhead, however, override the advantages of slice thickness focusing and annular arrays are not popular, at present, for clinical scanning.

6. *Endocavity Transducers*

Many soft tissue structures are not imaged easily from the skin surface. The female reproductive anatomy and the male prostate gland are two good examples. Years ago, obstetrical and gynecological ultrasonic scans were performed by giving the patient several large glasses of water and then using

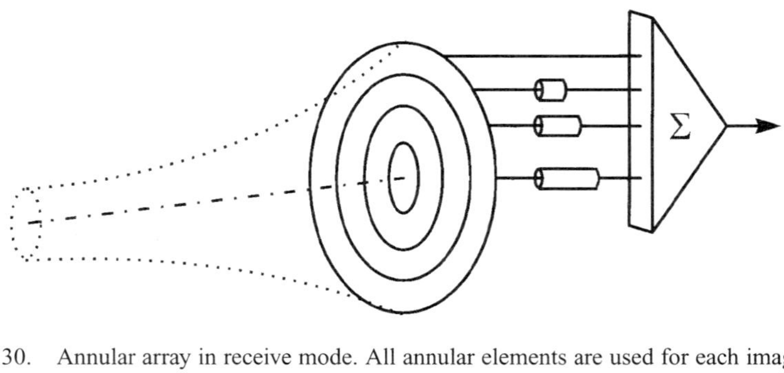

FIG. 30. Annular array in receive mode. All annular elements are used for each image line. The time delays of the echo signals from each element are depicted as cylinders whose length is proportional to the time delay. The time delayed element signals are summed in a summing circuit to produce the echo wavetrain received from the unsteered, focused beam pattern shown. The annular array must be mounted in a mechanical sector scanner head for image generation.

the filled urinary bladder as an acoustic window. Besides being very uncomfortable for the patient, the "normal anatomy" being viewed was distorted due to the space occupying urinary bladder. The development of transvaginal transducers that can be inserted into the vagina and placed next to the anatomy of interest greatly improved these structural images.

The transducers are mounted on long-handled probes (see Figure 31). Some are rotating or oscillating small single elements while others are linear or convex arrays. Some endocavity transducers are "front looking" or "endfire" transducers with a sector image format projecting straight out from the probe tip. Others use array transducers to produce scan planes perpendicular to the probe axis (transverse scans) or parallel to the probe axis (longitudinal scans).

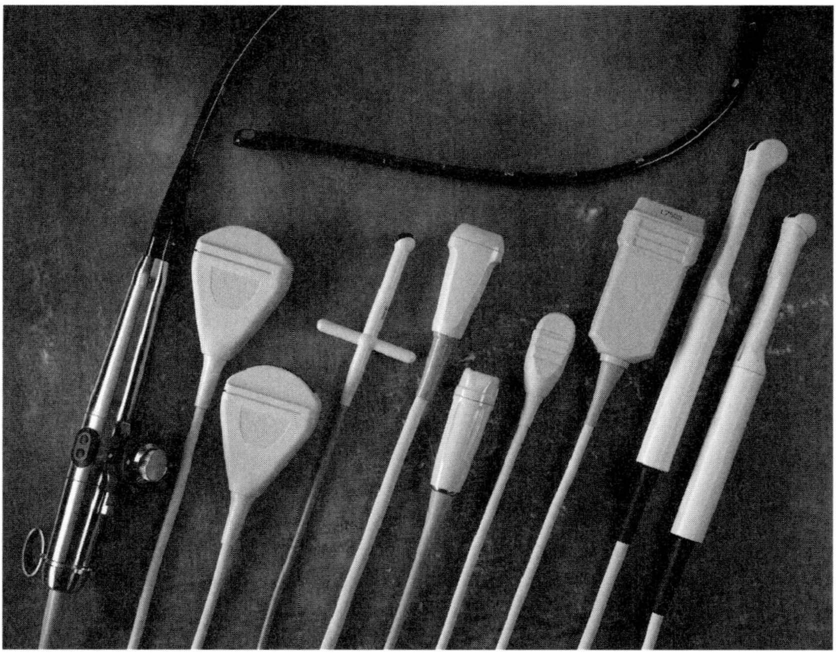

FIG. 31. Transducers available for specialized ultrasonic imaging exams. On the far left is a transesophageal probe with a side-looking phased array (no beam steering) at its tip. The probe is inserted into the esophagus and positioned at the level of the heart. The phased array is mechanically swept through a 180° sector to produce a cross-sectional cardiac image. The next two transducers are convex arrays. The transducer with the T-shaped handle is a 1.8-MHz cardiac nonimaging transducer for CW Doppler studies; next are two phased array transducers. The convex array transducer with a small radius of curvature is for pediatric studies. After the linear stepped array transducer is a transrectal probe with an almost front-looking convex array for prostate scanning. The last is a transvaginal probe with a front-looking convex array. (Photograph courtesy of Hewlett-Packard Company.)

Since the transducer can be placed close to the anatomy of interest, high-frequency transducers can be used. Image quality is improved because (1) the high frequencies produce well-focused ultrasonic beams and (2) the small field of view increases the image line density due to shorter image depths that permit higher *PRF*s and more image lines (Eq. (15)). For a fixed number of image lines, the smaller the image field of view, the higher the image line density (Goldstein, 1987).

Transrectal transducers are also used to search for cancer in the prostate gland, which is situated near the rectum. When suspicious areas are found in the prostate, an ultrasonically guided needle biopsy can be used to obtain small tissue samples for histological analysis.

7. *Intraoperative Transducers*

Special transducers have been developed for examination of exposed patient tissue during surgery. These transducers usually have high frequencies because the transducer can be placed close to the tissues of interest. They are generally sterilized using gas sterilization techniques. Figure 32 shows a typical intraoperative transducer.

C. Array Beam Patterns

The beam pattern of a uniform, linear multielement array is due to the coherently interfering contributions of the elements in its aperture. For a quasi-CW case with tone-burst transmission, it is found by analogy with the optical diffraction grating (Born and Wolf, 1975). The pulse-echo signal strength in its geometric focal plane—taking into account the fact that the ultrasonic waves generated from each element are in phase and polarized normal to the element's front face*—is for equal transmit and receive apertures

$$Sig(\theta) = \left[\frac{\sin\left(\frac{N\pi p}{\lambda}(\sin\theta - \sin\theta_0)\right)}{\sin\left(\frac{\pi p}{\lambda}(\sin\theta - \sin\theta_0)\right)}\right]^2 \cdot \left[\frac{\sin\left(\frac{\pi w}{\lambda}\sin\theta\right)}{\frac{\pi w}{\lambda}\sin\theta}\right]^2 \cdot \cos^2(\theta), \quad (38)$$

*The longitudinal polarization of a thickness-expander source of ultrasonic waves causes a $\cos\theta$ reduction in longitudinal wave amplitude in a fluid medium. Most optics and acoustics texts ignore this polarization term because the angles involved are usually quite small. However, in this case due to the large angles involved, it cannot be neglected. Experimental measurements of the angular beam pattern of narrow strip transducers well isolated (air gaps) from adjacent elements verify the presence of the $\cos\theta^2$ term (Selfridge et al., 1980 and von Ramm and Smith, 1983)

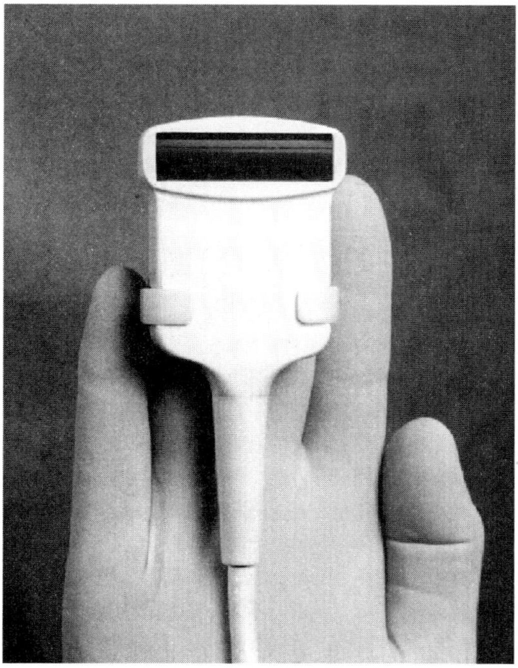

FIG. 32. Intraoperative transducer. This 7.0-MHz linear stepped array transducer is side-looking and comes with a plastic clip so it can be attached to gloved fingers with the array just below the finger tips. A surgeon can manually positioned it inside an incision by touch. (Courtesy of Diasonics Ultrasound.)

where N is the number of elements in the aperture, p is the array pitch, w is the width of an individual element, θ is the beam angle from the aperture center, θ_0 is the beam-steered angle of the beam central ray, and λ is the acoustic wavelength in the propagation medium. Np is equal to the aperture width D. The slice-thickness focus will not be considered here.

The first term in Eq. (38) is due to the periodicity of the array elements and will be called the *periodicity term*. The second and third terms in Eq. (38) are the unfocused, far-field beam pattern of one of the identical elements. Since they modulate the amplitude of the periodicity term, they will be called collectively the *envelope terms*.

1. Unsteered Beam Pattern

Figure 33 demonstrates the unsteered ($\theta_0 = 0$) beam pattern of a multielement array where $p = 1.5\lambda$. The sine of the beam angle is the independent variable

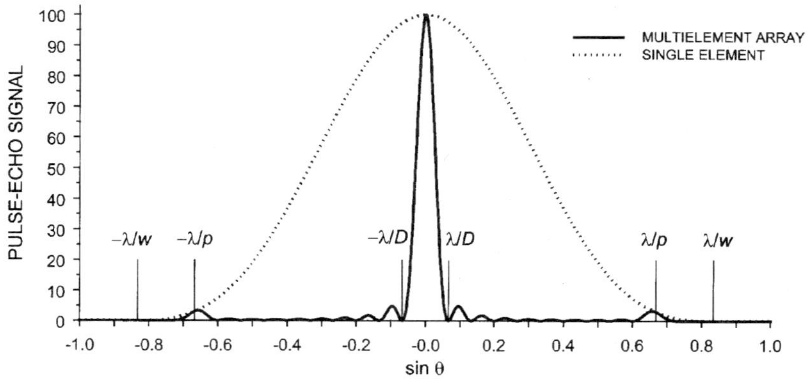

FIG. 33. Unsteered pulse-echo beam pattern of a 10-element, 5-MHz array radiating into a nonattenuating medium with $c = 1540$ m/s. The independent variable is the sine of the beam angle θ, which covers the full angular width of $\pm 90°$. $p = 1.5\lambda$ and $w = 0.8p$. The solid line is the array beam pattern and the dotted line is the envelope terms of Eq. (38) multiplied by N^2 to normalize the two curves.

covering the full beam angle range of -1 to 1 ($-90° \leq \theta \leq 90°$). Centered at 0 ($\theta = 0°$) is a beam pattern consisting of a large main lobe surrounded by side lobes. At $\sin\theta = \pm\lambda/p = \pm 0.67$, the beam pattern is replicated (although at lower amplitude due to the envelope terms). These replicas are called grating lobes and are due to constructive interference at the CW frequency. There is some small amount of ultrasonic signal received from all angles.

The half-width of the main lobe (in terms of $\sin\theta$), λ/D, is determined by the first zero of the periodic term, which occurs when the argument of its numerator sine term is equal to π. When converted to linear distance in the geometric focal plane, the full spot dimension is given by Eq. (17). The closest side-lobe amplitudes and beam echo amplitude dynamic range are almost identical with the rectangular aperture sinc function side lobes in Eq. (9). This is predicted by the periodic term when $\sin\theta$ is small enough that the denominator sine can be replaced by its argument, causing the periodic term to be proportional to a sinc function.

The main-lobe amplitude is proportional to N^2. This is predicted by the periodic term at angles near $0°$, where $\sin\theta$ is small enough that both sine terms can be replaced by their arguments. This main-lobe amplitude also can be predicted on a physical basis. In reception from the focal point, each of the N elements produces the same output signal (Eq. (6)) and all these signals are summed—whereas with a single element transducer the contributions from all areas of the front surface would average to the same signal received by each element. In transmission to the focal point, for identical element driving voltages acoustic reciprocity ensures that the ultrasonic pressure at the focal point is proportional to N.

Grating lobes permit reception of echo signals from beam angles different from the transducer line of sight (along the main lobe). There are three ways to counter this undesirable situation for unsteered beams. The first is to have nonuniform spacing of the array elements (Steinberg, 1976). The second is to notice that if the array elements were contiguous, $w = p$ and the zeros of the first envelope term would cancel the grating lobes. So if w is close to p in size (small gaps between elements), the grating lobes will be fairly low in amplitude (as in Figure 33). The third, more effective, way of eliminating grating lobes for unsteered beam patterns is to require that $p < \lambda$. Then, the first grating lobe is pushed out beyond $\pm 90°$ because the argument of the denominator sine in the periodic term will never reach $\pm\pi$.

2. *Steered Beam Pattern*

Figure 34 demonstrates the beam pattern if the multielement array of Figure 33 is beam steered 30°. While the main lobe has been displaced to $\sin\theta = 0.5$, one grating lobe (which is always $-\lambda/p$ less) has been pulled

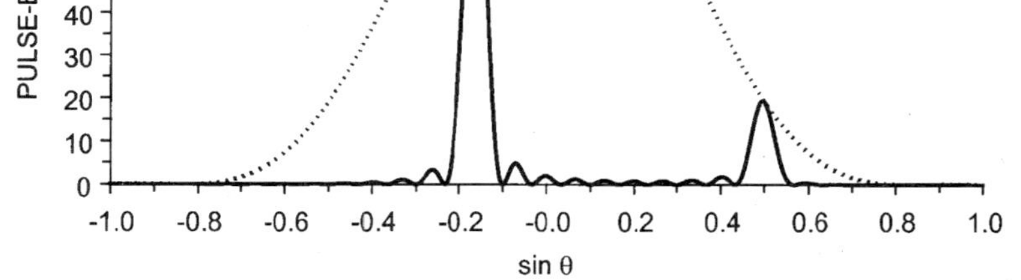

FIG. 34. Pulse-echo beam pattern of the array in Figure 33 beam steered 30° to the right. The independent variable is the sine of the beam angle θ, which covers the full angular width of $\pm 90°$. The solid line is the array beam pattern and the dotted line is the envelope terms of Eq. (38) multiplied by N^2 to normalize the two curves. The main lobe is seen at the sine of 30°, 0.5, and a large first grating lobe is seen at $0.5 - \lambda/p = -0.17$. A second grating lobe at -0.83 has been effectively eliminated by the first zero of the envelope terms.

to $\sin\theta = -0.17$. Due to the stationary envelope terms, this grating lobe is higher in amplitude than the beam-steered main lobe. So clearly grating lobes are much more of a problem with beam steering, and neither having w close to p in size nor having different size transmit and receive apertures will help in reducing grating-lobe amplitude.

Reducing p so that the argument of the denominator sine in the periodic term will never reach $\pm\pi$ is the only way to eliminate the first grating lobe for a uniformly spaced beam-steered array. For a beam-steered angle θ_0, the condition on p is

$$p < \frac{\lambda}{1 + \sin\theta_0}. \tag{39}$$

Note that p has to be sufficiently less than shown in Eq. (39) to also eliminate the close side lobes of the grating lobe. To be certain of eliminating the first grating lobe for all beam-steered angles (up to $\sin\theta_0 = 1$), the condition on p is $p < \lambda/2$ for the highest frequency in the array bandwidth.

Beam steering is an essential component of phased array operation. The phased array footprint (aperture) must be small to fit in available acoustic windows, which means they are designed with $p < \lambda/2$. Linear stepped arrays, on the other hand, usually do not beam steer and their footprints (apertures) are large, so with 128 elements they usually were designed with $\lambda \leq p \leq 1.5\lambda$.

Early high-resolution arrays had up to 128 elements. Newer high-resolution arrays typically have up to 512 elements. For the same aperture sizes, this fourfold increase in elements leads to a sixteenfold increase in pulse-echo signal strength for more tissue penetration (24 dB more penetration or $12/f$ cm more penetration at f MHz for tissue with an attenuation of 1 dB/cm-MHz). Also, the array pitch is four times smaller, which aids in eliminating grating lobes for the higher frequencies presently in use.

In the case of short ultrasonic pulses with wide frequency bandwidths, the grating lobes will be smeared out in space with a commensurate reduction in amplitude due to their frequency dependence. Since this spatial smearing is proportional to the distance of the lobe from $\sin\theta = 0$, the main lobe is hardly affected. The ratio of the peak of the nth grating lobe to the peak of the main lobe for broadband pulses has been calculated to be (Macovski, 1979)

$$\text{Ratio} = \begin{cases} \dfrac{m}{Nn}, & \text{for } \dfrac{m}{Nn} < 1 \\[2ex] 1, & \text{for } \dfrac{m}{Nn} \geq 1, \end{cases} \tag{40}$$

where n is the grating-lobe order and m is the number of center frequency cycles in the pulse. So for very short pulses and high N, the grating lobes can be substantially lower in amplitude than in the quasi-CW case.

Medical ultrasonic multielement beam patterns consist of a high-amplitude, narrow main lobe surrounded by off-axis signal known as clutter. The clutter level consists of side, grating, spurious lateral, and phase quantization lobes (Section IV.G) along with electrical noise pickup and element cross talk.

3. *Element Cross Talk*

Electrical or acoustic coupling between the elements of an array transducer can produce many undesirable effects. The electrical coupling can be due to unshielded cables or capacitive interactions between elements. If the capacitive coupling between nearest-neighbor elements has no phase shifts, it can cause a change of element electrical input impedance and an increase in effective element size (Schafer and Lewin, 1984). Increasing the effective element size will reduce the angular spread of the envelope terms in Eq. (38) and limit beam-steering operation. This type of electrical coupling should be kept below 30 dB to achieve broad acceptance angle arrays (Kino and DeSilets, 1979). If the capacitive coupling also has 90° phase shifts between adjacent elements, then the beam pattern can be dramatically altered, producing a minimum at 0°, with strong coupling (Kino and DeSilets, 1979).

Acoustic coupling arises from lateral mode generation in the elements as well as surface and Lamb waves in the backing layer and front matching layers, respectively. These latter types of coupling can be reduced by having the saw cuts (which define the elements) extend into the backing, tilted with respect to the array front surface, and of different depths to destroy the symmetry about each element (Dias, 1982). Also, it is recommended that a metal foil be placed over the array face to stiffen the structure, but this foil should be less than 25 μm in thickness to minimize Lamb wave effects (Larson, 1981).

D. Zone Focusing

Only the focal zone portion of a focused ultrasonic beam pattern is narrow enough to be effective in medical imaging. Use of the complex interference pattern near-field and the wide far-field beam portions leads to reduced image spatial and contrast resolution. In analog static imaging, the single-element transducers were weakly focused to lengthen the focal zone and minimize this problem. Modern imaging equipment with fast digital circuitry uses zone focusing to eliminate the undesirable portions of the beam pattern.

1. Selectable Zone Focus

The first attempt to improve image quality at all depths was selectable zone focusing, shown for a linear stepped array in Figure 35(a). For each image line there were three possible focused beam patterns: a near-depth focus, a mid-depth focus, or a far-depth focus. The operator selected the focal depth for each image with a three-way switch. High image quality occurred only at the selected image depth.

2. Composite Transmit and Receive Focus

A further refinement eliminated the undesirable portions of each depth's focused beam pattern. As shown in Figure 35(b) with three depth zones, for each image line the transducer sequentially transmits the three different focal patterns. After each transmission, only the echoes from that beam's focal zone are acquired. Then the acquired focal zone echoes are combined to form an effective beam pattern that is a composite of the three separate transmissions and partial receptions (see Figure 35(c)). Some image processing was used to eliminate any visual seams or "zipper patterns" in the resulting image. Now the image was focused at all image depths. Keeping the number of lines per frame constant to preserve spatial resolution, the trade-off was a commensurate reduction in frame rate (Eq. (15)).

3. Dynamic Receive Focus

Once an ultrasonic beam is transmitted, its focal properties cannot be changed. During echo reception, however, it is possible to sequentially change the array element time delays such that the receive beam pattern is continuously focused at the depth of origin of the echoes being received (Morgan *et al.*, 1978). In other words, the receive focus moves outward at half the acoustic velocity (time differentiate Eq. (14)). This switching of array element time delays is performed in discrete steps using fast digital circuitry.

Dynamic receive focus is accompanied by dynamic aperture and dynamic apodization (Maslak, 1985). That is, along with changing the focus time delays, as echoes are being received the aperture is increased stepwise (element by element) such that the receive beam spot size remains approximately constant (Eq. (16)). Also, the apodization function is changed commensurate

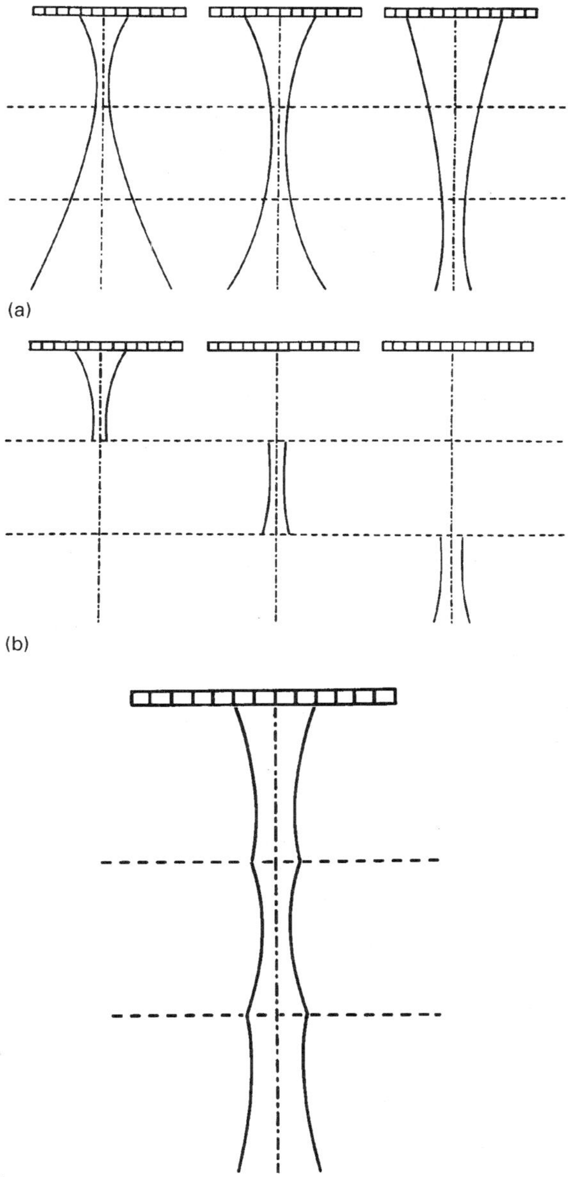

(a) (b) (c)

FIG. 35. Zone focusing. (a) Selectable transmit zone focusing with three beam patterns, each designed to optimally focus in one of the three depth zones. Note that the array aperture size increases with the required depth of focus. (b) Here, for each image line only the optimally focused portions of the pulse-echo beam pattern are acquired to form the composite transmit and receive beam pattern shown in (c). (Goldstein (1988). Reprinted by permission of John Wiley & Sons, Inc.)

with image depth and aperture size. In gray-scale imaging, the *f*-number is usually kept at $f/2$ during dynamic receive focus. For deep focuses the aperture will finally be limited by the physical array size, and then the receive focus *f*-number and beam spot size will increase with depth.

Since the pulse-echo beam pattern is the product of transmit and receive beam patterns, it is narrowed considerably by the use of dynamic receive focus with a commensurate lengthening of its effective focal zone (Thomenius, 1996).

4. Composite Transmit/Dynamic Receive Focus

Modern high-resolution ultrasonic imaging equipment uses a combination of successive transmit focus line segments in depth, each with dynamic receive focus to form a composite beam pattern for each image line—much like Figure 35(c), only narrower. The operator can select either one of many possible depth transmit focuses in a selectable dynamic receive zone focus mode for highest image frame rate or a composite transmit/dynamic receive focus mode for improved image quality over the whole image (or a large area of the image) at a commensurately reduced frame rate. For some transducers, two or three transmit focus line segments are combined in a composite transmit/dynamic receive focus combination, which can be shifted in position along the image lines. The trade-offs between area of high image quality and frame rate are different for each specific clinical exam and operator control of these image variables is essential.

The trade-off between image line density and frame rate stated in Eq. (15) assumes a constant transmit *PRF*. In practice, image frame rate has steadily increased over time. One improvement was altering the *PRF* in the composite transmit/dynamic receive focus mode so that immediately after the last echo was received from one line segment the transducer was fired for reception of the next line segment. This new mode of operation (presumably) acquired the line segments sequentially down each image line with some dead time between receptions that reduced the magnitude of range-ambiguous echoes (Goldstein *et al.*, 1989). It is possible to sequentially acquire line segments with sufficient lateral displacement (different image lines) to eliminate range-ambiguous echoes (von Ramm and Thurstone, 1976). Also, the transmit aperture firings could be timed so that there is no dead time between line segment receptions.

Another recent increase in image frame rate comes from the parallel processing of image lines possible with digital signal processing. Two different types of parallel processing have been proposed. One uses a wide transmit beam and multiple signal processing chains (such as shown in Figures 27 and 28) connected to each element in receive (Shattuck *et al.*, 1984). Thus, several receive beams may be synthesized simultaneously using the same element receive data. The trade-off for this type of parallel processing is a resultant widening of the pulse-echo beams leading to decreased image resolution and sensitivity. Another type of parallel processing is achieved by simultaneously transmitting and receiving several narrow beams (Snyder, 1989) (Mallart and Fink, 1992). The trade-off for this type of parallel processing is increased circuit complexity and the possible heating of the transducer structure if the acoustic matching of the transducer is not sufficient. Parallel processing is especially useful for color flow Doppler signal processing where multiple samplings of each color pixel are necessary to produce an accurate estimate of blood flow velocities (Section V.D).

One manufacturer uses strongly focused transmit beam patterns with short focal zones. By dividing each image line into a large number of transmit focal zones, an effective "transmit dynamic focus" can be achieved along with receive dynamic focus. Due to image frame rate considerations, the transducer must be fired as rapidly as possible and parallel processing is likely used. The line segments are acquired sequentially in horizontal image strips. Transmit focus beam parameters are kept constant for each horizontal strip and then changed for the acquisition of the next horizontal strip, which reduces digital circuit complexity. This rapid firing scheme is susceptible to range ambiguity artifacts. But due to the highly focused transmit beams, outside of the short focal zones the incident pulse intensity is reduced sufficiently to eliminate practically all range-ambiguous echoes from the image.

The design of composite focused beam patterns for medical imaging uses all of the principles described above. While these principles are simple in concept, their implementation in commercially available equipment would not be possible without digital hardware and software. Some general details concerning the digital signal processing used are presented in Section IV.F. Although the beamforming process is high-tech, a low-tech piece of coat hanger wire or paperclip can be used to determine the spatial characteristics of linear stepped, convex, and vector array beamforming (Goldstein *et al.*, 1989).

E. Transmitter

Transmitting circuits have changed from the simple shock pulses used in early analog equipment. The array elements, with nominal electrical impedances of

50 ohms, are now "driven" by a short burst of square waves with voltage amplitudes of 2 to nearly 200 volts (Quistgaard, 1997). These lower voltages are necessary due to voltage restrictions on the electronic switches used to access individual elements in arrays (Maginness, 1979). By varying the frequency, shape, and duration of the square wave burst, any portion of the broadband array's frequency spectrum can be used in the generation of each line segment (Persson, 1981).

The simplicity of the switching circuitry needed to generate the square wave burst is of vital importance. Each element in the multielement array must have its own transmitter circuit, and equipment cost could rapidly increase if more complicated driving circuitry were utilized.

F. Digital Beamforming

The hardware implementation of the various beamforming concepts discussed in Sections IV. B, C, and D has evolved from hybrid (digitally controlled analog circuits) to fully digital hardware. Figure 36 demonstrates a hypothetical delay-sum digital beamforming scheme, which will be discussed in some detail below. Note in Figure 36 that the complex digital circuitry from the array element to the summing and apodization bus (SAB) must be present for each array element. Other necessary circuit duplication is discussed below.

1. T/R Switch

Modern ultrasonic equipment utilizes fast electronic transmit/receive (T/R) switches to isolate the receiving circuitry from the transmit signal (Harmuth, 1979). These T/R switches permit better electrical impedance matching between the array elements and the receiving circuitry than the previously used diode clipping circuits.

2. Logarithmic Amplifier

The amplification function can be performed by both preamplifiers and/or amplifiers. Although the amplifiers used do not necessarily have to be logarithmic, they will be collectively represented here as a logarithmic amplifier since their gain is calibrated in dB. Some analog signal amplification is necessary for signal compression reasons due to the limitations of present analog-to-digital converters (see below). Operator-selected swept gain control curves usually control the analog amplification process.

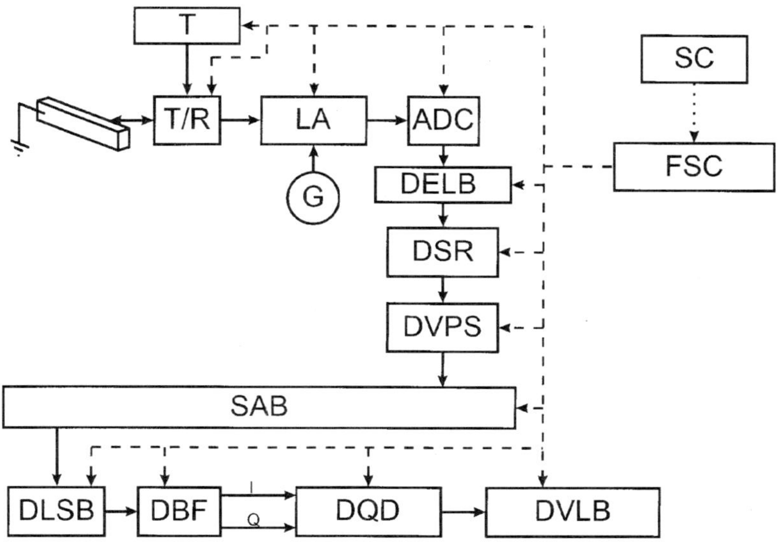

FIG. 36. Digital beamforming. A hypothetical delay-sum digital beamforming scheme is demonstrated. Solid lines indicate ultrasonic data flow and dotted and dashed lines indicate control. The system clock (SC) drives the focus and scan controller (FSC), which determines the beamforming scheme for the line segment being acquired and controls the beamforming process. T: transmitter, T/R: transmit/receive switch, LA: log amplifier, G: operator gain controls, ADC: analog-to-digital converter, DELB: digital element line buffer, DSR: digital shift register (for time delays), DVPS: digital vernier phase shifter, SAB: summing and apodization bus, DLSB: digital line segment buffer, DBF: digital bandpass filter, I: in-phase quadrature component, Q: out-of-phase quadrature component, DQD: digital quadrature detector, DVLB: digital video line buffer.

3. *Analog-to-Digital Conversion*

Sampling analog-to-digital converter (ADC) chips transform the rf echo signal from analog to digital form. The cost and performance of the ADCs is crucial to system performance. The ADC conversion speed (sampling rate) determines not only the digitized echo signal frequency bandwidth but also the accuracy of the digital waveform reconstruction. The Nyquist sampling theorem states that the ADC sampling rate must be at least twice the highest frequency in the echo bandwidth to avoid objectionable frequency aliasing (Woodward, 1953). However, for a delay-sum digital beamformer the sampling rate must be 5 to 10 times the highest frequency in the echo bandwidth to achieve accurate waveform reconstruction from the time delayed digitized array element signals (Mucci, 1984) (see also Section IV.G.2).

The number of bits in the digitized rf echo signal determines its dynamic range. Each bit represents a factor 2 in echo amplitude or 6 dB. To adequately convert analog rf echo signals to digital form, the ADC sampling rate and number of bits must be very high, due to signal reconstruction and dynamic range considerations. The availability of reasonably low-cost high-speed ADCs has been a limiting factor in beamformer development. The commercial introduction of 20- to 40-MHz ADCs with 8 and 10 bits has made possible the present generation of digital beamformers (Thomenius, 1995).

A 10-bit ADC produces digitized rf echo signals with a 60-dB dynamic range. The received analog rf echo signals have over 100 dBs of dynamic range. The logarithmic amplifier is, therefore, needed to compress the large received dynamic range echo signal down to the ADC dynamic range.

The delay-sum digital beamformer is only one of many possible digital beamformer designs. The delay-sum, partial-sum, interpolation, interpolation with complex sampling, shifted-sideband, discrete Fourier transform, and digital phase shift beamformers have been described in the literature (Mucci, 1984). Each has its advantages and disadvantages with respect to ADC sampling rate, memory requirements, and computational complexity. Some only require ADC sampling rates consistent with the Nyquist sampling criterion. It is likely that the digital beamformers used in modern high-resolution ultrasonic imagers have not been reported in the scientific literature. The delay-sum beamformer is presented here for pedagogical purposes only and is not meant to represent any commercially available model.

4. *Digital Element Line Buffers*

The digital element line buffers (DELB) store the digitized rf echo amplitude samples from the acquired line segments prior to subsequent digital processing. While in the DELB, the line segment data may be linearly or nonlinearly processed for many proprietary purposes. There are usually two DELBs associated with each ADC; they operate in ping-pong fashion such that one is acquiring the digitized samples while the other is either being digitally processed or supplying its stored data for subsequent digital processing.

5. *Digital Time Delays*

The digitized line segment data from each array element must be phase shifted for aperture beam steering and focusing prior to summation with apodization correction. Here this has been divided into two operations: a digital shift register (DSR) for coarse time delays and a digital vernier phase shift multiplication (at the mid-band frequency) for fine time delays (Steinberg, 1992).

6. *Digital Signal Processing*

The apodization and summation of all the array element signals produces the line segment echo signal, which is stored in the digital line segment buffer (DLSB). In equipment with true logarithmic compression before summation, the output of the summer more closely approximates their product rather than their linear sum (Thurstone and von Ramm, 1973). This multiplicative processing in receive produces an improvement in image resolution, a reduction of side-lobe amplitude, and an improved *FZ*. By digital techniques the DSLB rf echo data is bandpass filtered to remove unwanted low and high noise frequencies outside the transmitted frequency bandwidth. The bandpass filtering can be changed with depth to compensate for the loss of high frequencies due to the frequency-dependent tissue attenuation.

In most high-resolution equipment the digital bandpass filter has a complex multiplicative factor that, along with bandpass filtering, mixes the echo signal down to an intermediate frequency or to baseband (zero frequency). This digital mixing produces a set of quadrature signals, the I component (in phase) and a Q component (90° out of phase) that are processed separately and simultaneously (see Section V.C.4). Next, digital quadrature detection (DQD) is performed by calculating the square root of the sum of the squares of the I and Q signals at each instant of time. The resultant echo amplitude line segment signal is stored in a digital video line buffer (DVLB). The DVLB signal may be digitally processed for swept gain signal compression or further processed linearly or nonlinearly before being combined with other line segments along the same transducer line of sight to produce one line in the image frame.

An advantage of complex digital quadrature detection is that, in effect, two samples have been obtained per cycle. Thus, the rf signal only has to be sampled at the highest frequency in the transducer frequency bandwidth to satisfy the Nyquist theorem (Hildebrand and Brenden, 1972).

Since most digital processing schemes are linear, their sequence may be varied. The precise order of operations, ADC sampling rate, digital word size, exact filter design, etc. determines the quality of the final image and is closely held proprietary information.

In terms of data flow, the modern digital equipment functions differently from the analog systems depicted in Figure 16. In the analog systems, the data flowed from transducer to image memory in a synchronous, continuous flow. In modern digital equipment, the data moves in an asynchronous, discontinuous fashion. Many signal processing functions are duplicated so that one line segment can be acquired while a previously acquired one is processed. Much of the processing is performed in parallel and then combined. The speed of sound in tissue still limits the rate at which acoustic information can be acquired by the transducer, but the increased signal processing capability and speed has increased the amount of processing performed on the acoustic data and the resultant image quality.

7. *Parallel Processing*

The parallel processing described in Section IV.D.4 leads to further duplication of digital hardware and signal processing. In the parallel processing scheme where a wide transmit beam is used and several image lines are generated from the receive aperture array elements, the duplication begins after the digital element line buffer (DELB) in Figure 36. Several different digital phase shift circuits use this data to generate the separate image lines. Each line is produced from a separate summation and apodization bus and is processed in a separate set of digital circuits (bottom horizontal row of digital functions in Figure 36). In the parallel processing scheme where several narrow transmit pulses are generated simultaneously, the same number of receive apertures and associated digital circuitry are processing echo information in parallel.

8. *ASICs*

It is evident from the above general description that there is a great deal of redundant digital signal processing being performed in beamforming and other system digital operations. Early commercially available microprocessors were not fast enough to process the required large volume of data (Maslak, 1985). Thus, application specific integrated circuits (ASICs) have been developed so that these digital signal processing operations may be performed rapidly and efficiently using digital hardware (O'Donnell, 1988) (Robinson and Mo, 1992). Present-day high-resolution digital equipment typically perform over 3 billion operations per second.

Early on, Raytheon Company developed an off-the-shelf microprogrammable digital beamformer incorporating the partial-sum concept (Martin, 1974). A special-purpose swept gain amplifier integrated circuit (IC) has been developed by a major IC manufacturer (Analog Devices, Peabody, MA).

However, for the most part, each manufacturer has designed its own ASICs and had them specially fabricated by commercial IC manufacturers. This lack of standardization in ultrasonic imaging equipment is due to proprietary concerns but must be recognized also as a characteristic stage in the maturation of any field. Only when the optimum signal processing for every clinical imaging situation is known will there be standardization in medical ultrasonic equipment.

As the gate counts of ASICs have increased over the years, the number of processing channels for multielement arrays have risen (Thomenius, 1995). Presently, digital time delay ASICs with over 160,000 transistors are used in high-resolution equipment. One challenge in developing an ASIC is to design the hardware signal processing to be as flexible as possible so that modifications, system upgrades, and new signal processing schemes may be implemented through changes in their control software.

Recent advances in microprocessor technology have exponentially increased their functionality and speed while reducing size and cost. So now, for many digital operations after the beamformer, software control— using these new microprocessors—has replaced hardware control. Besides the reduction in hardware size and cost, software control permits ready modifications and upgrades on site and thus extends the useful life of the system hardware engine.

9. *Slice-Thickness Focus*

The multielement array beam patterns shown in Figures 27 and 28 for phased and linear stepped arrays have rectangular symmetry due to their rectangular array apertures and are electronically focused only in the in-plane dimension. The out-of-plane dimension or slice-thickness direction is focused mechanically either by a cylindrically curved wear surface lens or by segmenting the individual array elements and positioning them on a circular arc (while still functioning as a unit electrically).

The fixed slice-thickness focus leads to a variation of slice thickness with depth. Partial volume effects occur when more than one tissue type is contained in each voxel (three-dimensional pixel), and the echo signal obtained from the voxel (and placed in the corresponding image pixel) is an average of the contained tissues. Clearly, the smaller the voxel in the slice-thickness direction, the less the reduction in image contrast resolution due to partial volume effects. Up until the present, multielement arrays had mechanical slice-thickness focuses with optimum image quality only in the slice-thickness focal zone. This situation is changing rapidly. The new transducers will be discussed in Section VI.F.

G. Quantization Errors

Multielement arrays are sampled apertures with the samples at each location (element) obtained sequentially in time. The transmission and reception from each independent element along with the use of digital circuitry lead to the great flexibility in beamforming and signal processing described above. The trade-off for this increased signal flexibility is that all sampling must be performed properly. Often, due to design constraints, the sampling is inadequate and the resulting sampling errors will either limit system performance or introduce spurious noise signals into the image.

1. Spatial Sampling

The number of elements (spatial samples) in the array aperture determine beam focal properties and beamwidth. The spatial sampling restrictions have been derived using CW plane wave, Fraunhofer diffraction theory (Whittingham, 1991). The restrictions arise because of two factors. First, a plane wave of wavelength λ at the edge of a focused beam pattern (with a beam half-angle θ) interacts with the array equivalent to a wave of wavelength $\lambda/\sin\theta$ traveling across the array surface. And second, the Nyquist sampling theorem states that the phased elements of the array can only transmit or receive waves whose wavelengths have been sampled twice per cycle.

So for adequate spatial sampling per the Nyquist theorem,

$$2p < \frac{\lambda}{\sin\theta}$$

or

$$\sin\theta_{\max} = \frac{\lambda}{2p}. \tag{41}$$

For arrays where $p < \lambda/2$, this condition is met for all angles θ. For arrays where $p > \lambda/2$, Eq. (41) limits the maximum beam half-angle and the maximum effective aperture leading to an effective minimum spot size of (Eq. (17))

$$d_{\min} = \frac{F\lambda}{a_{\text{eff}}} = \frac{\lambda}{\sin\theta_{\max}} = 2p, \tag{42}$$

where a_{eff} is half of the effective array aperture. In other words, with these arrays, increasing the aperture will only decrease the spot size until the aperture exceeds the effective aperture defined by $\theta_{\max}$ and then the spot size will remain $d_{\min}$.

For steered beams, the above analysis can be extended by noting that beam steering by θ_0 will change the angles that the focused beam pattern edges make with the array normal to $\theta_0 - \theta$ and $\theta_0 + \theta$. As θ increases, the edge with the latter angle will undergo undersampling first, which leads to a modified form of Eq. (41):

$$\sin(\theta_{\max} + \theta_0) = \frac{\lambda}{2p}. \tag{43}$$

Phased arrays with $p < \lambda/2$ will not be affected by these sampling error restrictions. But linear stepped arrays can be affected. In the unsteered case for $p = \lambda$, $\theta_{\max} = 30^\circ$ corresponding to $fn = 1$. And for $p = 1.5\lambda$, $\theta_{\max} = 19.5^\circ$ corresponding to $fn = 1.5$. Gray-scale imaging equipment with 128-element linear stepped arrays was designed with $fn = 2$ focuses to avoid these restrictions. Newer linear stepped arrays with 512 elements will be able to utilize stronger $fn = 1$ focuses at the same frequencies.

In Doppler peripheral vascular studies of the carotid arteries, the linear stepped array beam must be beam steered away from normal incidence to these arteries to avoid a Doppler angle of 90° with the flowing blood. The transmitted Doppler frequency is at the low end of the array passband to minimize tissue attenuation, so p is closer to λ and Eq. (43) then mandates that $\theta_0 + \theta = 30^\circ$. The 128-element arrays use a 20° beam-steering angle that leaves a maximum focused beam half-angle θ of 10° corresponding to $fn = 2.88$, which is lower (stronger focus) than the typically used $fn = 4 - 6$ Doppler focus (Section V.C.1). With the new 512-element linear stepped arrays, much higher Doppler frequencies can be used, or at the same frequencies the Doppler beam-steering angles may be larger.

The curvature of the front face of convex arrays causes the ultrasonic rays from (to) each element directed to (or coming from) a point in space to have different angles with the normal to the convex array face at the element. If $p > \lambda/2$, for a focused beam with a focused beam half-angle $\theta_{\max}$, Eq. (43) applies with θ_0 now representing the angle between the normal to the convex array face at the aperture edge and the focused beam central ray. So small radii of curvature convex arrays not only have the disadvantage of small maximum apertures (twice the radius of curvature) but for $p > \lambda/2$ the effective aperture may be even smaller than the maximum possible due to the modified Eq. (43).

Another type of potential spatial sampling error concerns image line spacing. According to the Nyquist sampling theorem, for the image to attain the full resolution of the focused beam pattern, there must be two image lines spanning the smallest lateral spatial wavelength that can be resolved by the focused beam. Typically, this smallest spatial wavelength is taken as the lateral resolution distance (von Ramm and Smith, 1983). As was seen in Section II.F.1, there is some arbitrariness in the definition of lateral resolution. For line spacing considerations many favor the Rayleigh criterion, which leads to a small lateral resolution distance and an image line spacing that will tend to spatial oversampling thus avoiding the aliasing difficulties associated with spatial undersampling (Goodman, 1968).

2. *Phase Delay*

Temporal sampling of the returning rf echo signals leads to a reconstructed echo rf wavetrain that closely approximates the actual wavetrain if the Nyquist sampling theorem is obeyed. However, when the sampled wavetrains from each element are phase shifted and summed, the accuracy of the final focused and beam-steered array received wavetrain depends on phase delay errors.

Analog phase delay circuits and digital shift registers are quantized with a minimum time delay, δt_{min}. At a pulse center frequency f_c, for each element signal this results in a minimum phase delay of Φ_{min} and the following relation between the ideal phase delay (for a given focus and beam-steered angle), Φ_I, and the actual circuit phase delay of Φ_A (von Ramm and Smith, 1983):

$$\Phi_A = n\Phi_{\text{min}} = \Phi_I + \Phi_E \qquad |\Phi_E| < \pi, \tag{44}$$

where n is an integer. Φ_E, the phase error, is a function of δt_{min} and Φ_I varies with time in dynamic receive mode. (The time dependencies are omitted here.) For certain focus and beam-steered angle combinations, Φ_E can have a periodic time component that results in a grating-like effect and places an anomalous grating lobe in the angular response (beam profile) of the array. This anomalous grating lobe may lead to image artifacts and certainly will reduce the beam echo amplitude dynamic range (c.f. Figure 8). Apodization techniques will not reduce the anomalous grating-lobe amplitude (Beaver, 1977).

A CW study of these phase quantization grating lobes found that the ratio of the rms amplitude of these lobes to the main-lobe amplitude is (Peterson and Kino, 1984)

$$\left[\frac{1 - \operatorname{sinc}\left(\frac{1}{\mu}\right)}{N \operatorname{sinc}^2\left(\frac{1}{\mu}\right)}\right]^{1/2} \approx \frac{\pi}{\mu\sqrt{6N}} \qquad \text{for} \quad \mu \gg 1, \tag{45}$$

where μ is the number of temporal samples per signal period ($\mu = 2$ is the minimum permitted). So increasing μ and N will decrease their relative amplitude. Broadband simulation studies have demonstrated that the phase quantization grating lobes are greatly reduced in amplitude from those predicted for CW systems and tend to decrease in amplitude with increasing bandwidth (Magnin *et al.*, 1981).

For the broadband digital beamformer presented above, the digital time delays (Section IV.F.5) were implemented with a digital shift register for course delays and a digital vernier phase shift multiplication at the pulse midband frequency for fine time delays. These time delays only will be accurate at the midband frequency and will have random errors at other frequencies. Analysis of the phase quantization grating lobes produced revealed that their relative amplitude depends on the product Bt_0, where B is the pulse rf bandwidth and t_0 is the sample interval (t_0^{-1} is the sampling rate) (Steinberg, 1992). Only for very small Bt_0 values will the rms amplitude of the phase quantization grating lobes be negligible; i.e., the sampling rate should be 4 to 10 times the bandwidth. The bandwidth and not the midband frequency determines the minimum sampling rate because the information content of a wave is proportional to its bandwidth and independent of its midband frequency (Steinberg, 1992).

3. *ADC Performance*

The amplitude of these anomalous grating lobes also depends on the fineness of quantization in the ADC used to digitize the echo signal rf wavetrain, i.e., on its number of bits. The average amplitude quantization error between the actual rf signal amplitude and the ADC digital representation of this amplitude is one-half the value of the least significant bit in the digital word.

In a CW study for uncorrelated amplitude quantization errors between the N-element channels in the aperture, the ratio of the rms value of the phase quantization grating-lobe amplitude to the maximum digital main-lobe amplitude was found to be (Peterson and Kino, 1984)

$$\frac{1}{2^b\sqrt{3N}}, \tag{46}$$

where b is the number of bits. Then the average value of the phase quantization grating lobe is down from the main lobe by

$$-10\log(N) - 6b - 4.8\,\text{dB}.$$

With the broadband digital beamformer presented above for partially correlated amplitude quantization errors between the N-element channels in the aperture, the average value of the phase quantization grating-lobe power to the maximum digital main-lobe power was found to be approximately (Steinberg, 1992)

$$\frac{1}{2^{2b}N_{\text{eff}}}, \tag{47}$$

where $0 \leq N_{\text{eff}} \leq N$ but not further specified. Then the average value of the phase quantization grating lobe is down from the main lobe by

$$-10\log(N_{\text{eff}}) - 6b\,\text{dB}.$$

Independent of the degree of correlation between the amplitude quantization of the summed N-element signals, increasing N and/or b will reduce the average amplitude of the phase quantization grating lobes. This makes sense physically because increasing these parameters brings the quantized signals closer to their original analog values.

H. Digital Scan Converter

The digital scan converter is the portion of the digital signal processing circuitry that assembles the acquired line segments into an image format, performs signal processing on the line segments or the image data, and then converts the digitally stored data to analog form to drive the display TV.

1. Line Buffers

The digital video line buffer (DVLB) in Figure 36 is the first step in the scan conversion process. Depending on previous signal processing, the line segment data could be digitally compressed using swept gain or compensated for individual transducer nonuniform axial sensitivity (Schorum and Fidel,

1977). Also, compensation for nonanalytic circuit response could be performed here. DVLBs from adjacent image segments can be processed together in a spatial filtering operation to enhance or suppress low or high spatial frequencies in the combination before depositing them in the stored image.

An important signal processing step is the echo amplitude transfer curve shown in Figure 37. Image contrast resolution is dependent on image echo amplitude resolution. Echo amplitude resolution depends on the slope of the transfer curve. Curve **a** in Figure 37 has constant slope so adjacent display shades of gray represent equal differences in swept gain compressed echo amplitude. Curve **b** has a high slope for low-amplitude echoes and a low slope for high-amplitude echoes. Since the low-amplitude echoes are spread over a larger range of image shades of gray, small differences in their amplitudes are more resolvable in the image. The opposite is true for high-amplitude echoes. Curve **c** had better echo amplitude resolution for high-amplitude echoes. Since low-amplitude echoes from soft tissue parenchyma contain a great deal of diagnostic information, curve **b** is the best clinical choice of the three shown.

Each clinical imaging situation (type of scan) requires a different transfer curve, depending on its unique echo amplitude distribution. The appropriate transfer curve is usually determined at a clinical test site by trial and error. The stored transfer curves may be applied to the echo data at many possible

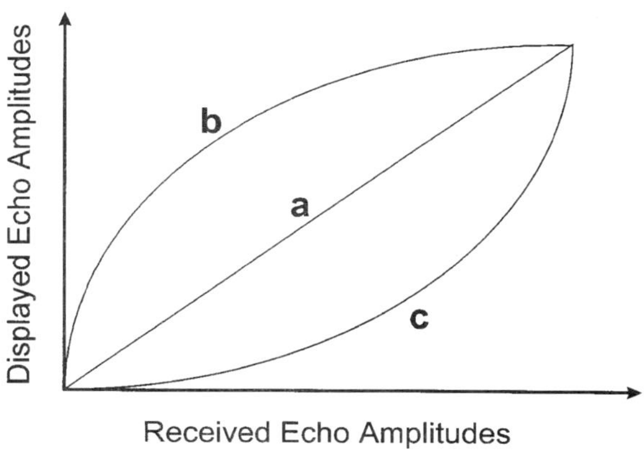

FIG. 37. Echo amplitude transfer curve. The received echo amplitudes have been compressed using swept gain. The echo amplitudes are displayed as image shades of gray. Curve **a** is linear and has equal echo amplitude resolution for all echoes. Curves **b** and **c** are nonlinear and their echo amplitude resolution is proportional to their local slope.

locations in the signal processing chain. It would seem most efficient to apply the transfer curve to the digitized data in the DVLB.

2. *Pixel Fill-In Algorithms*

There are always unsampled pixels in each image frame. The sector image format has a great many unsampled pixels, especially at large image depths where the image line density is lower. These unsampled pixels are filled in for image cosmetic reasons.

Early scan converters waited until the horizontal TV raster lines were read out of the stored image data and then used a simple one-dimensional interpolation fill-in (Park and Lee, 1984). Later, more sophisticated bilinear interpolation was adopted (Larsen and Leavitt, 1980). Here unsampled pixels were filled in by interpolating the sampled data both axially and in angle for sector images.

3. *Image Contrast*

The full echo amplitude dynamic range of the data stored in the digital scan converter does not have to be displayed. If it were, then the image would have low gray-scale image contrast. That is, echoes of almost the same amplitude would not be perceivably different in the image. To produce more gray-scale image contrast to aid in resolving echo amplitudes, the display echo amplitude dynamic range can be reduced. Typically the operator has a choice of a display echo amplitude dynamic range of 30 to 70 dB. Smaller display echo amplitude dynamic ranges are used either to produce an image with more contrast or to eliminate electronic or side-lobe fill-in noise in the image. When the display echo amplitude dynamic range is reduced, the overall gain control on the logarithmic amplifier determines which portion of the received echo amplitude dynamic range is displayed (i.e., low- or high-amplitude echoes).

4. *Freeze Frame*

Once the operator has adjusted all of the scan parameters to obtain a diagnostic quality image, he or she either makes a hardcopy or saves the image to a digital archiving system (Section VI.A). To assure that the proper image will be processed, the operator freezes the system operation, the image data acquisition ceases, and the frozen, stored image in the digital scan converter produces a steady-state image display. Once assured, the operator can then proceed to take hardcopy or save the image.

Freezing the image permits image post-processing (image annotation or calculations—Section IV.J) to be performed on the stored image prior to making a hardcopy or saving the image. It also permits multiple copies of the same image to be obtained on different forms of hardcopy or to be saved and have hardcopy taken (images for referring physicians or patients).

5. *Frame Averaging*

Another feature of high-resolution equipment is the ability to average over an operator-selected number of image frames while displaying a real-time image. The purpose of this temporal averaging of spatial data is to reduce the effects of image speckle. Since the operator's hand and/or the patient's tissue moves slightly during the scanning process, successive image frames have differing speckle patterns. Temporal averaging causes a spatial averaging over these different speckle patterns.

Since the integration time of the eye is 0.2 seconds (Davson, 1963), high frame rates will cause a similar perceived spatial averaging. The frozen image and its subsequent hardcopy, however, will not convey this information unless frame averaging is employed.

6. *Nonlinear Signal Processing*

The transfer curves **b** and **c** in Figure 37 are examples of nonlinear signal processing to enhance the image echo amplitude resolution of specific ranges of echo amplitudes. Another example would be nonlinear spatial filtering of the stored image data. The problem is that the spatial frequency filtering required for specular and diffuse echoes is entirely different. Specular echoes have high amplitudes, are beam direction dependent, and have little speckle since they are due to organized, constructive interference from specific tissue geometries. Diffuse echoes have low amplitudes, are beam direction independent (for the most part), and exhibit image speckle since they are due to interference from a collection of randomly positioned reflectors (Section II.G). Clearly, specular echoes will be enhanced by a high spatial frequency filter operation (or none at all), whereas diffuse echoes will be enhanced (speckle reduced) by a low spatial frequency filter operation.

A solution to this dilemma is a nonlinear spatial frequency filtering operation in which spatial filtering is applied selectively to the stored echo amplitudes. All low-amplitude echoes (below a determined threshold value) are filtered with a low spatial frequency filter and all high-amplitude echoes are filtered by a high spatial frequency filter. This sort of adaptive filtering is readily implemented using digital circuitry.

7. Ciné Loops

In high-resolution equipment, digital memory is used to store a number of image frames in a first-in-last-out manner. When the display is frozen by the operator, so is this image chain. These image frames can be stored or replayed as a closed temporal loop to view repetitive motion such as the cardiac cycle. Or the stored image frames can be sequenced through one at a time to choose an appropriate image for archiving or hardcopy (Figure 38).

8. Zooms

For each transducer used, the size of the image field of view in the patient is operator selectable over a predetermined range. Magnified, small field-of-view images always start from the patient's skin surface. While scanning a patient, it may become necessary to get a magnified view of an image portion at a lower depth. In this case the zoom function is used. Initially there were two types of zooms: read zooms and write zooms. The read zooms were magnified portions of the stored image data. But since the pixel size was also magnified, these images were blocky in appearance and did not contain any new diagnostic information.

The write zoom is much more useful. Here, the operator selects the portion of the displayed image to be magnified. Then, only this image portion is scanned and presented over the full display face. The number of display pixels remains the same but represents smaller voxels in the patient. Also, a different set of beam-steering angles for phased arrays (or half- or quarter-stepping for linear stepped arrays) can be used to place more image lines in the smaller field of view to increase the image line density. The write zoom presents more diagnostic information in terms of a more detailed view of the anatomy in question (note that the sample volume has not changed in size).

I. IMAGE DISPLAY

The ultrasonic image presents diagnostic medical information at the time of the patient examination. Its hardcopy (or digital archived file) has medical/legal importance because it documents not only the patient medical information but also the ultrasonic examination parameters and signal processing

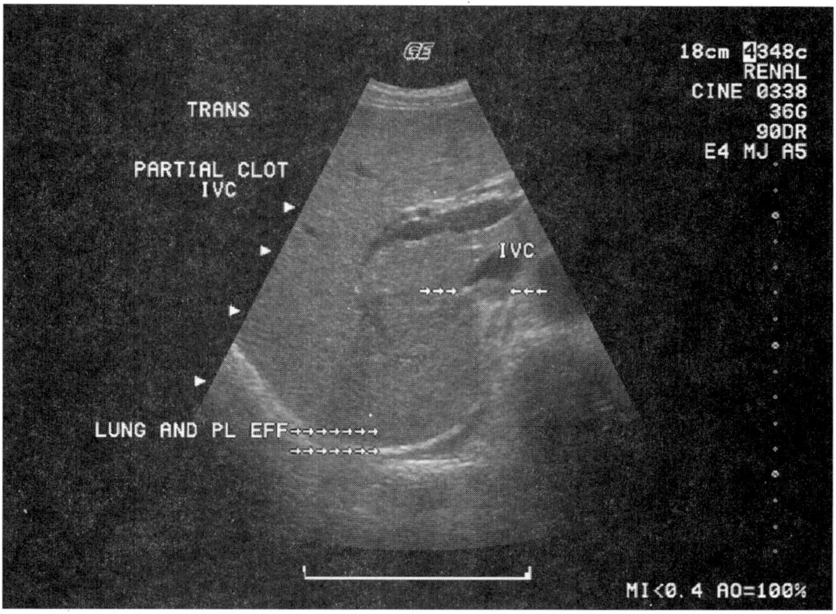

FIG. 38. Convex array abdominal scan. The equipment is being operated in a renal mode (see the upper-right alphanumerics). The white carets on the left side of the image denote the focal lengths of the multiple transmit focuses in the composite transmit/dynamic focus utilized. A thrombus in the inferior vena cava (IVC) is seen and denoted with image annotation. Pleural effusion is also seen and denoted with image annotation. (An air-filled lung causes the diaphragm to be a perfect reflector [part specular and part diffuse] and the fill-in seen below the white curved diaphragm just below the liver is artifactual. When there is liquid in the lung against the diaphragm, it is partially transmitting and the liquid itself can be seen in the image distal [further along the ultrasound line] to it.) This image is number 338 in a 340-image, ciné loop (third line down in the upper right and just on the right on the ciné loop indicator on the bottom). The mechanical index (MI) is < 0.4 (Section IV.I.1) with full acoustic output (AO). Note the unavoidable image degradation on its left edge due to the small asymmetric array apertures there. (Courtesy of GE Medical Systems.)

used. The typical digital display format is 640×480 pixels with 256 shades of gray (8 bit digital resolution of echo amplitude).

1. *Alphanumeric Fields*

A portion of the image display on the top or side is used for an alphanumeric field that presents date and time, patient information, scan information, and signal processing information. The patient information includes the patient's name and identification number (the patient information has been removed from all clinical images presented here for medical/legal reasons). The scan information includes transmit power, transducer model number, frequency bandwidth selected, image distance scale, number and position of transmit zones (in a composite transmit/dynamic receive multizone focus), and frame rate. The signal processing information includes the tissue-specific mode chosen, the degree of frame averaging, display echo amplitude dynamic range, transfer curve used, and results of any image processing.

Presently, a new type of information relating to the safety of ultrasonic imaging is being added to the image. The FDA, under the Medical Device Act of 1968, has been limiting the acoustic output of transducers. Recent imaging advances, however, such as Doppler color flow imaging (Section V.E) require higher acoustic outputs. A recent agreement permits the use of higher acoustic outputs if an index, related to safety, is displayed for the operator to monitor (AIUM, 1992). The two indices proposed are the thermal index (TI), related to the potential for tissue damage due to heating, and the mechanical index (MI), related to the potential for tissue damage due to cavitation effects.

2. *Gray-Scale Invert*

The presently accepted display convention is a white-on-black gray-scale scheme. The image background is black and successively higher echo amplitudes are encoded with lighter shades of gray. If desired, a black-on-white gray-scale scheme can be implemented with the push of a button.

3. *Image Invert*

The presentation of ultrasonic cross-sectional images has been standardized for ease in image interpretation (AIUM, 1986). Transverse cross-sectional images are always presented looking from the patient's feet toward the head. Longitudinal cross-sectional images are always presented with the patient's head to the left and the feet to the right. Tactile markings on the transducer case make it easy for operators to properly orient the transducer to comply with these standards. Sometimes, however, images are obtained incorrectly or transducer geometry forces an incorrect image. In these situations the images can be right–left inverted with the push of a button.

4. *Pseudo-Color Display*

An important component of image contrast resolution is visual perception of the image shades of gray. The eye is most sensitive in the orange (555 mμ) when it is adapted for cone vision (Adler, 1965). When a pseudo-color display is implemented, the shades of gray are converted to orange with various degrees of color saturation (white added to the orange). These "shades of orange" are visually more pleasing to view and produce some aid in echo amplitude differentiation.

5. *Duplex Displays*

Sometimes two different types of ultrasonic data displays are combined in the same image (see Section V.C.7). Examples are an M-Mode display along with a gray-scale image showing the highlighted M-Mode line and a Doppler spectral display along with a gray-scale image showing the location of the Doppler range gate (Section V.D). Usually, the duplex display contains a gray-scale image map, which aids in positioning the precise anatomical location of ultrasonic data acquisition for a specific ultrasonic signal processing operation and documents this location for medical and/or legal reasons.

J. Image Post-Processing

Any procedure performed on the frozen image data is called image post-processing. The result of the post-processing procedure usually is displayed in the image so that it is also captured in the hardcopy or digital archiving.

1. *Image Annotation*

Tissue identification labels may be typed onto the image in selected locations for pedagogic reasons or for publication purposes (Figure 38). Or a freehand trace function can be used to outline the shape of an organ or the interior of a fluid-filled structure (such as a cardiac chamber).

2. *Image Measurements*

Important diagnostic information can be obtained from measurements made on the frozen image data. Spatial measurements concerning the size of a lesion or tissue structure are a common example. Linear distance measurements are made using digital calipers. The calipers are sets of pixel patterns (crosses, squares, diamonds, etc.) that are moved to specific image locations by the operator. The equipment then computes the image distance (in pixel lengths) between the calipers' center pixels and converts the result to a linear distance using the image magnification (pixel real-space dimensions).

Usually, multiple pairs of digital calipers can be implemented simultaneously (Figure 39).

Tissue area measurements can be made by using a trace function to outline the tissue structure of interest. The equipment then computes the number of enclosed pixels and converts the result to an area using the image magnification. Or, separate orthogonal linear distance measurements are made of the tissue structure using digital caliper sets and the equipment computes the area of an elliptical model of the tissue shape. The trace function can be used similarly to compute the circumference of a tissue structure.

Tissue volume measurements can be made directly by summing the area measurements made in a series of uniformly spaced scan planes and multiplying by the interplane distance. Also, a set of three orthogonal digital

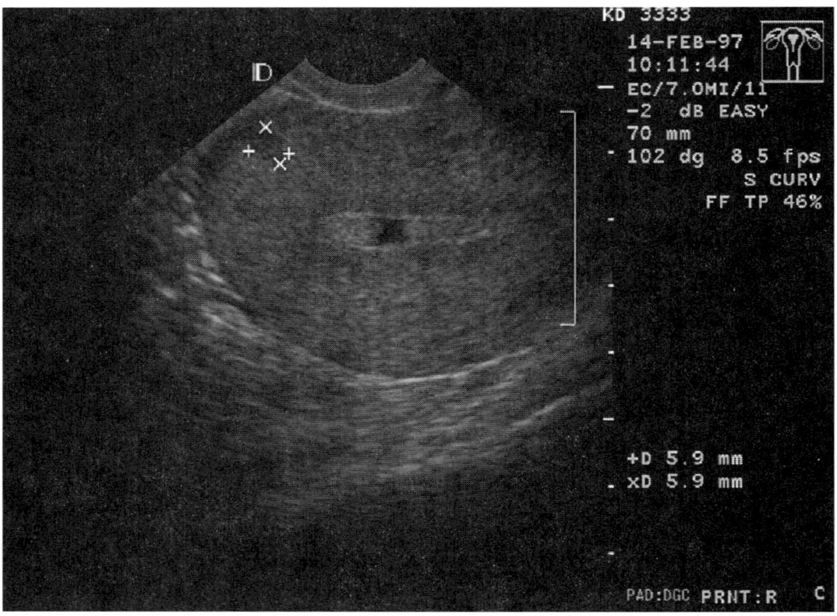

FIG. 39. Transvaginal image of a 6-mm uterine lesion. The small pictogram in the upper right of the display indicates optimization for an ob/gyn scan. The square bracket on the right of the gray-scale image defines the range of depths over which the "confocal imaging" multizone focus is performed. The longer this range of depths, the lower the image frame rate. A 7.0-MHz transvaginal probe was used to obtain this uterine image with good contrast resolution demonstrating the lesion in the upper left. The lesion dimensions are spanned by digital calipers, and the measurement results are displayed in the lower right. (Courtesy of Diasonics Ultrasound.)

caliper measurements (taken from two orthogonal scan planes) can be used with an ellipsoidal model to compute the tissue volume. Modern equipment can even perform the volume measurements in real-time (Figure 40, color plate).

One common use of distance measurements is in obstetrics. An important problem is the estimation of fetal gestational age. Various medical researchers have experimentally determined and published the relationship between fetal gestational age and certain fetal anatomic dimensions in ultrasonic images. The distance variables in these standard charts include bi-parietal diameter

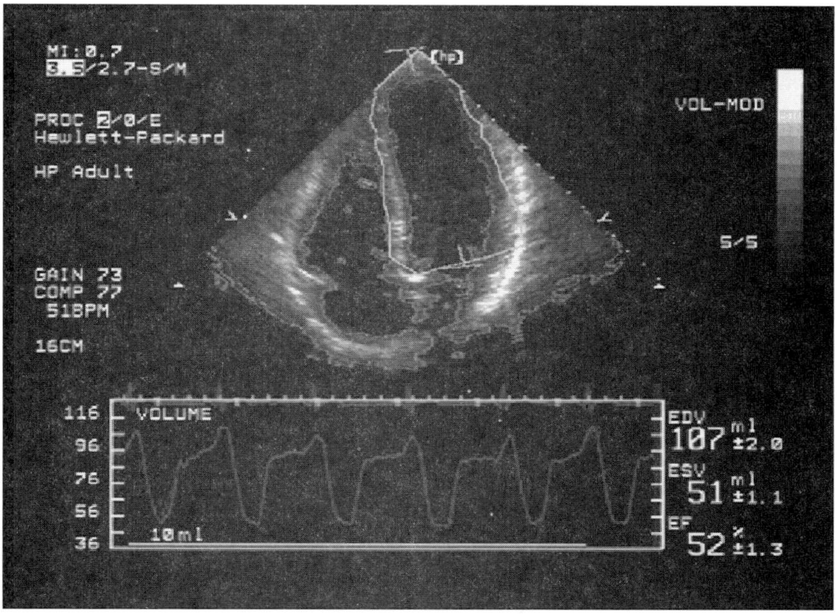

FIG. 40. Phased array cardiac scan. In this equipment the borders of the cardiac chambers are automatically located in real-time by an edge detection algorithm and a red line denoting the detected myocardium/blood boundary added to the image. The operator uses a freehand trace function (green line around rightmost chamber) to outline the left ventricle. The equipment, using an ellipsoidal model for the chamber volume, automatically calculates this chamber volume in real-time and displays it in the graph shown below the scan. The end diastolic volume (EDV) [high point of the curve], end systolic volume (ESV) [low point of the curve], and ejection fraction (EF) [difference between the two], averaged over five operator-selected cardiac cycles, are continually computed and displayed. The cardiac rate of 51 BPM (beats per minute) is automatically calculated from the measured EKG (displayed at the top border of the graph) and displayed on the left side above the graph. The mechanical index (MI) is 0.7 (AIUM, 1992). (Ultrasound image courtesy of Hewlett-Packard Company.)

(distance between the two parietal bones in the fetal skull), femur length, crown–rump length (head to tail distance of an early fetus), abdominal circumference, and head circumference. All of these charts (some anatomic dimensions have more than one chart due to multiple publications in the literature) are stored in the equipment in the form of look-up tables. After a fetal anatomic distance measurement is performed, a standard chart may be chosen and the equipment automatically includes the estimated fetal gestational age, including estimated error, on the image or in an equipment-generated report.

Another type of image measurement is region-of-interest (ROI) measurements. For example, once an ROI is denoted by using the trace function, the equipment will compute and display a plot of the enclosed echo amplitude histogram (number of pixels displaying each value of echo amplitude). Other types of ROI echo amplitude or tissue texture computations have been installed in ultrasonic imaging equipment over the years. The goal has been ultrasonic tissue characterization — the noninvasive identification of specific tissue types or disease states based solely on ultrasonic image information. As of yet, no tissue characterization scheme has ever been clinically accepted.

K. Human Engineering

While modern high-resolution ultrasonic equipment is at the cutting edge of imaging technology, it will be used in a sometimes-hostile hospital environment by nontechnical clinical end users. All of the technical wizardry incorporated in the equipment will be for naught if it cannot be used easily and efficiently under severe clinical pressures. Thus a great deal of time and effort is expended by manufacturers to determine the needs of the clinical user and to engineer the equipment to meet those needs in a consistent and reliable manner. Some of the human engineering features of the equipment are listed below to present the breadth of user-related problems.

Modern ultrasonic equipment has been redesigned from the bottom up and is a far cry from the modified oscilloscope shown in Figure 14. Figure 41 shows a modern scanner. The scan controls (buttons, dials, and switches) have been laid out on the control panel so that the ones most used are very conveniently positioned (the operator scans the patient with the transducer in one hand and the other on the control panel). The scan controls are also backlit so that they will be easy to see in a semidark scan room. The control panel has mounts for the many different transducers needed. It is hermetically sealed to protect the electronics underneath from the occasional coffee spill or

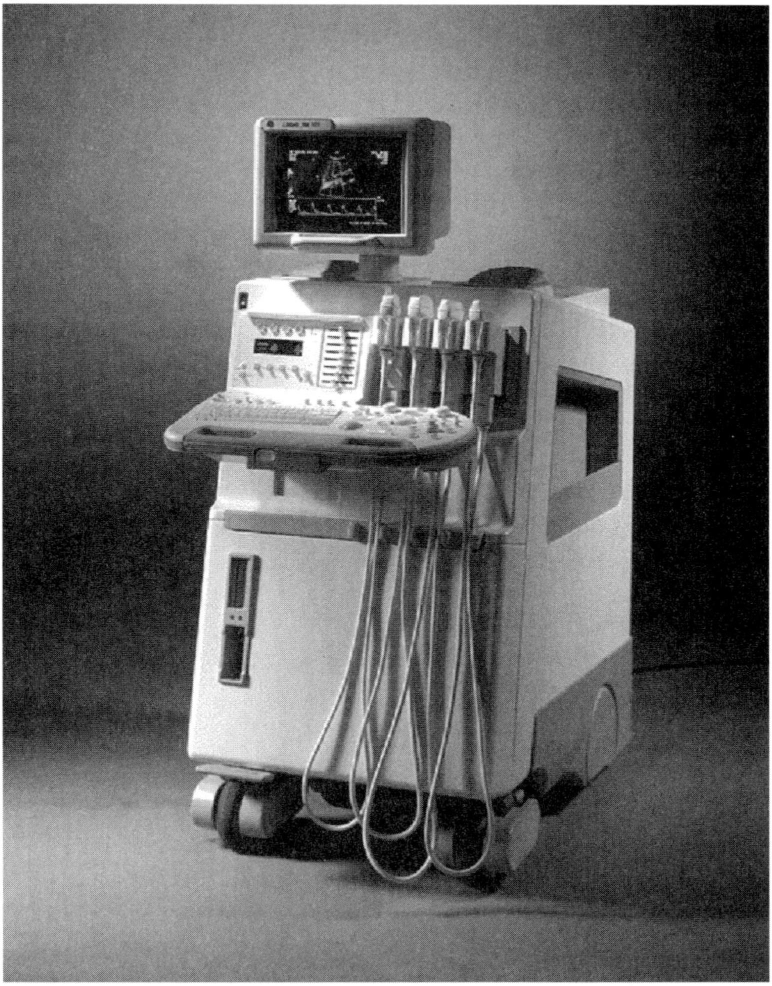

FIG. 41. Modern high-resolution ultrasonic scanner. This portable computer platform has been engineered for ease of use in the ultrasound clinic. The keyboard platform moves up and down and can be positioned for maximum operator comfort. To the right of the keyboard are logically grouped backlit controls and a roller ball for ease of use in a semidark room. Above the keyboard are slide-pot swept gain controls and a set of soft keys that drive a display for menu control of the scanner functions. One of the four transducers connected to the mainframe can be selected by software control. On bottom left is an optical disk drive (and a disk storage bay below it) that can store up to 300 black-and-white images or 100 color images. An internal hard disk can store up to 1500 black-and-white images or 500 color images. A 14-inch high-resolution TV monitor is mounted between stereo speakers that are used for audio presentation of Doppler signals. (Courtesy of GE Medical Systems.)

from patient liquids, and it can be raised or lowered to a height convenient for any operator.

Line voltage conditioning and power isolation circuits are included in the mainframe to protect the system electronics from the traditionally noisy hospital power circuits. The casters underneath have independent suspensions for shock absorption and are designed for easy movement and control when the scanner must be moved for portable patient examinations. They are also electrically conductive to avoid static electricity buildup.

Multielement transducers are connected to the mainframe by a bundle of microfine coaxial cables terminating in a low insertion force connector (Quistgaard, 1997). The microfine coaxial cables are essential to minimize operator muscle fatigue, and the connectors must be well shielded to avoid EMF pickup. Different pins on each connector are shorted so that the system recognizes which transducer has been selected and can load the correct signal processing software.

The equipment must be user-friendly. A patient information menu is filled out initially to be included with the archived patient images. Then a menu system is used to select the examination anatomical imaging mode and other scan parameters. The operator can develop his or her own scan parameters and save the new scan mode for future selection.

L. System Operation

The ultrasonic equipment is designed for maximum reliability. When it is powered up, a several-minute system self-check is initiated in which major circuit functions are verified for proper operation. If any difficulties are encountered, an appropriate error code is stored and a display screen requests a service call—either immediately or within several days, depending on the severity of the difficulty. Certain diagnostic routines are also run in the background to check for faults while the system is idle.

The service technician has access to a more comprehensive level of tests through a dedicated system diagnostics interface. All field repairs are made at the board level. The clinical user can purchase a service contract, which includes periodic preventive maintenance (PM). During the PM the technician inserts a special PM board into the system that runs system diagnostics on approximately 90% of the system's digital and analog functions.

Included with most service contracts is a guarantee of 95% or greater system up-time. This is quite important since a down machine not only affects the ultrasound clinic's schedule and revenue but also each scheduled patient's time and welfare. With periodic PMs, most high-resolution equipment can easily meet these up-time goals. Some manufacturers have included modems in their equipment so that troubleshooting can be performed remotely by telephone before dispatching a service technician.

System upgrades and the addition of newly purchased features are performed by updating the system operating software. On some equipment this requires new boards with updated read-only memory (ROM). On others the software update is performed with the aid of floppy disks. Systems with internal modems can be upgraded remotely.

V. Doppler Imaging

Doppler imaging is one of the newest and fastest-growing applications in medical ultrasonics. The clinical need for blood flow measurements, the various types of Doppler equipment, their physical principles of operation, and their limitations will be outlined here. The required signal processing and instrumentation are very complex and only can be covered briefly.

A. Doppler Imaging Goals

1. Ultrasonic Reflections from Blood

The primary target for medical Doppler applications is the red blood cell (RBC). Blood flow within the heart and the vascular system is observed by reflecting ultrasound off the moving blood and detecting the resulting Doppler-shifted frequencies (DSFs). Blood, however, is anything but a perfect ultrasonic reflector. To ultrasonic waves, blood appears as a fluid with lots of small reflecting particulates inside. Complicating things is the fact that these particulates can interact with one another and produce clots and thrombi that change the reflection characteristics of blood (Guyton, 1991). Table 3 shows some of the physical characteristics of whole blood *in vivo*.

Blood makes up about 7% of the total body weight (Physical Chemistry etc., 1984). It is a distributed tissue composed of erythrocytes or red blood cells (RBCs), leukocytes or white blood cells, and thrombocytes or platelets. These particulates are suspended in a plasma thick with proteins (albumin, immunoglobulins, and various carrier proteins). RBCs make up the largest population of cells, averaging 5.2 million/mm^3 for men and 4.5 million/mm^3 for women (Physical Chemistry etc., 1984). The RBCs are small, discoid cells 7- to 10-μm wide and about 2-μm thick. Their small size and low reflectivity

TABLE 3

PROPERTIES OF BLOOD

Specific gravity	1.0595
Relative viscosity (18°C) *in vitro*	4.75 centipoise
Relative viscosity (18°C) *in vivo*	2.3 to 2.75 centipoise
Red blood cell concentration (male)	5.2 Million/mm^3
Red blood cell concentration (female)	4.5 Million/mm^3
RBC dimensions	discoid shape, 7–10 microns diameter, 2 microns thick
Effective scattering unit size	30–50 microns
White blood cell concentration	5000 to 9000/mm^3

make individual RBCs very poor reflectors. In groups, however, they can form larger aggregates that become scattering bodies that change with the hematocrit (Hanss and Boynard, 1979). Even larger aggregates called Rouleaux formations also occur (Bloom and Fawcett, 1968). These formations are often seen in gray-scale images of the larger veins.

Figure 42 demonstrates the acquisition of Doppler information from moving blood in the vascular system. Transducer B has a smaller Doppler angle with the flowing blood than transducer A, so its measured DSFs will be higher (Eq. (38)). It is important to realize that the Doppler measurement yields RBC cluster velocities in the lumen of the vessel and not the important clinical information of blood volume flow rate in ml/sec.

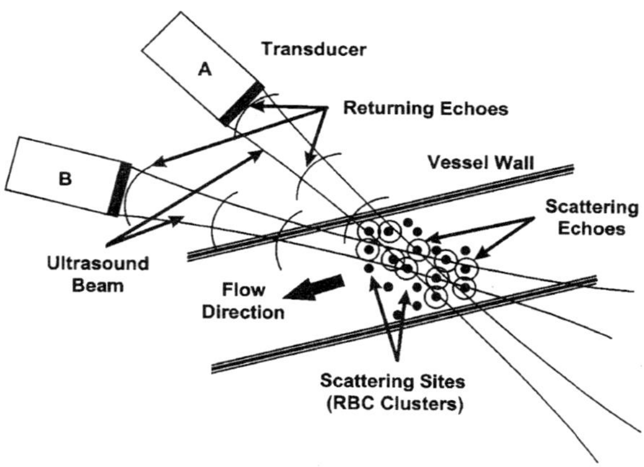

FIG. 42. Doppler signal acquisition. The Doppler ultrasound beam looks at flowing RBCs in ever-changing clusters from a variety of angles. Changing the angle to the vessel changes the Doppler shift frequencies. Changes in cluster size changes the amplitude of the Doppler signals.

2. Cardiovasular Events

Blood movement depends on the energy imparted to it by the beating heart and the vascular channels that deliver blood to all the living cells in the body. Clinically, there is interest in any reduction of blood flow to the body's living cells. A reduction in regional blood flow that may not sustain cellular or tissue functions is called *ischemia*. If the flow reduction goes to zero (arterial *occlusion*), the formerly supported (now nonsupported) cells die, producing an *infarction*. Therefore, from a Doppler point of view, the interest is in the mechanical energy source of blood delivery (the heart) and the means of delivery (the vessels). Doppler imaging goals have been developed with these aspects of blood flow in mind. The imaging goals represent signal processing steps within a typical Doppler signal chain.

a. Determining the Doppler Signal Source. The peripheral vascular system is a set of parallel vascular beds, each supplied with roughly the same amount of hydrodynamic energy. The blood flow pattern through each of these vascular beds depends on the physiology of the local vascular system. Different vessels, then, have different specific flow patterns that represent specific vascular resistance and characteristic vascular tone (the amount of smooth muscle tension) within each vascular bed. For example, the common carotid artery, internal carotid artery, and external carotid artery are all connected together yet each has a different flow pattern (Roederer *et al.*, 1984). Progressive disease will change the flow patterns in each. Consequently, it is important to uniquely identify the vascular source of the echo signals being used for Doppler signal processing. The identification of the vessel establishes what a clinician expects to see in the flow pattern under normal and diseased conditions.

b. Form of Flow Over Time. The sequence of flow pattern events over time reveals how the pulse wave is shaped as it enters and passes through a particular vascular bed. The presence of disease can and will change this flow pattern over time. The ability to see these flow pattern changes and compare them against the expected flow patterns permits the detection of vascular disease. Comparisons can take the form of calculations that depend on the shape of the flow pulse. Parameters include the flow acceleration, deceleration, peak-to-peak values, peak systolic frequency or velocity, and averages of peak flow values over the cardiac cycle.

c. Frequency Content Over Time. When red blood cells enter an ultrasonic beam traversing a vessel, they return not a single frequency but many frequencies. These many frequencies come from the velocity gradients within the vascular compartment that the ultrasound beam is interrogating. Major stream lines will have highest velocities, with linear velocities steadily decreasing toward the walls of the vessels. Examining the frequency content of the Doppler signals returning to the signal processing system presents a picture of the degree of organization within the vascular compartment. In general, frequency broadening represents the range of velocities that may be found in both normal flow patterns and those resulting from disease. The rule here is: *Spectral broadening suggests disease, but does not uniquely indicate disease.*

d. Direction Over Time. Each heartbeat sends out a hydraulic pulse that travels along the vascular system, from the aorta to all the smaller vessels. As the pulse moves more peripherally, it tends to steepen and become larger while the average pulse pressure systematically decreases (Guyton, 1991). The damped pulse becomes smoother as it passes through the smaller arteries and capillary beds, which ultimately removes all cardiac pulsations. Flow in the venous system tends to be steady, modulated by the respiratory system and local events such as skeletal muscle contraction and the presence of venous valves. Generally, flow should run from larger to the smaller arteries. Significant blood flow in the opposite direction suggests a medical problem. The seriousness of the situation is expressed in the timing of the reversal. For example, pulse reflections and vascular tone from high-resistance vessels change both the pulse pressure and the flow response, producing a period of flow in the opposite direction. A low-resistance vascular bed will not have the same sort of reflection event and flow will occur throughout the cardiac cycle. Figure 43 shows the basic characteristics of these flow patterns. The spectral Doppler display in Figure 43 is explained in Section V.C.6.b. With these Doppler imaging goals in mind, the functional organization of various Doppler technologies are now presented.

B. CW Doppler Systems

The simplest Doppler ultrasound technology is the handheld, continuous wave (CW) Doppler device. This is an ever-present tool for the vascular clinician performing a variety of tasks, ranging from primary patient vascular examinations to measuring blood pressure. In more sophisticated forms, CW Doppler can be used to measure high blood velocities in various cardiac valve stenoses or to estimate the cardiac output. Figure 44 shows some commercial examples of this versatile tool.

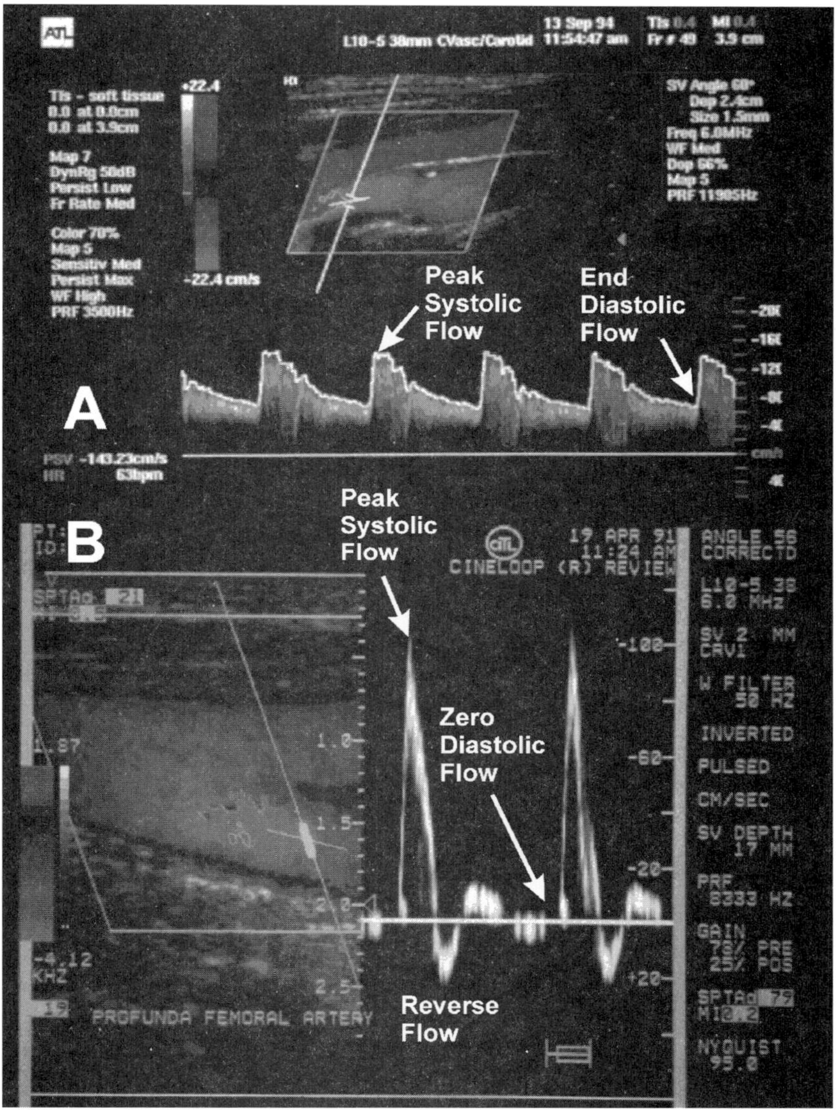

FIG. 43. Examples of flow for low- and high-resistance vessels. The upper waveform (A) shows a low-resistance, internal carotid artery with flow throughout the cardiac cycle. The CFI portion shows carotid flow from image right to left. The brighter portion of the spectral waveform tracks the maximum frequency. The lower image (B) shows high-resistance flow in a femoral artery. The high resistance causes reflected pressure waves that produce reversed flow in late systole, and a small forward flow rebound. This pattern is called triphasic flow. The CFI portion shows flow from image left to right. Note in both images that the Doppler spectrum vertical axis is calibrated in velocity due to the use of the flow angle indicator line at the range gate position. (Courtesy of ATL Corporation.)

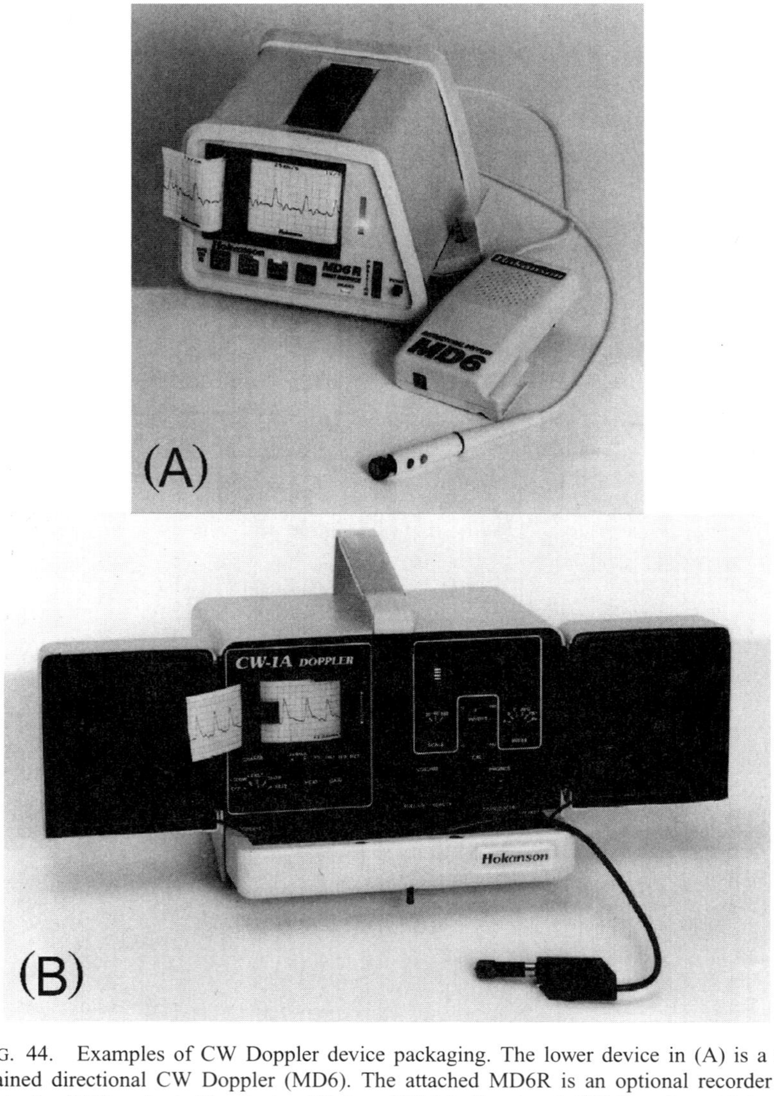

FIG. 44. Examples of CW Doppler device packaging. The lower device in (A) is a self-contained directional CW Doppler (MD6). The attached MD6R is an optional recorder that records the MD6 output. The device (B) is a CW-1A directional CW Doppler with built-in speakers separated to present forward (left speaker) and reverse (right speaker) flow. The recorder can show peak bidirectional, unidirectional, or average flow. Several flow parameters are shown in a digital readout. (Courtesy of D. E. Hokanson, Inc.)

1. Transducer Characteristics

CW operation requires two separate transducers — one transmitting ultrasound, the other receiving the echoes. They are often made from a circular wafer that is cut down the middle into two Ds, which are then slightly separated (Kikuchi, 1978). Because the individual Ds do not have circular symmetry, they do not have cylindrically symmetric beam patterns. The close proximity of the two Ds means that a "leakage" signal from the transmitter constantly enters the receiver along with the much smaller RBC echoes (Baker and Daigle, 1977). A slight cant of the two Ds forms an intersection of their beam patterns in the patient, as shown in Figure 45. The region of overlap is called the *region of sensitivity* (Baker *et al.*, 1978), and only enclosed RBCs will produce DSFs. Applying an external lens or using internal focusing can sharpen this region even further. A trade-off for the simple CW transducer design is that RBC range resolution is not possible, so it is sometimes difficult to determine the Doppler signal source.

CW transducers are typically undamped with a narrow bandwidth. When it is a handheld device, a CW system requires a low-energy power source to set the transducer vibrating at a stable frequency. Voltages can vary from 10 VAC to 60 VAC. Regardless of the operating voltage, an undamped transducer resonating at its natural frequency creates the most efficient system.

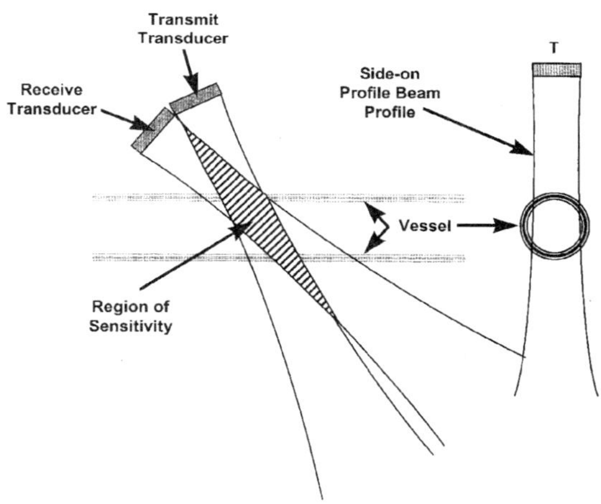

FIG. 45. Continuous wave Doppler scanning geometry. The continuous operation of CW Doppler means only RBCs in the intersection of the transmit and receive beam patterns (region of sensitivity) produce DSFs. CW Dopplers are easy to use, but have no range discrimination. The T represents the transducer structure.

2. *Instrumentation*

As Figure 46 demonstrates, an internal master oscillator (MO) sets the transducer into vibration at a known transmit frequency, f_0, and provides a *coherent* signal reference within the system for detecting the DSF. The system compares the received echo signal, f_r (a Doppler-shifted frequency) and the transmitted frequency, f_0. The DSF is, of course, the difference between f_0 and f_r. An alternative means of comparison is to use the leakage signal as a reference, which forms an *incoherent* detection scheme (Evans, 1985). The frequency comparison (Figure 46) requires little more than a mixing circuit, which produces four output frequencies: f_r, f_0, $f_r + f_0$, and $f_r - f_0$. A suitable low-pass filter removes all but the difference frequency, which is the DSF. The detected audio can appear as an audio output (speaker or headphones) for simple audio analysis, or it can undergo further spectral analysis for presentation as a spectrum on a CRT display or a small paper printer.

Comparing an internal reference against the received echo signal is an easier task than measuring the absolute value of f_r (Atkinson and Woodcock, 1982). The detector (functional demodulator) shifts the received signals from

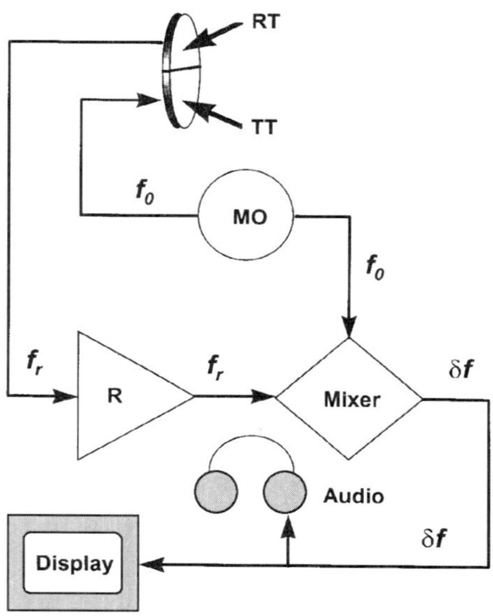

FIG. 46. CW Doppler signal processing. RT and TT: receive and transmit transducers, respectively, MO: master oscillator, f_0: carrier frequency, f_r: receive signal, δf: Doppler shift frequencies, R: receiver.

frequency-modulated rf to baseband audio with no frequency offset. In so doing, however, the system loses its ability to show directionality, that is, to show motion toward the transducer as different from motion away from the transducer. In general, then, simple CW Doppler systems cannot indicate flow direction over time.

Flow directionality can be added to CW Doppler with a commensurate increase in equipment complexity. It can be displayed within a single channel by introducing an audio frequency offset (Atkinson and Woodcock, 1982). An alternative means of detecting and following the directionality of flow is the quadrature phase detector (QPD), which will be examined in detail in Section V.C.4.

C. Coherent PW Doppler Systems

A coherent pulsed wave (PW) Doppler system sends out a tone-burst of ultrasound with a pulse length τ_P that follows the transducer focused beam pattern shown in Figure 9. The primary operational purpose of the PW Doppler system is to identify uniquely the source of the DSFs in the heart or the vascular system and determine flow direction over time. By pulsing the transmitted wave, echo ranging can be used to locate and interrogate cardiac chambers or specific vessels. This activity requires a signal processing design different from the CW system. Figure 47 shows the functional organization of PW Doppler system.

1. Transducer Characteristics

Since PW Doppler is a pulse-echo measurement, a gray-scale imaging transducer (single element or multielement) can be used. However, there are two differences in its Doppler mode: Tone-burst transmission must be used for improved frequency resolution (Eq. (2)) and the beam focus must be weaker. In gray-scale imaging, the beam patterns are highly focused to improve image spatial and contrast resolution. In PW Doppler, accurate frequency measurements require minimal spectral broadening, and highly focused beam patterns can cause excessive spectral broadening.

This spectral broadening can be understood physically from two equivalent viewpoints (Newhouse *et al.*, 1980), one considering the Doppler signals received with a finite array aperture and the other considering the time of flight of the RBCs through the focused beam. With a finite aperture transducer, each portion of its front face (or each element in a multielement array) has a different Doppler angle with the direction of blood flow and a

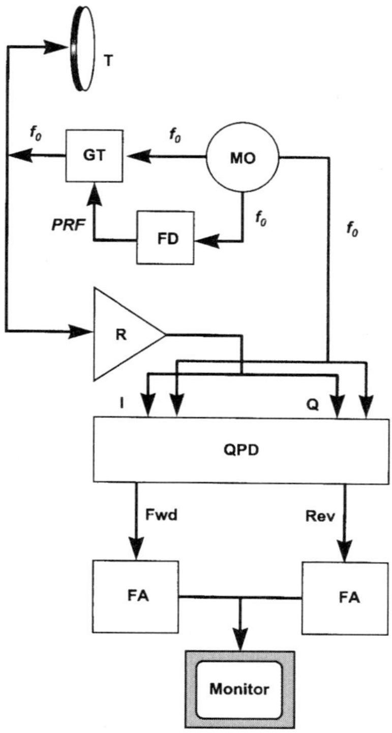

FIG. 47. Pulse Wave (PW) coherent Doppler system. A coherent PW Doppler has a pulse repetition frequency (*PRF*) that is a subharmonic of the carrier frequency. Comparing the carrier frequency against the receive signal in a quadrature phase detector (QPD) permits the depiction of motion direction. T: transducer, MO: master oscillator, FD: frequency divider, GT: gated transmitter, R: receiver, I and Q: in-phase quadrature channels, respectively, Fwd and Rev: forward and reverse flow channels, respectively, FA: frequency analysis.

different detected DSF (Eq. (32)). The larger the aperture, the greater the spread of Doppler angles and the detected DSFs. Or, the finite time of flight through the narrowest beamwidth (spot size Eq. (17)) leads to a commensurate uncertainty in the measured DSF (Eq. (2)). The stronger the focus, the smaller the spot size and the larger the frequency uncertainty.

For a rectangular aperture array with a blood vessel at the focal point of a Doppler beam, the fractional broadening of the detected DSF is given by (Newhouse *et al.*, 1980)

$$\frac{\Delta f_{DS}}{f_{DS}} = \psi \tan \theta_D = 2 \tan \theta_D \tan^{-1}\left(\frac{1}{2fn}\right), \tag{48}$$

where ψ is the full beam spread angle, *fn* is the focus *f*-number, and θ_D is the Doppler angle of the beam central ray. Figure 48 presents the predictions of Eq. (48) for beam focuses *fn* of 2, 4, 6, and 8. For the commonly used peripheral vascular, linear stepped array beam-steered Doppler angle of 70° (90° − 20°), the typical gray-scale *fn* of 2 produces excessive spectral broadening and an *fn* of 4 to 6 is preferred for DSF measurements.

2. *Coherent Transmitter*

As suggested by the derivation of the pulse-echo Doppler equation in Section II.I, detecting the DSF requires phase detection signal processing. This requires that a coherent PW transmitter send out a tone-burst of ultrasound with a known frequency and timing. To obtain that explicit relationship, the MO sends its CW signal to the transmitter gate (Figure 47). In addition, a frequency divider receives the MO signal and creates a *subharmonic* of the MO frequency, which becomes the system *PRF*. The *PRF* trigger opens the gate, letting the MO excite the transducer into vibration. The transmitter gate then closes after allowing a fixed number of cycles to reach the transducer.

3. *Range-Gated Receiver*

The transmitted ultrasonic tone-bursts interact with tissue reflectors, whose echoes return to the transducer between transmissions. Doppler echo ranging measurements can be performed by only accepting echo signals from a selected range of tissue depths using a range gate (RG). The operator selects

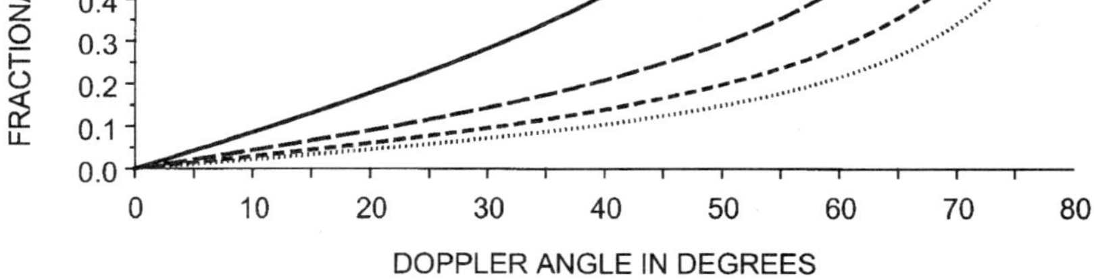

FIG. 48. Fractional broadening of Doppler signal with beam focus and Doppler angle.

the Doppler signal source by placing the RG in any vessel of interest (see Section V.C.7). At this operator-selected depth (time), the rf amplifier opens, letting the echo signals reach the subsequent signal processing. After a certain time τ_{RG}, the RG closes. The tone-burst length, τ_p, and the RG duration determines the range of tissue depths interrogated in the PW Doppler DSF measurement (Newhouse *et al.*, 1980):

$$\Delta z = \frac{c}{2}(\tau_P + \tau_{RG}). \tag{49}$$

4. *Signal Detection*

The PW Doppler measurement obtains one DSF sample per transmission and the Nyquist sampling theorem mandates that the Doppler *PRF* must be twice the highest DSF to be measured. These restrictions limit the maximum target velocity that can be properly measured in PW Doppler. To avoid range ambiguities, the RG signal must be received before the next subsequent transmission, so the maximum value of the PRF is

$$PRF = \frac{c}{2z_{RG}}, \tag{50}$$

where z_{RG} is the RG depth in tissue. Substituting the maximum DSF that can be measured ($\frac{1}{2}$ of Eq. (50)) into Eq. (33) yields

$$V_{max} = \frac{c^2}{8f_0 z_{RG} \cos\theta_D}. \tag{51}$$

To some extent the operator can increase V_{max} by choosing a low carrier frequency or a large Doppler angle. In modern PW Doppler equipment with complex Doppler signals, the Nyquist sampling limit changes to a *PRF* equal to the highest DSF to be measured. This doubles the V_{max} stated in Eq. (51).

Some of the RBCs in the PW Doppler RG sample volume have *forward* motion (toward the transducer) and others have *reverse* motion (away from the transducer). To measure blood flow direction over time, these two motions have to be separated in the PW Doppler signal processing. A quadrature phase detection (QPD) method (Atkinson and Woodcock, 1982) commonly used in modern equipment is presented in Figure 49. This figure is a more detailed view of the QPD in Figure 47.

Consider a returning echo $E(t)$ from the RG sample volume:

$$E(t) = A\cos(\omega_0 t + \varphi_c) + B_f \cos(\omega_f t + \varphi_f) + B_r \cos(\omega_r t + \varphi_r). \tag{52}$$

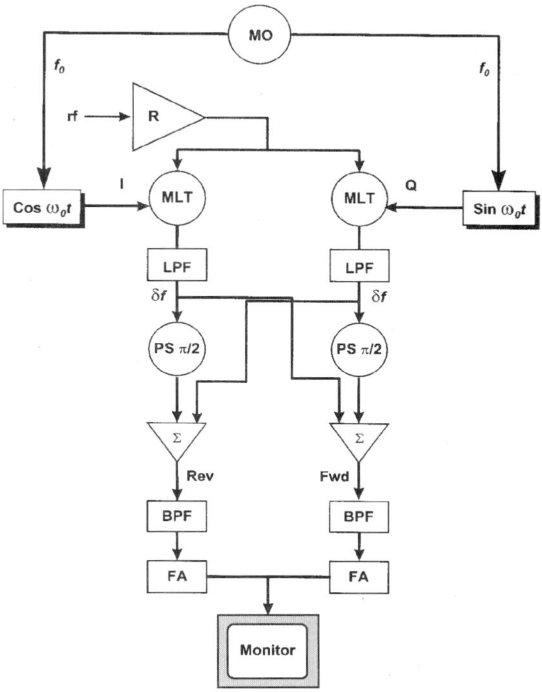

FIG. 49. Coherent Doppler with a quadrature phase detector (QPD). The basic function of a QPD coherent system begins with injection of cosine and sine functions where ω_0 is the carrier angular frequency. By using appropriate phase shifts (PS), signals can be summed (Σ) to form forward (Fwd) and reverse (Rev) flow channels. rf: radio frequency input signal, R: receiver, MO: master oscillator, f_0: carrier frequency, PS: phase shifter, MLT: multiplier mixer, LPF: low-pass filter, δf: Doppler shift frequency, BPF: bandpass filter, FA: frequency analyzer.

The first term is the clutter signal from stationary echo sources and the last two terms arise from moving echo sources. A and B are signal amplitudes and ω is an angular frequency with the subscripts representing the transmitted signal [0], forward motion [f], and reverse motion [r]. φ is a phase angle with subscripts representing clutter [c], forward motion [f], and reverse motion [r]. The QPD signal processing sets up two rf signal channels that are identical but separated by a 90° phase shift. The unshifted channel is the *in-phase* or I channel. The other is the *quadrature* or Q channel. The I channel signal is set up by inserting a $\cos \omega_0 t$ signal into its mixer and the Q channel signal is set up by inserting a $\sin \omega_0 t$ into its mixer, where ω_0 is the angular frequency of the system MO. The mixing produces sum and difference frequencies. The sum frequencies are removed by low-pass filtering, leaving the difference frequency terms

$$\mathrm{I}(t) = \frac{1}{2}[A\cos(\varphi_c) + B_f\cos(d\omega_f t + \varphi_f) + B_r\cos(d\omega_r t - \varphi_r)] \tag{53}$$

and

$$\mathrm{Q}(t) = \frac{1}{2}[A\sin(\varphi_c) + B_f\sin(d\omega_f t + \varphi_f) - B_r\sin(d\omega_r t - \varphi_r)], \tag{54}$$

where $d\omega$ is the absolute value of the Doppler shifted angular frequency of the echoes.

The clutter signal terms will be ignored since they are removed from the signal processing by various filtering techniques. Expressing the I and Q channel signals in terms of cosines gives

$$\mathrm{I}(t) = \frac{1}{2}B_f\cos(d\omega_f t + \varphi_f) + \frac{1}{2}B_r\cos(d\omega_r t - \varphi_r) \tag{55}$$

and

$$\mathrm{Q}(t) = \frac{1}{2}B_f\cos\left(d\omega_f t + \varphi_f - \frac{\pi}{2}\right) + \frac{1}{2}B_r\cos\left(d\omega_r t - \varphi_r + \frac{\pi}{2}\right), \tag{56}$$

where the first terms are due to forward motion and the second terms are due to reverse motion. This representation demonstrates that the Q channel lags the I channel by $\pi/2$ for forward motion and the Q channel leads the I channel by $\pi/2$ for reverse motion. To take advantage of these phase shifts, the I and Q channels are further phase shifted by $\pi/2$ to produce

$$\mathrm{I}_{\pi/2}(t) = \frac{1}{2}B_f\cos\left(d\omega_f t + \varphi_f + \frac{\pi}{2}\right) + \frac{1}{2}B_r\cos\left(d\omega_r t - \varphi_r + \frac{\pi}{2}\right) \tag{57}$$

and

$$\mathrm{Q}_{\pi/2}(t) = \frac{1}{2}B_f\cos(d\omega_f t + \varphi_f) + \frac{1}{2}B_r\cos(d\omega_r t - \varphi_r + \pi). \tag{58}$$

Finally, to produce the forward motion signal, $\mathrm{I}(t)$ and $\mathrm{Q}_{\pi/2}(t)$ are summed:

$$\mathrm{I}(t) + \mathrm{Q}_{\pi/2}(t) = B_f\cos(d\omega_f t + \varphi_f). \tag{59}$$

To produce the reverse motion signal, $\mathrm{I}_{\pi/2}(t)$ and $\mathrm{Q}(t)$ are summed:

$$\mathrm{I}_{\pi/2}(t) + \mathrm{Q}(t) = B_r\cos\left(d\omega_r t - \varphi_r + \frac{\pi}{2}\right). \tag{60}$$

5. Signal Processing

Many different analog and digital circuits have been developed to analyze the frequency content of the forward and reverse motion signals (Evans *et al.*, 1989). At present, the preferred method of accurately extracting these DSFs is the *fast Fourier transform* (FFT). In digital signal processing, this decomposition is available through the *discrete Fourier transform* (DFT), which works for aperiodic digital signals of finite length (Lynn and Fuerst, 1994). An FFT is not a single process, but a set of algorithms that are specialized for extracting the DFT frequency components in a particular environment and at a speed suitable (hence fast) to the flow of information. If necessary, the output of the QPD is converted to digital form and then sent into an FFT circuit through a buffer. The FFT algorithm, which is typically a mix of software and hardware, is then applied. Since the FFT frequency components are orthogonal (each frequency component is independent of all the others (Davis, 1963)), truncating the number of components does not affect the accuracy of the DSF extraction process. The FFT output is a set of frequency components of varying frequency and amplitude. An output buffer then prepares the signals for presentation as an audio signal or in a Doppler spectral display.

6. Signal Presentation

a. Audio. The traditional Doppler output has been a composite audio output, either to a speaker system or headphones. If the DSF detection is directional, then the output is in *stereo*, with forward flow in one speaker or headphone, reverse flow in the other. This is still the case, even on the most sophisticated equipment. Although Doppler sounds are very qualitative, they can reveal a great deal about vascular flow events without the use of more sophisticated display techniques.

b. Doppler Spectral Display. The RG frequency components obtained from the FFT are displayed as a function of time. Its CRT presentation is three dimensional. As shown in Figure 50, the y-axis presents the component frequencies, the x-axis is time, and the amplitudes of the component DSFs are presented in gray-scale with the highest amplitudes represented by the lightest shades of gray.

Because the FFT output is digital, the y-axis is a collection of frequency bins, each representing a small range of frequencies. Since Doppler signal processing must detect changes in phase between successive pulse-echo RG

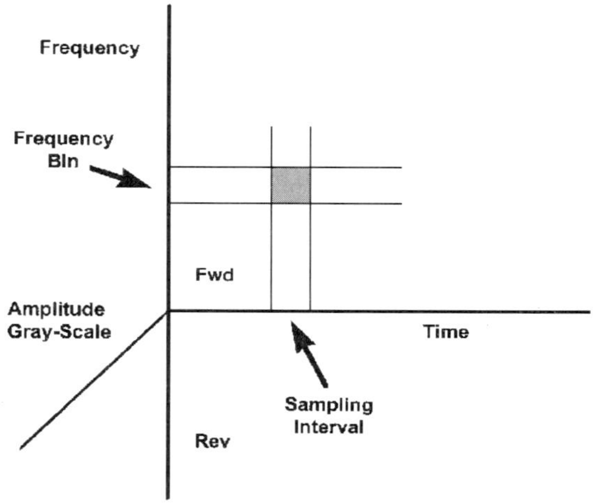

FIG. 50. Display elements of a Doppler spectrum. A time varying spectrum based on the fast Fourier transform (FFT) expresses frequency in small frequency bin intervals for each sampling interval. The amplitude of the frequency component appears as gray scale.

samples, the FFT input encompasses many transmit pulses over time intervals from a few milliseconds to as long as 300 milliseconds. The longer the FFT input time interval (more samples), the better the DSF signal extraction — i.e., the more sensitive the amplitude detection and the better the *y*-axis frequency resolution (Eq. (2)) (i.e., more frequency bins). In general, the trade-off is speed versus accuracy. In modern equipment these signal processing choices are available to the user, who can select the required level of FFT accuracy.

A difficulty in Doppler measurements is that echoes from RBC clusters are 40 to 60 dB lower in amplitude than soft tissue echoes. So the removal of soft tissue "clutter" from the Doppler spectrum is of prime importance. (Sometimes these clutter echoes are due to voluntary or involuntary tissue motion and sometimes due to involuntary transducer motion.) Two sorts of frequency filtering schemes can be used to help separate the clutter from the DSFs representing the flow events. The first is a frequency- and amplitude-dependent filtering scheme, as shown in Figure 51. This filter, known as a clutter canceller, works on the assumption that low-frequency, high-amplitude signals are from tissue, while high-frequency, low-amplitude signals represent blood flow. The second filtering scheme is an application of "wall filters" that approximately threshold the frequencies (Figure 51), functioning as a high-pass filter. (They were developed to remove the low frequencies due to vessel

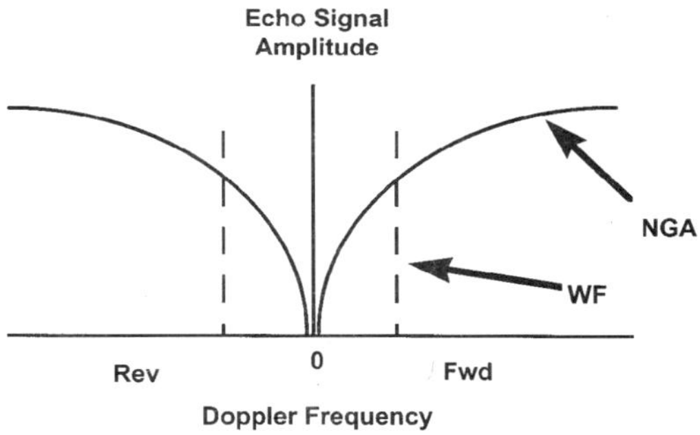

FIG. 51. Clutter signal separation. Doppler signal processing attempts to separate flow from stationary signals by testing frequency and amplitude. Clutter signals can be separated from the Doppler signal using a canceller based on a nonlinear gain amplifier (NGA) with the frequency dependence shown. Wall filters (WF) that threshold frequency produce a simpler separation. The WF is shown here schematically with a very well-defined cut-off frequency. Commercial WFs do not have a well-defined cut-off frequency due to the trade-off between filter cut-off frequency sharpness and time response (Eq. (2)).

wall motion.) The clutter canceller technique provides a shading to the frequency rejection process that is not available to the wall filters. Clearly, a system with wall filters set too high may not show the diastolic (slow) flow in a vessel that could be an indication of disease or the recovery of a vessel to an occlusion.

The Doppler spectral display permits the determination, at the RG anatomical location, of the form of flow over time, its frequency content over time, and the flow direction over time. The accurate determination of the Doppler signal source became possible with the development of duplex imaging systems.

7. *Duplex Imaging Systems*

A duplex imaging system is a CW or PW Doppler device coupled with a gray-scale imager to form a common imaging system (presenting both displays). The goal is to show unambiguously the source of the Doppler signals using the anatomy in the gray-scale image coupled with the flow pattern in the Doppler spectral display. The early forms of duplex imaging centered on coupling CW Doppler with a linear array and PW Doppler coupled with a mechanical sector scanner. These were "outrigger" systems in which the Doppler transducer was mounted alongside the imaging transducer with the Doppler beam intercepting the gray-scale scanning plane. Optical encoders read the position of the Doppler transducer, producing a line (Doppler-line) and cursor on the screen to show the Doppler beam and RG positions.

Integrating the Doppler signal processing with the gray-scale image meant using the same transducer for these very different functions. An early and successful application was the so-called M/Q system that combined an M-mode display with Doppler signal processing (the Q or flow portion of the system). Because of the narrow bandwidth of the transducer for Doppler applications, axial resolution for the M/Q system was relatively poor. This was not considered a problem since the primary purpose of the system was Doppler flowmetry. A strong point for the M/Q system was its simultaneous imaging—that is, the system used the same echo signal for both amplitude (M-mode) and Doppler signal processing. The *PRF* was the same for both the PW and imaging tasks, and both functions could be implemented at the same time.

Things became more complicated when integrating a mechanical sector scan (with rotating transducers) gray-scale image with PW Doppler. The system function depended on stopping the transducer rotation suddenly and through the use of a position encoder, steering the transducer along a Doppler-line into the correct position to obtain Doppler signals from a cardiovascular site. The position of the RG moved along the Doppler-line with an operator adjustment on the scanhead. During the Doppler acquisition period, the gray-scale image was frozen with no image update. The operator pushed on a small lever to position the Doppler-line and twisted the end of the lever to position the RG.

The multielement array provides a more amenable platform for duplex imaging. Electronic positioning is faster and easier, with no mechanical inertial problems to overcome for beam positioning. The Doppler-line can assume any of the fixed beam positions, and a RG control sets the position of the sample volume. Also, the Doppler-line beam pattern can be changed to a weaker focus to reduce spectral broadening effects (Section V.C.1). Because the scan and Doppler-line are under electronic control, the gray-scale image can be regularly refreshed between periods of Doppler sampling without disrupting the Doppler information significantly.

The phased array is a good transducer for echocardiography and some vascular applications, but a linear stepped array or a convex array is far better for abdominal, pelvic, and vascular scanning. Duplex imaging with a linear stepped array requires either a separate, external transducer or internal beam steering (with a weaker focus) to form the Doppler portion of duplex image. The convex array requires no beam steering. The Doppler-line needs only lateral positioning, a weaker focus, and RG positioning.

In duplex imaging systems, the DSFs presented in the Doppler spectral display may be replaced by RBC or blood flow velocities using Eq. (33). The operator can display a short line, called the flow angle indicator, centered on the RG in the gray-scale display. By rotating the flow angle indicator until it is parallel to the RBC flow direction (usually taken as the vessel axis), the operator specifies the Doppler angle (since the equipment now knows the two lines — flow direction and Doppler-line — that define this angle). When the flow angle indicator is displayed, the vertical axis of the Doppler spectral display changes from frequency to velocity. As the velocity axis scale is changed, the equipment prevents the operator from choosing velocities greater than the V_{max} given by Eq. (51) to prevent aliasing.

There is a velocity measurement error associated with using the flow angle indicator since it has a fixed angular increment, $\delta\theta$. The largest error occurs when the vessel axis lies halfway between two possible flow angle indicator settings. Then, the erroneous Doppler angle is

$$\theta_I = \theta_D \pm \frac{\delta\theta}{2}, \tag{61}$$

and using θ_I in Eq. (33) results in a percentage velocity error of (Goldstein, 1991)

$$VelErr = \left(\frac{\cos\theta_D - \cos\theta_I}{\cos\theta_I}\right)100. \tag{62}$$

The predictions of Eq. (62) are presented in Figure 52 for $\delta\theta/2$ values of 0.5°, 1°, 3°, and 5°. These results and those presented in Figure 48 demonstrate the importance of using the smallest possible Doppler angle in performing flow measurements.

Before the widespread use of color flow imaging (Section V.D), the inability to perceive flow patterns stimulated a number of analytical tools that operated on the single LOS RG spectrum. The goal was to find an accurate, fast, angle-independent means of detecting the presence of vascular stenosis and major changes in vessel compliance. The detection of stenotic disease depended on increased blood flow velocities and the presence of spectral broadening. Complicating the desire for the simple calculation for all situations was a wide range of vascular resistances found even under normal circumstances and the periodic changes in vessel resistance with physiological changes or compensation to disease. The result was a number of indices and ratios that became part of the equipment software-based calculations from the

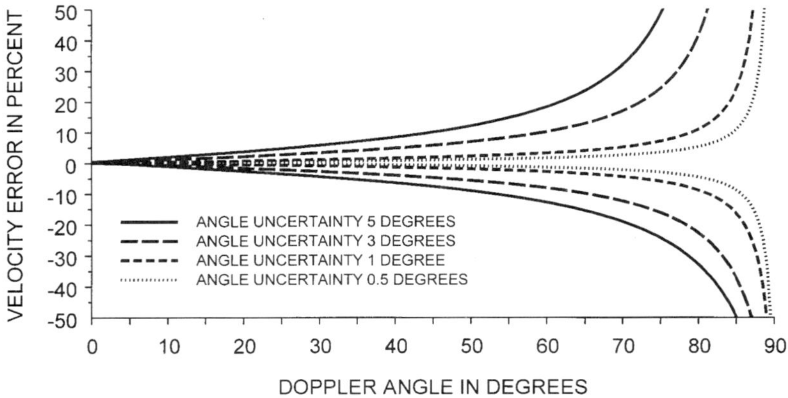

FIG. 52. Effect of Doppler angle error on velocity measurement. Figure 43 shows the flow angle indicator line in position to estimate the Doppler angle.

Doppler spectrum. These calculations, in turn, needed a number of parameters that were extracted from the spectrum.

The spectral parameters for calculations included (Powis and Schwartz, 1991) (1) the maximum DSF, with the peak systolic value as a specific value; (2) the minimum DSF; (3) the mean or first-moment average frequency; (4) the median value between maximum and minimum frequencies; (5) the mode value, representing the largest number of RBCs traveling at the same velocity; and (6) the systolic acceleration and deceleration. Determining the maximum and minimum values usually requires some consideration for noise and the amplitude of the frequency component. Noise may have high frequencies but typically exhibits low amplitudes.

Blood flow is a three-dimensional process that can be better appreciated as an image rather than a Doppler spectral display from a single point. Thus the direct logical extension from duplex imaging is color flow imaging.

D. COLOR FLOW IMAGING

A color flow image is a combination of two tissue maps: a gray-scale map presenting either the local anatomy of the heart or that surrounding blood vessels of interest and a color map presenting the results of many Doppler flow measurements performed over the full gray-scale field of view or a selected ROI (see Figure 53 on p. 169) (Ferrara and DeAngelis, 1997). The cardiac chambers or the blood vessels appear black on the gray-scale map because of the low amplitude of RBC cluster echoes. The regional Doppler measurements fill in these gray-scale echo-free areas with blood flow measurement results encoded in color for ready viewer identification. The gray-scale map is composed of echo amplitude and time delay measurements using many gray-scale lines of sight (LOS). The color map is composed of many blood flow measurements made using many Doppler LOSs, each interrogated along its full length by many contiguous RGs.

Color flow imaging (CFI) means the real-time depiction of soft tissue in a gray-scale, two-dimensional image with the simultaneous depiction of blood flow within the vessels in color *regardless of how the system develops flow information*. Thus, CFI can result from a multirange-gate Doppler signal processing scheme, a time domain analysis (TDA) that directly measures the motion of RBC clusters within an image, or the simple comparison of successive image frames. The Doppler-based systems can employ any number of analytical techniques that may be synchronous (simultaneous extraction of amplitude, phase, and frequency information from echo signals) or asynchronous (nonsimultaneous extraction of amplitude, phase, and frequency information from echo signals). And, of course, CFI means the flow information arrives in a real-time format. As early as 1978 (Brandestini and Forster, 1978) the pieces for the CFI system were coming together conceptually, but it took the speed of the modern microprocessor to bring the information to the image at real-time speeds.

CFI represents a direct extension of duplex imaging to multiple sampling sites rather than a single site. The earlier M/Q system combined a single Doppler sample site with the M-mode image to localize the Doppler signal sources. The next step was to simultaneously process motion information from several Doppler RGs on each LOS, superimpose it on the M-mode image, and collect information from many different LOSs by moving the Doppler transducer. These data were color encoded and overlaid onto a frozen gray-scale image (Brandestini and Forster, 1978). The only thing lacking was a real-time presentation of both the gray-scale image and the color Doppler flow map. The current commercial CFI technologies utilize two sources of motion information: the Doppler effect and a direct measurement of RBC cluster displacement using time domain analysis.

1. Doppler-Based Systems

Each Doppler sample within the vascular system produces a range of DSFs representing the various scattering site velocities within the vascular compartment and spectral broadening due to beamwidth. CFI requires a compression of this information because the color encoding can show only a single color for each Doppler map image pixel. A compressed value would be either a spectral mean, median, or mode value at that pixel, with further modulation of the color to represent the spread of Doppler frequencies (i.e., spectral variance).

The Doppler portion of the CFI system needs to *dwell* for a period of time on an LOS to detect the DSFs. This *dwell time* or *ensemble time* can range from 4 to 24 pulse–listen cycles for each LOS. The longer the dwell time, the better the frequency resolution (Eq. (2)) and the more sensitive the Doppler, but the image frame rate slows. Doppler-based CFI is divided into two technical approaches: synchronous and asynchronous signal processing. In both systems, the signal processing extracts echo signal amplitude, phase, and frequency. The difference is whether or not that processing is on the gray-scale echoes.

a. Synchronous Signal Processing. A synchronous CFI system simultaneously processes gray-scale and DSF information from the same echo signals with the same spatial resolution in both image maps (Powis, 1994). It uses the same transducer (or multielement aperture) for both amplitude and Doppler signal processing. The echo signals from each sample site are simultaneously processed in separate signal channels, one for amplitude and the other for Doppler. Because the system requires directional information, the Doppler signal processing establishes I and Q channels and finally QPD to produce forward and reverse directional channels. Because the system gathers Doppler information synchronously with the gray-scale image, the beam focuses are typically selected for a balance between gray-scale imaging needs and Doppler, which at times may not be ideal for either one.

b. Asynchronous Signal Processing. Asynchronous signal processing separates the amplitude and Doppler information gathering events. The system uses separate transmitters, beams, sampling intervals, scanning geometries, and at times different frequencies for gray-scale imaging and Doppler. Figure 53 shows the sampling geometry for the separate image maps. The gray-scale map is set up in a conventional rectangular field of view with the beam focusing perpendicular to the linear stepped array. The Doppler map steers the beam to provide a Doppler angle with the vessel flow (Powis, 1986). These two maps need to be interweaved in a sequence that can provide sufficient spatial resolution for both and dwell time for Doppler. Running high

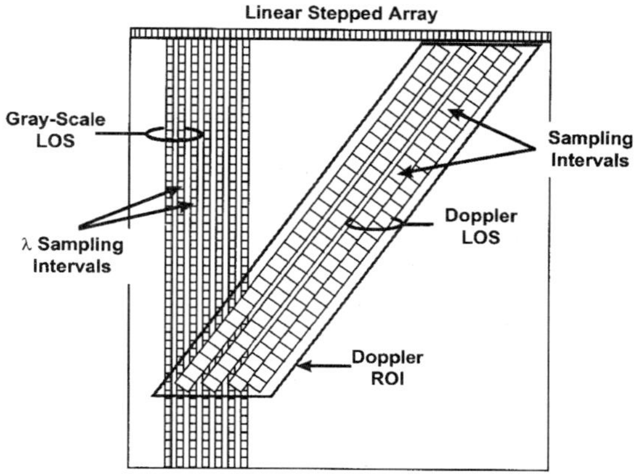

FIG. 53. Linear stepped array CFI geometry. Separate sampling intervals, beam focus, and beam steering are used for the gray-scale and Doppler maps of the composite image. The Doppler carrier and gray-scale carrier frequencies can be different. The region of interest (ROI) can be controlled to improve the CFI frame rate. LOS is an image line of sight.

resolution simultaneously for both maps can dramatically reduce the combined image frame rate. An effective method of keeping the frame rate up is to limit the region of Doppler mapping. By forming a smaller ROI, the system can speed up the composite image formation. In addition, the operator can decrease the number of LOSs that comprise the ROI, increase the sampling intervals along each LOS, or double up the data gathering by parallel processing, that is, by transmitting a broad Doppler beam and receiving echo signals from two adjacent LOSs.

Cardiac CFI centers on the phased array with its sector field of view. From an apical or subcostal scanning window, the flow patterns within the heart are well within a 20° Doppler angle or less to the central sector LOS. In contrast, vascular CFI centers on the linear stepped array, which normally produces a rectangular field of view perpendicular to the array face. Because the vessels of interest are generally parallel to the skin surface, a conventional linear stepped array places the LOS perpendicular to the flow. Early synchronous CFI used a triangular plastic, water-filled wedge standoff to obtain a suitable Doppler angle to the vessel. Figure 54 shows how the wedge set the Doppler scanning angle. The present method of obtaining a suitable Doppler angle is beam steering.

Many early multielement arrays used for CFI were designed for conventional gray-scale imaging so their elements were too large to support beam

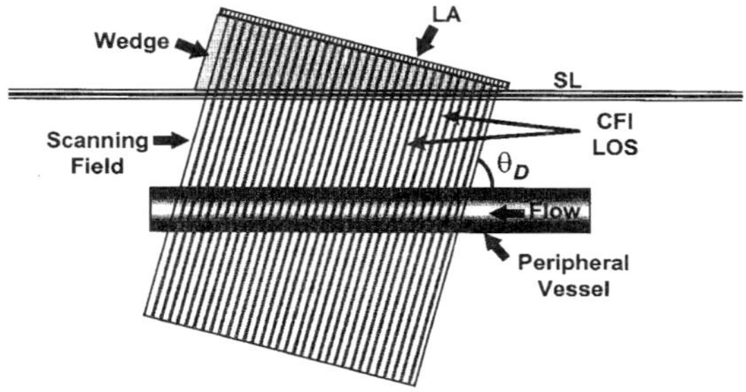

FIG. 54. Mechanical wedge standoff for a Doppler angle. Because peripheral blood vessels are generally parallel to the skin surface, linear stepped arrays without beam steering can use a wedge standoff to obtain a nonperpendicular Doppler angle with the blood flow. LSA: linear stepped array, SL: skin line, θ_D: Doppler angle, LOS: lines of sight.

steering at the array (Doppler) carrier frequency (see Section IV.C.2). As a consequence, several manufacturers lowered their Doppler frequency to support moderate beam steering. Later array improvements reduced the element width and increased their number. Beam steering has its trade-offs, however, which include a reduced ability to focus the ultrasound beam and the production of side lobes and grating lobes (Section IV.C.2). In clinical situations involving high-velocity jets in the heart, PW Doppler cannot easily quantify the flow velocities without high-frequency aliasing (Eq. (51)). Changing the Doppler function from PW to CW and steering the beam using phase delays can remove the aliasing problem.

2. *Time-Domain-Analysis–Based Systems*

An alternative to utilizing the Doppler effect is to shift the analysis out of the frequency domain into the time domain. Called time domain analysis (TDA), this technique focuses on the ability to identify a "piece of blood" (RBC cluster) and follow its movement over time. By measuring distance and knowing the time interval between transmissions, the system directly calculates the regional blood velocity. (Note that TDA simultaneously extracts gray-scale and RBC cluster velocity information from the same focused beam.) This technology rests on some central assumptions about the tissue–ultrasound interaction and the behavior of moving blood: (1) The speckle pattern from an RBC cluster is sufficiently unique that it can be separated from all other surrounding speckle patterns. (2) The speckle pattern of an RBC cluster does not change enough over very short time periods to prevent its reidentification by the machine. As it turns out, these assumptions are sufficiently true (Trahey *et al.*, 1988) to allow a new and entirely different CFI approach.

a. Ideal TDA System. The ideal TDA system first collects a portion of a gray-scale image speckle pattern in a memory buffer. The system then performs a regional sampling of the image one frame later, sampling the space surrounding the original data collection. The recognition of the speckle sample comes from a digital cross-correlation between regions of image speckle and the stored information in the buffer. The interrogation has some functional limits, including a limit on the size of the original sample. The sample must be large enough to have some uniqueness to its speckle pattern and yet small enough to represent a useful region of flow, even in smaller vessels. Tracking the movement of a piece of tissue requires a two-dimensional interrogation of the image and resolution of both horizontal and vertical velocity components. The image must be divided into a set of speckle sampling sites with a follow-up spatial interrogation for the recorded speckle over a region large enough to match the highest velocity being measured. By setting the threshold for the degree of cross-correlation, some alterations can occur in the moving blood and a useful color flow image can still be obtained. And the analysis need not be broken into small regions. At least one developed system has examined gray-scale speckle with succeeding frames and used Boolean algebra to produce CFI (Gardiner and Fox, 1989).

b. Color Velocity Imaging. The current commercial application of TDA is called color velocity imaging or CVI (CVI/CVI-Q Primer, 1994). It relaxes cross-correlation requirements by limiting the sampling and testing only to the image LOS (motion toward or away from the transducer). CVI is a bit like Doppler in that it is angle dependent. It permits the images to appear at a real-time rate while remaining within the limits of available microprocessors. The early CVI transducers used a wedge on a linear array to establish a viewing angle. Current CVI applications depend on beam steering to obtain a good viewing angle. Key to its difference from Doppler is that CVI does not have to contend with high-frequency aliasing. Increased blood velocities only require increasing the range of interrogation along the line of sight. A spatial form of aliasing does occur for TDA, however. At the heart of TDA is the requirement for the speckle pattern to remain unique. Unfortunately, speckle patterns can repeat themselves in the near-field of multielement transducers (Wagner *et al.*,

1988), and this can open the door to unexpected errors in velocity calculations.

Although CVI is a non-Doppler means of imaging the blood flow pattern, it does not remove the need for conventional Doppler duplex or triplex (CVI and a single-point Doppler spectral display) operation. The preponderance of clinical evaluations in the vascular system require a Doppler spectral display and to make similar determinations CVI systems also need PW Doppler capability.

3. Color Encoding Schemes

The process of encoding flow information into color utilizes the three basic properties of color: hue, brightness, and saturation. By changing any one or all of these parameters, specific information can be imparted to the image viewer. The CFI information to be encoded includes: the direction of motion, the average frequency within a sample site, and some measure of spectral broadening within the sample site frequencies.

Hue represents the perceived color, usually expressed as either a frequency or as a wavelength. Changing a color from red to green to blue represents a change in hue. In CFI systems, the flow directions (forward and reverse) are usually expressed as different colors, e.g., red and blue. The assignment is typically chosen to show arteries red and veins blue, with additional color modulation to impart other information.

Brightness is the intensity or luminance of a color. In CFI systems, the brightness of a color is often purposefully reduced when it appears close to a vessel wall. This smoothes the transition between the color and the gray scale representing the surrounding soft tissues.

Color saturation is a measure of color purity. White light is the presence of all colors. If a color is saturated 100%, then it contains no other colors. Decreasing the color saturation means adding more whiteness to the color. Changing color saturation from a 100% red to a 10% red means the color will change from a "deep" red to a light pink.

CFI color encoding begins with directionality—usually red (or blue) represents the closing velocity (or its negative). Within each image sampling pixel, the system makes a determination of the average frequency, which can be a mean, mode, or median value. As the average frequency value increases, the color can be modulated in hue or saturation. Changes in hue may not move toward the same color endpoint. For example, Figure 55 (color plate)

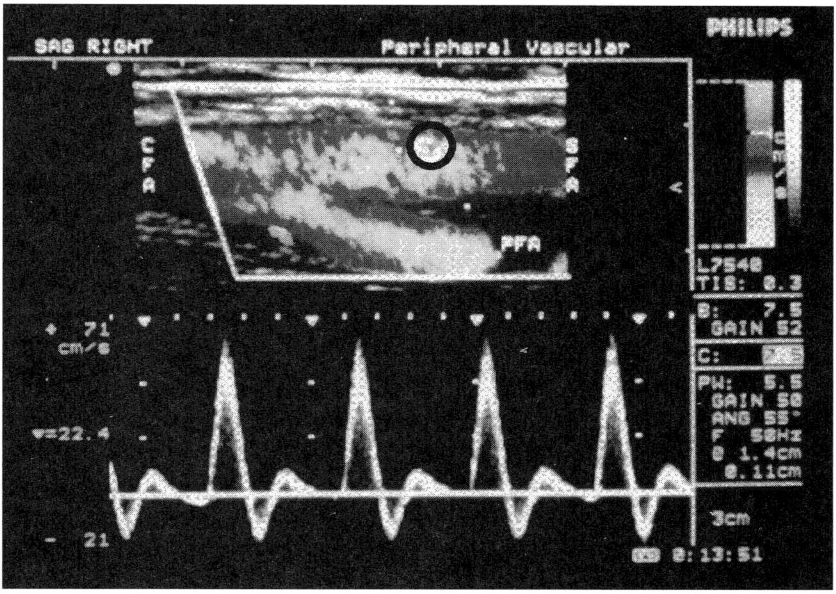

FIG. 55. CFI hue encoding. This example of color velocity Imaging of the femoral artery uses color hue (red to yellow) to encode flow away from the transducer. The color bar (image right) also shows that flow toward the transducer would be blue. The image incorporates both CVI (time domain analysis) for the CFI and duplex Doppler to produce the flow spectrum. The black circle surrounds the location of the Doppler range gate and angle corrector. (Courtesy of Philips Ultrasound International.)

shows a color encoding that runs from red to yellow for one direction and blue to aqua in the other direction. In contrast, using saturation to show higher frequencies easily demonstrates higher frequency patterns within the vessel. Figure 56 (color plate) shows an encoding of frequency to color saturation. As the frequencies increase, the color becomes whiter.

Doppler CFI revolves around a frequency measurement and a velocity calculation. In some systems, the CFI display expresses velocities without an angle correction. These values represent the closing velocity values. Including the Doppler angle into the calculation estimates the true velocity. Often, the system shows some measure of spectral broadening in the Doppler CFI color encoding using a further modulation of hue. Figure 55 shows a hue change toward green as an expression of spectral broadening.

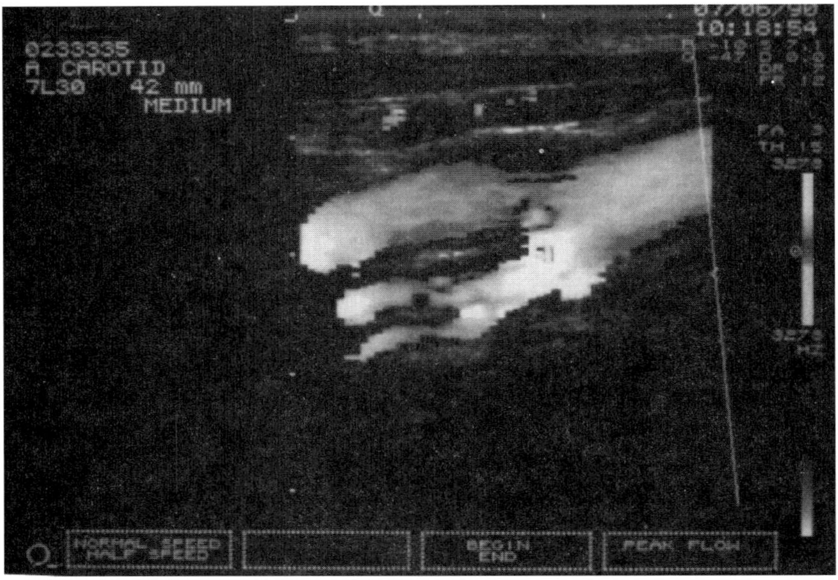

FIG. 56. CFI saturation encoding. The color bar on the image right shows the color code. Red is flow away from the transducer; blue is flow toward it. The image shows a common carotid artery bifurcating into the internal (upper) and external (lower) carotids, with a further branching off the external carotid (lowest vessel). (Used with permission of R. Knighton 11/5/97.)

4. Power Doppler Imaging

In a practical sense, Doppler-based CFI has three limitations that can affect its clinical application: (1) In the frequency domain, random noise looks like flow events. (2) High frequency aliasing occurs when a DSF exceeds the Nyquist frequency. (3) The detection and portrayal of flow is angle dependent. These limitations all derive from the frequency domain analysis.

An alternative to showing the average DSF in color is to encode the average power in the Doppler signal in color (Rubin *et al.*, 1995). This calculation can be carried out at the autocorrelator level used to derive the average DSF. The color encoding indicates the presence of Doppler signals (color or no color) and the amount of power within the signal using a change in color hue, brightness, or saturation.

The improvements are fourfold: (1) Power Doppler color encoding makes noise appear significantly different. Because the average power of noise is so low, it appears as a constant background color rather than speckling random events. In addition, low-frequency signals with a lot of power are more easily portrayed. (2) Because the signal processing is power dependent and frequency independent, no aliasing occurs even for very high velocities. (3) The portrayal of the signal is mostly angle independent (the low-frequency noise suppression filters (see Figure 51) produce some angle dependence at large Doppler angles). (4) The stronger echo signals from contrast media do not produce color blooming outside the vessel walls. The stronger contrast media reflections are simply encoded as larger amplitudes. Figure 57 (color plate) demonstrates a power Doppler image.

The trade-offs for this improved sensitivity to the presence of flow are slow data acquisition and no indication of flow speed or direction. In addition, relative tissue motion from transducer movement, heartbeat pulsations, and respiratory motion all create spurious color additions to the image that do not represent flow. The response to this problem by industry has been the development of more sophisticated methods of CFI "flash suppression."

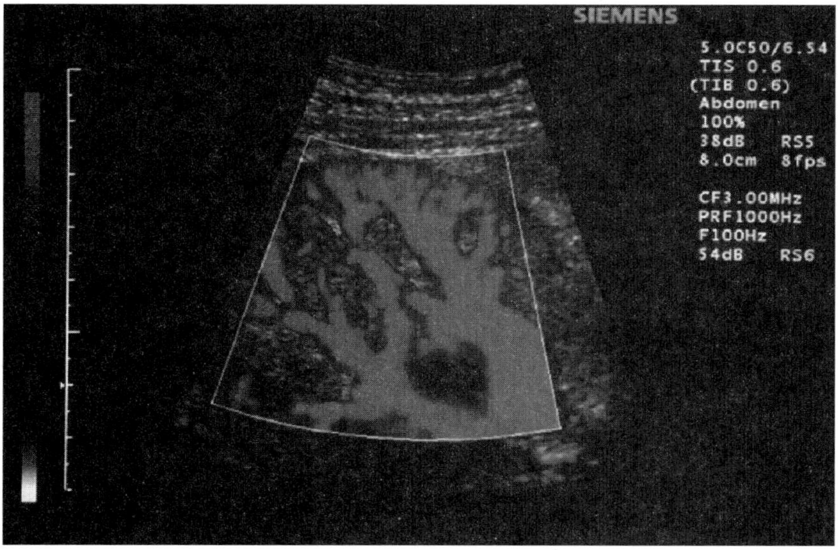

FIG. 57. Power Doppler color flow image. Power Doppler encodes the amplitude of the Doppler signals into color rather than the frequency. The color bar on image left shows the amplitude color encoding. The image shows flow in the renal vessels, both arteries and veins. (Used with permission of R. Knighton 11/5/97.)

VI. Recent Developments

Improvements in medical ultrasonic imaging cover all aspects of its use. Some are basic developments in ultrasonic science or digital electronics, others are concerned with its day-to-day clinical use. Some are evolutionary and others revolutionary. A few of the newest developments are summarized here. Some are promising preliminary studies whereas others are being incorporated currently into high-resolution equipment.

A. PAC Systems

Digital archiving offers improvements in several different areas of medical imaging. It is cost-effective, since expensive hardcopy need be made only for the most diagnostic images and those exported to referring physicians and patients. Thus the large film library maintained in radiology departments (costly due to personnel and film silver content) is replaced by a collection of optical disks. It is efficient, since patient images are permanently stored in digital form and can be accessed to be reviewed or copied at any time. (In contrast, stored photographic hardcopy patient images can "wander" onto someone's desk or be in photography for slide production and thus not be available for comparison to more recent images.) And it permits the rapid digital transmission of images across large distances by phone or microwave transmission.

Picture archiving and communications systems (PACs) are now commercially available and are part of the rapid digitization of all of medical imaging (Huang, 1993). Ultrasound manufacturers initially opted for the development of closed PACs systems that stored only the images produced on their equipment. The adoption of standardized image formats is now permitting cross-platform storage and transmission of ultrasonic images.

B. Contrast Agents

Due to ever-present speckle patterns and the slight differences in ultrasonic parameters of normal and diseased tissue, it is very difficult to identify tissue structures and/or disease in ultrasonic images. By trial and error, various methods of increasing the image contrast between tissues have evolved. These usually involve introducing substances into the body that accumulate selectively in certain tissues and modify their ultrasonic parameters. Contrast-agent-induced changes in tissue scattering cross-section, attenuation coefficient, or acoustic velocity have been used to improve image contrast.

The most commonly used contrast agent is microbubbles of air (less than 5 μm in diameter) introduced into flowing blood. The microbubbles can be as simple as free bubbles in an agitated saline solution, but they usually have a shell of albumin or sugar for stability (de Jong, 1996). A concentration of microbubbles increases blood echogenicity (scattering cross-section) due to acoustic impedance mismatching. Microbubbles have proven to be valuable in the investigation of blood perfusion or cardiac function.

Another contrast agent that shows promise is perfluorocarbons, a colloidal suspension (Mattrey, 1989). These enhance ultrasonic backscatter in tissue as well as blood. Again, acoustic impedance mismatches of about 30% between tissue or blood and the perfluorocarbons (with a density of 1.9 g/mL and an acoustic velocity of 600 m/s) are responsible for the increased backscatter. Many other compounds introduced into the body have demonstrated improvements in the visualization of certain tissues or disease states (Goldberg *et al.*, 1994).

Encapsulated air bubbles have resonant frequencies determined by their radii, the ambient fluid pressure and density, and the bubble surface tension (de Jong *et al.*, 1992). They exhibit a nonlinearity in their scattering cross section and, therefore, produce harmonics in the scattered ultrasonic waves. This phenomenon is utilized in Doppler harmonic imaging where the bubbles are insonified at a fundamental transmission frequency and their echoes are passband detected at its first harmonic, twice the fundamental frequency (Schrope and Newhouse, 1993). By detecting only the first harmonic echo signals from the bubbles, any tissue motion echoes are suppressed and the Doppler contrast is improved considerably. The only trade-off is that the SNR improvement depends on the bandwidth of the transmitted ultrasonic waves (Chang *et al.*, 1995). The narrower the bandwidth, the greater the Doppler contrast improvement.

Harmonic imaging has been introduced into gray-scale imaging as well. This mode of operation takes advantage of the nonlinear properties of soft tissue (Section VI.C). Since the transmitted and first harmonic signals have wide bandwidths, there is some crossover between the two so that some tissue information is present in the image even in the absence of nonlinear effects. The amount of crossover is tissue specific and is a proprietary image quality parameter. Figure 58 demonstrates the improvement in image quality for an obese patient from one manufacturer's implementation of harmonic imaging. This imaging is useful for hard-to-image patients since the

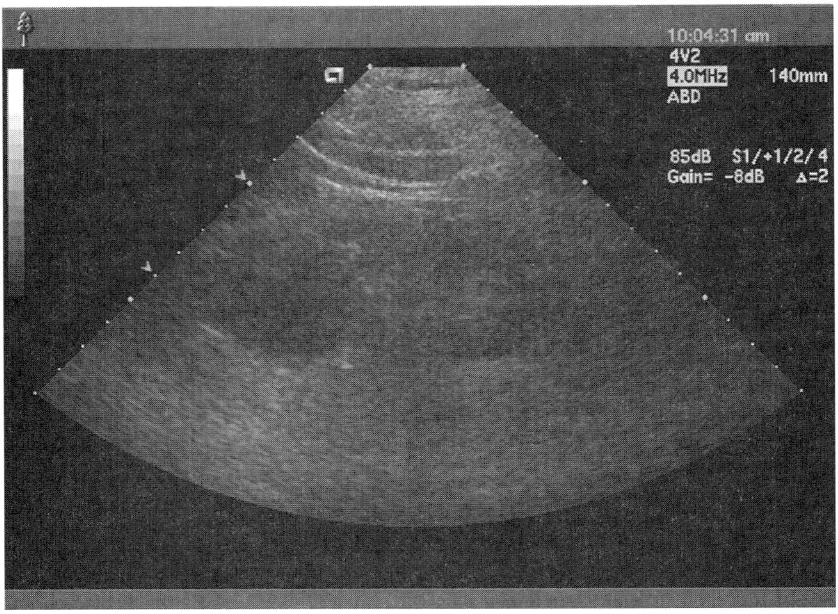
10:04:31 am
4V2
4.0MHz 140mm
ABD
85dB S1/+1/2/4
Gain= -8dB Δ=2

(a)

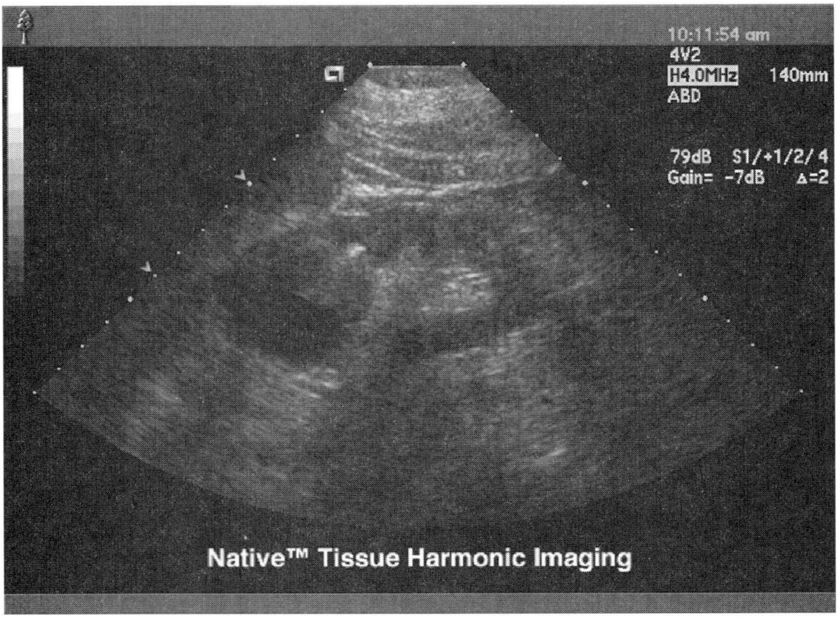
10:11:54 am
4V2
H4.0MHz 140mm
ABD
79dB S1/+1/2/4
Gain= -7dB Δ=2
Native™ Tissue Harmonic Imaging

(b)

low-frequency transmitted signal penetrates better and has less scatter for small reflectors.

C. Nonlinear Propagation

All the theory concerning tissue ultrasonic properties and beam focusing previously presented was under the assumption of finite-amplitude, linear wave propagation. Actually, it is well known that high-amplitude transmitted medical ultrasonic waves undergo nonlinear transmission in tissue (Muir, 1980). Nonlinear ultrasonic waves have amplitude-dependent acoustic velocities (Beyer, 1969). This causes waveform distortion, shock wave formation, and harmonic generation (Muir and Carstensen, 1980). The nonlinear parameters of tissue have been studied for some time (Law *et al.*, 1985).

It has been demonstrated recently that harmonically distorted ultrasonic transmit beam patterns are narrower and have lower side-lobe levels and shorter pulse lengths than undistorted fundamental frequency beams (Ward *et al.*, 1997). So if nonlinear shock wave formation can be demonstrated not to cause any bioeffects in tissue (NCRP, 1983), the high-pressure amplitudes used in ultrasonic imaging equipment will produce only benefits for the patient.

D. Adaptive Focusing

Recent research has demonstrated that tissue acoustic velocity is neither uniform in specific tissue types nor constant from tissue to tissue. Ultrasonic wave distortions produced by the abdominal wall (Sumino and Wagg, 1991) and secondary wavefronts generated by bone and cartilage in the chest wall (Hinkelman *et al.*, 1997) have been measured.

These wavefront distortions will not only defocus ultrasonic beams designed under the assumption of a uniform and well-known tissue acoustic

---

FIG. 58. Gray-scale harmonic imaging. (a) A renal vector array image of a 5′5″, 265 lb 65-year-old patient with a thick body wall. (b) The same patient anatomy scanned using harmonic imaging. The large complex renal mass is now clearly seen. Although the harmonic image is not as cosmetically pleasing as regular gray-scale images obtained with thinner patients, its improved spatial and contrast resolution provide important diagnostic information for these clinically important, difficult-to-image patients. The trade-off for harmonic imaging is increased acoustic output — an increase in MI from 1.0 to 1.9 in the harmonic image. (Images provided by Acuson Corporation.)

velocity, but can even cause double images of single targets. In other remote sensing fields, well-defined point targets are surrounded by a uniform unattenuating medium so that selected targets can be used as "beacon" signals to enable adaptive focusing corrections. However, in medical ultrasonics the tissue is continuous, nonuniform, and attenuating, so it is not possible to select ideal point target "beacon" signals during a clinical examination.

To get around this difficulty, many investigators have been studying adaptive focusing schemes using the ubiquitous low-amplitude speckle signals. In one early scheme, the nonuniformities in acoustic velocity were measured and corrected using a cross-correlation technique between adjacent array elements (Flax and O'Donnell, 1988 and O'Donnell and Flax, 1988). In another, speckle signal brightness was maximized by iteratively phase shifting each array element's signal in the summing circuit (Nock *et al.*, 1989). In still another, wavefront distortions were compensated for by backpropagation of the wavefront using the angular spectrum method (Liu and Waag, 1994). Ultrasonic image quality will be improved for difficult patients when adaptive focusing is commercially available, but the solution must be computationally simple for application in real-time equipment.

E. Intraluminal Imaging

Many different types of small, high-frequency transducers have been mounted on catheters and inserted into arteries and veins to visualize their walls and any plaque formation. The limited space creates many difficulties in transducer design and focusing. A new approach is the use of synthetic aperture imaging (Karaman *et al.*, 1995). In synthetic aperture imaging, a large aperture (with its small spot size, Eq. (16)) is synthesized by translating a small active aperture (transmit or receive) over a multielement array. The usual difficulty of patient motion during synthetic aperture scan acquisition is minimized with the small multielement array resting inside of the vessel.

F. Slice-Thickness Focusing

The use of mechanical slice-thickness focusing in multielement arrays causes a reduction in image contrast resolution outside of the limited slice-thickness focal zone. Electronic focusing in the slice-thickness direction is just now becoming commercially available, because of transducer difficulties (more complicated element electrode patterns with electrical contacting) and circuit difficulties (a drastic increase in the number of parallel signal processing channels required). With more experience in multielement transducer fabrication and ASIC design, however, most manufacturers are now tackling this important problem. The present arrays—which focus and beam steer electronically in-plane and focus mechanically out-of-plane—are called 1D arrays. A 2D array structure would have out-of-plane electronic focusing and beam steering capabilities.

A simple interim solution presently being used is to divide the long, thin linear array elements into a small number of elements (up to eight) in the slice-thickness direction with individual electrical contacts to all subelements. Then, a limited dynamic slice-thickness focus is attained with dynamic aperture (and possibly dynamic apodization) applied to these subelements (the mechanical cylindrical lens is still used). This is called a 1.25D array. By not having time delays in the slice-thickness direction, the required circuit complexity is greatly reduced. Image contrast resolution at shallow image depths is improved due to the smaller slice-thickness aperture used at these depths.

A better solution is a 1.5D array, which has electronic focusing in the slice-thickness direction but has no beam steering (Wildes *et al.*, 1997). In one feasibility test, the existing 256 equipment signal channels were divided into a shorter-than-usual linear array with a limited number of elements in the slice-thickness direction (Daft *et al.*, 1994). One interesting result of this study was the ease with which unusual anatomy could be found with the 1.5D array. With the dual benefits of increased visualization of low-contrast objects and reduced scan time (while producing superior images), it is clear that it is only a matter of time before most high-resolution ultrasonic imaging equipment routinely focuses electronically in the slice-thickness direction. One manufacturer is now claiming to have commercial 1.5D array transducers with 1024 active channels for in-plane and out-of-plane focusing.

G. 3D Imaging

3D ultrasonic imaging refers to the interrogation of a tissue volume and its display in a 3D format. Some 3D schemes acquire the ultrasonic data as a series of adjacent planes, while others utilize 2D multielement arrays that can beam steer and focus over the entire interrogated tissue volume (Masotti and Pini, 1993; Greenleaf, 1995).

The large amount of required information puts a premium on data acquisition speed. Some of the parallel processing schemes described in Section IV.F.7 were developed for 3D imaging. The problems associated with 3D data acquisition will be solved by the evolution of ultrasonic digital circuitry. The main difficulty with 3D imaging is the proper display of the data. With the use of a threshold function, high-contrast tissue structures (such as ultrasonic cardiac images) can be easily visualized in a 3D display. But since most ultrasonic images have lower contrast, a single threshold cannot be used to suitably remove unwanted information from the 3D image. Many volume-rendering techniques are being studied at present in an effort to properly visualize the low-contrast information in 3D images (Steen and Olstad, 1994).

A simple, practical method of 3D imaging utilizes existing ultrasonic equipment and an add-on mechanical means to acquire a series of indexed scan planes (Fenster and Downey, 1996). The 3D display is unique as well, since it presents the block of volume data on the CRT display and allows the operator to select any 2D plane cutting through the block for visualization. Advantages of this technique include rapid acquisition (~10 sec) of the 3D data, user-friendly data manipulation, and the ability to visualize 2D image planes that are impossible to acquire directly (due to scan plane positioning limitations and the out-of-plane directionality of the incident ultrasonic pulses).

H. Panoramic Imaging

Although the development of real-time imaging has greatly improved the diagnostic image quality and utility of medical ultrasonics, the image field of view is limited by the multielement transducer aperture—unlike the global images possible in analog static images. One manufacturer has developed a real-time technique for acquiring static panoramic images up to 60 cm long by moving the transducer parallel to its scan plane. As the real-time image frames are acquired, they are divided into several smaller subimage regions that are rapidly analyzed for movement of individual image features from frame to frame. The results of the image feature motion analysis are used to translate and rotate the newly acquired real-time image data so that it is properly registered spatially when it is inserted into the static image map being generated. Fuzzy logic control is employed to increase simplicity and flexibility and to give robust performance even in the presence of large amounts of image noise (Klir and Folger, 1988). Figure 59 demonstrates this technique.

FIG. 59. Panoramic image of renal transplant. (a) A standard linear stepped array view of a renal transplant (placed in the abdomen). (b) A 12-cm wide field of view panoramic image obtained by slow lateral movement of the same transducer demonstrating liver (on left) and transplanted kidney with a fluid collection below it. The mechanical index (MI) is 0.7 (AIUM, 1992). (Courtesy of Siemens Medical Systems, Inc., Ultrasound Group.)

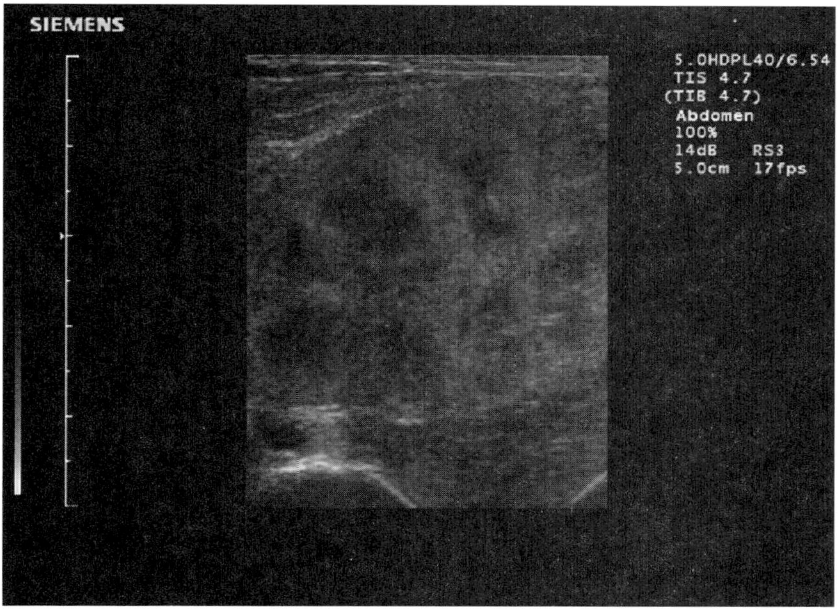
SIEMENS
5.0HDPL40/6.54
TIS 4.7
(TIB 4.7)
Abdomen
100%
14dB RS3
5.0cm 17fps

(a)

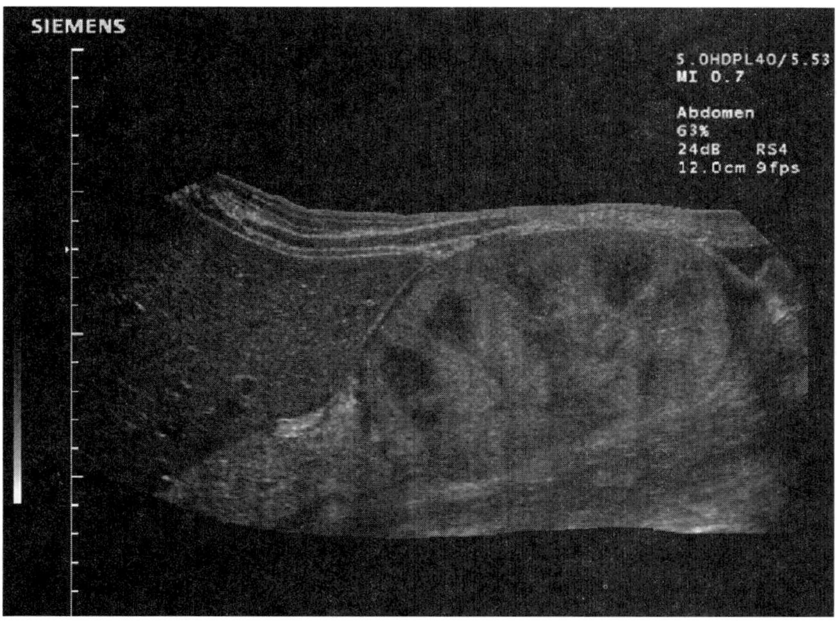
SIEMENS
5.0HDPL40/5.53
MI 0.7
Abdomen
63%
24dB RS4
12.0cm 9fps

(b)

VII. Summary

The field of medical ultrasonic diagnostics is much more comprehensive and includes more types of instrumentation and new research than described in this chapter. Space considerations mandated just a description of the basic equipment. Reviews and recent articles were used as references whenever possible to aid the readers in understanding this complex field. Our apologies to those investigators whose important work was not cited. This chapter was not intended to be a critical review of the progress made in this field but merely as an introductory description of the use of ultrasonic waves and digital technology in medical imaging. Due to the very competitive nature of this field and proprietary considerations, only a simple description of the use of digital technology was possible.

It is certain that more evolutionary and hopefully some revolutionary improvements in this equipment will be forthcoming. Certainly, medical ultrasonic imaging will finally mature. Perhaps its maturation will include the ability to separate the acquired raw data into two patient tissue maps, one of tissue attenuation coefficients and one of tissue reflection coefficients.

References

Adler, F. H. (1965). "Physiology of the Eye." Mosby, Saint Louis, Missouri.

Aero-Tech Reports (1980). Multiple matching layer theory and application. *KB-Aerotech*, Vol. 1., No. 5., Lewiston, Pennsylvania.

AIUM (1986)."Standard Presentation and Labeling of Ultrasound Images." American Institute of Ultrasound in Medicine, Bethesda, Maryland.

AIUM (1992). "Acoustic Output Measurement and Labeling Standard for Diagnostic Ultrasound Equipment." American Institute of Ultrasound in Medicine, Bethesda, Maryland.

AIUM (1994)."Medical Ultrasound Safety," American Institute of Ultrasound in Medicine, Bethesda, Maryland.

Arditi, M., Foster, F. S., and Hunt, J. W. (1981). Transient fields of concave annular arrays. *Ultrasonic Imaging* **3**, 37–61.

Atkinson, P., and Woodcock, J. P. (1982). "Doppler Ultrasound and Its Use in Clinical Measurement." Academic Press, London.

Augustine, L. J., and Anderson, J. (1979). An algorithm for the design of transformerless broadband equalizers of ultrasonic transducers. *J. Acoust. Soc. Am.* **66**, 629–635.

Baker, D. W., and Daigle, R. E. (1977). Noninvasive ultrasonic flowmetry. *In* "Cardiovascular Flow Dynamics and Measurements" (N. H. C. Hwang and N. A. Normann, eds.). University Park Press, Baltimore.

Baker, D. W., Forster, F. K., and Daigle R. E. (1978). Doppler principles and techniques. *In* "Ultrasound: Its Applications in Medicine and Biology," Vol 1 (S. P. Wolsky and A. W. Czanderna, eds.). Elsevier Scientific Publishing, New York.

Bamber, J. C. (1993). Speckle reduction. *In* "Advances in Ultrasound Techniques and Instrumentation" (P. N. T. Wells, ed.), Churchill Livingstone, New York, pp. 55–67.

Barber, F. E., Baker, D. W., Nation, A. W. C., Strandness, D. E. Jr., and Reid, J. M. (1974). Ultrasonic duplex echo-doppler scanner. *IEEE Trans. Biomed. Eng.* **BME-21**, 109–113.

Barnett, S. B., Rott, H.-D., ter Haar, G. R., Ziskin, M. C., and Maeda, K. (1997). The sensitivity of biological tissue to ultrasound. *Ultrasound in Med. & Biol.* **23**, 805–812.

Beaver, W. L. (1977). Phase error effects in phased array beam steering. *In* "1977 IEEE Ultrasonics Symposium Proceedings," IEEE, New York, pp. 264–267.

Berlincourt, D. A, Curran, D. R., and Jaffe, H. (1964). Piezoelectric and piezomagnetic materials and their function in transducers. *In* "Physical Acoustics," Vol. 1A (W. P. Mason, ed.), Academic Press, New York, pp. 170–270.

Beyer, R. T. (1969). "Physical Acoustics." Academic Press, New York.

Bloom, W., and Fawcett, D. W. (1968). "A Textbook of Histology," 9th Edition. W. B. Saunders, Philadelphia.

Bom, N., Lancee, C. T., and Honkoop, J. (1971). Ultrasonics viewer for cross-sectional analysis of moving cardiac structures. *Biomed Eng.* **6**, 500–511.

Born, M., and Wolf, E. (1975). "Principles of Optics," 5th Edition. Pergamon Press, Oxford.

Brandestini, M. A., and Forster, F. K. (1978). Blood flow imaging using a discrete-time frequency meter. *In* "1978 IEEE Ultrasonics Symposium Proceedings." IEEE, New York, pp. 348–352.

Brice, J. (1993). "Diagnostic Ultrasound Market Trends and Clinical Practice in the U.S., Diagnostic Imaging Technology Report," Vol. 2, No. 1. Miller Freeman, San Francisco.

Chang, P. H., Shung, K. K., Wu, S.-J., and Levene, H. B. (1995). Second harmonic imaging and harmonic Doppler measurements with Albunex®. *IEEE Trans. UFFC* **42**, 1020–1027.

Chen, X., Schwarz, K. Q., and Parker, K. J. (1993). Radiation pattern of a focused transducer: A numerically convergent solution. *J. Acoust. Soc. Am.* **94**, 2979–2991.

Christensen, D. A. (1988). "Ultrasonic Bioinstrumentation." John Wiley & Sons, New York.

Cornsweet, T. N. (1970). "Visual Perception." Academic Press, New York.

"CVI/CVI-Q Primer" (1994). Philips Ultrasound, Santa Ana, California.

Daft, C. M. W., Thomas, L. J., Smith, L. S., Lewandowski, R. S., Leue, W. M., Rigby, K. W., Chalek, C. L., and Hatfield, W. T. (1994). A 1.5D transducer for medical ultrasound. *In* "Proc. 1994 IEEE Ultrason. Symp." IEEE, New York, pp. 1491–1495.

Davis, H. F. (1963). "Fourier Series and Orthogonal Functions." Allyn and Bacon, Boston.

Davson, H. (1963). "The Physiology of the Eye." Little, Brown and Co., Boston.

de Jong, N. (1996). Improvements in ultrasound contrast agents. *IEEE Eng. Med. Biol.* **15**, 72–82.

de Jong, N., Hoff, L., Skotland, T, and Bom, N. (1992). Absorption and scatter of encapsulated gas filled microspheres: theoretical considerations and some measurements. *Ultrasonics* **30**, 95–103.

DeClercq, A., and Maginness, M. G. (1975). Adaptive gain control for dynamic ultrasound imaging. *In* "Proc. 1975 IEEE Ultrason. Symp." IEEE, New York, pp. 59–63.

Desilets, C. S., Fraser, J. D., and Kino, G. S. (1978). The design of efficient broad-band piezoelectric transducers. *IEEE Trans. Son. & Ultrason.* **SU-25**, 115–125.

Dias, J. P. (1982). An experimental investigation of the cross-coupling between elements of an acoustic imaging array transducer. *Ultrasonic Imaging* **4**, 44–55.

Erickson, J. J., and Brill, A. B. (1970). Digitization of ultrasonic images. *Radiology* **95**, 589–593.

Evans, D. H. (1985). Doppler signal processing. *In* "Cardiovascular Ultrasonic Flowmetry" (S. A. Altobelli, W. F. Voyles, and E. R. Greene, eds.). Elsevier, New York.

Evans, D. H., McDicken, W. N., Skidmore, R., and Woodcock, J. P. (1989). "Doppler Ultrasound—Physics, Instrumentation and Clinical Applications." John Wiley & Sons, Chichester.

Faran, J. J. Jr. (1951). Sound scattering by solid cylinders and spheres. *J. Acoust. Soc. Am.* **23**, 405–418.

Fenster, A., and Downey, D. B. (1996). 3-Dimensional ultrasound imaging: a review. *IEEE Engin. Med. Biol.* **15**, 41–51.

Ferrara, K., and DeAngelis, G. (1997). Color flow mapping. *Ultrasound in Med. & Biol.* **23**, 321–345.

Flax, S. W., and O'Donnell, M. (1988). Phase-aberration correction using signals from point reflectors and diffuse scatterers: basic principles. *IEEE Trans. UFFC* **35**, 758–767.

Fry, W. J., Leichner, G. H., and Okuyama, D. (1968). Ultrasonic visualization system employing new scanning and presentation methods. *J. Acoust. Soc. Am,* **44**, 1324–1388.

Gardiner, W. M., and Fox, M. D. (1989). Color flow imaging through analysis of speckle motion. *Radiology* **172**, 866–868.

Goldberg, B. B., Liu, J.-B., and Forsberg, F. (1994). Ultrasound contrast agents: a review. *Ultrasound Med. & Biol.* **20**, 319–333.

Goldstein, A. (1976). A computerized ultrasound processing, acquisition, and display system. Final Report, NSF Grant GJ-41682.

Goldstein, A. (1981). Range ambiguities in real-time ultrasound. *J. Clin. Ultra.* **9**, 83–90.

Goldstein, A. (1987). Pertinent physics of an optimal examination. *In* "The Use of Transrectal Ultrasound in the Diagnosis and Management of Prostate Cancer" (F. Lee and R. D. McLeary, eds.). Alan R. Liss, New York, pp. 31–48.

Goldstein, A. (1988). Ultrasonic imaging. *In* "Encyclopedia of Medical Devices and Instrumentation" (J. G. Webster, ed.), Vol 4. John Wiley & Sons, New York, pp. 2803–2823.

Goldstein, A. (1991). Performance tests of Doppler ultrasound equipment with a string phantom. *J. Ultrasound Med.* **10**, 125–139.

Goldstein, A., and Clayman, W. (1983). Particle image resolution test object. *J. Ultrasound Med.* **2**, 195–209.

Goldstein, A., Ophir, J., and Templeton, A. W. (1974). A computerized ultrasound processing, acquisition and display (CUPAD) system: research in ultrasound image generation. *In* "Ultrasound in Medicine" (D. White and R. E. Brown, eds.), Vol. 1. Plenum Press, New York, pp. 475–480.

Goldstein, A., Ophir, J., and Templeton, A. W. (1975). Research in ultrasound image generation: a computerized ultrasound processing, acquisition and display (CUPAD) system. *In* "Acoustic Holography" (N. Booth, ed.), Vol. 6. Plenum Press, New York, pp. 57–70.

Goldstein, A., Ranney, D., and McLeary, R. D. (1989). Linear array test tool. *J. Ultrasound Med.* **8**, 385–397.

Goodman, J. W. (1968). "Introduction to Fourier Optics." McGraw-Hill, San Francisco.

Goss, S. A., Johnston, R. L., and Dunn, F. (1978). Comprehensive compilation of empirical ultrasonic properties of mammalian tissues, I. *J. Acoust. Soc. Am.* **64**, 423–457.

Goss, S. A., Johnston, R. L., and Dunn, F. (1980). Comprehensive compilation of empirical ultrasonic properties of mammalian tissues, II. *J. Acoust. Soc. Am.* **68**, 93–108.

Greenleaf, J. F. (1995). 3D and tomographic ultrasound. *In* "Medical CT and Ultrasound: Current Technology and Applications" (L. W. Goldman and J. B. Fowlkes, eds.). AAPM, College Park, Maryland, pp. 267–284.

Guyton, A. C. (1991). "Textbook of Medical Physiology," 8th Edition. W. B. Saunders, Philadelphia.

Haller, M. I., and Khuri-Yakub, B. T. (1992). Micromachined acoustic matching layers. *In* "New Developments in Ultrasonic Transducers and Transducer Systems," Vol. 1733 (F. L. Lizzi, ed.). SPIE Proceedings, Bellingham, pp. 72–77.

Hanss, M., and Boynard, M. (1979). Ultrasound backscattering from blood: hematocrit and erythrocyte aggregation dependence. *In* "Ultrasonic Tissue Characterization II" (M. Linzer, ed.). National Bureau of Standards Special Publication 525. U.S. Government Printing Office, Washington, D.C.

Harmuth, H. F. (1979). "Acoustic Imaging with Electronic Circuits." Academic Press, New York.

Hildebrand, B. P., and Brenden, B. B. (1972). "An Introduction to Acoustical Holography." Plenum Press, New York.

Hinkelman, L. M., Szabo, T. L., and Waag, R. C. (1997). Measurements of ultrasonic pulse distortion produced by human chest wall. *J. Acoust. Soc. Am.* **101**, 2365–2373.

Hossack, J. A., and Auld, B. A. (1992). Improving transducer performance using multiple active layers. *In* "New Developments in Ultrasonic Transducers and Transducer Systems," Vol. 1733 (F. L. Lizzi, ed.). SPIE Proceedings, Bellingham, pp. 284–296.

Huang, H. K. (1993). Ultrasound picture archiving and communication systems. *In* "Advances in Ultrasound Techniques and Instrumentation" (P. N. T. Wells, ed.). Churchill Livingstone, New York, pp. 141–150.

Hubelbank, M. (1970). Computer enhancement of ultrasound images. Sc.D. Dissertation, Dept. Electrical Engineering. MIT, Cambridge, Massachusetts.

Hunt, J. W., Arditi, M., and Foster, F. S. (1983). Ultrasound transducers for pulse-echo medical imaging. *IEEE Trans. Biomed. Eng.* **BME-30**, 453–481.

Insana, M. F. (1995). Sound attenuation in tissue. *In* "Medical CT and Ultrasound: Current Technology and Applications" (L. W. Goldman and J. B. Fowlkes, eds.). Advanced Medical Publishing, Madison, Wisconsin, pp. 15–33.

Insana, M. F., and Hall, T. J. (1993). Quality management of ultrasound diagnosis. *In* "Advances in Ultrasound Techniques and Instrumentation" (P. N. T. Wells, ed.). Churchill Livingstone, New York, pp. 161–181.

Jensen, J. A. (1991). A model for the propagation and scattering of ultrasound in tissue. *J. Acoust. Soc. Am.* **89**, 182–190.

Karaman, M., Li, P.-C., and O'Donnell, M. (1995). Synthetic aperture imaging for small scale systems. *IEEE Trans. UFFC* **42**, 429–442.

Kikuchi, Y. (1978). Transducers for ultrasonic systems. *In* "Ultrasound: Its Applications in Medicine and Biology," Vol. 1 (S. P. Wolsky and A. W. Czanderna, eds.). Elsevier Scientific Publishing, New York.

Kimmelman, B. A. (1988). "Medical Diagnostic Ultrasound: A Retrospective on Its 40th Anniversary." Eastman Kodak, Rochester. Copies may be obtained from Dr. B. B. Goldberg, chairman of the AIUM Archives Committee.

Kino, G. S., and DeSilets, C. S. (1979). Design of slotted transducer arrays with matched backings. *Ultrasonic Imaging* **1**, 189–209.

Kinsler, L. E., Frey, A. R., Coppens, A. B., and Sanders, J. V. (1982). "Fundamentals of Acoustics," 3rd Edition. John Wiley & Sons, New York.

Klir, G. J., and Folger, T. A. (1988). "Fuzzy Sets, Uncertainty and Information." Prentice-Hall, Englewood Cliffs, New Jersey.

Kossoff, G. (1966). The effects of backing and matching on the performance of piezoelectric ceramic transducers. *IEEE Trans. Son. & Ultrason.* **SU-13**, 20–30.

Kossoff, G. (1972). Improved techniques in ultrasonic cross-sectional echography. *Ultrasonics* **10**, 221–229.

Kossoff, G. (1979). Analysis of focusing action of spherically curved transducers. *Ultrasound in Med. & Biol.* **5**, 359–365.

Kossoff, G., and Garrett, W. F. (1972). Ultrasonic film echography in gynecology and obstetrics. *Obst. and Gyn.* **40**, 299–309.

Larsen, H. G., and Leavitt, S. C. (1980). An image display algorithm for use in real-time sector scanners with digital scan conversion. *In* "Proc. 1980 IEEE Ultrason. Symp." IEEE, New York, pp. 763–765.

Larson, J. D. (1981). Non-ideal radiators in phased array transducers. *In* "Proc. 1982 IEEE Ultrason. Symp." IEEE, New York, pp. 673–683.

Law, W. K., Frizzell, L. A., and Dunn, F. (1985). Determination of the nonlinearity parameter B/A of biological media. *Ultrasound in Med. & Biol.* **11**, 307–318.

Leopold, G. (1974). Gray scale imaging. *19th Annual Conference of the American Institute of Ultrasound in Medicine*, Oct. 8–10, Seattle, Washington.

Liu, D.-L., and Waag, R. C. (1994). Correction of ultrasonic wavefront distortion using backpropagation and a reference waveform method for time-shift compensation. *J. Acoust. Soc. Am.* **96**, 649–660.

Lord, A. E. Jr. (1966). Changes in velocity of an elastic pulse owing to geometrical diffraction. *J. Acoust. Soc. Am.* **40**, 163–171.

Loupas, T., McDicken, W. N., Anderson, T., and Allan, P. L. (1994). Development of an advanced digital image processor for real-time speckle suppression in routine ultrasonic scanning. *Ultrasound in Med. & Biol.* **20**, 239–249.

Lucas, B. G., and Muir, T. G. (1982). The field of a focusing source. *J. Acoust. Soc. Am.* **72**, 1289–1296.

Lynn, P. A., and Fuerst, W. (1994). "Introductory Digital Signal Processing with Computer Applications," Revised Edition. John Wiley & Sons, New York.

Macovski, A. (1979). Ultrasonic imaging using arrays. *Proc. IEEE* **67**, 484–495.

Maginness, M. G. (1979). Methods and terminology for diagnostic ultrasound imaging systems. *Proc. IEEE* **67**, 641–653.

Magnin, P. A., von Ramm, O. T., and Thurstone, F. L. (1981). Delay quantization error in phased array images. *IEEE Trans. Son. & Ultrason.* **SU-28**, 305–310.

Mallart, R., and Fink, M. (1992). Improved imaging rate through simultaneous transmission of several ultrasound beams. *In* "New Developments in Ultrasonic Transducers and Transducer Systems," Vol. 1733 (F. L. Lizzi, ed.). SPIE Proceedings, Bellingham, pp. 120–130.

Martin, W. (1974). "The Microprogrammable Beamformer." Technical Report, Raytheon Co.

Maslak, S. H. (1985). Computed sonography. *In* "Ultrasound Annual 1985" (R. C. Sanders and M. C. Hill, eds.), Raven Press, New York, pp. 1–16.

Masotti, L., and Pini, R. (1993). Three-dimensional imaging. *In* "Advances in Ultrasound Techniques and Instrumentation" (P. N. T. Wells, ed.). Churchill Livingstone, New York, pp. 69–77.

Mattrey, R. F. (1989). Perfluoroctylbromide: a new contrast agent for CT, sonography and MR imaging. *Am. J. Roentgenol.* **152**, 247–252.

McDicken, W. N. (1991). "Diagnostic Ultrasonics." 3rd Edition. Churchill Livingstone, Edinburgh.

Morgan, C., Trought, W. S., Clark, W. M., von Ramm, O. T., and Thurstone, F. L. (1978). Principles and applications of a dynamically focused phased array real time ultrasound system. *J. Clin. Ultra.* **6**, 385–391.

Morse, P. M., and Ingard, K. U. (1968). "Theoretical Acoustics." McGraw-Hill, New York.

Moshfeghi, M. (1987). Side-lobe suppression for ultrasonic imaging arrays. *Ultrasonics* **25**, 322–327.

Mucci, R. A. (1984). A comparison of efficient beamforming algorithms. *IEEE Trans. ASSP* **ASSP-32**, 548–558.

Muir, T. G. (1980). Nonlinear effects in acoustic imaging. *Acoust. Imag.* **9**, 93–109.

Muir, T. G., and Carstensen, E. L. (1980). Prediction of nonlinear acoustic effects at biomedical frequencies and intensities. *Ultrasound in Med. & Biol.* **6**, 345–357.

NCRP (1983). "Biological Effects of Ultrasound: Mechanisms and Clinical Implications." NCRP Report No. 74. National Council on Radiation Protection and Measurements, Bethesda.

Newhouse, V. L., Furgason, E. S., Johnson, G. F., and Wolf, D. A. (1980). The dependence of ultrasound Doppler bandwidth on beam geometry. *IEEE Trans. Son. & Ultrason.* **SU-27**, 50–59.

Nock, L., Trahey, G. E., and Smith, S. W. (1989). Phase aberation correction in medical ultrasound using speckle brightness as a quality factor. *J. Acoust. Soc. Am.* **85**, 1819–1833.

O'Donnell, M. (1988). Applications of VLSI circuits to medical imaging. *Proc. IEEE.* **76**, 1106–1114.

O'Donnell, M., and Flax, S. W. (1988). Phase-aberation correction using signals from point reflectors and diffuse scatterers: measurements. *IEEE Trans. UFFC* **35**, 768–774.

O'Neil, H. T. (1949). Theory of focusing radiators. *J. Acoust. Soc. Am.* **21**, 516–526.

Oakley, C. G., and Marsh, P. (1992). Development of 1-3 ceramic-air composite transducers. *In* "New Developments in Ultrasonic Transducers and Transducer Systems," Vol. 1733 (F. L. Lizzi, ed.). SPIE Proceedings, Bellingham, pp. 274–283.

Ophir, J., and Goldstein, A. (1977). The principles of digital scan conversion and its application to diagnostic ultrasound. *In* "Ultrasound in Medicine," Vol. 3B (D. White and R. E. Brown, eds.). Plenum Press, New York, pp. 1707–1713.

Ophir, J., and Maklad, N. F. (1979). Digital scan converters in diagnostic ultrasound imaging. *Proc. IEEE* **67**, 654–664.

Park, S. B., and Lee, M. H. (1984). A new scan conversion algorithm for real-time sector scanner. *In* "1984 IEEE Ultrason. Symp." IEEE, New York, pp. 723–727.

Persson, H. W. (1981). Electric excitation of ultrasound transducers for short pulse generation. *Ultrasound in Med. & Biol.* **7**, 285–291.

Peterson, D. K., and Kino, G. S. (1984). Real-time digital image reconstruction: a description of imaging hardware and an analysis of quantization errors. *IEEE Trans. Son. & Ultrason.* **SU-31**, 337–351.

Physical chemistry, composition of blood, hematology, somatometric data (1984). *In* "Geigy Scientific Tables" (C. Lentner, ed.). CIBA-Geigy Limited, Bask, Switzerland, pp. 205–207.

Powis, R. L. (1986). Angiodynography, a new real-time look at the vascular system. *Appl. Radiol.* **15**, 55–59.

Powis, R. L. (1994). Color flow imaging. *Radiographics* **14**, 415–428.

Powis, R. L., and Schwartz, R. A. (1991). "Practical Doppler Ultrasound for the Clinician." Williams & Wilkins, Baltimore.

Quistgaard, J. U. (1997). Signal acquisition and processing in medical diagnostic ultrasound. *IEEE Signal Processing Magazine* **14**, 67–74.

Ranalli, R. (1975). U.S. Patent 3 964 661, February 1975.

Robinson, A. L., and Mo, J. H. (1992). Application of microelectronics and microfabrication technology to ultrasound imaging systems. *In* "Proc. 1992 IEEE Ultrason. Symp." IEEE, New York, pp. 681–691.

Roederer, G. O., Langlois, Y., and Strandness, D. E. (1984). Comprehensive noninvasive evaluation of extracranial cerebrovascular disease. *In* "Noninvasive Diagnosis of Vascular Disease" (F. B Hershey, R. W. Barnes, and D. S. Sumner, eds.). Appleton Davies, Inc., Pasadena, California.

Rose, A. (1973). "Vision, Human and Electronic." Plenum Press, New York.

Rubin, J. M., Adler, R. S., Fowlkes, J. B., Spratt, S., Pallister, J. E., Chen, J. F., and Carson, P. L. (1995). Fractional moving blood volume estimation with power Doppler imaging. *Radiology* **197**, 183–190.

Schafer, M. E., and Lewin, P. A. (1984). The influence of front-end hardware on digital ultrasonic imaging. *IEEE Trans. Son. & Ultrason.* **SU-31**, 295–306.

Schorum, S., and Fidel, H. (1977). The Pho-Sonic SM: a computer controlled image forming system. *In* "Ultrasound in Medicine," Vol. 3B (D. White and R. E. Brown, eds.). Plenum Press, New York, pp. 1433–1441.

Schrope, B., and Newhouse, V. L. (1993). Second harmonic blood perfusion measurement. *Ultrasound in Med. & Biol.* **19**, 567–579.

Sears, F. W., and Zemansky, M. W. (1957). "University Physics," 2nd Edition. Addison-Wesley, Reading, Massachusetts.

Selfridge, A. R., Kino, G. S., and Khuri-Yakub, B. T. (1980). A theory for the radiation pattern of a narrow-strip acoustic transducer. *Appl. Phys. Lett.* **37**, 35–36.

Shattuck, D. P., Weinshenker, M. D., Smith, S. W., and von Ramm, O. T. (1984). Explososcan: a parallel processing technique for high speed ultrasonic imaging with linear phased arrays. *J. Acoust. Soc. Am.* **75**, 1273–1282.

Smith, S. W., Pavy, H. G. Jr., and von Ramm, O. T. (1991). High-speed ultrasound volumetric imaging system—part 2: parallel processing and image display. *IEEE Trans. UFFC* **38**, 109–115.

Smith, S. W., Wagner, R. F., Sandrik, J. M., and Lopez, H. (1983). Low contrast detectability and contrast/detail analysis in medical ultrasound. *IEEE Trans. Son. & Ultrason.* **30**, 164–173.

Smith, W. A. (1992). New opportunities in ultrasonic transducers emerging from innovations in piezoelectric materials. *In* "New Developments in Ultrasonic Transducers and Transducer Systems," Vol. 1733 (F. L. Lizzi, ed.). SPIE Proceedings, Bellingham, pp. 3–26.

Snyder, R. A. (1989). Ultrasonic imaging system utilizing two or more simultaneously active apertures. European patent # 0.335.578 A2.

Steen, E., and Olstad, B. (1994). Volume rendering of 3D medical ultrasound data using direct feature mapping. *IEEE Trans. Med. Imag.* **13**, 517–525.

Steinberg, B. D. (1976). "Principles of Aperture and Array System Design." John Wiley & Sons, New York.

Steinberg, B. D. (1992). Digital beamforming in ultrasound. *IEEE Trans. UFFC* **39**, 716–721.

Sumino, Y., and Wagg, R. C. (1991). Measurements of ultrasonic pulse arrival time differences produced by abdominal wall specimens. *J. Acoust. Soc. Am.* **90**, 2924–2930.

't Hoen, P. J. (1982). Aperture apodization to reduce the off-axis intensity of the pulsed-mode directivity of linear arrays. *Ultrasonics* **20**, 231–236.

Thomenius, K. E. (1995). Instrumentation for B-mode imaging. *In* "Medical CT and Ultrasound: Current Technology and Applications" (L. W. Goldman and J. B. Fowlkes, eds.). AAPM, College Park, Maryland, pp. 67–83.

Thomenius, K. E. (1996). Evolution of ultrasound beamformers. *In* "1996 IEEE Ultrason. Symp." IEEE, New York, pp. 1615–1622.

Thurston, F. L., and von Ramm, O. T. (1973). A new ultrasound imaging technique employing two-dimensional electronic beam steering, *In* "Acoustical Holography," Vol. 5 (N. Booth, ed.). Plenum Press, New York, pp. 249–259.

Thurston, R. N. (1960). Effect of electrical and mechanical terminating resistances on loss and bandwidth according to the conventional equivalent circuits of a piezoelectric transducer. *IRE Trans. Ultrason. Eng.* **UE-7**, 16–25.

Torrence, K., and Ranalli, R. (1975). U.S. Patent 3 864 660, February 1975.

Trahey, G. E., Hubbard, S. M., and von Ramm, O. T. (1988). Angle independent ultrasonic blood flow detection by frame-to-frame correlation of B-mode images. *Ultrasonics* **26**, 271–276.

Turner, J. A., and Weaver, R. L. (1995). Time dependence of multiply scattered diffuse ultrasound in polycrystalline media. *J. Acoust. Soc. Am.* **97**, 2639–2644.

von Ramm, O. T., and Smith, S. W. (1978). A multiple frequency array for improved diagnostic imaging. *IEEE Trans. Son. & Ultrason.* **SU-25**, 340–345.

von Ramm, O. T. and Smith, S. W. (1983). Beam steering with linear arrays. *IEEE Trans. Biomed. Eng.* **BME-30**, 438–452.

von Ramm, O. T., and Thurstone, F. L. (1976). Thaumascan: improved image quality and clinical usefulness. *In* "Ultrasound in Medicine," Vol. 2 (D. White and R. Barnes, eds.). Plenum Press, New York, pp. 463–464.

Wagner, R. F., Insana, M. F., and Smith, S. W. (1988). Fundamental correlation lengths of coherent speckle in medical ultrasonic images. *IEEE Trans. UFFC* **35**, 34–44.

Wagner, R. F., Smith, S. W., Sandrik, J. M., and Lopez, H. (1983). Statistics of speckle in ultrasound B-scans. *IEEE Trans. Son. & Ultrason.* **30**, 156–163.

Ward, B., Baker, A. C., and Humphrey, V. F. (1997). Nonlinear propagation applied to the improvement of resolution in diagnostic medical ultrasound. *J. Acoust. Soc. Am.* **101**, 143–154.

Wells, P. N. T. (1974). The receiver in the pulse-echo system. *In* "Ultrasonics in Medicine" (M. de Vlieger, D. N. White, and V. R. McCready, eds). Exerta Medica, Amsterdam, pp. 30–36.

Wells, P. N. T. (1977). "Biomedical Ultrasonics." Academic Press, London.

Whittingham, T. A. (1991). Resolution and information limitations from transducer arrays. *Phys. Med. Biol.* **36**, 1503–1514.

Wildes, D. G., Chiao, R. Y., Daft, C. M. W., Rigby, K. W., Smith, L. S., and Thomenius, K. E. (1997). Elevation performance of 1.25D and 1.5D transducer arrays. *IEEE Trans. UFFC* **44**, 1027–1037.

Wilhelmij, P., and Denbigh, P. (1984). A statistical approach to determining the number density of random scatterers from backscattered pulses. *J. Acoust. Soc. Am.* **76**, 1810–1819.

Woodward, P. M. (1953). "Probability and Information Theory, with Applications to Radar." McGraw-Hill, New York.

Yokoi, H., and Ito, K. (1972). Ultrasonic diagnostic equipment with color display unit for simultaneous tomogram method. *Toshiba Rev.* **76**, 13–21.

Yokoi, H., and Ito, K. (1973). Computer aided ultrasonic diagnosis equipment for simultaneous tomogram method. *Toshiba Rev.* **84**, 1–12.

Zemanek, J. (1971). Beam behavior within the nearfield of a vibrating piston. *J. Acoust. Soc. Am.* **49**, 181–191.

— 3 —

Nondestructive Testing

EMMANUEL P. PAPADAKIS, PH.D.

Quality Systems Concepts, Inc., 379 Diem Woods Drive, New Holland, PA 17557-8800

ISBN 0-12-477923-9 $30.00

I. Introduction and Orientation

Nondestructive testing (NDT) is defined loosely as all the methods of testing an object to ensure that it is fit for service without damaging it and making it unfit for service. The presupposition is that certain classes of mechanisms that would make an object unfit for service can be detected by nondestructive applications of physics embodied in electronic devices.

Nondestructive testing is an amalgam of three inseparable aspects: methods, instruments, and intelligence. Methods are developed by intelligent people using theory and instruments for experiments. Then tests based on the methods are carried out either by people using instruments or by automated systems. Intelligence is required for the interpretation of the output of the instruments. The intelligence may be supplied directly by a certified operator (ASNT, 1988) or indirectly by artificial intelligence "trained" by a certified operator (Papadakis and Mack, 1997). Since expendable materials and devices are used, NDT also can be described in the quality sense as a process incorporating the Four Ms: men, materials, methods, and machines (Scherkenbach, 1986).

The term *nondestructive testing* is used in this chapter because this term describes what is actually done in the real world. Other terms, such as *nondestructive evaluation* and *nondestructive inspection* are also used. (The vocabulary can change as rapidly as the philosophy of management. For example, at one organization "testing" became an unacceptable term to a new corporate vice-president who thought the term was routine, repetitive, and not good enough for his research and development directorate. To accommodate his outlook, the name of this department was changed from "Nondestructive

Testing" to "Nondestructive Inspection Technology" because "technology" fit his image and "Inspection" seemed a more sophisticated term for the function the technology was to perform. A few years later, the Deming philosophy (Deming, 1982) was adopted and "inspection" became absolutely taboo in deference to one of Dr. Deming's 14 Points. The name of the group was changed again, eliminating "inspection.") However, nondestructive testing is the recognized generic nomenclature, consistent with the name of the leading technical society in this field, the American Society for Nondestructive Testing.

In the field of nondestructive testing, ultrasonics is commonly thought of as a subfield or a method (a family of methodologies). In other words, ultrasonic testing (UT) is one of the Big 5 testing methods recognized by the nondestructive testing profession, along with other methods or specialized fields. (The other fields and methods are growing so that in the not-too-distant future there may be a Big 7 or a Big 11; change is the only constant. In particular, acoustic emission (AE) as a method has always been treated separately from UT because it arose and matured later and was passive instead of active with respect to radiant mechanical energy in the ultrasonic range. [Some practitioners claim it is not nondestructive because AE arises from crack growth under excess stress.]) Thus it requires a reorientation of thinking to speak of nondestructive testing as a subheading under ultrasound instead of the inverse.

Nevertheless, nondestructive testing is one of the disciplines or professions that utilizes ultrasound. In the context of commercially successful instruments and devices, nondestructive testing has been the beneficiary of many ideas common to the field of ultrasound. Special-purpose developments and technology transfer to manufacturers in the NDT field have resulted in the commercialization of these common ideas (and some special ones) into salable items.

The manufacturers are generally of two types: (1) large corporations where NDT has been a necessary sideline resulting in some special salable product, and (2) relatively small NDT specialty companies.

The customers are the users of NDT equipment. These users can be categorized into four broad types: (1) manufacturers using NDT methods to ensure the quality of the manufactured product, (2) users of the manufactured product for maintenance or periodic inspection, (3) NDT service companies working for either of the above or government units, and (4) others such as university or government laboratories. Some users require much equipment—one engineer at an aircraft manufacturer (category #1) claimed in 1971 to have $4,000,000 worth of ultrasonic transducers (sensors, search units) in an array of drawers in a laboratory area and still needed some of a particular external shape to fit into a groove in a particular aircraft part. Other users, of course, get along on a bare-bones budget.

NDT is analogous to medical diagnostic ultrasound: Medical diagnostic ultrasound is NDT of the living body. As such, medical ultrasound is much better known to the general public than is NDT. People may have their own bodies tested but not realize that the brake calipers in the cars they drive and the wing spars of the planes they fly in are tested also. The customer of the medical manufacturer may be the hospital, but the visibility of the doctor to the medical end user is much higher than, for instance, the visibility of the technician in the hangar of the major airline. On the other hand, NDT can get beyond the NDT customer to be seen by end user: In 1967 a mechanic in a car dealership used a dye penetrant (another NDT Big 5 Method) to prove that the cylinder head of my car was cracked. In short, NDT does for airplane wings what a bite wing does for your teeth—finds the holes. Simplistic, but expressive of part of reality.

What is the range of what can be done, and hence what is the range of instruments and devices that have come to commercialization? These questions are addressed in the next section.

II. Principles of NDT

A. General View of Ultrasonics in NDT

As mentioned previously, nondestructive testing (NDT) comprises all methods of testing an object to ensure that it is fit for service without damaging it and making it unfit for service. Ultrasonics and other regimes of physical acoustics such as sonic resonance provide the methods considered in this chapter. The presupposition is that a class of mechanisms that would make an object unfit for service can be detected by nondestructive applications of physical acoustics embodied in electronic devices, instruments, and systems.

Ultrasonics, of course, is sound above the range of human hearing. Ultrasound in NDT is an active radiation method, meaning that there is a source of ultrasound sending ultrasonic energy into the object to be tested. The ultrasonic radiation is then received, at least in part, by a receiver after traversing the object in a preassigned path. The resulting sequence of signals is displayed or processed for some kind of synthetic display or decision mechanism.

B. Production and Reception of Ultrasound

Consider the most generic type of ultrasonic radiating element, a piezoelectric plate with electrodes on both sides. (Other types are also considered in Chapter 2, Vol. 24.) It may be typically 0.5 inches (1.27 cm) in diameter and several thousandths of an inch (a fraction of a millimeter) thick. The thickness defines half a wavelength of the ultrasound to be generated if the plate vibrates in a free-free bulk mode. The wavelength is in the material of the piezoelectric plate, of course, and is related to the ultrasonic frequency f and the ultrasonic velocity v in the piezoelectric material by

$$\lambda = v/f \tag{1}$$

The piezoelectric plates are either cut from piezoelectric crystals or formed from ferroelectric ceramics that are poled (electrically polarized) in the proper directions. The useful cuts and directions are specified for two types of waves, longitudinal and shear (transverse). Longitudinal plates vibrate with particle motion in the thickness direction and generate longitudinal waves propagating normal to their major faces. (See Figure 1.) Shear plates, on the other hand, vibrate with particle motion in one direction in the plane of the major faces and generate shear waves also propagating normal to their major faces. (See Figure 2.) To produce ultrasonic beams from such plates, the lateral dimensions must be many wavelengths. Perusal of Figures 1 and 2 will indicate that there are some shear forces at the perimeter of the longitudinal plates and some pressures at the perimeter of the shear plates to satisfy the clamped boundary conditions. In practice, these are of minor consequence. For more details concerning piezoelectricity and piezoelectric plates, see Berlincourt *et al.* (1964), Cady (1946), IEEE (1987), Jaffe and Berlincourt (1965), Jaffe *et al.* (1971), Mason (1950), Mattiatt (1971), and Meeker (1996).

Piezoelectric elements are reciprocal. An applied voltage generates a deflection, and an impinging stress generates a voltage. This physical condition leads to the use of piezoelectric elements, typically plates, as transducer from electrical signals to stress signals (waves) and from stress waves to electrical signals. In other words, the piezoelectric elements can be used as transmitters and receivers for stress waves. Lindsay (1960) has labeled this subject of useful stress waves "mechanical radiation." In NDT, the term *transducers* refers to piezoelectric plates with backing and frontal elements to modify their vibration characteristics. These assemblies are potted inside cases to protect them and to provide a means for gripping them by hand or for mounting them in systems. These potted transducers are sometimes referred

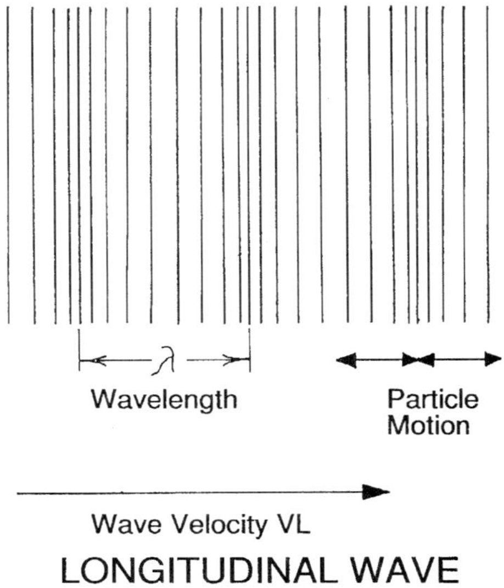

FIG. 1. Longitudinal wave directions of propagation and particle motion. The strain is actually of the order of 1/1,000,000.

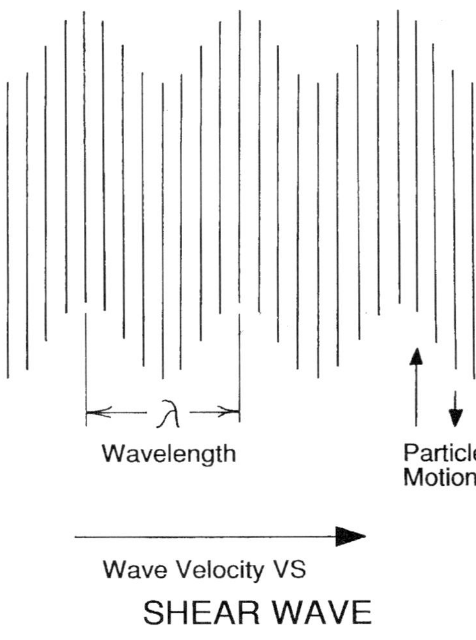

FIG. 2. Shear wave directions of propagation and particle motion. The strain is actually of the order of 1/1,000,000.

to as *search units*, although this nomenclature is disappearing from use. Transducers will be treated more thoroughly in Section III.D Chapter 2, Vol. 24, which is devoted to their analysis.

Piezoelectric plates many wavelengths in diameter generate beams of ultrasound when they are caused to vibrate by an electric field applied between their electrodes. The beams are not confined to cylinders but spread because of the finite size of the plate source (Roderick and Truell, 1952; Seki, Granato, and Truell, 1956; Papadakis, 1959, 1963, 1964, 1966, 1971, 1972, 1975; Papadakis and Fowler, 1971; Benson and Kiyohara, 1974; ASNT 1959, 1991). Sometimes the spreading is useful and sometimes it is deleterious. The spreading can be corrected for, sometimes rigorously and sometimes approximately. In NDT, the amplitude of signals is sometimes corrected for distance approximately by a factor called ADC, the amplitude distance correction. As in rigorously computed beam-spreading (ultrasonic diffraction from single apertures) corrections, the ADC depends on frequency, distance, piezoelectric plate diameter, and the velocity in the material supporting propagation. The ADC is electronically built into flaw detection instruments, which will be treated in Section III.B.

C. Instruments and Scan Displays

A piezoelectric transducer attached to a transmitting and receiving instrument is shown schematically in Figure 3. A wave packet a few wavelengths long generated by the pulser is shown traveling as an idealized nonspreading beam in an idealized workpiece. The wave packet will travel to the back face of the workpiece and be reflected back to the transducer. Any discontinuities in the beam area will produce other reflections. All these will be received, amplified, and displayed.

1. *A-Scans*

A modern instrument may be designed with a computer for data analysis and storage. The display of a generic instrument such as this would be a plot of the amplitudes of the received echoes on the *Y*-axis versus time subsequent to the input pulse on the *T*-axis. This is known generically as an A-Scan. Time on the *T*-axis is proportional to the propagation distance z in the workpiece. The constant of proportionality is the ultrasonic velocity. Diagrammatic representations of such displays are shown in Figure 4. If this velocity is known or assumed, the A-Scan can be converted electronically into a thickness gage.

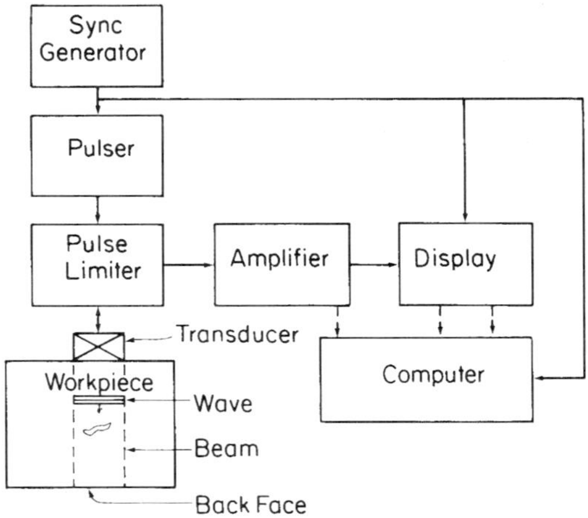

FIG. 3. Diagram of an idealized ultrasonic instrument to generate, transmit, receive, and display ultrasonic signals in a workpiece. (From E. P. Papadakis, "Ultrasonic Instruments for Nondestructive Testing," *Encyclopedia of Acoustics*, M. J. Crocker, editor. Copyright © 1997 John Wiley & Sons, Inc. Reprinted by permission of John Wiley & Sons. Inc.)

2. *B-Scans*

In the B-Scan, the time on the *T*-axis is still proportional to the distance z in the workpiece. The distance along the *Y*-axis of the scan is proportional to distance y in the workpiece normal to the propagation direction z. This dimension is achieved by moving the transducer laterally or sweeping it in an arc or using a phased array of transducer elements. The display is a gray scale with brightness proportional to the amplitude of any reflections occurring at that y-z coordinate in the workpiece. The B-scan, often used in medicine, was described at length in Chapter 2.

3. *C-Scans*

In the C-scan, the transducer is swept rapidly back and forth in the x-direction while being stepped in the y-direction after each sweep. The repetition rate of the pulses is rapid compared with the x sweep rate. This method is most often used in a tank of liquid, as shown diagrammatically in Figure 5. The x-y display is a gray scale with amplitude proportional to the reflection amplitudes within the specimen in the liquid tank. The returning signals are preprocessed by a gate such that the large reflections at the front and back surfaces are

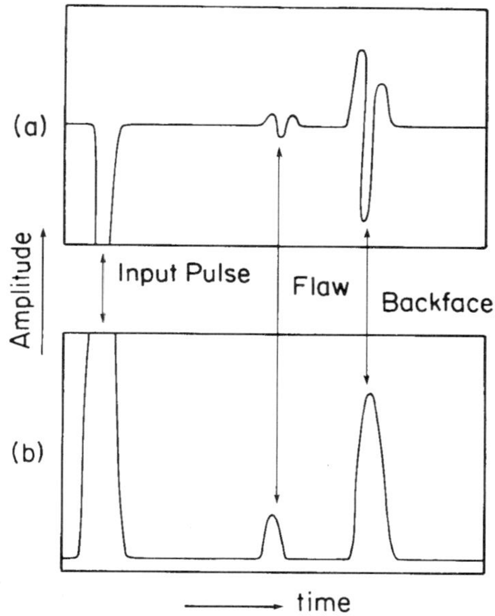

FIG. 4. Display of signal amplitude versus time as might be seen on a flaw detection instrument analog display. (a) rf representation of a short broadband pulse. (b) Rectified and detected representation. This configuration of amplitude versus time is called an A-scan. (From E. P. Papadakis, "Ultrasonic Instruments for Nondestructive Testing," *Encyclopedia of Acoustics*, M. J. Crocker, editor. Copyright © 1997 John Wiley & Sons, Inc. Reprinted by permission of John Wiley & Sons, Inc.)

eliminated from the display. C-scans may be either reflection-type (as described) or through-transmission. In through-transmission, a collinear receiving transducer is swept past the back face of the specimen. C-scans will be described further in Sections III.E and III.F.

D. Sonic Resonance

Sonic resonance is a technique in which the sonic or ultasonic energy is imparted to a workpiece and the transducer is then used only as a receiver. The workpiece vibrates at its natural frequencies, which are altered by deleterious conditions in the workpiece. The analysis of the resonance frequencies permits the detection of nonconforming parts. These systems will be treated in Section III.H despite the fact that discontinuities can be detected by resonance also.

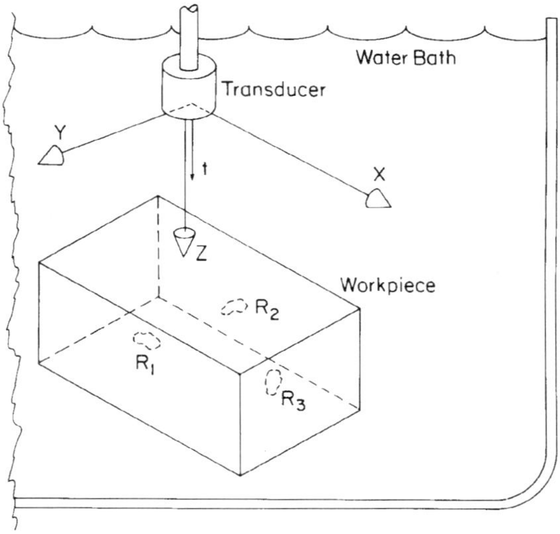

FIG. 5. Generic sketch of a C-Scan. The transducer gantry sweeps back and forth along *X* while stepping in short increments along *Y* at each sweep. The pulse repetition rate and the ultrasonic velocities permit complete coverage of the interior of the part during this sweep sequence. Electronic gates within the receiver stage permit the selective viewing of the interior of the part without interference from its surfaces. (From E. P. Papadakis, "Ultrasonic Instruments for Nondestructive Testing," *Encyclopedia of Acoustics*, M. J. Crocker, Editor. Copyright © 1997 John Wiley & Sons, Inc. Reprinted by permission of John Wiley & Sons, Inc.)

E. Two Times for Testing in a Product's Life Cycle

It is necessary to recognize that there are two distinctly different times in a product's life cycle when testing may be performed: during manufacture and after it has been put to use. In addition, the testing may be quite different in the two time frames.

1. *Manufacture*

While the product is being manufactured, the testing may be of incoming stock, partially fabricated parts, or completed objects. The testing may be motivated by cost considerations, by previous experience, or by "what if" thought experiments (sometimes known as Failure Mode and Effect Analyses or FMEAs (Ford Motor Company, 1979)). Sometimes the tests are formalized to optimally meet government requirements such as (but not limited to) the HMVSS (Highway Motor Vehicle Safety Standards) series. To assure that minimal amounts of nonconforming product reach users, NDT and statistical process control (SPC) are both used. The purpose of NDT and SPC in such safety situations is to attempt to limit any occurring field failure to being unique in the universe. Basically, the manufacturer is required to do whatever a reasonable person would do to assure that no such flaws reach the field.

However, everyone is familiar with recalls of such products as automobiles or jet engines to fix possible flaws in families of parts. Sometimes NDT can be used to find and salvage good parts out of such families of parts during recalls. Although this use of NDT is on manufacturing flaws, it is done after the parts had been put into use.

2. *Maintenance*

After the product has been put into use, testing may occur at any point over the entire lifetime of the product. Some products survive without testing whereas others must be tested periodically for certain types of deleterious conditions. NDT contributes to such testing. In fact, such testing during the lifetime of a product is the better-known application of NDT in contradistinction to the testing during manufacturing. At the present time, there is a heavy emphasis on extending the useful lifetime of objects beyond their design lifetimes, using NDT to ascertain whether deleterious conditions exist that could be repaired.

F. Two Types of Deleterious Conditions

It is also necessary to recognize that there are two entirely different types of deleterious conditions that could make a part unfit for service: discontinuities and inadequate material properties. The NDT approach for the two types of conditions are radically different.

1. *Discontinuities*

The most prevalent and dangerous type of discontinuity is a crack, which could grow under cyclic stress and cause a part to rupture during use. Another type is a hole below the surface of a material, unseen in the feed stock, which might interfere with the operation of a part after machining. A hole might accidentally connect two drilled fluid passages, mixing oil and water, for instance. Or a hole might be opened up when a surface is machined, making that wear surface inadequate for its intended purpose. Still another type of discontinuity is a localized subsurface solid anomaly that could cause damage. "Hard Alpha" in titanium is one example that can lead to cracks in forged parts used in high-stress environments. Another example is hard regions such as carbides or sand inclusions in castings, which can destroy machine tool bits. (Carbides are sometimes detected by methods applicable to material properties. See below.) The possible list of discontinuities is very long, and the motivations for finding them vary from barely economical cost savings to destruction or even death avoidance.

2. *Inadequate Material Properties*

The other deleterious condition is inadequate material properties. Engineers are fully familiar with yield strength, ultimate tensile strength, hardness as measured by penetrators, case depth of deliberately surface-hardened parts, wear resistance, impact breaking energy, and a host of other properties. One or more properties are specified for parts to make them capable of the desired performance in the mechanism in which they are designed to function. If the properties in the fabricated parts are inadequate, the parts will not perform adequately. If the properties of the feed stock are inadequate, then resources will be wasted processing it. Standard old-line tests for all these material properties are destructive (as with tensile machines) and therefore must be statistical only; it is not permissible to break all of production.

Fortunately, nondestructive testing methods have been discovered and developed to measure certain material properties of some very important materials. The methods involve the relationship between some parameter that is measurable by nondestructive means and the material property of interest. In the case of ultrasound and sonic resonance, the parameter could be the ultrasonic velocity, the ultrasonic attenuation, the resonance frequency, the damping, or possibly the backscattering amplitude on an A-scan. The "relationship" may be either a true mathematical function or a correlation. The capabilities of nondestructive testing in the realm of the measurement of material properties have been utilized in some industries but are not widely known by the engineering community at large. However, in many cases, the nondestructive testing method may be more accurate than the old-line test methods, against which it is forced to compete by the engineering establishment. Where NDT is used for material properties, the parts can be measured 100% by electronic instruments that are compatible with automated parts handling and decision making. The decisions can be made by simple Go/No-Go circuitry or by sophisticated artificial intelligence (AI). In addition, testing can be done at any convenient step in the production process.

An example of an easily understood functional relationship is the measurement of yield strength in steel by Brinnell Hardness Number (BHN) indentation with a hardened steel ball (Lysaght, 1949). The BHN test is considered nondestructive if it can be performed on a part in a location that will not present problems such as a stress riser, a bump in a bearing surface, or a leak under a seal. The BHN test is performed by applying a weight to the ball touching the horizontal surface of the workpiece. The workpiece yields, leaving a shallower circular crater or "caldera." The diameter of the caldera for a given weight is controlled by the yield strength of the workpiece, so the BHN number derived from this diameter by a single-valued function is functionally related to the yield strength.

G. Test Methods and Criteria

1. Flaws

a. General. In industry, the goal is to find flaws that will cause detrimental effects within the next time period. This "time period" may be the time until the next scheduled inspection or the entire lifetime of the part if the part is designed to last longer than the mechanism it will be in. Several NDT methods are available for achieving this goal.

b. Reflection. The most common test for flaws is reflection. In Figure 3, a flaw was shown diagrammatically in the beam path. This flaw would provide a reflection on an A-scan as in Figure 4. A variant is *pitch-and-catch*, in which two transducers are almost collinear, aimed at a region within a part, and then one transmits and the other receives. The amplitude of the flaw reflection as well as some other characteristics (such as its frequency spectrum) could be used to approximate its size, shape, and severity. Learning programs in artificial intelligence can be applied to such problems.

c. Through-Transmission. With transducers placed on both sides of a part and aimed at each other along a common axis, one can perform *through-transmission* to find flaws. The signal passes through the part directly and only once. Reflections are disregarded. The amplitude of the transmitted signal is reduced by the presence of flaws. *Through-transmission* principally to find disributed flaws such as porosity in castings and poorly bonded regions in fiber-reinforced composites. The method is useful where a reflected signal could be noisy due to the presence of many nondeleterious reflectors such as the grains in a metal or the fibers in a composite.

d. Acousto-Ultrasonics. This method is a variant of through-transmission. The two transducers need not be aimed directly at each other. The signal to be analyzed is not just the first-arrival pulse from the input pulse but also includes multiple internal reflections, longer paths due to beam spreading and refraction, and so on. See Duke (1988) for many examples of this relatively modern technique pioneered by Vary (1978).

e. Acoustic Emission. This technique relies on the reception of signals generated within a material or structure when stresses are applied. The stresses may cause cracks to propagate releasing strain energy or parts to move with attendant friction noise. If the stress is pressure, a leak may emit high-frequency sound.

f. Inverse Problem. The "forward problem" of finding the characteristics of a reflection from a known shape are mathematically difficult but tractable. However, the "inverse problem" of finding the shape of a reflector from its reflections, even at many angles of incidence, is still basically unsolved. The question of corrective action in the presence of a reflection from a potential flaw that may or may not be of a detrimental size or shape is still a management decision rather than a purely scientific one.

g. Probability of Detection. Given a particular method such as reflection and a set of parameters such as ultrasonic frequency, part size, part shape, and material attenuation, reflectors of a relatively small size will not be detected at all while reflectors of a relatively large size will definitely be detected. An intermediate range may be detected or missed. The intermediate range becomes probabilistic. The term for the detectability of a reflector versus its size is *probability of detection* (POD). As the probability of detection itself has errors, one must express the probability desired as, for instance, the probability of detecting 90% of the flaws of this size 95% of the time. As in all statistical determinations, there is a probability of calling a nonconforming part good (missing the flaw) and a probability of calling a conforming part bad (interpreting the signal from a benign reflector as serious). A generic curve for probability of detection is given in Figure 6. The probabilistic nature is shown by the inclusive curves one standard deviation away from the main function.

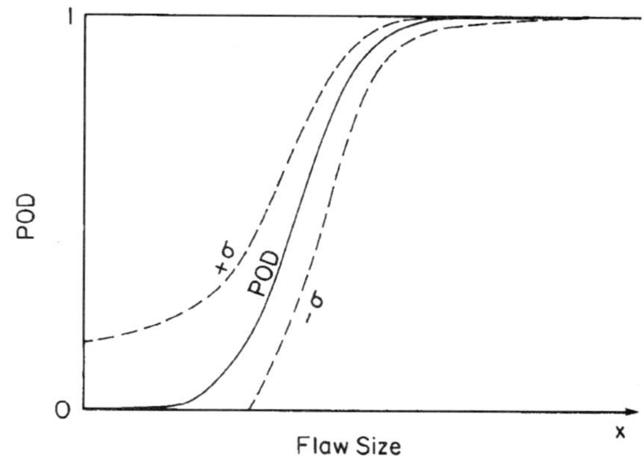

FIG. 6. Generic diagram of probability of detection (POD). The probability is plotted versus flaw size for the test parameters being used. A reject level is chosen to allow the acceptance of a minimal number of nonconforming parts while permitting the rejection of many more conforming parts. The former is not desirable from a performance point of view; the latter is not desirable from a cost point of view.

Suppose that the critical flaw size has been determined for the part in use. A larger flaw will fail the part before the next inspection. A smaller flaw probably will permit the part to survive until the next inspection. (Some specifications require that the part last through two inspection periods in case the flaw is missed the first time.) In Figure 7, the critical flaw is superimposed upon the probability of detection curve (Papadakis, 1992). An "ideal" technique would have a step function POD, the vertical dashed line. The real technique produces the two shaded areas. Area FA is false accepts (where nonconforming parts are falsely accepted by the test) and area FR is false rejects (where good parts are falsely rejected by the test). These areas could be potentially augmented if one were to take into consideration the probabilistic nature of the reliability of the POD curve as in Figure 6. The ramifications of FA and FR are as follows: FA parts are likely to fail before the next inspection, producing danger and undue expense. FR parts produce the expense of early, unneeded repair.

The ratio of FA to FR can be adjusted by developing a more sensitive test (to lower FA) or by recalibrating to yield less sensitivity (to lower FR). Management generally wishes to minimize FR within a set of constraints placed upon FA. If expense were the only criterion, then the adjustment would be simple. However, FA frequently entails danger to humans and/or regulations by government. Note all the MIL SPECS, HMVSS rules, NRC

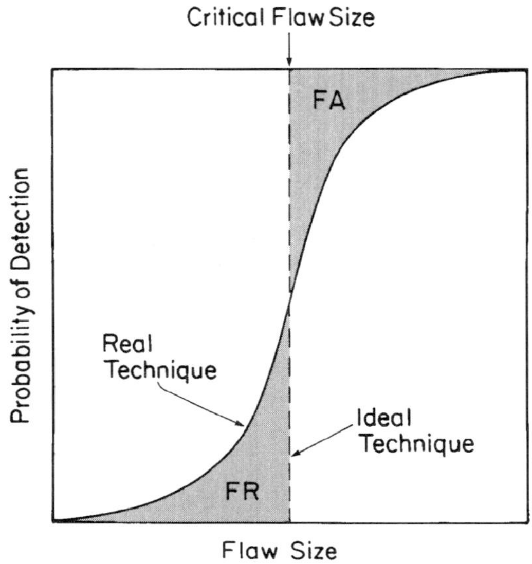

FIG. 7. Probability of detection curve with the critical flow size shown. The net result is that some nonconforming parts are falsely accepted (FA) and some good parts are falsely rejected (FR). The ratio of FA to FR can be adjusted by increasing or decreasing the sensitivity of the test, if possible. (*Materials Evaluation*, 1992. Used by permission.)

regulation, FARs from the FAA, and so on. Also note the well-known examples of catastrophes: For example, the crash of United Air Lines Flight 232 in 1989 near Sioux City, Iowa, was caused by a flaw in a titanium alloy engine turbine disc. Research is still underway to solve this problem from the metallurgical angle and from the NDT angle—ultrasound is a prime candidate for the latter. For another example, the sinking of the nuclear submarine USS Thresher was caused by faulty brazing of piping leading to the exterior of the hull. After the fact, John E. Bobbin of Branson Instruments demonstrated to the Navy that ultrasonic inspection of such brazed pipe could have detected the poor brazing (Bobbin, 1974).

2. *Material Properties*

a. Methods. In general, there are four types of ultrasonic measurements used for measuring material properties:

1. *Ultrasonic Velocity.* This is the most common and most successful method. Velocity, being a function of the moduli of elasticity and the

density, correlates with (or may be a function of) many changes in materials. As these changes are brought about by processes applied to the materials, the resulting velocities can be used to monitor the processes and ensure the quality of the outputs. In some cases the changes in ultrasonic velocity are major, leading to definitive tests. An outstanding example is the use of velocity to quantitatively measure the graphite shape in cast iron (Papadakis, 1984). In other cases the changes in ultrasonic velocity may be so small that other fluctuations may mask the ultrasound results.

2. *Sonic Resonance.* Sonic resonance has recently been gaining wider acceptance. The resonances in a body depend on size, shape, the moduli, and the density. Sonic resonance shows the same type of results as ultrasonic velocity, but also gives information on flaws that may change the resonance frequencies. Although it is claimed that resonance methods test the whole body whereas velocity methods only test the small volume in the ultrasonic beam, the resonance test is not uniform; it is weighted toward the regions of the antinodes in strain where the strain is maximum.
3. *Ultrasonic Attenuation.* This method is sensitive to the grain size in polycrystalline solids and to the structure within the grains of metals. Attenuation in this regime is principally a scattering problem. Because of difficulties in making measurements in the factory regime, simultaneous competing causes of attenuation, and cost factors, the attenuation method has not found wide acceptance. One notable exception is the ultrasonic test for spot weld integrity (Mansour, 1988). (Transducer innovation was particularly critical to this test; see Section III.D.3.) Another exception is the use of attenuation to detect sludge in water streams in sewage treatment plants for the control of pumping cycles (Envirotech, 1973, 1974). Other uses of attenuation are discussed in Chapter 4. Interestingly, the velocity method for assuring the nodularity of graphite in ductile cast iron was discovered serendipitously while engineers were trouble-shooting an attenuation test for the nodularity circa 1960 (Torre, 1986).
4. *Ultrasonic Backscattering.* In polycrystalline metals, the scattering from the grains depends on the grain size and the crystalline substructure of the grains. When either parameter is changed by a process, the scattering changes. Scattering shows up as noise on the baseline of an A-Scan, for instance. Mathematical methods can be used to analyze this

noise for information. One straightforward example is the low scattering from an induction-hardened case on steel where the core is soft or normalized and presents much higher scattering. The noise beginning on the A-Scan baseline as the wave enters the core permits a measurement of the case depth. While this is true in principle, the measurement is not used because eddy current response in steel provides a better measurement of case depth and hardness (Stephan, 1983).

b. Correlations and Functions. This section gives fundamental background information on certain relationships that can exist between NDT-measurable variables and engineering parameters (Papadakis, 1993a). Often certain engineering parameters are defined by destructive measurements while the engineering community would like to be able to measure the parameters nondestructively. Other times the engineering parameters have been found to relate to certain measurables so that the NDT measurement is to be the third in a string of parameters. Take, for example, the yield strength of nodular iron, which has been found to be predictable from optical measurements of the degree of nodularity. The engineering community wanted an NDT measurement of the degree of nodularity, so ultrasonic velocity testing was developed as step 3 to correlate with the optical measurement, step 2, which then correlated with yield strength, step 1, the parameter of primary interest. More information on this problem will be given later, but it is mentioned here to lead into the discussion of the use of correlations. Although there is no such instrument sold as "The Ultrasonic Correlator," this discussion is vital to the understanding of some other instruments and measurements examined in this chapter.

As the reader knows from algebra, a function is a relationship in which

$$y = f(x). \tag{2}$$

Over the domain of x, there will be defined values for y over a range. One can say that y is caused by x. On the other hand, in the case of a correlation x may be the principal cause of y, but there may be other causes as well. It is said that a portion of the variance in y is explained by x. In a scientific experiment, one could hold all the other causes constant and measure the function $y = f(x)$. In an engineering situation (e.g., production in a factory with the attendant variables uncontrolled to a degree), the points will be scattered about the function if x is taken as the principal variable. Consider, in Figure 8, that the function $y = f(x)$ is a straight line for convenience. Consider also that both variables, y and x, may be affected to different (unknown) degrees by a third

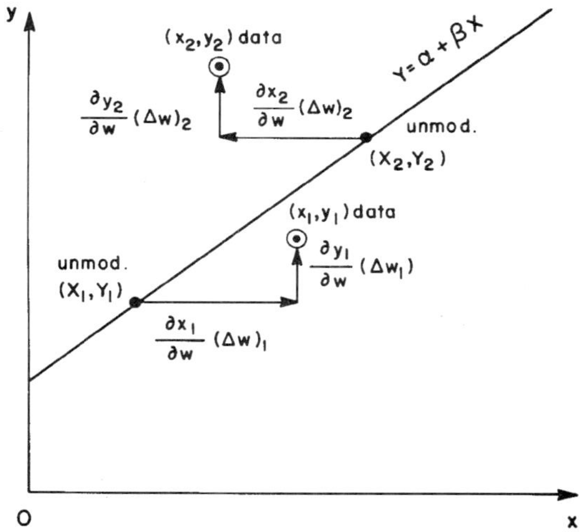

FIG. 8. The underlying function of y versus x denoted by the sloping line is modified by variability in the ordinates and abscissas of its points by another variable, w. The result is a spread distribution of points appearing to be a correlation, not just a function with errors. (*Materials Evaluation*, 1993. Used by permission.)

variable, w, a fourth variable z, and so on. Now consider only the perturbations from variables w. If the variable w were held constant at some baseline value, then $y = f(x)$ would be found and the points (X_i, Y_i) would be found at the positions shown at the unmodified positions labeled as "unmod."

In the context of engineering, however, a process has brought a part to its present state. The generic "fishbone" diagram of a process showing all the potential influences as drawn by quality professionals (Scherkenbach, 1986) is shown in Figure 9. Each branch of the fishbone diagram itself has many branches, leading to a plethora of w-type variables. There is no reason to believe *a priori* that either x or y are free from variability brought about by the other variables such as w. The changes in x and y from the unmodified baseline values are

$$\Delta x = (\delta x/\delta w)\Delta w \tag{3}$$

and

$$\Delta y = (\delta y/\delta w)\Delta w \tag{4}$$

The resulting points as measured and plotted onto Figure 8 are (x_i, y_i). As can be seen, these points may be close to the original line or far from it. One or the

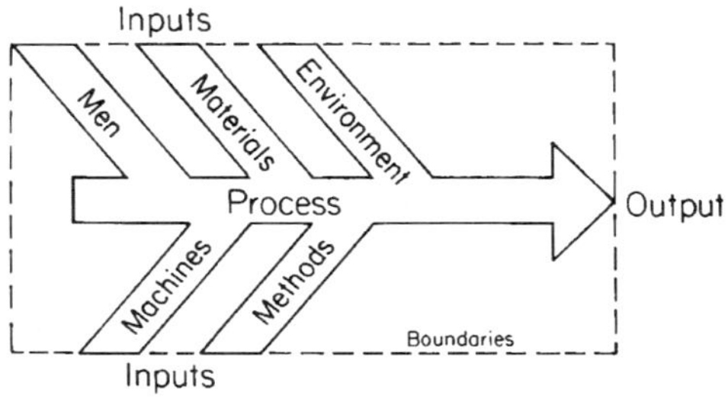

FIG. 9. Diagram descriptive of an industrial process. All five classes of inputs can introduce variability into the output by inadequate uniformity and controls. (*Materials Evaluation*, 1993. Used by permission.)

other of y or x may be more greatly influenced by w, z, and any other subsidiary variables.

When many data points are accumulated, the result is a clustering of points about the line rather than a set of points on the line. One might visualize a cluster describing the shape of a cigar or an ellipse, but rather jagged. A curve, in this case a line, through the points is needed to use as a predictor of the physical property y from the NDT variable x. If one were asked to "eyeball" the best line through the points, one might sketch something like the smoke path through the center of the cigar or the major axis of the ellipse.

However, lines thus defined are not derived when the usual least-squares formulas are applied to the points to find a regression line. The statistical regression line always has a slope lower than the "eyeball" line. This result occurs because the regression formulas are derived under the assumption that the running variable x is absolutely accurate and that all the variability resides in y. However, Figure 8 shows that the subsidiary variables w, z, etc., produce variability in both x and y in the engineering situation. Thus, the "eyeball" best fit is actually better in the engineering production milieu.

An algorithm has been derived (Papadakis, 1995) to find the equivalent of the major axis of the ellipse of points by doing successive regression calculations and then rotations of coordinates to make the regression line of calculation (n) the new x-axis for the next regression line of calculation $(n + 1)$ until the rotations become infinitesimal. It was shown that the line derived in this fashion is a better prediction line than the first regression done in the ordinary way.

Figure 10 deals with the problem presented by preexisting specifications and recommended practices. Suppose that a process such as represented in Figure 9 is applied in a factory to a part. This factory process introduces into the part some physical determinant that changes the part. (An example could be heat treatment.) This physical determinant can have various effects, represented by internal processes in the part and labeled Process #1, Process #2, and Process #3 in the diagram. All these go on during the factory process as a whole, and they produce changes in certain intensive properties of the part. As shown in Figure 9, Process #1 results in the physical property desired (Box 1). Process #2 results in some property that can be utilized to develop a slow or destructive test method (Box 2). Assume that this method, being low-tech relative to more modern NDT, was developed first and finalized into specifications and recommended practices by standards organizations or individual companies. Process #3 results in another property that is later developed into a fast, electronic, and nondestructive test method (Box 3). The NDT test challenges the slow or destructive test. In the hierarchy of the acceptance of standards, the NDT test frequently must show a correlation with

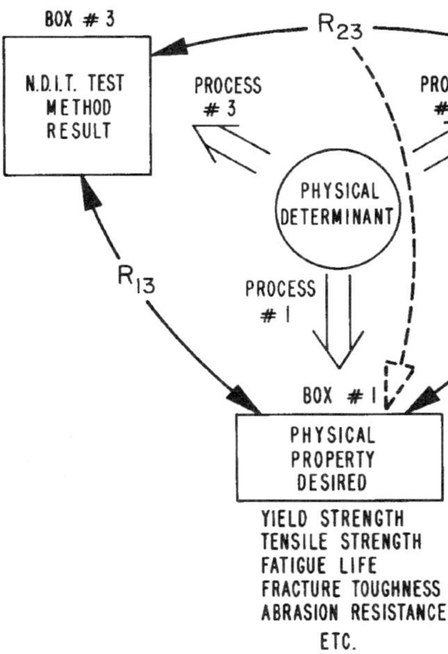

FIG. 10. Three measurables are hypothesized as resulting from the process in Figure 9 as intensive properties of the part being made by the process. In Box #1 is the physical property desired. In Box #2 is a measurable adopted in the past to use as a predictor of the property in Box #1. In Box #3 is a recently discovered measurable proposed to supplant that in Box #2. The three measurables are connected by correlation coefficients R_{12}, R_{23}, and R_{13}. It is true mathematically that $R_{13} > R_{12} \times R_{23}$. (*Materials Evaluation*, 1993. Used by permission.)

the earlier standardized test to win acceptance. However, this is not rational; the NDT test should be applied directly to the physical property desired in Box 1.

Between each box is a correlation coefficient R_{ij}. To go from the physical property desired to the old test and then from the old test to the NDT test, the resultant correlation coefficient is

$$R = R_{12} \times R_{23} \tag{5}$$

To go directly from the physical property desired to the NDT test, the correlation coefficient is R_{13}. It is well-known from statistics that

$$R_{13} > R_{12} \times R_{23} \tag{6}$$

for any permutation of the three coefficients, so the correlation in Eq. (5) forced by adherence to the old standard is never as great as the correlation that could be found by starting afresh with the NDT test and the physical property desired.

Correlations and functions with errors can be used as predictors of physical properties in quality control scenarios. Consider Figure 11, which shows a function and the 95% statistical reliability limits (Papadakis, 1982, 1993a). Data points fall inside the jagged curves beyond these limits to a small finite degree. To use the correlation for prediction, the minimum acceptable value of the design parameter, $Y_{\min}$, is drawn to intersect the lower 95% limit. That intersection defines the set-point for the test, $X_{\min}$. Values below $X_{\min}$ indicate that a part should be rejected. A few nonconforming parts in Area A are accepted and constitute Type II errors of the test. At the same time, a few good parts in Area B are rejected, constituting Type I errors of the test.

Type II errors as represented by Area A are more benign on the average than Type II errors made by statistical process control because the nonconformity of a part falsely passed by SPC, i.e., between periodic samplings, can be of any severity whereas the severity of the nonconformity possible in Area A is minimal. SPC cannot catch the "wild card," whereas 100% NDT inspection can.

The false rejects in Area B should be compared with the false reject rate of the particular set of SPC statistical criteria (Run Rules) being used. For the four Run Rules in the Western Electric formulation (Western Electric, 1956), the false reject rate is 1.0%.

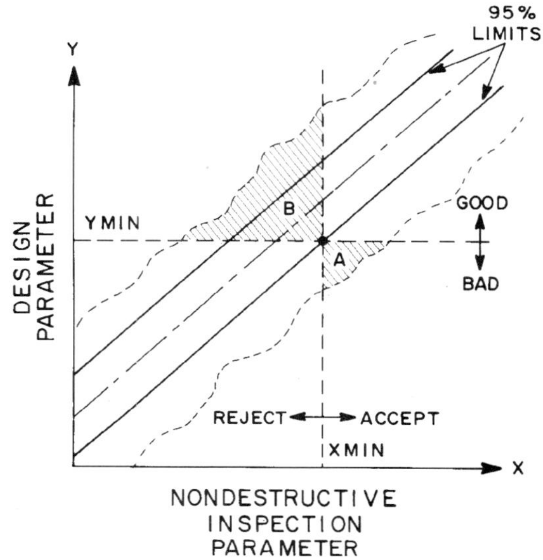

FIG. 11. Diagram of accept–reject levels of an NDT test for materials properties relative to the acceptable vs conconforming level of performance of a material. Type I and Type II errors occur. The nonconforming material accepted by the NDT test (Area A) deviates from acceptability by a minimal amount. (*Materials Evaluation*, 1993. Used by permission.)

III. Instruments and Systems

A. General

Instruments and systems are at the core of nondestructive testing. Without them, all the erudite theories and elegant experiments in the laboratory are just laboratory curiosities. An instrument or a system can solve a problem. Successful instruments can solve enough problems for enough customers that it becomes worth building and marketing the instruments. In the field, a salesman can convince a potential customer that some piece of commercial apparatus can solve his or her problem, thus turning a potential into an actual customer. The sale, an arms-length transaction, multiplied many times over throughout the industrial world, provides the funding for all the salesmen, manufacturers, laborers, technicians, designers, engineers, managers, and scientists who participate in the NDT "food chain." Some tax money even comes into the chain through various forms of subsidies.

In this book the interest is in instruments and systems with successful arms-length sales transactions in volume. Various classes of instruments and systems will be explored and examples will be given. The treatment will be thorough but not complete or exhaustive; this chapter is neither a catalog nor a buyers' guide. One should look to commercial advertising sources for a more complete coverage of instruments, manufacturers, and suppliers. See, for instance, the annual Buyers' Guide issue of *Materials Evaluation*, an official journal of the American Society for Nondestructive Testing. In addition, specific instruments are mentioned in this book only as examples; mention of an instrument does not necessarily imply that the author advocates the use of the instrument. Mention of the capabilities of an instrument is only a reiteration of the manufacturer's claims and is not necessarily an endorsement. (See the References for the author's personal knowledge of some items.)

B. Flaw Detectors

1. *Historical*

Some flaw detectors were in use in defense industry plants in World War II. They were tube-type electronic instruments with oscilloscopes borrowed from radar and were as large as full-size refrigerators. Transducers were generally lithium sulfate, a fragile and water-soluble crystal, which was encapsulated for protection. Quartz was also used. Note that this time frame was previous to epoxy resins.

Figure 12 is a picture of a comparable instrument that was in use at the University of Michigan before 1945 (Firestone, 1945, 1946; Firestone and Frederick, 1946). Firestone called this instrument the "Supersonic Reflectoscope"—the terminology had not been standardized as yet to denote flight faster than sound as "supersonic" and the sound higher than human hearing as "ultrasonic." The word *Reflectoscope* indicated pulse-echo operation with one transducer and an oscilloscope display (A-Scan). The pulses and echoes were several cycles of the rf coming from pulse-excited tuned circuits.

This instrument was commercialized by the Sperry Products Company as the Sperry Reflectoscope, and many copies were sold to defense and other industries. Sperry used it along with eddy current inspection for testing rail *in situ* with its Sperry Rail Cars, which ran under contract on rail lines across the country doing tests. (Interjecting a bit of oral history, as early as 1938 I saw these self-propelled railroad cars running occasionally on the New York Central main line from Albany to New York. As a child I called them "Funny Face" because of the massive sergeant's stripes painted on the front. I was told

FIG. 12. Photograph of a WW II vintage ultrasonic pulse-echo flaw detector, the Supersonic Reflectoscope, in a physics laboratory at the University of Michigan (*Materials Evaluation*, 1983. Used by permission.)

then that the cars were delivering U.S. mail; only later [in 1953] when I applied for a summer job at Sperry did I learn about the testing function.)

Soon after the Firestone work, the U.S. Army Watertown Arsenal funded work at Brown University on elastic interactions with solids. This work resulted first in a large laboratory test set for ultrasonic attenuation and changes in velocity (Roderick and Truell, 1952), and then in a relatively compact tabletop instrument for convenient ultrasonic attenuation measurements. There were two salient features of the electronics in this instrument. The first was that the output pulse was a tunable pulsed rf burst with an envelope shaped like a Gaussian bell curve, giving the narrowest bandwidth for a given pulse length. The second was that the oscilloscope display was alternatively the echo train (rectified) and then a calibrated decaying exponential curve with adjustable time constant for convenient measurement of attenuation between pairs of echoes.

The name given to the Brown instrument was the Ultrasonic Attenuation Comparator. The name stuck when the instrument was commercialized by the Sperry Products Company as the Sperry Ultrasonic Attenuation Comparator. A CRO photograph of the echoes with the superimposed exponential curve is shown in Figure 13. Sperry sold a few of these instruments, but the attenuation method did not find wide acceptance in NDT because of experimental difficulties and interfering phenomena in the causation of attenuation. Torre (1986) was using a similar attenuation instrument on

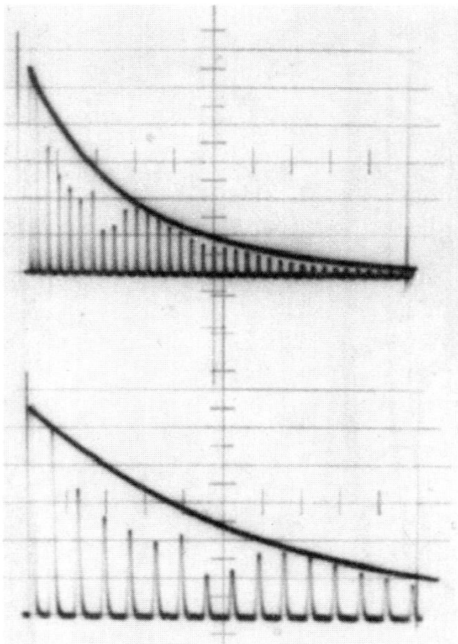

FIG. 13. Oscilloscope photographs of echoes in NaCl taken using a Sperry Ultrasonic Attenuation Comparator. Special plating on the transducer plate magnified the diffraction effects by perturbing the behavior of the initially piston source. (Reprinted with permission from E. P. Papadakis, "Diffraction of Ultrasound in Elastically Anisotropic NaCl and in Some Other Materials," *J. Acoust. Soc. Amer.* **35**, 490–494, 1963.)

nodular iron at General Motors when he discovered that ultrasonic velocity provided a better test. One notable success of the Sperry Ultrasonic Attenuation Comparator was a test devised at the Watertown Arsenal Laboratories that detected a soft condition in hardened and tempered 40-mm cannon barrels (Orner, 1958). The test was multidisciplinary, using eddy current response and ultrasonic attenuation. With flat transducers and couplant on the cylindrical gun tubes, excellent multiple echoes were seen from the bore, and relative attenuation could be measured precisely. A batch of 20,000 cannon barrels for the Northrop F-89 Scorpion night figher were tested in a Georgia warehouse in August, 1958. The barrels had been improperly quenched after austenitizing by the supplier. Supporting scientific work was performed on the steel material (Papadakis, 1960). An advanced version of the attenuation instrument was reported in the literature (Chick *et al.*, 1960). Semiquantitative attenuation measurements as part of a test for spot weld integrity will be mentioned below.

2. Modern

"Modern" is a relative term. For flaw detectors, "modern" began with the tabletop flaw detector with a block diagram somewhat like Figure 3 but without the computer. Everything including the display, a cathode ray oscilloscope, was in a single chassis. The echoes were rectified and detected before being displayed. Commerical oscilloscopes could not display the rf wave forms in those days. Some instruments had military specifications—some had to continue to function after being dropped from a certain height; others had to be able to be carried down the conning tower of a submarine. Requirements often included filters for the cooling fan so that the instrument could operate in a dusty factory environment without air conditioning. The practical aspects of commercializing instruments go on and on and include such electronic features as gates to pick out certain signals and amplitude-distance-correction (ADC) time-variable amplification in the receiver stage to correct (approximately) for beam spreading.

One instrument available in 1970, the Sonic Mk IV from Sonic Instruments in Trenton, New Jersey, is shown in use in Figure 14. It has all the characteristics mentioned above, including the required size for the conning tower. In the figure, the operator is shown in a development laboratory of an automotive supplier's plant, testing spot welds on bumper reinforcement bars for the 5-mile-an-hour bumpers specified by the U.S. federal government for introduction in 1974. The testing method was developed at the Budd

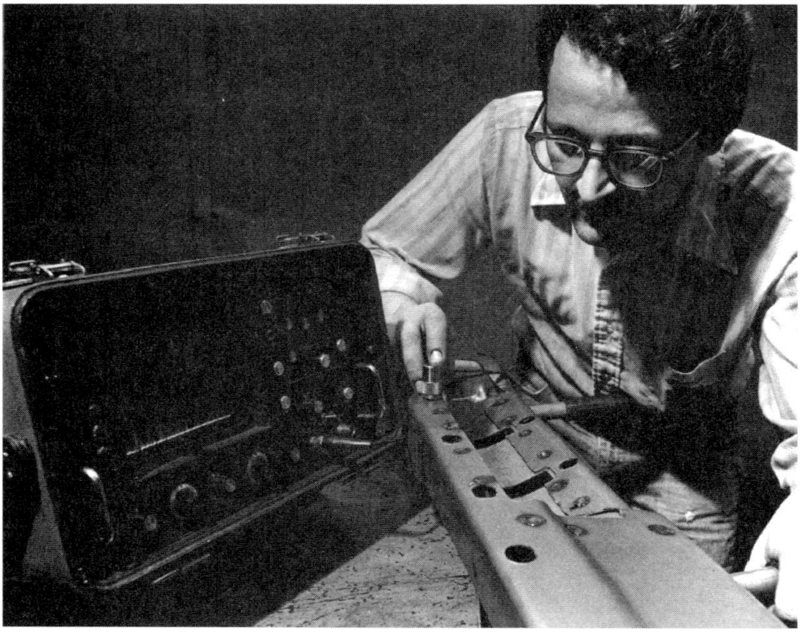

FIG. 14. Flaw detection instrument of the 1970–1975 era being used to test spot welds. (Ford Motor Company, 1975. Used by permission.)

Company in Philadelphia and was adopted by the Ford Motor Company. The hot-rolled sheet steel in the reinforcement bars was 0.120 to 0.160 inches thick (3 to 4 mm). The method involved instruments, transducers, attenuation, and thickness gaging, all of which are in different sections. Because the principal breakthrough in the test method depended on the transducer design, the full explanation of the method will be given in Section III.D on transducers. A second transducer breakthrough (Mansour, 1988) allowed the method to be extended to thin-gage cold-rolled sheet steel down to 0.023 inches thick (sheet steel for automobile bodies).

Up to this point in time, the controls on the flaw detection instruments were simply a combination of analog controls for the oscilloscope portion, the pulser-receiver portion, and special-purpose gates and ADC. The next step was to add computer memory coordinated with a special-purpose panel of buttons to control the instrument in a repeatable way so that an operator could exactly duplicate the settings at a future date. One such instrument, the Epoch II from Panametrics, Inc., is shown in Figure 15. Note the touch pad of color-

FIG. 15. Flaw detection instrument, the Panametrics Epoch II, introduced circa 1985. Features in this generation of instruments were controls and computer memory to permit accurate resetting of test parameters. Some accessories are shown. (From E. P. Papadakis, "Ultrasonic Instruments for Nondestructive Testing," *Enclyclopedia of Acoustics*, M. J. Crocker, editor. Copyright © 1997 John Wiley & Sons. Inc. Reprinted by permission of John Wiley & Sons, Inc.)

coded and labeled controls on the right and the oscilloscope face on the left with digital information displayed below it. The display is clear to an operator, but not very clear in this picture. The picture also shows some specialized transducers and two gage blocks for testing the operation of the inspection system (instrument, transducer, and angle-beam foot or wedge operating on the principle of refraction). At this point in time, with systems being all solid-state electronics, battery packs were built into some instruments, permitting them to operate 4 hours or even 8 hours in the field before recharging.

Next came miniaturization. A handheld instrument weighing less than 3 lb (1.4 kg) is shown in use on a pipeline in Figure 16. This model, the DuPont QFT-2, was available in 1989. Two-dimensional digital electronic displays made this step down in size and weight possible vis à vis the cathode ray tubes of earlier instruments. The miniaturized instrument was special-purpose and did not displace the larger models for less strenuous operations. However, small instruments with large capabilities have been developed for practical

FIG. 16. Handheld miniature flaw detector being used on a major job. Two-dimensional digital displays made this size reduction possible. (From E. P. Papadakis, "Ultrasonic Instruments for Nondestructive Testing," *Encyclopedia of Acoustics*, M. J. Crocker, editor. Copyright © 1997 John Wiley & Sons, Inc. Reprinted by permission of John Wiley & Sons, Inc.)

purposes. The convenient-sized instrument from Panametrics, Inc.—the Epoch III-B circa 1996, with flat digital display and the color-coded touch pad controls—is shown in Figure 17. In the figure, a wedge-mounted transducer is being used in the development laboratory to interrogate a weldment. In Figure 18, the same instrument is shown in industrial use on a large pipe fitting.

The next stage of development, currently in process, is to put the flaw detection circuitry into a computer as one or just a few cards. The monitor becomes the digital oscilloscope part of the time, the set-up controller with user-friendly software part of the time, and the report writer for the results at the end of the test. With the PRINT SCREEN command, anything can be output as hardcopy. The computer interface can allow the computer memory to store great numbers of sets of results for later correlations and other

FIG. 17. A portable flaw detection instrument, the Epoch III-B, with full range of computerized control capabilities, touch-panel control buttons, and two-dimensional screen. (Panametrics, Inc. Used by permission.)

analyses. One such flaw detection instrument, the USPC 2100 from Krautkramer circa 1997, is shown in Figure 19. Some details from the manufacturer's catalog are shown in Figures 20 and 21. (As previously mentioned, the use of these catalog pages is not meant as a sales promotion for the instrument; other manufacturers also offer flaw detection systems built on a computer platform. The advertising material shown is simply very lucidly written and quite informative.)

The next stage of development should involve artificial intelligence (AI). To date, all instruments, despite their computer interfaces or computer-based architecture, depend on the operator to interpret the presence or absence of flaws, set the gate height, and so on, even if the instrument is then set on an

FIG. 18. The instrument of Figure 17 in use in an industrial setting on a large pipe junction. (Panametrics, Inc. Used by permission.)

automatic mode to monitor product or production. AI will, in principle, supplant the operator. Operating models to accommodate the intelligence of human beings must be worked out (Papadakis and Mack, 1997).

C. THICKNESS GAGES

Thickness gaging is a form of flaw detection used when the flaw is corrosion that removes thickness from a structure. This application will be explained further in Section III.F.3.

Thickness gages have evolved from bulky resonance instruments with large analog output scales to compact pulse-echo instruments about as large as a cellular telephone. The read-outs are digital displays of thickness to three or four digits with comparable accuracy. A typical modern thickness gage, the Model 25DL from Panametrics, Inc., is shown in Figure 22. A modern thickness gage usually utilizes a single highly damped transducer in the pulse-echo mode. The time is measured between the "main bang" input pulse and

FIG. 19. Computer-based nondestructive testing instrument mounted in an industrial case and being used next to its parts-testing fixture. The flaw detection circuitry is one card in the computer chassis at the bottom of the rack. The monitor is visible at the top. (Krautkramer Branson. Used by permission.)

the first echo from the specimen being measured for thickness. The conversion from time to thickness comes by way of the ultrasonic velocity in the material of the part. This velocity is entered into the computer memory of the gage by a setup program before the measurements are made; the velocity can be stored for future use, as can data about the transducer, so that setups can be repeated conveniently. Some instruments have the facility to store information about several transducers and recall the information for different tests requiring different transducers.

Two other transducer configurations are in use. One is the delay line transducer, which has a short delay line attached to its face. (See the

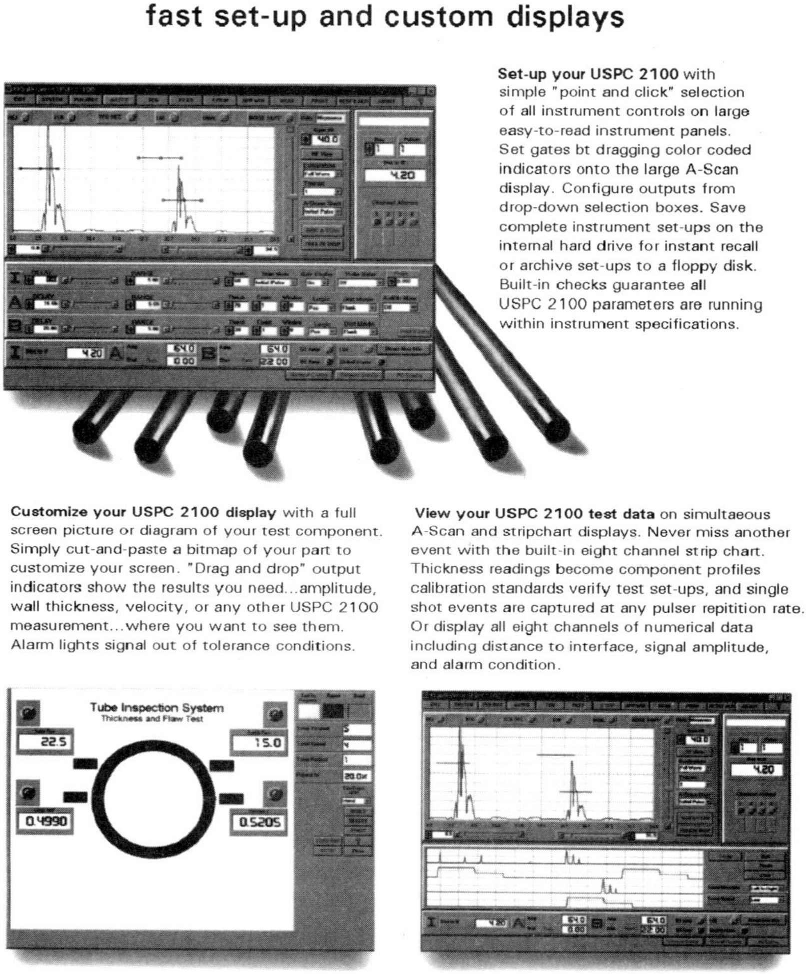

FIG. 20. Setup and display features of the instrument in Figure 19. (Krautkramer Branson. Used by permission.)

illustrative drawing (Papadakis, 1996) in Figure 23.) With this configuration, very thin specimens can be measured accurately. The time delay used for the measurement is either from the interface echo to the first back echo in the specimen or from the first back echo to the second back echo. (See Figure 24.) Another configuration sometimes used is "pitch-and-catch," in which two transducers are mounted in the same housing and aimed to converge at a certain useful range (as were the wing guns of WW II fighter planes). One

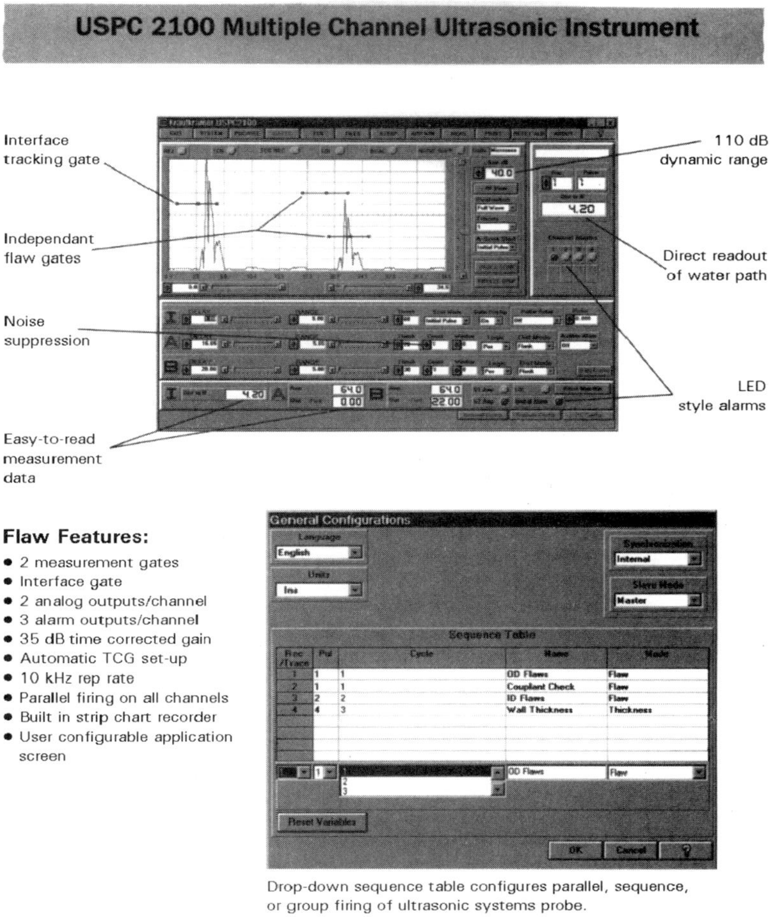

FIG. 21. Features and displays of the instrument of Figure 19 set up in its flaw detection mode. (Krautkramer Branson. Used by permission.)

transducer transmits and the other receives. This configuration is useful in materials with very high attenuation, where the receiver gain must be set very high.

The degree of computerization in the typical thickness gage can be duplicated, of course, in the computer-based flaw detector instrument shown in Figure 19, so thickness-gaging software is offered with it. Using that software, the displays are as shown in Figure 25.

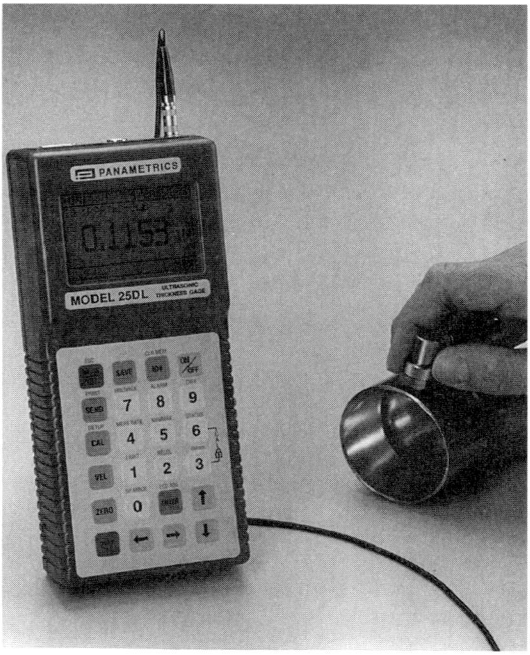

FIG. 22. A modern ultrasonic thickness gage. With the ultrasonic velocity in the material to be tested known and input into the memory of the gage, the instrument will measure the thickness using the pulse-echo technique. The round-trip transit time through the material is measured and translated into thickness by use of the velocity. (Panametrics, Inc. Used by permission.)

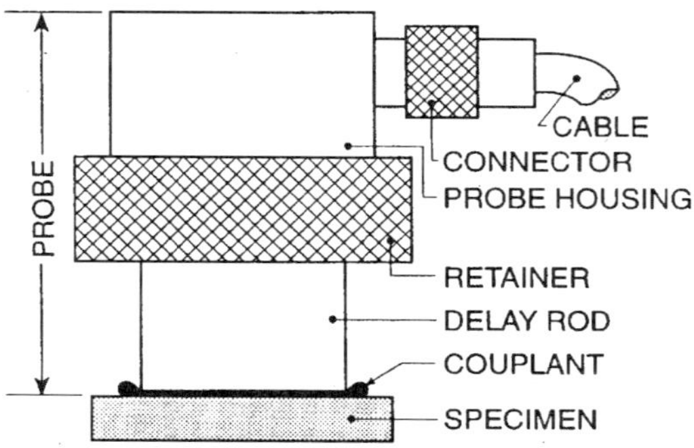

FIG. 23. Diagram of a thickness-gaging transducer with a delay line to permit measurement of very thin materials. (Society for Engineering Mechanics. Used by permission.)

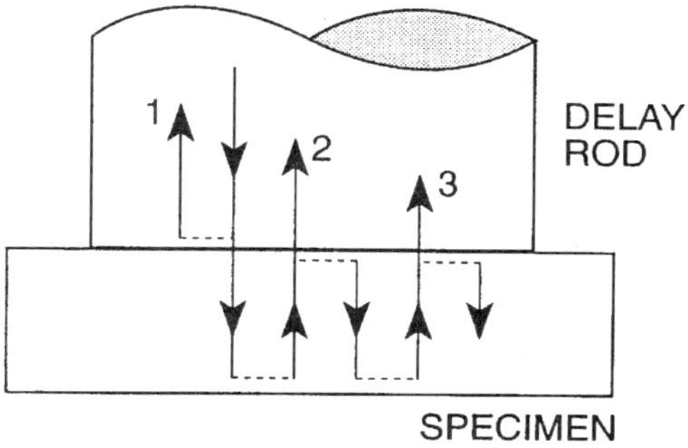

FIG. 24. Diagram of the wave path of the echoes that can be used in measurements with the delay line transducer. Either echoes 1 and 2 or 2 and 3 can be chosen. (Society for Engineering Mechanics. Used by permission.)

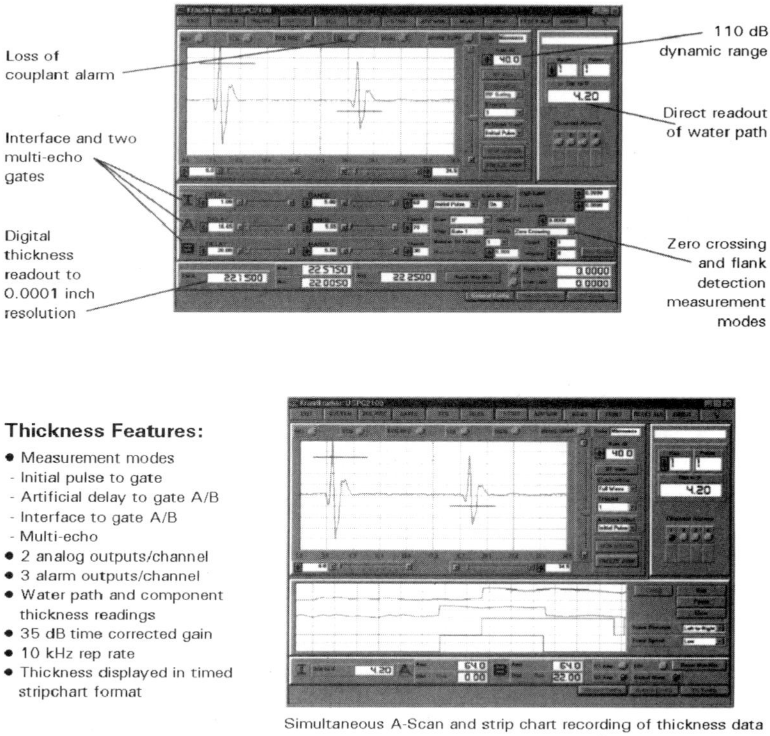

FIG. 25. Computer display of the instrument in Figure 19 when using thickness-gaging software. (Krautkramer Branson. Used by permission.)

D. NDT Generic Transducers

1. Construction

As mentioned before, transducers are treated as a separate topic in the chapter devoted to them. However, they need to be mentioned here since they are "the eyes and the ears" of the NDT ultrasonic industry. A generic transducer is pictured in Figure 26 (Papadakis, 1983). The active element is a monolithic piezoelectric plate, shown as XTAL in the diagram. This may be a piezoelectric crystal, poled ferroelectric crystal, or a poled ferroelectric ceramic. (Nonmonolithic elements are treated in Chapter 2, Vol. 24.) To achieve broadband operation and eliminate extraneous echoes, an acoustically absorbent and electrically conductive backing, B, matched in acoustic impedance to the piezoelectric element is bonded to the piezoelectric plate, which is plated on both sides. The high-voltage lead and the ground strap are bonded in place as shown. This ground strap configuration modifies the field pattern of the radiating transducer only slightly from the ideal piston source (again, see Chapter 2, Vol. 24). The other elements, including the protective wear plate, are self-explanatory.

Transducers come in various sizes and frequencies and can have a variety of external configurations to accommodate different uses. Some have easy-to-

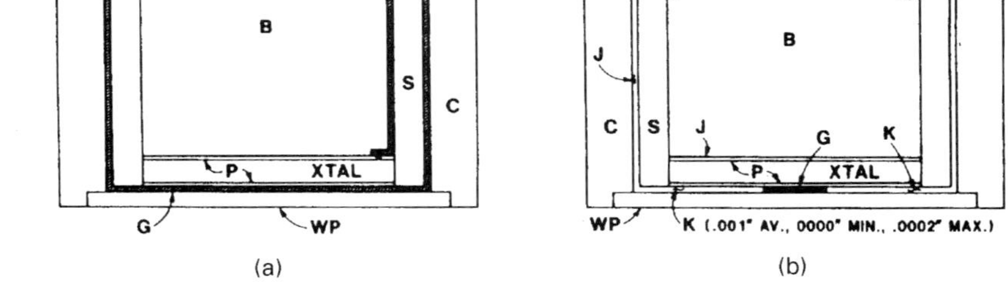

FIG. 26. Transducer construction diagram showing the components. XTAL: piezoelectric element, P: plating, WP: wear plate, G: ground strap, B: backing, S: insulating shield, C: case, T: top cover, HV: high-voltage lead, E: electrical connector. The bond lines or layers J and K are epoxy. Views (a): side elevation; View (b): front elevation. (*Materials Evaluation*, 1983. Used by permission.)

grip case covers whereas others are coaxial and waterproof for immersion applications. All transducers of this type require a couplant to transmit the ultrasound into a workpiece. The couplant may be a thin layer of liquid or a tank of liquid in the case of immersion. (See Figure 5.)

2. *Angle Beams*

Most transducers are built with longitudinal wave active elements. These radiate normal to their wear plate surfaces into the workpieces. To generate shear waves at an angle to the surface for flaw detection, angle blocks are used between the transducers and the workpiece. The transducer is clamped with couplant and mechanical fasteners to the angle block, and the angle block uses added couplant to touch the workpiece. When a longitudinal wave impinges upon a boundary at an angle other than normal incidence, the boundary conditions require a longitudinal solution and a shear solution in the second medium (Mason, 1958). Only the shear solution in the workpiece is desired, to avoid ambiguity.

To achieve this by design, the ultrasonic longitudinal wave velocity in the angle block is chosen to be lower than the longitudinal velocity in the workpiece. Under these conditions, the block can be manufactured with an angle that gives total internal reflection to the longitudinal wave and a propagating solution for the shear wave at some desired angle, such as 70° or 45° from the normal. The combination of transducers and angle blocks yield angle beam transducers that are widely used in detecting flaws in engineering materials. The beam of ultrasound from one such angle beam transducer is shown in Figure 27. The method used was a photoelastic method in Lucite® using strobed light synchronized with but delayed from the ultrasonic pulse. A variety of commercial transducers are shown in Figures 28 and 29.

3. *Spot Weld*

Of special note is the type of transducer used to test spot welds, a methodology alluded to earlier but left to this section because of its total dependence on the development of specialized transducers. One fundamental fact of spot weld fabrication is that even the best surface is not flat but bears an indentation. The transducer face must conform to this variable shape. The shape leads to the requirement for a flexible membrane face on a compartment filled with a liquid under some pressure to force conformity of the membrane to the spot weld face. This requirement then means that the transducer will

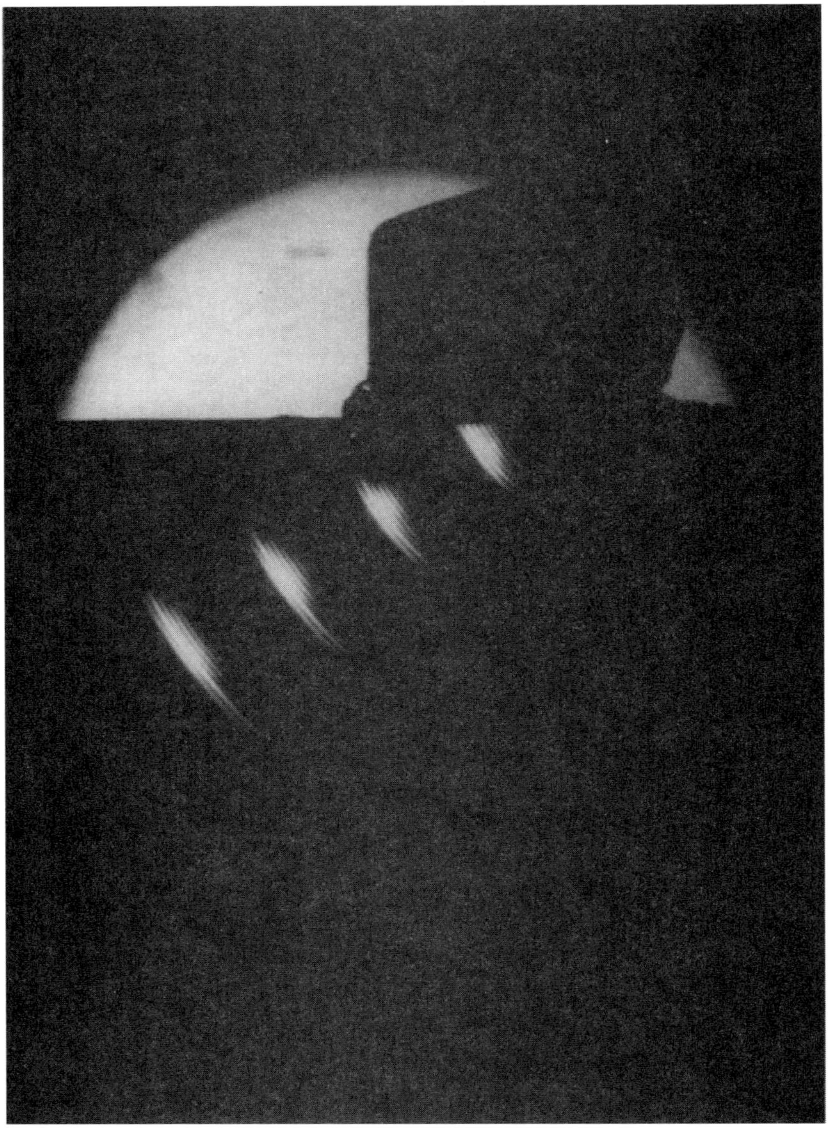

FIG. 27. The radiation beam of pulses from an angle beam transducer radiating into Lucite® as illuminated by a synchronous pulsed photoelastic method. (R. C. Wyatt, Central Electricity Generating Board, U.K. Used by permission.)

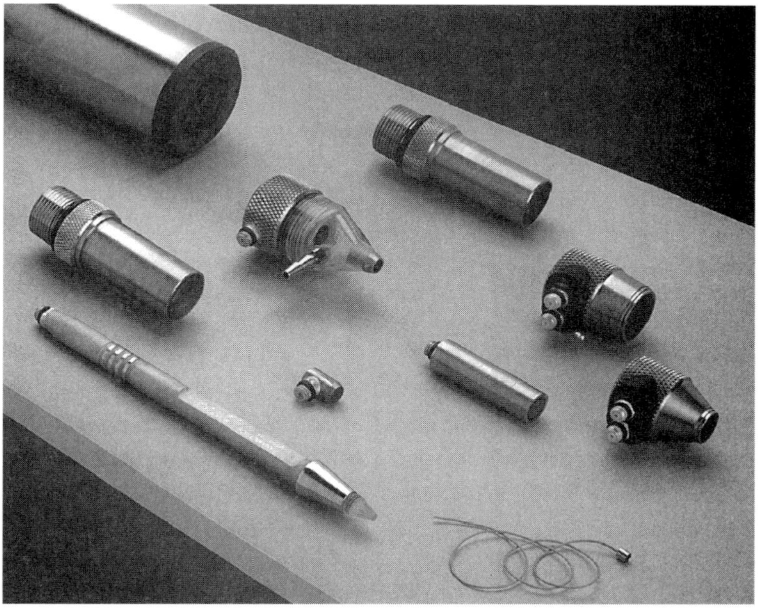

FIG. 28. Several commercial transducers of various configurations. (Panametrics, Inc. Used by permission.)

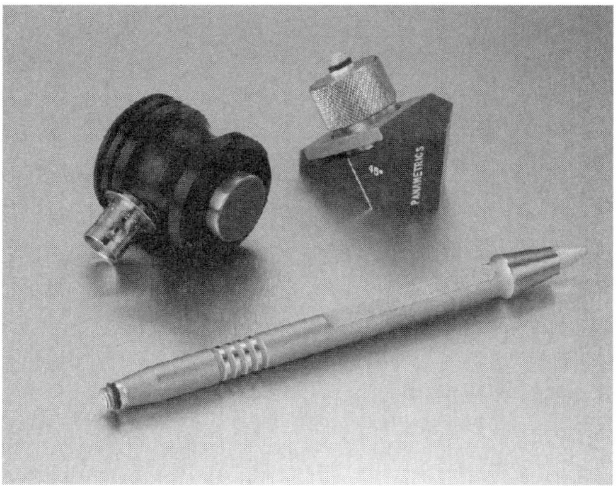

FIG. 29. Several commercial transducers of various configurations. (Panametrics, Inc. Used by permission.)

have a delay line in front of the active element and that this delay line is a liquid column.

The commercial transducers used in the thick-gage sheet metal work had a pumped water source and a small hole in the membrane to let flowing water provide the couplant (Papadakis, 1976). A drawing of the transducer structure with the water column is shown in Figure 30. These transducers were supplied by Automation Industries, the successor to the ultasonic side of the Sperry Products Company. In use, the transducer is placed on the spot weld area, as shown in the figure. The pulse travels through the water and the rubber membrane (which is an excellent acoustic impedance match) and impinges upon the metal. Part of the wave enters the metal and suffers multiple reflections at the free surfaces. Some of the wave energy reenters the water column at each echo from whatever source. The spot weld nugget does not represent a free surface, as the nugget material matches the parent material in acoustic impedance but has much higher attenuation. The spot weld test has

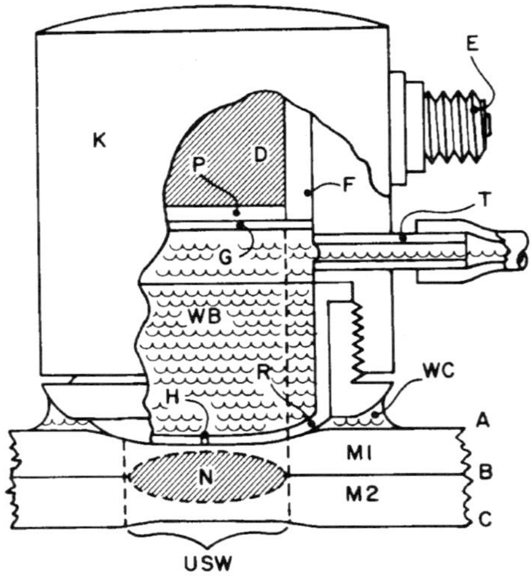

FIG. 30. Sketch of water column transducer on sheet metal with a weld nugget N. Highly attenuated multiple echoes beetween surfaces A and C are desired. B is the interface between metal sheets M1 and M2. In the transducer, P is the piezoelectric element, G the protective layer, D the damping backing, F the encapsulating material, WB the water buffer (water column delay line), T the water inlet tube, R the rubber membrane, E the electrical connector, and K the case. Some water can escape through the hole H to become the water couplant WC. The ultrasonic waves travel in the beam labeled USW. The desired conditions are (1) no reflections from B (i.e., wide nugget) and (2) high attenuation echoes from C (i.e., thick nugget). (Academic Press, Mason XII. Used by permission.)

two parts: (1) If the nugget is wide enough, the reflections reentering the water column come only from the double thickness of the metal; if the nugget is too narrow, single-thickness echoes are visible from the perimeter of the ultrasonic beam. (2) If the nugget is thick enough, the attenuation of the double-thickness echoes is above a minimum threshold. A so-called stick weld, which is unacceptable because of a very thin nugget, may show only double-thickness echoes but inadequate attenuation. The test was developed to conform to the automotive standards for spot welds (Ford Motor Company, 1972) and may vary from industry to industry.

The same principle applies to thin-gage spot weld testing (Mansour, 1988). The development of a high-frequency transducer (15–20 MHz) with a captive water column by Mansour permitted the extension of the spot weld test to steel as thin as 0.023 inches. A picture of the transducer and its parts is shown in Figure 31. The circular membrane is held inside the knurled ring by the cylindrical body, which screws up against the O-ring touching the membrane. The transducer element with backing, etc., is in the T-shaped portion with the Microdot® connector. To assemble the structure, the knurled ring, membrane, O-ring, and cylindrical body are screwed together first. Then the cylindrical

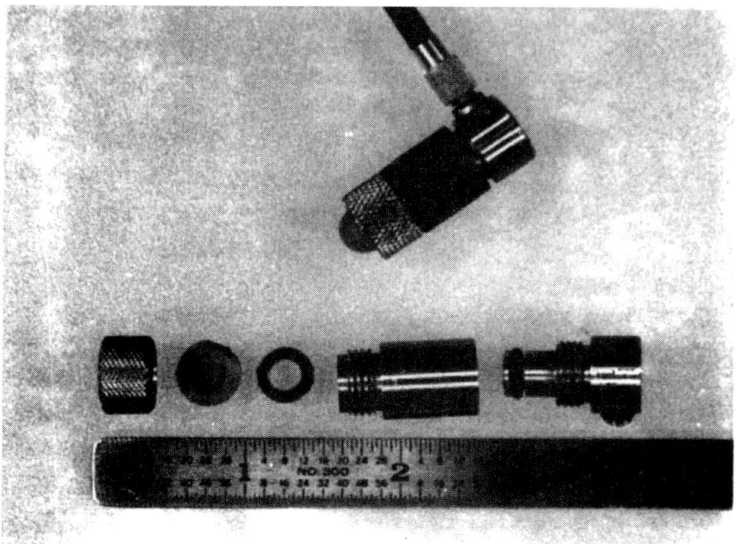

FIG. 31. Assembled and exploded view of a transducer with a captive water column for spot weld testing in thin-gage steel. The water column is held in by the hemispherical slightly pressurized membrane. (*Materials Evaluation*, 1988. Used by permission.)

body is filled with water. After that, the transducer is inserted and screwed into place. Its frontal O-ring holds the water as a captive water column, causing the membrane to bulge as in the upper assembled picture. Extra couplant is used in the test.

Some years ago, operators from nine stamping plants at the Ford Motor Company were trained in this testing procedure, and the test was implemented in those plants. Within two years of the inception of this testing, the net savings from the elimination of two-thirds of the destructive audits and from more rapid quality information acquisition was $2 million per year.

E. C-SCANS

1. Early Models

An initial explanation of C-Scans was given in Section II.C.3. The C-Scan presents a picture of a slice of the workpiece. The slice is normal to the transducer beam and does not include the material very close to the top and bottom surfaces of the workpiece because the reflections from the surfaces would swamp the small reflections from flaws. The unwanted major reflections are deleted by an electronic gate. The display is proportional to the amplitude.

This explanation is simplistic considering the versatility of the C-Scan and the capability of its modern versions. For instance, if one set the gates to include the top surface, one could check on the presence or absence of drilled holes. Or, one could obtain transit time outputs rather than amplitude outputs so that the surface contours could be mapped or velocity anomalies could be detected. Furthermore, X-Y operation is not the only directional set of coordinates; one could use R-Θ for the ends of cylinders and Z-Θ motion for the perimeters of cylinders.

An early commercial version of a C-Scan system from Automation Industries is diagrammed in Figure 32. Its transducer holder traversed back and forth rapidly along X while it was stepped in small increments along Y at each X-traverse. The output was an X-Y recorder using special paper. An electric arc pen was scanned over sensitive paper synchronously with the X-Y scan of the transducer in the liquid tank. The electric pen was either "on" or "off," corresponding to an echo larger in amplitude than a set threshold. Interestingly, some of the pioneering work done on primitive equipment like this has stood the test of time and has proved to be of fundamental value. For instance, the ground strap across the face of the piezoelectric element in commercial ultrasonic NDT transducers (Figure 26) was detected (Mansour, 1979) with the ball reflector scheme shown in Figure 32. This experimental work is explained further in Chapter 2, Vol. 24. From these early beginnings, the C-Scan experienced improvements in two directions.

a. Topological Improvements. One direction of improvement was topological. The transducer mechanism was made aimable by adding multiple axes of gimbaling in the mechanism holding the transducer. Thus, the transducer beam could be held normal to a complicated surface while the traversing mechansims were driven by stepper motors controlled by computer programs. The programs could be written to give adequate coverage for the entire surface and the material beneath the surface. This facility also included holding the beam at a definite angle to the surface in Eulerian coordinates to send in shear waves at particular angles for interrogation of a part. With dual systems, two transducers could be used in through-transmission in parts with nonparallel faces. Learning programs for curve following have been introduced.

b. Display Improvements. The other direction of improvement was display. The use of computers resulted in gray-scale images on monitors instead of on-off electrically written *X-Y* plots. Up to 60 dB of amplitude could be easily

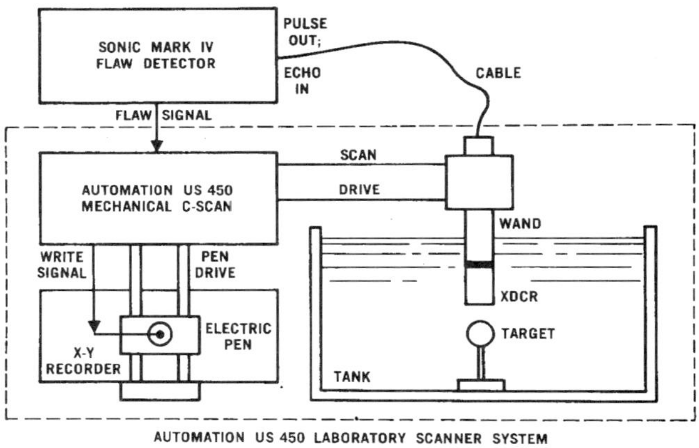

FIG. 32. Diagram of an early commercial C-Scan. An electric arc pen was scanned over sensitive paper synchronously with the *X-Y* scan of the transducer in the liquid tank. The electric pen was either "on" of "off" corresponding to an echo larger in amplitude than a set threshold. The setup with the ball target shown was for the evaluation of transducers, a nonstandard application. (*Materials Evaluation*, 1979. Used by permission.)

shown on a gray scale. The successor to the gray scale is pseudocolor, where colors on a color monitor can be equated to the ultrasound intensity level. However, as every manufacturer seems to use a different pseudocolor protocol, the operator must learn the scheme of each particular brand of software in use. Often, computer monitors are used for the setup, echo display, and monitoring functions.

2. *Computer Versatility*

The use of computer information control and manipulation control makes it possible to do B-Scans as well as C-Scans in what used to be thought of as a C-Scan environment. As there is no intrinsic limitation on the size of tanks, manipulation arms, etc., the C-Scans in this section flow naturally into large installations, which are treated below. With the aiming capabilities, the C-Scan manipulators can accommodate bubblers to aim at curved surfaces, a subject also treated below. Some very precise C-Scans with focused probes can be classed as ultrasonic microscopes (see Chapter 5 of Vol. 24). At the risk of encroaching upon territory beyond the scope of C-Scans, we show here illustrations of modern systems supplied by Panametrics, Inc., and Sonix, Inc. in Figures 33 and 34, respectively. Of course, other configurations and systems from still other suppliers are available.

F. Large Installations

1. *General*

Large installations are characterized by both size, complexity, and the massive character of the facilities needed to run them. Large installations exist in manufacturing facilities where large parts are fabricated and in maintenance facilities if large parts can be brought to the installation. Otherwise, testing instruments are brought to the large structures and scanned over them by manual or automated means.

2. *Tanks*

A large installation at Douglas Aircraft is pictured in Figure 35. Shown is an immersion tank large enough to test the wing spars of large aircraft and the aluminum plates that are to be machined into very large parts. Tanks like these have been used for many years and continue in use with updated electronics and manipulation gear for C-Scans, B-Scans, and complex shape-following

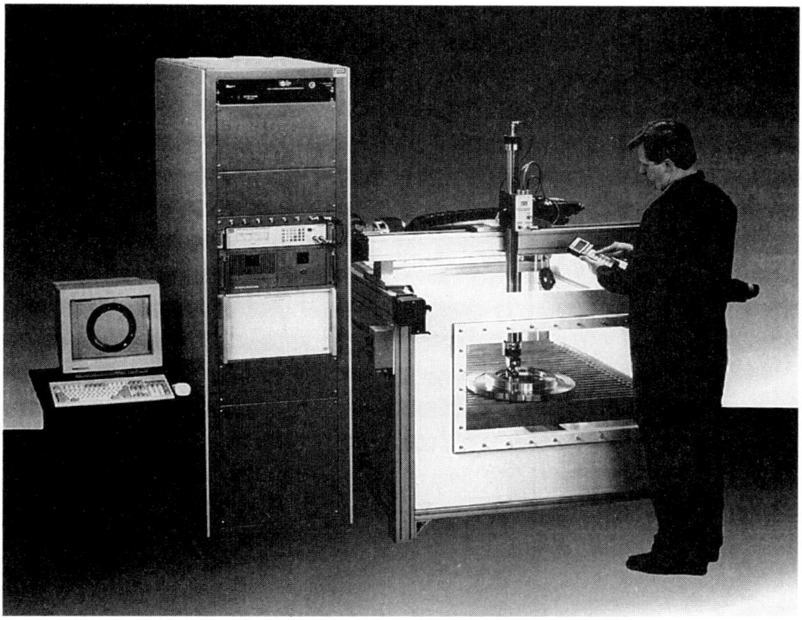

FIG. 33. A modern C-Scan with multiaxis manipulation and computer control and display. This system has many versatile features, including the learning module in the operator's hand that teaches the system to follow curves. The system can also be supplied with two manipulators and can do B-Scans. (Panametrics, Inc. Used by permission.)

programs. The material must be tested prior to the very complex machining operations in the aircraft industry to ensure that the material is sound and that the machining time will not be wasted. Fabricated structures with welds must also be tested.

The facilities associated with these large installations include large high-bay buildings, high capacity cranes, water treatment equipment, and access roads for transport both within the factory and outside. Large computers are used to analyze and store massive amounts of data as well as to run the operations.

While the large tanks like the one shown in Figure 35 are individually made in custom job-shops and the lengths of the *X*-, *Y*-, and *Z*-traverse mechanisms may be made-to-order, these large facilities qualify for inclusion in this book because the concept is universal and has been sold many times over to a variety of customers. The multiaxis manipulators and control computers of several NDT manufacturers can be installed on the traverse mechanisms and chosen to best fit the needs. NDT manufacturers may choose to be the prime contractor for an entire system including the operating tank, gantries, and

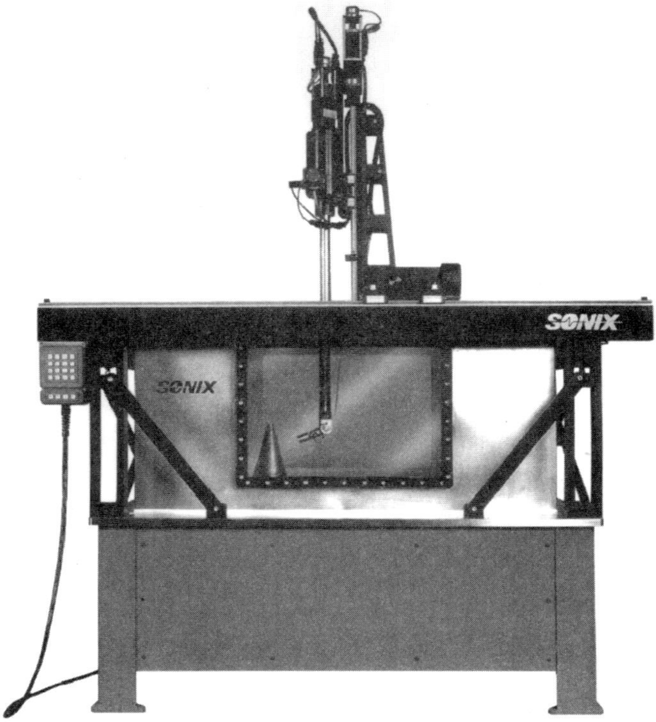

FIG. 34. A modern C-Scan with a single scanning bridge (manipulator) for several axes. The system is also built in a model with two manipulators and can perform B-Scans as well as C-Scans. (Sonix, Inc. Used by permission.)

whatever, or they may choose to be only suppliers of equipment. Business is complex. One large company purchasing NDT systems always chose a factory machine maker accomplished at automated parts handling as the prime contractor and then let the prime contractor (also knowledgeable in NDT) choose the NDT instrument vendor and assemble the entire machine/ instrument package. The reader should refer to the November and December 1984, issues of *Materials Evaluation* for examples of automation applied to large systems.

3. Bubblers (Squirters)

A bubbler, sometimes known as a squirter, is a hydraulic structure mounted in front of the face of an ultrasonic transducer. Water enters the bubbler under

FIG. 35. A very large immersion tank for ultrasonic testing in the aircraft industry. The tube-type electronics attests to the length of time ultrasonic testing has been successful and commercial in important applications. (From E. P. Papadakis, "Ultrasonic Instruments for Nondestructive Testing," *Encyclopedia of Acoustics*, M. J. Crocker, editor. Copyright © 1997 John Wiley & Sons, Inc. Reprinted by permission of John Wiley & Sons, Inc.)

pressure at the base near the transducer and flows away from the transducer through a constricting nozzle. The axis of the nozzle is concentric with the centerline of the transducer beam. The nozzle is a laminar flow configuration so that the ultrasonic beam travels in the water without perturbation of its phase front. The nozzle is aimed at the part to be tested. The water, leaving the nozzle at high velocity, travels to the part and acts as a delay line to carry the ultrasonic wave to the part. The stream of water provides a self-couplant to the part. Echoes can travel back up the stream of water in pulse-echo or through the part to the stream of another bubbler on the other side of the part for through-transmission. To illustrate bubbler action, a photograph of two horizontally opposed bubblers made by Sonix, Inc., scanning a small section of composite material is shown in Figure 36 . In Figure 37, the photograph by McDonnell Douglas (Boeing) shows the precision of bubbler alignment as an almost-artistic splash of a galaxy of droplets.

FIG. 36. Two concentric horizontally opposed bubblers used for through-transmission on a section of composite. (Sonix, Inc. Used by permission.)

Because the bubblers can be mounted on computer-controlled manipulation mechanisms, the water streams can be directed properly to utilize refraction to penetrate curved shapes with nonuniform thickness. The curve-following facility is used extensively in the aircraft industry for testing composite structures for skins and control structures of aircraft. The through-transmission mode is very useful for detecting unbonded areas in these structures, which are adhesively bonded and must be structurally sound to perform aerodynamically. One system in use on a composite part at the Boeing Commercial Airplane Company is shown in Figure 38. Frequently banks of bubblers are mounted together to cover more area, like a so-called paintbrush transducer. Adjacent mounting can be used when the curvature in that direction is minimal. Another large facility using bubblers on aircraft parts at McDonnell Douglas (Boeing) is shown in Figure 39 using the AUSS-V

FIG. 37. Two concentric horizontally opposed bubblers showing the precision of the aiming manipulation to produce an almost symmetrical "galaxy" of droplets at the impact of the two water streams. (McDonnell Douglas, Inc. Used by permission.)

system. An illustration of bubblers emphasizing their role in large installations is shown in Figure 40, again on aircraft parts at McDonnell Douglas (Boeing).

G. Portable Systems for Large Installations and Objects

1. *General*

Huge parts like submarine hulls and nuclear reactor containment vessels are tested with installations brought to the structure. These installations may be attached to robots or to "crawlers" that follow the surfaces and record the

FIG. 38. One example of a bubbler system used on a large composite part. Through-transmission is used with a contour-following wand system with multiple axis control. (From E. P. Papadakis, "Ultrasonic Instruments for Nondestructive Testing," *Encyclopedia of Acoustics*, M. J. Crocker, editor. Copyright © 1997 John Wiley & Sons, Inc. Reprinted by permission of John Wiley & Sons, Inc.)

positions of the transducers they are holding to perform the interrogating. While these systems make use of standard flaw detection instruments or at least the modules that comprise them, they are not treated at length in this chapter because they tend to be *ad hoc* systems built on the captive audience principle where a government agency contracts with a development facility for one or a few items. One scanning system being commercialized (NDT Update, 1997) is the MAUS (Mobile Automated Scanning) by McDonnell Douglas (Boeing). The system handles the scanning operations and uses either ultrasonics or eddy currents as the interrogation means. The next set of objects which in time probably will be found amenable to the crawler systems is aircraft skins during major maintenance checks. The captive audience principle is already evident in the research being conducted on this subject.

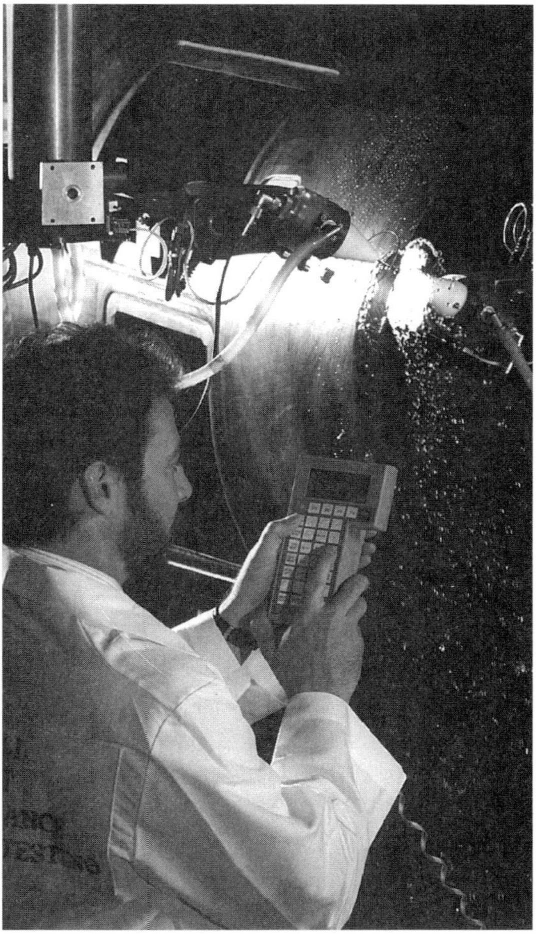

FIG. 39. Another example of a bubbler system (AUSS-V, Automated Ultrasonic Scanning System) used on a composite structure in an airframe. (McDonnell Douglas, Inc. Used by permission.)

2. *Tubes and Pipes*

A specialty field of growing importance is the testing of boiler pipes and tubes. These may be in power plants, petrochemical plants, and other facilities where pipes and tubes are subject to corrosion. Thinning of the tube wall due to corrosion is one flaw to be detected; the other is cracks. Private, competitive suppliers have already commercialized mechanisms for this testing.

FIG. 40. Bubbler system of Figure 39 positioned to test an aircraft part. This figure emphasizes the carriage and complex manipulation mechanisms and shows the system as a large installation. (McDonnell Douglas, Inc. Used by permission.)

Ultrasonic thickness gaging is the technique in use for corrosion. Ultrasound is also used for crack detection. The tubes are generally inaccessible from the outside, frequently being hexagonally close packed in large arrays with only a little liquid flow space between them and held in place by spacers and bulkheads that further restrict access. Therefore, the technology for the ultrasonic methods must accommodate viewing from the inside of the tube. As the tubes can contain liquids, an immersion ultrasonic technique is optimum.

One class of instruments in the field consists of an ultrasonic thickness gage with computer data processing and a search unit that carries an ultrasonic transducer and a mirror along the centerline of the tube. The transducer radiates axially. The beam is turned radially by the stainless steel mirror at 45° to the beam and coaxial with it. The mirror is rotated through 360° azimuthally as the search unit is scanned axially along the tube. The rates of axial motion, mirror rotation, and pulse repetition are appropriate to give 100% coverage to the tube wall.

Figure 41 shows a composite illustration of a tube wall thickness inspection system by IRIS Inspection Services. A heat exchanger with a bank of tubes is being tested. The airbrush drawing of the tube shows the self-centering test head inside with its ultrasonic transducer radiating outward from its rotating mirror. The test head is pushed through the tube by a stiffened cable and a "pusher" device, which can be reversed to pull. In addition, the illustration shows the electronic instrument with monitor, keyboard, and hardcopy printer. The instrument images the thickness of the tube around its circumference. The plots show four calibration grooves incised into the wall thickness. With three interchangeable test heads, tube diameters from 0.48 to 2.5 inches can be tested for thickness. Figure 42 shows the three test heads and the "pusher." Figure 43 is a detail view of one of the self-centering test heads.

Characteristics of an instrument by Anser, Inc., using essentially the same principle of the axial transducer and the rotating mirror are shown in Figures 44 and 45. The computer monitor display of the instrument in Figure 44 shows the thickness of the tube around a 360° circumference. Dents, pits, corrosion, and out-of-round can be detected by the instrument. In this display, the contour of a thinned area is shown. Both above and below this area, there is a gray stripe indicating lack of received signal because the slope of the sides of this pitted area were too steep to give a reflection back to the transducer. Figure 45(a) is a photograph of four ultrasonic probes; Fig. 45(b) is a mechanical drawing of the structure of a probe. The mirror is mounted with ball bearings and rotated by water flow through an integral turbine in a probe large enough for such a structure. One of the probes is flexible for curved tubes.

An instrument by Nerason, Inc., that has a flexible test head section is shown in Figures 46 and 47. Having the capability to traverse bends in U-tubes, this system can test tubing of ID as small as 22 mm and bend

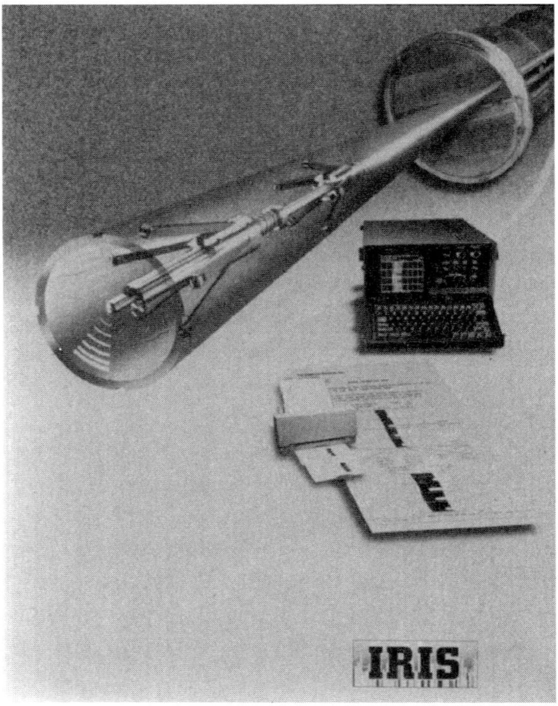

FIG. 41. Composite illustration of a tube wall thickness inspection system. A heat exchanger with a bank of tubes is being tested. The airbrush drawing of the tube shows the self-centering test head inside with its ultrasonic transducer radiating outward from its rotating mirror. The test head is pushed through the tube by a stiffened cable and a "pusher" device, which can be reversed to pull. In addition, the illustration shows the electronic instrument with a monitor, keyboard and a hardcopy printer. The instrument images the thickness of the tube around its circumference. The plots show four calibration grooves incised into the wall thickness. (IRIS Inspection Services, Inc. Used by permission.)

radius as small as 55 mm on the inside. Using the same principle as an axially directed transducer beam and mirrors, both at 45° and other angles, the instrument can measure wall thickness and detect the presence of flaws. Mirror angles can be chosen to cause the ultrasonic beam in the water in the tube to refract at the tube ID into Rayleigh waves, lamb waves, or shear waves for reflection from cracks. After the path back through the metal and the water, the reflections are picked up by pulse-echo circuitry.

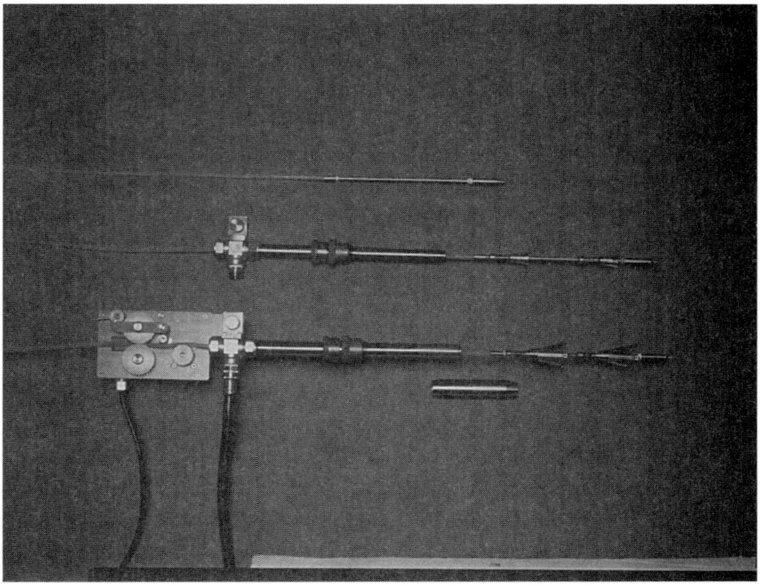

FIG. 42. Details of the "pusher" and three test heads for the instrument in Figure 41. The size range of measurement in the tubes is from 0.48 inch ID to 2.50 inch OD. (IRIS Inspection Services, Inc. Used by permission.)

H. MATERIALS PROPERTIES SYSTEMS

1. General

Materials properties systems using elastic wave motion fall into two categories, ultrasonic velocity and sonic resonance. While sonic resonance need not be ultrasonic, it is very relevant to this chapter. Omitting it would do the field a disservice.

2. Ultrasonic Velocity

After the landmark discovery by Torre and others at GM that ultrasonic velocity was monotonically related to nodularity, yield strength, and tensile strength in ductile (nodular) cast iron (Torre, 1986), the field of ultrasonic velocity for materials properties measurements came into its own and spawned many instruments. Independently, engineers at Norton Abrasives discovered that sonic resonance could be used to ensure the integrity of high-speed grinding wheels.

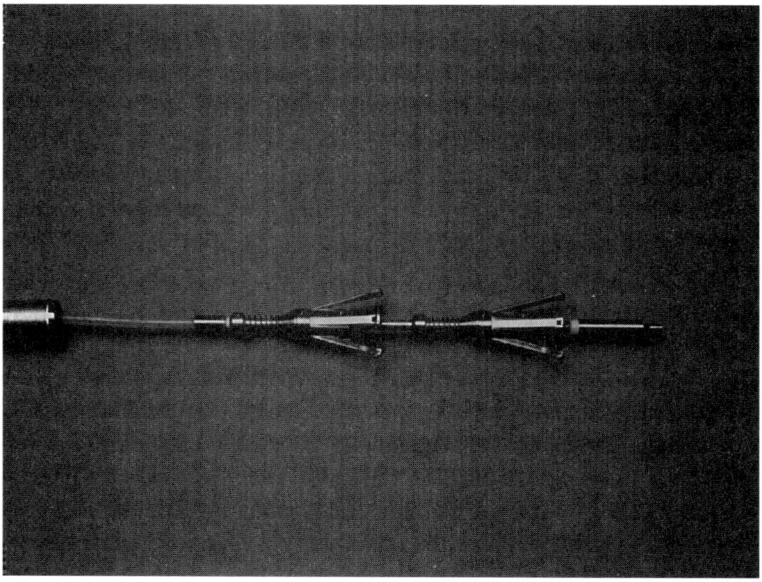

FIG. 43. Details of one of the test heads for the instrument in Figure 41. (IRIS Inspection Services, Inc. Used by permission.)

FIG. 44. Another instrument for corrosion testing in tubes and pipes. The computer monitor display shows the thickness of the tube around a 360° circumference. Dents, pits, corrosion, and out-of-round can be detected by the instrument. The contour of a thinned area is shown. Both above and below this area, there is a gray stripe indicating lack of received signal because the slope of the sides of this pitted area were too steep to give a reflection back to the transducer. (Anser, Inc. Used by permission.)

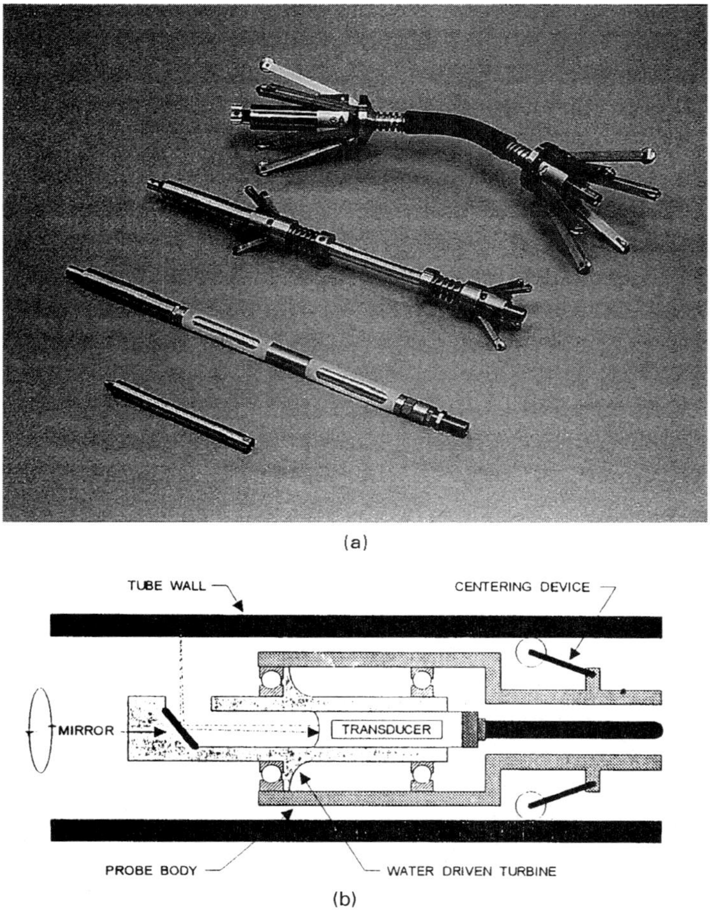

FIG. 45. Details of the probes for the instrument system in Figure 45. (a) Four ultrasonic probes. One of the probes is flexible for curved tubes. (b) Diagram of the structure of one probe. Note the rotation of the mirror by a hydraulic turbine on ball bearings. Anser, Inc. Used by permission.)

A graph of yield strength and tensile strength of nodular iron versus ultrasonic velocity is shown in Figure 48. The work represented by this recapitulation was done at the Ford Motor Company (Klenk, 1973) on tensile bars and velocity specimens cut from large castings. Data from over 130 parts is represented by the 95% confidence limits drawn in this graph. At this confidence, to assure 60 kpsi yield strength, velocity above 0.221 inches/ microsecond would be required. Generally, the iron foundry industry standard

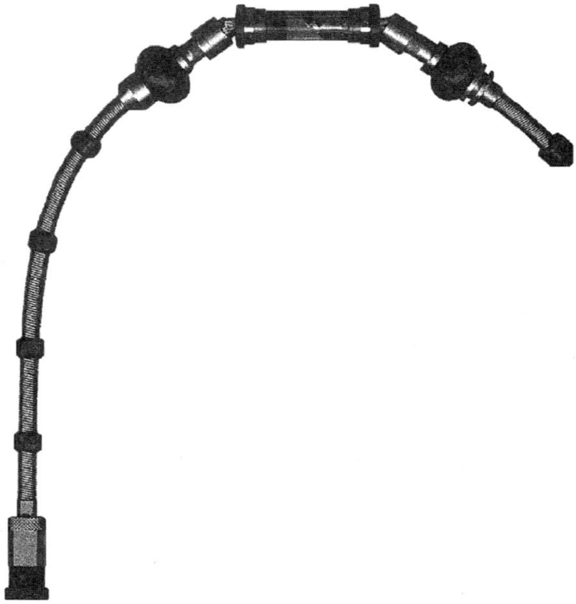

FIG. 46. An instrument for bent tubing to find corrosion and cracks. (Nerason, Inc. Used by permission.)

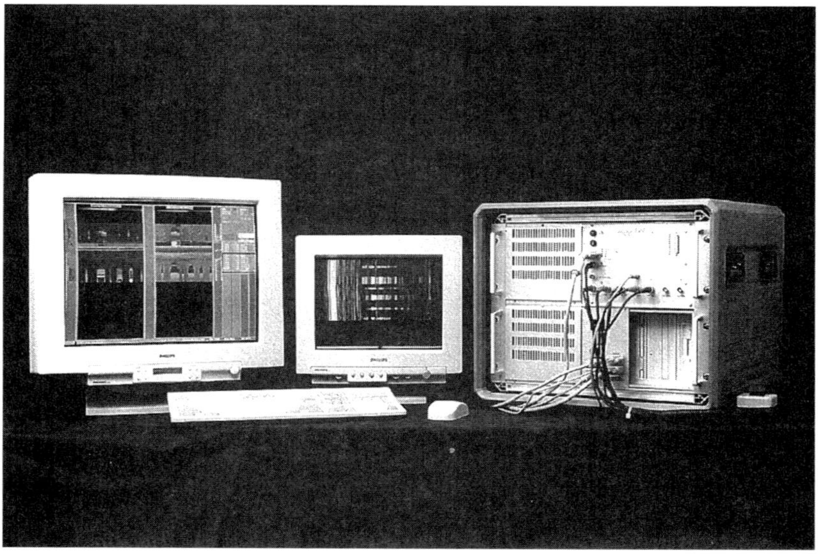

FIG. 47. Details of the instrument in Figure 46. (Nerason, Inc. Used by permission.)

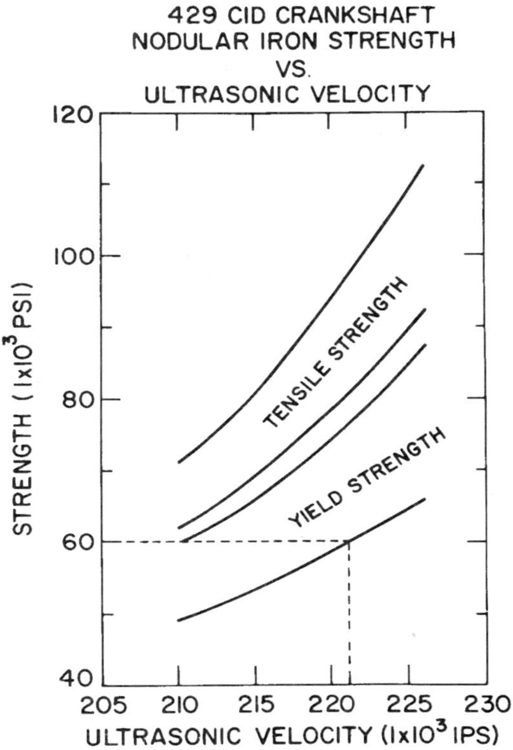

FIG. 48. Tensile strength and yield strength in a pearlitic nodular iron as a function of ultrasonic longitudinal wave velocity. The 95% confidence limits are drawn. Most of the spread lies in the tensile bar measurements of strength. It is believed that the data really represents functional relationships. (Academic Press, Mason XII. Used by permission.)

for excellent nodular iron is 0.223 inches/microsecond. The desired accuracy is therefore about 5 parts in 2000 or 1/4%.

The desire of industry to test castings before machining to save money (which would be wasted by processing faulty material) has led to equipment designs that are ingenious. However, these designs may embody some compromise relative to the idealized situation of testing machined specimens with flat and parallel faces. Castings generally have roughness corresponding to the sand-cast faces and a draft angle to allow for the sand mold to be pulled away from the master while making the mold. To accommodate these conditions and obtain readings by pulse-echo or through-transmission, the frequency must be relatively low, say, 5 MHz. The low frequency permits the penetration of rather large pieces of iron, possibly up to 5 inches. In only a few instances are the measurements made on machined pieces in a factory beyond the foundry. When measuring as-cast material, the exact thickness is not known so the measuring instrument must have the capability of finding thickness as well as travel time to find the velocity.

One manually operated portable ultrasonic instrument uses a caliper fitted with an LVDT gage for thickness. This arrangement was pioneered by Magnaflux, Inc. The caliper measures from the ultrasonic transducer face where it touches the workpiece to the point on the opposite face of the workpiece where the ultrasonic beam will impinge. Then the same transducer picks up the reflection in the pulse-echo mode. The pulse-echo travel time measured by the instrument and the LVDT measurement of thickness are then used in the instrument to compute velocity. This instrument has gone through several corporate owners and several generations of improvement. A photograph of the present version marketed by Centurion NDT is shown in Figure 49.

An alternative method is through-transmission in immersion. A diagram of the ray paths is shown in Figure 50. With the distance L known by construction, the three measurements of time, t_0, t_1, and t_2, provide sufficient data to calculate the specimen thickness d and the velocity v_M in the workpiece M (Papadakis, 1976).

A modification to this routine is to skip the time t_2 while instituting another step in which each transducer uses pulse-echo to find the time in water to the adjacent side of the workpiece M. The reader can work out the arithmetic.

Systems to use this measurement method are generally classed as large installations, being built around a large water tank and often having automated handling equipment attached as well as the ultrasonic instruments. Figure 51 is a photograph of a tank designed to test two automotive front wheel spindle supports, one right and one left, (colloquially known as steering knuckles) for both ultrasonic velocity and flaws. One spindle support is in place in the right-hand fixture. The three wands more or less parallel at the bottom of the picture hold pulse-echo flaw detection transducers aimed at areas of the workpiece to detect porosity if present. The two horizontally opposed wands at the top of the picture hold through-transmission transducers aimed at a segment of the workpiece to be tested for velocity. A fail-safe feature of this instrumentation is the calibration block in each fixture, which rotates up between the velocity transducers on a pivoted, counterbalanced arm when the workpiece is removed. This block can be seen in the left-hand jig where the workpiece has been removed. The calibration block is measured

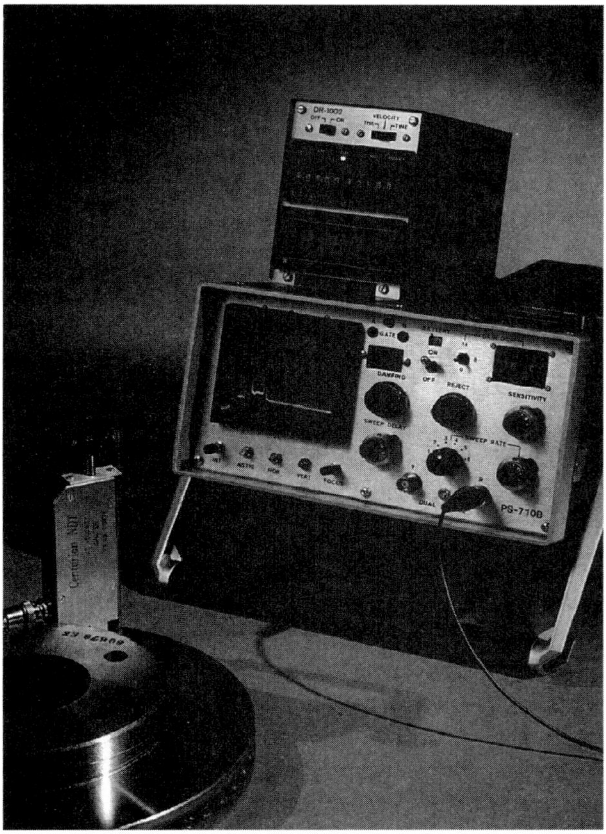

FIG. 49. Ultrasonic instrument with LVDT caliper for thickness being used to measure ultrasonic velocity. (Centurion NDT, Inc. Used by permission.)

every time before the next workpiece is inserted. The instrument must find the correct calibration velocity before being electronically enabled to continue measuring parts.

In Figure 52 a front view of the tank is shown with the Magnaflux electronic operating system for flaw detection and velocity measurements standing next to it. This equipment was state of the art in 1975. The thicknesses and the velocities of the two workpieces are read out on the illuminated displays on the bottom panel in the instrument case. In addition, reject signals indicating inadequate velocity are provided.

A Krautkramer model circa 1995 of a multichannel ultrasonic system to test for flaws and velocity is shown in Figure 53. Both the flaw and the velocity

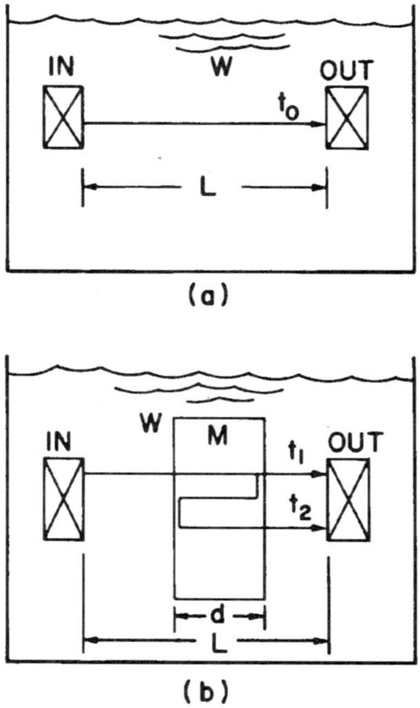

FIG. 50. Diagram of the ultrasonic signal paths used for the measurement of velocity in nodular iron. The time t_0 is measured in water (W) alone. Then the metal M is inserted, and two times t_1 and t_2 are measured. The velocity v_M in the metal can be derived without knowledge of the thickness d. (Academic Press. Mason XII. Used by permission.)

systems come in 4- or 8-channel models. The system comes in a NEMA-12 cabinet with crane hooks, attesting to its status as belonging in large installations.

The instrumentation mentioned earlier (Figure 19) as built on a computer platform has been configured to test for ultrasonic velocity. It is shown in use on ductile iron pipe in Figure 54. The electronics for the velocity test are in the bottom section of the rack in the illustration. All the "bells and whistles" needed to automate the test are in the software.

3. *Sonic Resonance*

Sonic resonance is related to ultrasonic velocity through the wavelength relationship in Eq. (1). A resonance will occur at a frequency at which an integral number of half-wavelengths fit into some dimension of the work-

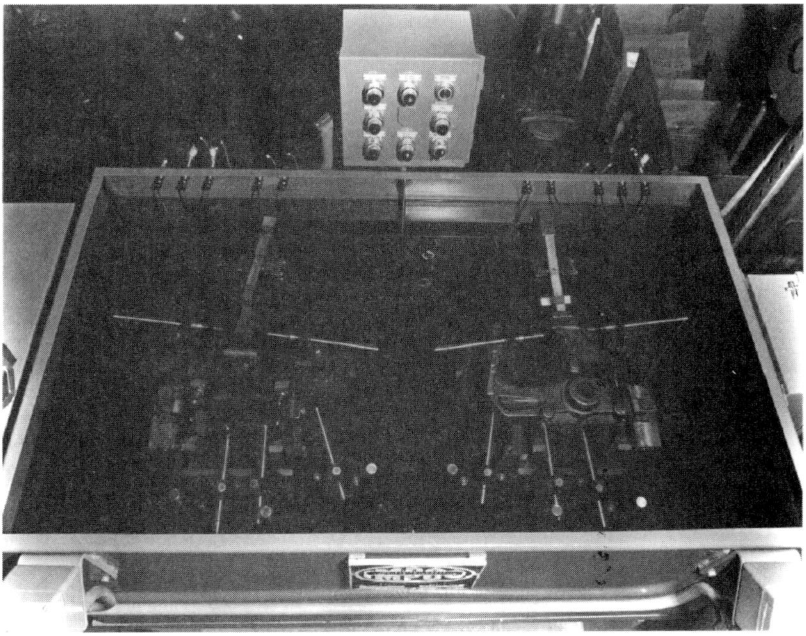

FIG. 51. Plan view of the interior of the ultrasonic test tank. The circular transducers aimed upward look for porosity by the pulse-echo method. The pair of transducers aimed horizontally measure the velocity in one portion of the casting. With no casting in place, a steel block rotates into position between these transducers for calibration. (Academic Press. Mason XII. Used by permission.)

piece. For complex shapes such as engine blocks for an internal combustion engine, the relationships are very complex and can be approached only by finite element analysis software for modal analysis. Both longitudinal and shear motion as well as flexure must be considered in complex parts that are excited by arbitrary forcing functions.

The desire is to use sonic resonance for quality assurance in the same sense as ultrasonic velocity is used. If the material has the best quality for the highest velocity as in nodular iron, then the set of resonance frequencies in a workpiece will each individually be highest for the best nodular iron. (Other grades of iron may be desired, however, with different velocity ranges.) To monitor the frequencies of parts as they are produced, resonance may be a better tool than velocity in the sense that resonances sample almost the entire workpiece except for areas around nodes that are not stressed (Papadakis and Kovacs, 1980). (See Figure 55.) The size must be held constant within

FIG. 52. Photograph of the ultrasonic velocity test set (bottom unit in rack) and the test tank for testing cast iron for strength parameters. (Academic Press. Mason XII. Used by permission.)

adequate tolerances and flashing must be removed from castings to stabilize the frequencies.

The elementary approach is to make some parts with the proper physical properties, particularly the moduli of elasticity, and then find the resonance frequencies. Then, one frequency can represent a mode of motion that will test the workpiece over areas of interest and that is separated adequately from other resonance frequencies. An electronic system can then be built or adapted to measure that frequency when a workpiece is excited into resonance. Some examples of such systems are given next.

One system in commercial use in the automotive industry (Kovacs *et al.*, 1984) is shown in Figure 56. This photograph in a development laboratory shows the commercial instrument by Datac, Inc., being adapted to the testing of cast nodular iron V-8 crankshafts. The lowest mode of longitudinal vibration was chosen for use both in V-8 and in I-4 crankshafts. The test was performed after the sprues and risers had been broken off and the flashing between the cope and the drag of the mold had been sheared off in the casting

FIG. 53. A multichannel ultrasonic velocity and flaw detection test set circa 1995. (Krautkramer Branson. Used by permission.)

plant. Flashing played the dual role of lowering the frequency by adding mass and raising the frequency by adding stiffness across each journal area. In the figure, a solenoid-driven impactor at the right end of the crankshaft (front in a rear-wheel drive car) set the part into oscillation. The oscillation was damped principally by internal friction. To achieve independence of the environment consistent with a heavy-duty factory environment, the design of the supports under and near the ends of the crankshaft was critical (See Figure 57.) The support at the right end was two chemical rubber stoppers or, alternatively, two commercially available rubber bumpers molded onto bolts. The support at the left end (fly-wheel end) was designed to rotate about two rubber mounting

FIG. 54. A current multichannel ultrasonic velocity test set built on a computer platform. This instrument is an adaptation of the instrument in Figures 19 through 21. The velocity circuity is at the bottom of the rack. (Krautkramer Branson. Used by permission.)

pivots and transmit the vibration frequency to an accelerometer (which is not seen because it is protected by the angle bracket at the left in the picture). The instrument measured the frequency and the damping. Originally designed to measure damping when GM was still working on damping and attenuation (Torre, 1986), the damping feature was used in the Kovacs *et al.* development to monitor deterioration of the mounting mechanisms. Details of the system are given in Kovacs *et al.* (1984).

A system installed in a casting plant for a manufacturing feasibility study is shown in Figure 58. This version of the system was manually loaded and hand-triggered by the two-handed OSHA switches at the right and left of the

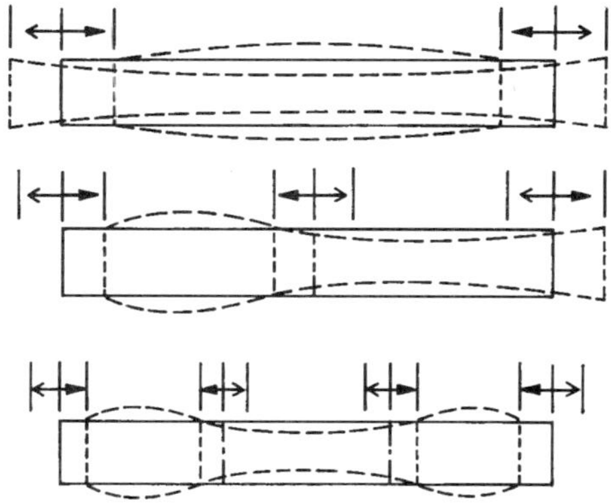

FIG. 55. Fundamental and first two overtones of longitudinal resonance in a bar. The strain is minimum near the nodes, so the resonance method will not test these volumes.

electronics. That feature was installed in anticipation of using more robust and energetic (potentially dangerous) solenoidal impactors on subsequent models.

The installation shown in Figure 59 is automated with a walking beam transfer mechanism to load and unload and a paint spray apparatus to mark rejected crankshafts. One drawback of this installation was the spurious vibrations introduced by the walking beam when loading the workpiece

FIG. 56. Commercial sonic resonance instrument and experimental setup to adapt it to the testing of cast nodular iron crankshafts. The instrument is a Datac Sonic Tester made by the Datac Division of Comtel, Inc. (Ford Motor Company. Used by permission.)

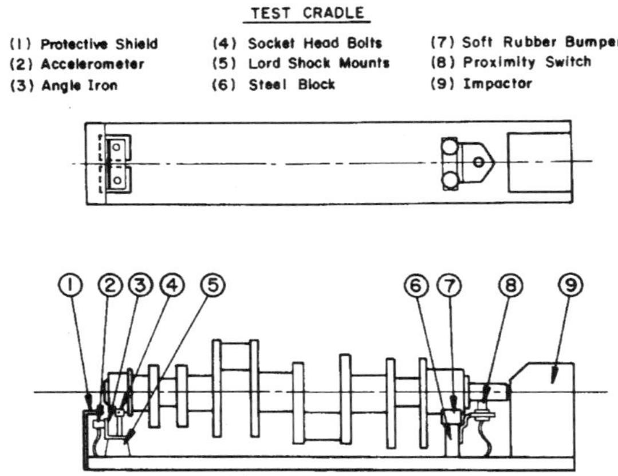

FIG. 57. Detail diagram of the test cradle of the instrument in Figure 56. (Ford Motor Company. Used by permission.)

FIG. 58. The system of Figure 56 set up in a casting plant for a manufacturing feasibility trial. (Ford Motor Company. Used by permission.)

FIG. 59. The walking beam automation for a ruggedized version of the system in Figure 56 installed in a casting plant for testing production. (Ford Motor Company. Used by permission.)

into its cradle. A time delay had to be utilized before the impactor was activated to start the ring-down of the workpiece to measure its resonance frequency. This time delay resulted in an excessive cycle time and the resulting ability to test only 600 parts per hour, whereas the casting plant wanted 850 per hour to achieve 100% testing in one shift. Compromise was necessary.

Still another system installed in a casting plant was inoperable until modified because initially its walking beam was welded to the walking beam output structure of the trim press for removing flash. Major intolerable vibrations were encountered from the trim press shear until the weld was cut and the feet of the sonic resonance section were isolated with absorbent pads. The difficulty would have been eliminated from the initial design if concurrent (simultaneous) engineering had been in vogue at the time.

Another sonic resonance electronic instrument that has been available for some years and has gone through updates is shown in Figure 60. An ASTM Standard Test Method specifically mentions this instrument (ASTM, 1994). (This issue of the Standard is a revision. The fundamental references cited date from 1961 and earlier.) This instrument was developed to ensure the integrity of grinding wheels by sonic resonance; hence the name, GrindoSonic. Figure 61 shows the instrument of Figure 60 in use in a laboratory. View (a) is of a scientist preparing to impact a dental material specimen with the microhammer in his right hand. The specimen is supported on two short pylons on the base plate being steadied by his left hand. The microphone is the cylindrical structure behind the specimen. View (b) shows details of the specimen, test stand, two microphones (on cables), and impactor (microhammer), which is the small steel ball on a thin rod to the left of the test stand. In another application, (Figure 62) a scientist is carrying out expeirments on a ceramic specimen with the handheld impactor, microphone, and the GrindoSonic instrument.

FIG. 60. The GrindoSonic Mk5i sonic resonance instrument determines the frequency of the vibrations created in an object after a shock (impulse) excitation. The instrument uses a microphone or a piezoelectric vibration detector to pick up the vibrations. (J. W. Lemmens, Inc. Used by permission.)

(a)

(b)

FIG. 61. Instrument of Figure 60 in use in a laboratory. (a) Scientist preparing to impact a dental material specimen with the microhammer in his right hand. The specimen is supported on soft foam strips on the base plate being steadied by his left hand. The microphone is the cylindrical structure behind the specimen. (b) Details of the specimen, test stand, microphone (below test stand), piezoelectric vibration detector (above test stand), and impactor (microhammer), which is a 4-mm ball bearing on a long thin plastic rod (left of the test stand). (American Dental Association Health Foundation; Paffenbarger Research Center; NIST-Bldg. 224, Polymers; Gaithersburg, MD 20899. (301) 975-4344/FAX (301) 963-9143/hmueller@paffenbarger.nist.gov. Photo supplied by J. W. Lemmens, Inc. Used by permission.)

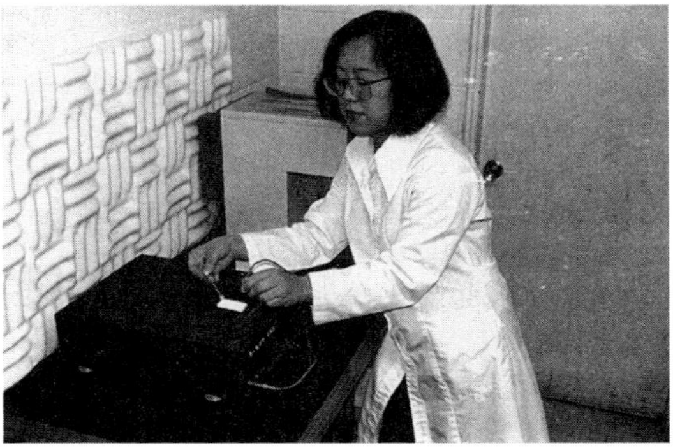

FIG. 62. Scientist carrying out a GrindoSonic measurement on a ceramic specimen with the handheld impactor and piezoelectric vibration detector. (LoTEC, Inc. Photo supplied by J. W. Lemmens, Inc. Used by permission.)

A modern sonic resonance system that has found acceptance in the iron foundry business is the ExperTest instrument from France. The instrument is shown in Figure 63(a) in use on an automotive front wheel spindle support. Other parts are shown in Figure 63(b). This instrument excites the workpiece by sound waves and detects the sonic resonance with a microphone. Both input and output are noncontacting and hence do not perturb the resonance frequency. Many factors do perturb the resonance frequency in addition to the metallurgy; these include cracks, voids, porosity, irregularity in welds or heat treatments, and other potential problems. In the context of the instrument's having "learned" the proper frequency on good parts, these other nonconforming conditions can be detected. The materials that can be tested are not limited to iron castings.

A complex computer-controlled and computer-analyzed instrument developed recently at a U.S. government laboratory (Migliori *et al.*, 1993) has been adapted for quality assurance. The technique is known as resonant ultrasonic spectroscopy (RUS) because of its ability to sweep through and analyze a large range of ultrasonic frequencies. In a part of a known simple shape, enough frequencies can be measured and inserted into enough equations in the computer to calculate all the elastic moduli of materials of nonsimple symmetry (such as orthorhombic). In the quality assurance mode, however, the practical version (Magnaflux MRI-100K) analyzes a few resonances for peak shape and relates the deviations from the ideal to nonconforming conditions. In the case of nodular iron, poor nodularity would lower the resonance frequency and broaden and lower the peak because of higher

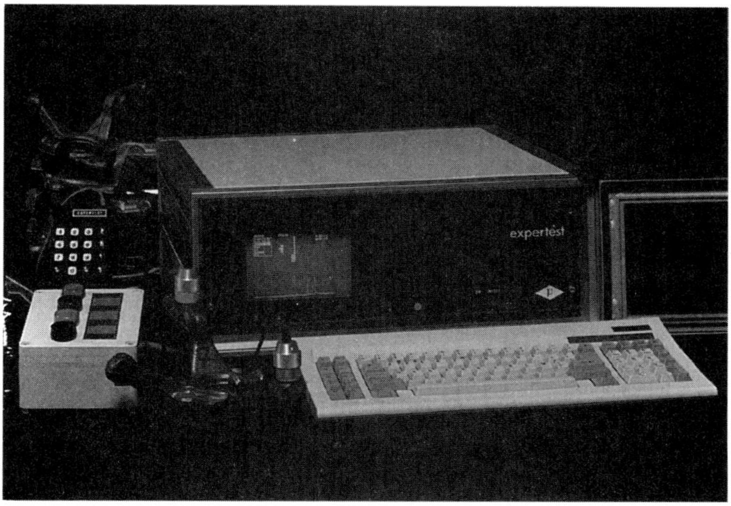

(a)

(b)

FIG. 63. ExperTest sonic resonance test instrument. The instrument is shown in (a) being set up to test an automotive front wheel spindle support. Other parts are shown in (b). This instrument excites the workpiece by sound waves and detects the sonic resonance with a microphone. Both input and output are noncontacting and hence do not perturb the resonance frequency. (Micrel/Hennebont, France. Photo supplied by Casting Consulting. Used by permission.)

damping. Various problems such as cracks, voids, porosity, inadequate heat treatment, and the like can be detected. Metals and ceramics can be tested. A photograph of the Magnaflux MRI-1000K is shown in Figure 64; a fixture for testing ball bearings is in the foreground. A dual system for the testing of industrial parts is shown in Figure 65(a). The fixture for one part and a grouping of several different automotive parts is shown in Figure 65(b).

An instrument that is indescribable in terms of pure resonance or pure propagating waves is shown in Figure 66. This is the Sondicator S2B made in former years by Automation Industries. A newer model based on similar principles is currently marketed by Staveley Industries. This instrument uses a probe with two point contacts for transmission and reception. However, the points are so close together at the frequency used (around 25 KHz) that the points are only a few degrees of phase apart instead of many wavelengths.

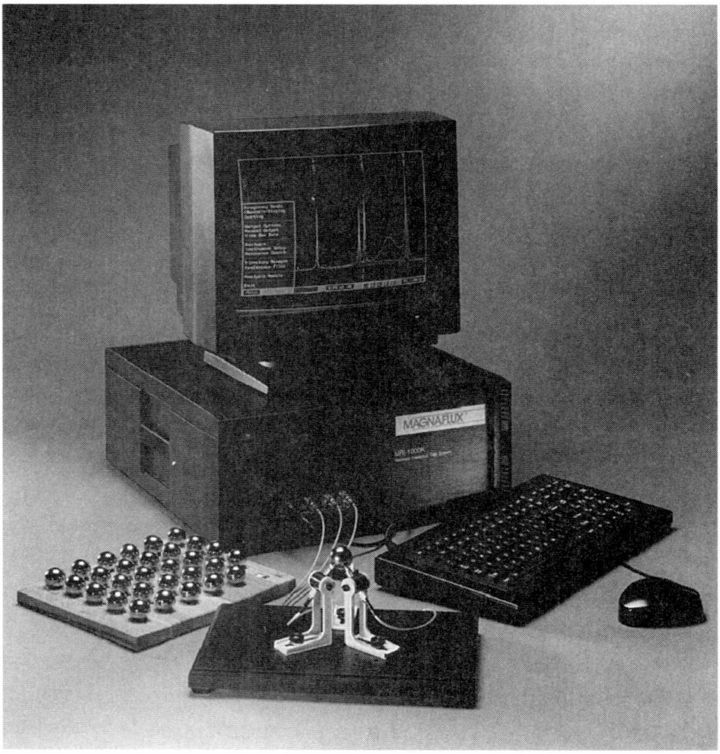

FIG. 64. Photograph of the Magnaflux MRI-1000K. A fixture for testing ball bearings is in the foreground. (Magnaflux, a Division of Illinois Tool Works, Inc. Used by permission.)

(a)

(b)

FIG. 65. (a) A system for the testing of industrial parts comprising two of the instruments in Figure 64 plus parts fixtures. (b) The fixture for one part and a grouping of several different automotive parts. (Magnaflux, a Division of Illinois Tool Works, Inc. Used by permission.)

The instrument is used as a bond tester for relatively thin materials so that the material thickness is a fraction of the distance between the points. The probe is placed on the top side of the top layer bonded. The wave motion is principally Lamb waves propagating away from the transmitting tip but also includes standing waves in the region of the tips and evanescent waves that would have imaginary propagation constants and thus never reach the radiation region away from the tips. The principle of operation is a Go/No-Go alarm based on a phase and amplitude "window" set relative to the input 25-KHz burst phase and amplitude. With the "window" set to accept a well-bonded specimen (calibration standard), the alarm is supposed to go off when signals from a disbonded region have a different phase and amplitude.

G. B. Chapman showed (Chapman, 1981) that the choice of the quality of the reference specimen was critical to the proper operation of the instrument in a high-visibility inspection task on lap joints in sheet molding compound (SMC). He found that if the best laboratory practice were used to produce the calibration standard, then almost all of production was rejected. On the other hand, most of production survived in the field. Thus, superlative calibration standards were inadequate. Chapman devised a statistical method to choose a constant level of mediocrity in survivable bond specimens to use as calibration standards. The ingenious statistical invention lifted a Manufacturing

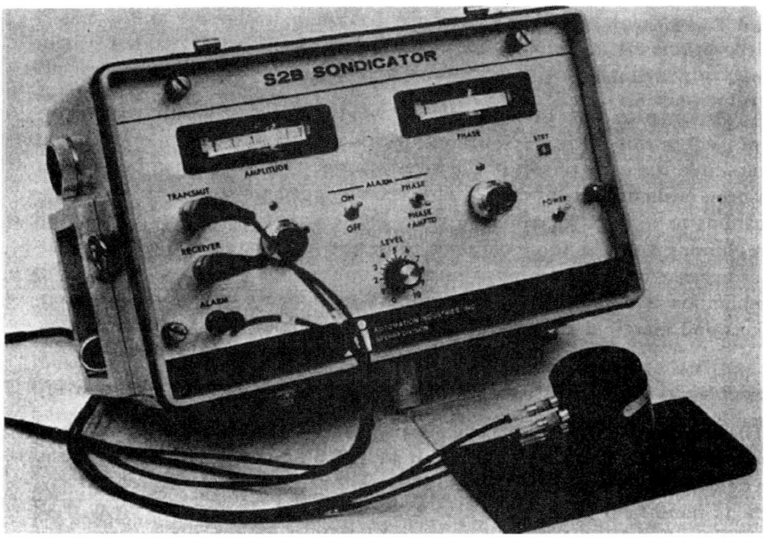

FIG. 66. A bond tester measuring phase and amplitude in a combination of resonance, traveling waves, and evanescent waves. (Copyright ASTM. Reprinted with permission.)

Feasibility Rejection that had been issued by the Ford Automotive Assembly Division against the use of adhesively bonded SMC and staved off a Stop Production order hanging like the Sword of Damocles over Ford heavy trucks. The invention also led to a specification for the quality of adhesive bonds in SMC (Ford Motor Company, 1980) where one had never existed before. As always, intelligence is needed to make use of any instrument.

IV. Summary

In a real sense there can be no summary to this chapter because commercializing research into salable NDT devices, instruments, and systems is an on-going and open-ended endeavor. While the researcher may stop with the publication of some new result, the business person has no place to stop. The researcher goes back to the lab but the business person must go forward into the real world and bear the burden and the anxiety of making the right decisions. The business person earns the daily bread for the whole "food chain" including the researcher, ultimately.

We have seen in this chapter how developments lead to improvements and bring forth new items to commercialize. New products succeed if they serve real customers and present them with value, which usually involves (1) a solution to some previously intractable problem or (2) a solution that is better in some sense than what is presently available. Salespeople and marketing managers who can move product as well as the engineers and expert technicians who order and operate equipment are part of the "food chain" just as the researchers and the electronics engineers and all the others who get ideas, make instruments, do applications, and so on. An egalitarian outlook on all the contributors to NDT is fostered by a study of the commercialization process, and the egalitarian outlook fosters appreciation of every individual.

As to instruments, the future will bring forth more built on computer platforms. Scan presentations in pseudocolor will be used as much as possible. Machine vision with artificial intelligence will be applied to interpret these images to increase reliability by "taking the people out of the loop." With machine processing of information (data) into a picture, the critera for the decision on acceptance or rejection may lie within the computational scheme somewhere previous to the generation of the actual image (Papadakis, 1993b, 1993c). Time could be saved by bypassing the image. Caution will be required; it has been seen that human intelligence makes the instruments operable where all else fails.

The only constant is change.

REFERENCES

ASNT (1959). "Nondestructive Testing Handbook," 1st Edition, Vol. 2, Sect. 44 (R. C. McMasters, ed.). Roland Press, New York, pp. 12–19.

ASNT (1988). Recommended Practice SNT-TC-1A, "Personnel Qualification and Certification in Nondestructive Testing," American Society for Nondestructive Testing, Columbus, Ohio.

ASNT (1991). "Nondestructive Testing Handbook," 2nd edition, Vol. 7 (P. McIntire, ed.). American Society for Nondestructive Testing, Columbus, Ohio.

ASTM (1994). C-1259-94, "Standard Test Method for Dynamic Young's Modulus, Shear Modulus, and Poisson's Ratio for Advanced Ceramics by Impulse Excitation of Vibration." American Society for Testing and Materials, W. Conshohocken, Pennsylvania.

Benson, G. C., and Kiyohara, O. (1974). Tabuluation of some integral functions describing diffraction effects in the ultrasonic field of a circular source. *J. Acoust. Soc. Amer.* **55**, 184–185.

Berlincourt, D. A., Curran, D. R., and Jaffe, H. (1964). "Piezoelectric and piezomagnetic materials and their function in transducers." *In* "Physical Acoustics: Principles and Methods," Vol. 1A, (W. P. Mason, ed.). Academic Press, New York.

Bobbin, John E. (1974). Unpublished talk on Detroit chapter of ASNT.

Cady, W. G. (1946). "Piezoelectricity." McGraw-Hill, New York.

Chapman, G. B. II (1981). A nondestructive method of evaluating adhesive bond strength in fiberglass reinforced plastic assemblies. *In* "Joining of Composite Materials, ASTM STP 749." (K. T. Kedward, ed.). Am. Soc. for Testing and Materials, W. Conshohocken, Pennsylvania, pp. 32–60.

Chick, B. B., Anderson, G., and Truell, R. (1960). Ultrasonic attenuation units and its use in measuring attenuation in alkali halides. *J. Acoust. Soc. Am.* **32**, 186–193.

Deming, W. E. (1982). "Quality, Productivity, and Competitive Position." Center for Advanced Engineeering Study, MIT, Cambridge, Massachusetts.

Duke, John C. Jr., Editor. (1988), "Accousto-Ultrasonics." Plenum Press, New York.

Envirotech (1973). PDS #2100 and #4100. National Sonics, Farmingdale, Long Island, New York.

Envirotech (1974). PDS #4101. National Sonics, Farmingdale, Long Island, New York.

Firestone, F. A. (1945). The supersonic reflectoscope for interior inspection. *Metal Progress*, 505–509.

Firestone, F. A. (1946). The supersonic reflectoscope, an instrument for inspecting the interior of solid parts by means of sound waves. *J. Acoust. Soc. Am.* **17**, 287–300.

Firestone, F. A., and Frederick, J. R. (1946). Refinements in supersonic reflectoscopy: polarized sound. *J. Acoust. Soc. Am.* **18**, 200–211.

Ford Motor Company (1972). Low carbon bare steel spot welding schedule standards. *In* "Welding Design and Reference Data, **WX-12**." June 1972. Manufacturing Standards, Manufacturing and Engineering Staff, p. 3. Ford Motor Company, Dearborn, Michigan.

Ford Motor Company (1979). "Potential Failure Mode and Effect Analysis: An Instruction Manual." Engineering and Research Staff. Ford Motor Company, Dearborn, Michigan.

Ford Motor Company (1980). "Nondestructive Inspection of Adhesive Bonds." Ford Laboratory Test Method FLTM BU 17-1, Product Quality Office, Manufacturing Staff. Ford Motor Company, Dearborn, Michigan.

IEEE (1987). ANSI/IEEE standard on piezoelectricity, Standard #176-1987. *Trans. IEEE-UFFC* **43**(5), 719–772.

Jaffe, H., and Berlincourt, D. A. (1965). Piezoelectric transducer materials. *Proc. IEEE* **53**, 1372–1386.

Jaffe, B., Cooke, W. R., and Jaffe, H. (1971). "Piezoelectric Ceramics." Academic Press, New York and London.

Klenk, R. (1973). Private communication.

Kovacs, B. V., Stone, J., and Papadakis, E. P. (1984). Development of an improved sonic resonance inspection system for nodularity in crankshafts. *Mater. Eval.* **42** (7), 906–916.

Lindsay, R. B. (1960). "Mechanical Radiation." McGraw-Hill, New York.

Lysaght, V. W. (1949). "Indentation Hardness Testing." Reinhold, New York, pp. 134–135, 151, 254–256.
Mansour, T. M. (1979). Evaluation of ultrasonic transducers by cross-sectional mapping of the near field using a point reflector. *Mater. Eval.* **37**(7), 50–54.
Mansour, T. M. (1988). Ultrasonic inspection of spot welds in thin-gage steel. *Mater. Eval.* **46**(4), 650–658.
Mason, W. P. (1950). "Piezoelectric Crystals and Their Application to Ultrasonics." Van Nostrand, New York.
Mason, W. P. (1958). "Physical Acoustics and the Properties of Solids." Van Nostrand, New York, pp. 22–27.
Mattiatt, O. E. (1971). "Ultrasonic Transducer Materials." Plenum Press, New York.
Meeker, T. R. (1996). Publication and proposed revision of ANSI/IEEE Standard 176-1987, 'ANSI/IEEE standard on piezoelectricity.' *Trans. IEEE UFFC* **43**(5), 717–772.
Migliori, A., Sarrao, J. L., Visscher, W. M., Bell, T. M., Ming Lei, Fisk, Z., and Leisure, R. G. (1993). Resonant ultrasonic spectroscopic techniques for measurement of the elastic moduli of solids. *Physica B* **183**, 1–24.
NDT Update (1997). "Portable Scanner Goes Commercial." Business Communications Company, Norwalk, Connecticut (August), p. 6.
Orner, J. W. (1958). Private communication.
Papadakis, E. P. (1959). Correction for diffraction losses in the ultrasonic field of a piston source. *J. Acoust. Soc. Am.* **31**, 150–152.
Papadakis, E. P. (1960). Ultrasonic attenuation in S.A.E. 3140 and 4150 steel. *J. Acoust. Soc. Am.* **32**, 1628–1639.
Papadakis, E. P. (1963). Diffraction of ultrasound in elastically anisotropic NaCl and some other materials. *J. Acoust. Soc. Amer.* **35**, 490–494.
Papadakis, E. P. (1964). Diffraction of ultrasound radiating into an elastically anisotropic medium. *J. Acoust. Soc. Am.* **36**, 414–422.
Papadakis, E. P. (1966). Ultrasonic diffraction loss and phase change in anisotropic materials. *J. Acoust. Soc. Am.* **40**, 863–876.
Papadakis, E. P. (1971). Effect of input amplitude profile upon diffraction loss and phase change in a pulse-echo system. *J. Acoust. Soc. Am.* **49**, 166–168.
Papadakis, E. P. (1972). Ultrasonic diffraction loss and phase change for broad-band pulses. *J. Acoust. Soc. Am.* **52** (Pt. 2), 847–849.
Papadakis, E. P. (1975). Ultrasonic diffraction from single apertures with application to pulse measurements and crystal physics. *In* "Physical Acoustics: Principles and Methods," Vol. XI, (W. P. Mason and R. N. Thurston, eds.). Academic Press, New York, pp. 151–211.
Papadakis, E. P. (1976). Ultrasonic velocity and attenuation: measurement methods with scientific and industrial applications. *In* Physical Acoustics: Principles and Methods," Vol. XII, (W. P. Mason and R. N. Thurston, eds.). Academic Press, New York, pp. 277–374.
Papadakis, E. P. (1982). Sampling plans and 100% nondestructive testing compared. *Quality Progress* **15**(4), 38–39.
Papadakis, E. P. (1983). Use of computer model and experimental methods to design, analyze, and evaluate ultrasonic NDE transducers. *Mater. Eval.* **41**(12), 1378–1388.
Papadakis, E. P. (1984). Physical acoustics and the microstructure of iron alloys. *Int. Metals Rev.* **29**(1), 1–24.
Papadakis, E. P. (1992). Inspection decisions based on costs averted. *Mater. Eval.* **50**(6), 774–776.
Papadakis, E. P. (1993a). Correlations and functions for determining nondestructive tests for material properties. *Mater. Eval.* **51**(5), 601–606.
Papadakis, E. P. (1993b). Imaging methodologies: making decisions before seeing the images. *Mater. Eval.* **51**(2), 120–124.

Papadakis, E. P. (1993c). Image interpretation for quality decisions at production line speeds. *Mater. Eval.* **51**(7), 816.

Papadakis, E. P. (1995). Theory and applications of improved regression analysis for determination of mechanical properties. *Mater. Eval.* **53**(5), 590–592.

Papadakis, E. P. (1996). Elastic moduli for F.E.A./F.E.M. from ultrasonic velocity. *Experimental Techniques* **20**(4), 21–24.

Papadakis, E. P., and Fowler, K. A. (1971). Broad-band transducers: radiation field and selected applications. *J. Acoust. Soc. Am.* **50** (Pt. 1), 729–745.

Papadakis, E. P., and Kovacs, B. V. (1980). Theoretical model for comparison of sonic-resonance and ultrasonic-velocity techniques for assuring quality in instant nodular iron parts. *Mater. Eval.* **38**(6), 25–30.

Papadakis, E. P., and Mack, R. T. (1997). Will artificial and human intelligence compete in NDT?' *Mater. Eval.* **55**(5), 570–572.

Roderick, R. L., and Truell, R. (1952). The measurement of ultrasonic attenuation in solids by the pulse technique and some results in steel. *J. Appl. Phys.* **23**, 267–279.

Scherkenbach, W. W. (1986). "The Deming Route to Quality and Productivity." ASQC Quality Press, Milwaukee, pp. 60, 105.

Seki, H., Granato, A., and Truell, R. (1956). Diffraction effects in the ultrasonic field of a piston source and their importance in the accurate measurement of attenuation. *J. Acoust. Soc. Am.* **28**, 230–238.

Stephan, C. H. (1983). Computer-controlled eddy current inspection of axle shafts for heat treatment. *In* "Computer Integrated Manufacturing—PED-Vol. 8." (M. R. Martinez and M. C. Leu, eds.) (Book No. H00288). American Society of Mechanical Engineers, New York.

Torre, R. (1986). Private communication.

Vary, A. (1978). Correlation of fibre composite tensile strength with ultrasonic stress wave factor. NASA TM-78846 and *J. Testing and Evaluation* **7**(4), 185–191 (1979).

Western Electric (1956). "Statistical Quality Control Handbook" (D. W. Thomas, *et al.* eds.). Western Electric, Newark, New Jersey, pp. 24–28.

Wyatt, R. C. (1975). Imaging ultrasonic probe beams in solids. *British J. Nondestructive Testing* **17**, (September), 133–140.

—4—

Industrial Process Control Sensors and Systems

LAWRENCE C. LYNNWORTH

Panametrics, Inc., Waltham, MA

VALENTIN MÁGORI

Institut für Meß- und Automatisierungstechnik, Universität der Bundeswehr München, Neubiberg, Germany

PHYSICAL ACOUSTICS, VOL. XXIII
ISBN 0-12-477923-9 $30.00

I. General Remarks on Ultrasonic vs Nonultrasonic Technologies and Sensors; Clamp-On vs Wetted Transducers and Sensors; Wireless Remote Sensing

In the world of *homo sapiens*, when one refers to an individual's "sixth sense," it is not likely that an *ultrasonic sensing* capability is being acknowledged. *Acoustic sensing*, however, ranks among the first five of our senses—most of us use ten octaves of its bandwidth as we go about our business or as we take pleasure in listening to birds or music. In this chapter, we will investigate both ultrasonic and acoustic sensing in their various industrial applications.

Industrial applications of sound waves may be anticipated at any frequency where a useful effect or useful sensing is obtainable in a practical manner. This chapter is limited to *sensing* applications primarily in the ultrasonic frequency range from 20 kHz to 2.45 GHz and where the pressure of the interrogating wave is very small compared to the ambient pressure. Also included, however, will be a few applications using audible sound waves, e.g. seismic sensor, an ocean temperature application using an initially high-amplitude carrier having a frequency of 57 Hz, and smokestack temperature and pressure measurements where the frequency of the interrogating pulse, created by an air blast, is on the order of 1 kHz.

Sensing may be *passive*, as simple as listening to the tone, spectrum, or intensity of sound created by flow; or it may be *active*, analogous to a bat's echolocation by the two-step process of transmitting and then receiving. The bat, and most ultrasonic instruments, do not merely receive an echo. They interpret the echo, deciding if the insect or flaw is large enough to demand some further action, or so small it can be disregarded. Bats use ultrasound to navigate, and probably to communicate, but they do not appear to have a high-intensity mode of operation to stun their prey. High-intensity ultrasonic applications such as cleaning, welding, reducing friction, and heating, while important in industry and in medicine, are beyond the scope of this chapter. The focus here is on sensing. (Table 1.)

To assist the engineer or scientist who is trying to select the most appropriate technology to sense a particular measurand (flow, temperature, etc.), several lists of reasons for selecting ultrasound over competing technologies have been compiled (see, for example, Asher (1997) and Lynnworth (1989, 1998)). These lists can be summarized by stating that one generally would select ultrasound when it yields a solution that is *better*, *faster*, or *cheaper* than the alternatives. *Better* can mean more accurate, less intrusive, easier and safer to install, or easier to use. *Faster* means that one can utilize the interaction of ultrasound with the measurand in question within the medium itself to obtain a faster response. Here the ultrasonic signal collects and carries information about the variables to be measured. In this sense, the ultrasonic waves are the true sensors and avoid inertial effects. For example, thermocouples and their sheath have thermal mass, in contrast to an ultrasonic thermometer based on speed of sound through a gas or liquid. A temperature sensor's thermal mass raises questions about thermal equilibrium and response time. Similar remarks apply to a comparison of a big turbine flowmeter's inertia versus that of an ultrasonic flowmeter. (However, if ultrasonic signal processing requires 1000 readings to be averaged, then even if each reading can be obtained in one ms, the effective response time is one second and perhaps no faster than the big turbine with its mechanical inertia.)

Determining whether ultrasonics is *cheaper* than competing technologies is not always a matter of a quick calculation. One must consider the demands of the technology on personnel and energy resources as well as the purchase price of the required instruments. The total price of a sensor may be taken as the sum of its purchase price, plus installation and maintenance costs. Maintenance costs should include an allowance for insurance, representing the economic risk associated with a sensor *not* doing its job with 100% reliability. (This includes not quite achieving the elusive zero false alarm rate.) For example, an ultrasonic flowmeter that exhibits virtually zero pressure drop (zero ΔP) in some instances provides significant energy savings to the user who might otherwise install an orifice plate, vortex shedder, or other ΔP device. On the technical side, the ultrasonic flowmeter might provide advantages of bidirectionality and linear response over a wide range of flow.

Clamp-on capability is often the main reason given for specifying an ultrasonic solution. Clamp-on flowmeters, for example, can be installed without cutting into the pipe, without emptying or depressurizing it to

TABLE 1.

DIFFERENT ULTRASONIC SENSOR PRINCIPLES IN DIFFERENT APPLICATION AREAS. MEASURING TASKS AND APPLICATION AREAS FOR ULTRASONIC SENSORS IN GASES (G, **G**) AND LIQUIDS (L, **L**).

	Ultrasonic Sensor Type											
Application Area	Presence, Approximation	Distance, Mutual Distance	Objects' and Positions' Recognition	Level of Solids & Powder, Liquid Level	Limit Level, also Through Vessel Wall	Density, Mass Flow	Flow by Carry-On Effect	Material Recognition, Concentration	Flow by US-Doppler	Motion by US-Doppler, Alarm	Particle Size and Distribution	Others
Manufacturing technology	**G**	**G**	G									
Machine fabrication	G	**G L**	G					**G**				
Chemical industry				**G L**	**L**	G L	**G L**	**L**			L	L
Petrochemical industry				**G L**	**L**	G L	**G L**	L			L	L
Vehicular technology	**G**		G							G		
Automobile technology		**G**					G			**G**		G
Water supply, waste water		**G**		**G**		L	**L**		**L**			**L**
Food industry				**G L**	**L**		**G L**	**L**			L	L
House, building technology	G	G	G	G L		G L	**G** L	G		**G**		G
Medical technology	G		G				G				L	**L**
Gas transmission, distribution						G	**G**					
Raw material technology				**G L**	**G**							

Bold Print: Existing Products

allow for a sensor to be installed. For pipes of small diameter, the sensor can be simply "snapped on" by an operator with no particular skill in coupling. However, what can be clamped on easily also might easily be installed backwards or be dislodged accidentally or maliciously.

On the other hand, sensors installed *into* a tank are automatically protected by that tank from outside influences. The tank acts as armor (although the sensor penetrates that armor when said sensor is invasive). Nevertheless, one of the reasons intrusive sensors are attractive is that they can be calibrated before or, if removed, after installation. Thus the performance of the invasive sensor is probably less influenced by the tank or process boundary than is the performance of a clamp-on device.

In Asher (1997), the differences between intrusive and noninvasive sensors are nicely illustrated. Apart from the question of intruding beyond or invading (breaching) a pressure boundary, one must also ask whether it is "intrusive" or perhaps "obnoxious" if a process has to be stopped, or whether a pipeline has to be emptied and scoured to install the sensor. In addition, atmospheric (meteorological), oceanic, or downhole sensors generate analogous but different questions. For instance, consider *hot tapping*, which will be discussed later in this chapter. Hot tapping is a widely used method of installing various sensors without interrupting the process, but where the pressure boundary (wall) of the pipe is certainly invaded by drilling or hole-sawing a hole through it. After the cutting tool is withdrawn, the sensor is installed to a depth that may or may not intrude beyond the original internal boundary (pipe ID). (For a hot-tap example, see Figure 1.))

Some pipes are too rough, badly corroded, or otherwise inappropriate for clamp-on. Here a better, more economical solution may be a mass-produced sensor installed by the same methods and with the same tools as are ordinarily used in a particular industry. Thus we find tens of thousands of ultrasonic energy meters installed in European buildings to monitor hot water usage and hundreds of thousands of ultrasonic gas flowmeters monitoring flow in small pipes that deliver methane to British households. The sensors, or transducers, used in these installations are *wetted*, meaning they touch the fluid. Installable devices can be serviced and recalibrated periodically just like many other nonultrasonic sensors.

There are combinations, sometimes called *hybrids*, that combine the principal advantages of wetted and clamp-on transducers. Here a wetted plug touches the fluid but the ultrasonic transducer is external, touching the dry side of the plug, and removable at any time from the plug that is permanently part of the pressure boundary. Sometimes the plug is elongated to many times

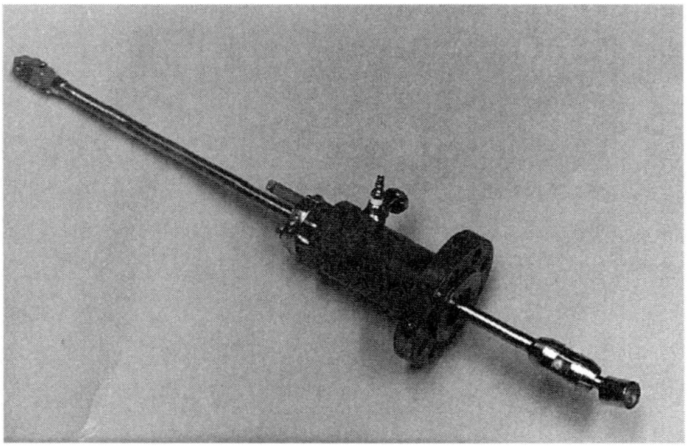

FIG. 1. Transducer for Daniel Ultra Tap™ Ultrasonic Flowmeter mounted on insertion tool. Insertion/extraction system uses line pressure and allows transducer change without interrupting line flow. Design is applicable for both 'hot tap' and spoolpiece type meters, i.e., flowmeter models shown in Figure 53(d), page 394. Photo courtesy Daniel Measurement and Control.

its diameter. In that case it is a waveguide and can buffer or thermally isolate the piezoelectric transducer from exposure to the high temperature of the fluid. One type of hybrid is the calibratable spoolpiece with clamp-on transducers for measuring flow. (A *spoolpiece* is typically a short section of pipe, flanged at each end, and containing two or more transducers appropriate to the paths that must be interrogated to measure flow or another measurand.) In the simplest example the inside of the spoolpiece is unbroken, round, and smooth like a section of a standard pipe. The spoolpiece may be said to be invasive with respect to the pipeline but the external transducers to not intrude. Hybrid flowmeters also exist where a spoolpiece provides a separate calibratable element. On its outside surface the transducers may be attached removably (using yokes or clamps) or permanently (using an adhesive or by welding). Elbows, valves, reducers, thermowells, fluid stratification, or other disturbances in industrial plants, lead to complicated flow patterns (see, for example, Yeh and Mattingly (1997) or Miller (1997)). For a more robust solution that integrates the flow profile even under nonideal unpredictable flow conditions, the calibratable spoolpiece's internal geometry still can be generally circularly cylindrical but it contains special reflective and refractive features. The reflective and integrating features alternatively can be achieved in a spoolpiece of rectilinear, e.g., square cross section. Another type of hybrid is the wireless sensing of a SAW (surface acoustic wave) or plate mode device. The SAW device is installed on a fixed or moving object. When interrogated, it reveals information about the identity of its carrier, or its environment. A recent review article on wireless remote ultrasonic sensing to which one of the authors contributed is Schmidt *et al.* (1996). This topic forms much of the subject matter in Section IV.

The wide range of measurands addressable by ultrasound is often divided into four process control categories (level, flow, temperature, and pressure) and another category served by analytical instruments. This grouping puts NDT instruments in the analyzer area, even though NDT instruments often are in a process control loop, as are densimeters or viscometers. See Figure 2. It will be understood that the same instrument can be used for process control or process monitoring—e.g., a gas meter for controlling a valve or for totalizing gas usage—provided the appropriate output signals or displays are present. Figure 3 is intended to remind us that merely because an ultrasonic approach can measure one or more measurands, there is no guarantee that it will be selected over competing technologies more familiar to the customer, or selected over nonultrasonic meters for which spares are already available and for which personnel are already trained. Ultrasound might be selected only if it is really better when judged by all the parameters appropriate to a given application.

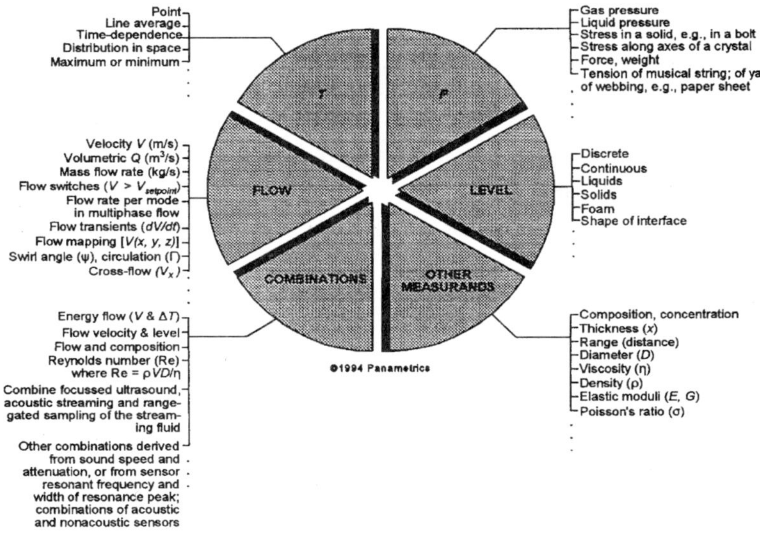

FIG. 2. Acoustic sensors and acoustic techniques measure pressure, temperature, flow, liquid and solids level, and other measurands in the process control field.

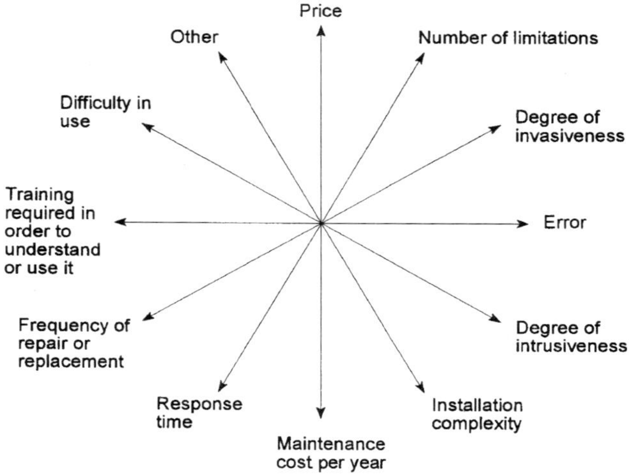

FIG. 3. Zeroing in on an ideal instrument. The origin is the target. The objective is to minimize each term so that the sum will be as close to zero as is practical. Three of the terms will be recognized as the inverses of "better, faster, cheaper" goals.

Because ultrasonic propagation responds to many variables (Figure 2), it offers on the one hand a way of providing a multivariable sensor system (flow, temperature, viscosity, etc.). But on the other hand, its proper use must avoid unwanted response to variables that in the case at hand might be considered *interfering* variables, e.g., flow *profile*, temperature *gradients*, or sources of attenuation *other* than viscosity.

Often, one has a choice of a number of different physical effects that can be used to sense a particular parameter or to obtain the sought information. For sensor applications, the effects have to be compared carefully and the principle with the best chances for a successful realization should be chosen. Therefore not only the technical properties but also (and often even more important) the economic aspects must be considered. Given the sensor's intended function in the system one should derive the properties of the sensor, e.g., accuracy, reproducibility, appropriate measuring range, and a stable, process-compatible housing technology. Important economically relevant objectives are, among others, ruggedness, long-term stability and an optimum cost-to-benefit ratio with respect to its anticipated operation in the overall system. Prior investments in competing technologies by the manufacturer and/or the user also play a role in deciding whether or not to choose the ultrasonic solution.

The costs of ownership to be optimized, apart from purchase price, includes expenses for installation, operation, maintenance, and replacement. Costs also include the impediments and circumstances caused by the sensor. These costs have to be balanced against the 'benefits of ownership', which mean better products with fewer expenses, less pollution, and sometimes even lower insurance rates as the process becomes more reliable and safer. Cost considerations include how the sensors will be handled by the customer. It is not reasonable to expect an average customer to be eager to become an expert in ultrasonic sensor technology. The customer's typical intention would be install it and then forget about it. In the vernacular of 1998, this is expressed as "plug and play."

In keeping with the theme for this volume, the presentation generally will concentrate on showing examples where particular ultrasonic instruments are used in recurring applications. Schematic diagrams are included, however, where it is felt that a *concept* may be better illustrated by a drawing than by a photograph of equipment. Some of the concepts presented, generally referred to as concepts or proposals, may not have been reduced to commercial practice or even to a laboratory prototype, as we go to press. The emphasis is on routine, rather than novel, one-of-a-kind situations. Technical explanations will cover the background, basic principles, and basic ideas, but not detailed mathematical analyses. More complete explanations are generally available elsewhere. In the case of flow, some applications have become so well developed that standards and guidelines exist, e.g., an ASME or AGA Recommended Practice (Anon., 1998), ISO tentative standards (Anon., 1996), or GERG 2 Guidelines. These documents contain technical explanations as well as references for more complete explanations. The reader is referred to the *latest versions* of such documents.

In contrast to, but also analogous to many other sensing technologies that sample distributions by multipoint methods, multipath ultrasound can look at large numbers of points arranged along several lines, as in the examples shown in Figure 4. Ultrasound can also participate in discrete multipoint sensing, or sense the general response of a large volume, e.g., measure the tone and/or rate of decay of sound in a resonant region. The latter parts of Figure 4 serve as samples of or an introduction to topics, techniques, or trends evident in the sensors literature shortly before press time. Readers interested in tracking trends in ultrasonic instrumentation may want to examine surveys or books from recent decades. Examples might be Bergmann (1954, general); Knapp Boetticher (1958; on flow); Babikov (1960, general); Frederick (1965, general); Lynnworth (1975, general; 1979, flow; 1989, general); Kažys (1986, general), Hamidullin (1989, flow and liquid level); Royer and Dieulesaint (1996, 1998, general); Ballantine *et al.* (1996, sensors); and Asher (1997, general). To compare ultrasonic vs nonultrasonic flowmetering technologies, see Miller (1996). Note that in short reviews of flowmeters the range of features and prices of ultrasonic and probably nonultrasonic meters cannot be explored in much depth.

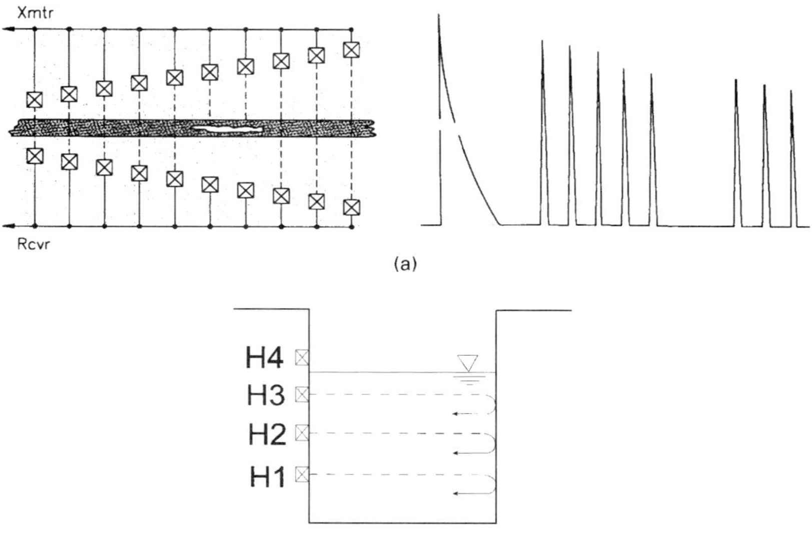

FIG. 4. Industrial applications of multipaths. (a) Ten parallel through-transmission paths, used to scan an immersed sheet for delaminations that are indicated by signal blockage. Adapted from Krautkrämer and Krautkrämer, p. 415 (1977). (b) Liquid level sensing by the pulse-echo method. (c) Top: Sound speed *c*-scan to determine temperature T distribution by a tomographic method. Only one of the eight scan patterns is shown by solid lines. Each of the other seven (dotted paths) originates from another transducer. Bottom: Proposed adaptation to flow imaging in a rectilinear duct. (d) End view of off-diameter tilted chord bidirectional scan, according to the GC (Gauss-Chebyshev) quadrature integration method, to determine flow in a pipe when the flow profile is unknown. Cavities correspond to vee paths obtained using thin transducers as described in Lynnworth *et al.* (1996). For the four-path GC arrangement shown, the volumetric flow rate is $Q = [0.138(V_1 + V_4) + 0.362(V_2 + V_3)]\pi D^2/4$. (e) Bidirectional simultaneous scan of chords of different lengths, analogous to Fig. 4(a). In (a), a *line* is all scanned at once, whereas here flow across the cross-sectional *area* is scanned along three chords all at once. Not all paths in the fluid would be necessarily of different length if the measurements were confined to one plane. It may necessary be to skew the paths, or to add shorter or longer buffers outside the fluid path, so that the received signals, including their ring-down, fall within separate range-gated transmission windows W_A, W_B, and W_C. The respective insertion losses are $(IL)_A$, $(IL)_B$, and $(IL)_C$.

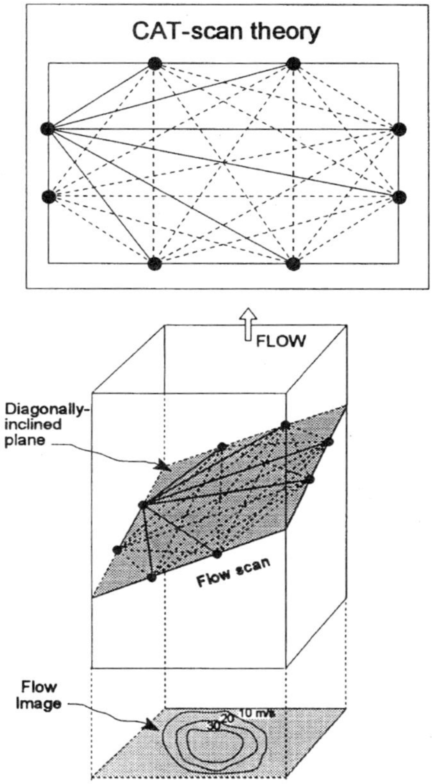

(c)

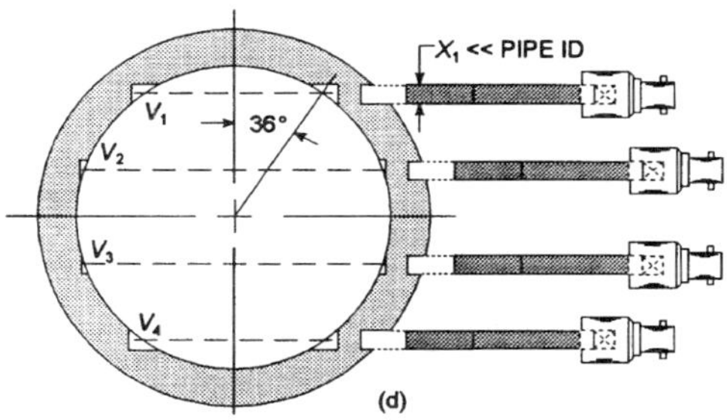

(d)

FIG. 4. *(continued)*

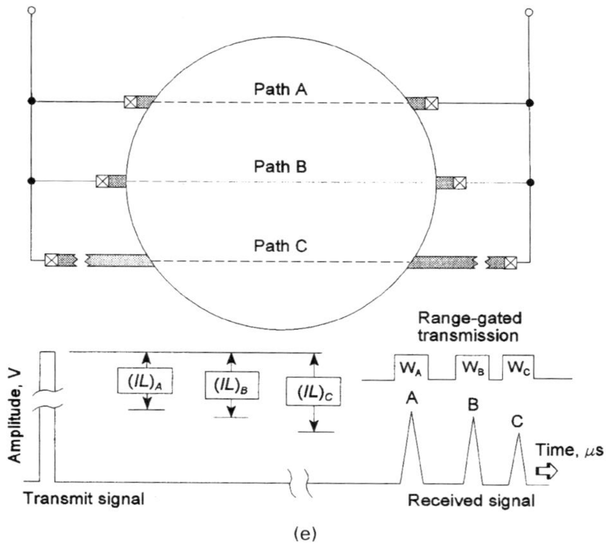

FIG. 4. *(continued)*

It will be recognized that there are important high-quantity installations of ultrasonics in areas not touched on in this chapter. For example, ultrasound is used in measuring fuel remaining on many large commerical airplanes. Aircraft ice detection sensors based on ultrasonics is an evolving technology. Proprietary, timing, or other concerns have prevented details from being included here.

A. ARCHITECTURE OF AN ULTRASONIC SENSOR SYSTEM

Sensors are the sensing organs of technical systems. They collect information about variables in the environment as well as on nonelectrical system parameters, and they provide the results as electrical signals. Sensors are an essential part of power generation and distribution systems, automated industrial processes, traffic management systems, and environmental and health maintenance systems. The development of sensors was stimulated by modern microelectronics. Relatively complex sensors, which previously were built up as "scientific instruments" (Bergmann, 1954), are now feasible as compact devices at low cost. The proliferation of control systems opened a high application potential for ultrasonic sensors (Lynnworth, 1989), which provide many benefits due to their intertialess operation principle.

Ultrasonic sonar sensors actively transmit acoustic waves and receive them later. This is done by ultrasonic transducers, which transform electrical signals to an ultrasonic wave and vice versa. Often it is possible to use the same transducer for both transmitting and receiving. On its path from the transmitter to the receiver, the wave becomes modified by the situation under investigation (see Figure 5). The ultrasound signal carries the information about the variables to be measured.

The task for the ultrasonic sensors is not merely to detect ultrasound. As *intelligent sensors* they efficiently extract the information carried by the ultrasonic signals with high significance (accuracy, resolution, repeatability). To achieve this high performance, the signals are processed, demodulated, and evaluated by dedicated hardware. Finally, the results are given to a computer for interpretation and classification. Thus, for example, not only a distance but a sequence of distances for recognizing objects can be measured (Schoenwald *et al.*, 1982). By the separation of partial echoes from so-called echo profiles, patterns are achieved for object recognition and lateral position identification (Bernst *et al.*, 1986). Algorithms based on models for the transducer performance, the ultrasonic signal propagation, and the interaction between the physical or chemical variables are employed in the time and frequency

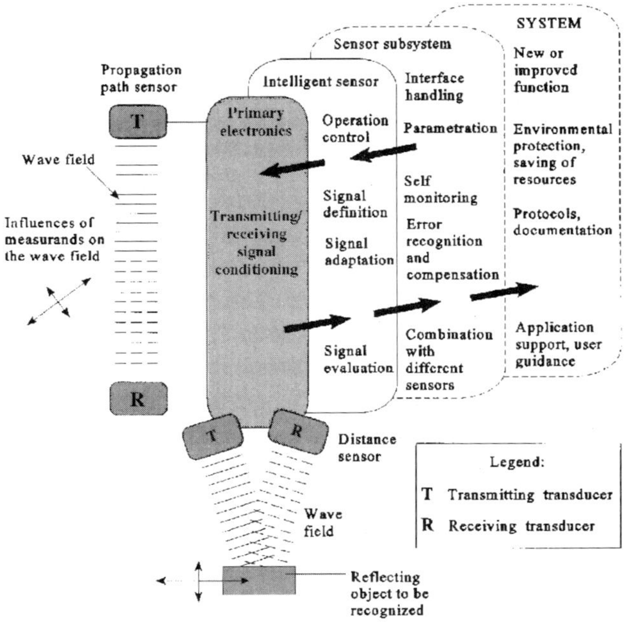

FIG. 5. Architecture of an ultrasonic sensing system.

domains (Lach and Ermert, 1992), including holographic techniques (Löschberger and Mágori, 1988; Knoll, 1991). By the employment of inverse filtered transmission signals, which are "designed" in accordance with a set of "master situations," a fast object recognition or position determination can be performed (Mágori, 1989). Previous knowledge or automatically "learned" experience can be used to improve the sensor's functionality, including *fuzzy logic*, *neural network*, or *nonlinear dynamical (chaos)* approaches (Hill, 1997). Ultrasonic sensors can be embedded in a control system, which uses additional sensors, combines information from the different sensors, handles the bus protocols, and initiates actions.

Depending on how the ultrasonic signal has been changed on its path from transmitting transducer to receiving transducer, *sonar sensors* can be classified as *distance sensors* (reflection) and *propagation path sensors*. Ultrasonic *distance sensors* use the travel time and amplitude of the received signal to derive the presence, distance, and type of a sound reflecting object. By the intelligent evaluation of their echoes objects become recognized and classified. Furthermore, lateral details can be recognized by the consideration of motions between the sensor and the object. In the case of *propagation path sensors* the effect of the tested variables on the ultrasonic transmission is evaluated. Here, the parameters that are affected are the speed of propagation, local changes of propagation properties (diffraction and refraction), directional and frequency dependence (anisotropy and dispersion), propagation attenuation, acoustic impedance, and scattering and waveguiding coefficients.

There are also passive *ultrasonic sensors*, which detect acoustic emissions emanating from objects. These emissions include flow-induced or leak-induced noise. Other sensor systems are sensitive to the change of the properties of oscillating transducers in response to contact with different media.

Figure 5 displays a general architecture of an ultrasonic sensor system, consisting of five layers:

1. The *physical/chemical interaction* layer, composed of the electroacoustic transducers that are used for ultrasonic wave transmission and reception and the interaction between the sound rays or sound field with the quantities to be detected, measured, or recognized.
2. The *primary electronics* layer, representing the electronic circuitry for generating the transmitting signals and amplifiers for the received signals. Fast-acting transmit/receive switches sequentially connect the ultrasonic transducer either to the transmitting or the receiving circuitry. The received signals are filtered and amplified before demodulation.

Distance-dependence of the echo amplitude can be equalized by an automatic distance-dependent (time-dependent) amplification control.

3. The *intelligent sensor layer*, where the received signals are evaluated by efficient sensor-specific algorithms, extracting the information impressed on the ultrasonic signal. On the other side, the form of the transmitted signal can be defined for a most significant sensoric interaction or for an easy and efficient evaluation.
4. The *sensor subsystem* layer, where the sensor becomes interconnected with other sensors, yielding more sensoric information than all the information obtainable with the individual sensor. Further, this layer comprises self-monitoring, error recognition, and error elimination strategies as well as interface handling.
5. The *system* layer is coordinated with the system in which the sensor is applied. It defines tasks to the sensor and utilizes information provided by the sensor to the system's advantage. By the achieved new or enhanced system functionality, for instance, energy can be saved and adverse influences to the environment can be minimized. The application and operation of the sensor can be made user-friendly by support given by the system. It could be important to get a very easy-to-use sensor that only needs to be turned on, with the necessary knowledge for installation and operation implemented in the system layer.

II. Industrial Process Control and Similar Applications

Of the four main process control categories shown in Figure 2, the two that contain the most variety of proven industrial applications, and also the greatest numbers of a particular instrument or sensor, are level and flow. Most of this chapter's examples are therefore in level and flow. Because of the variety of solutions, we include a number of schematics showing sensor and path configurations as well as photographs of commercially available equipment.

1. Different Ultrasonic Sensor Principles in Different Application Areas

Depending on the application, the designer will utilize different sensor principles or a combination of principles. The choice is governed by the factors given earlier. A simple summary statement would be that one seeks a strong and unique influence on the sensor's reading by the measurand, with no influence from interfering variables. Examples of the different ultrasonic or sonic principles available are illustrated by liquid level.

a. Liquid Level. The level of liquids in closed vessels such as tanks or pipes can be measured with audible and ultrasonic elastic waves. The methods currently in use in industry may first be divided into nonintrusive and intrusive categories, i.e., the sensor does or does not protrude beyond the process' natural (internal) boundary. Next the methods may be divided into invasive (boundary-breaching) or noninvasive (e.g., clamp-on); then divided according to whether the measurement occurs at one or more discrete levels or is continuous; further divided according to the type of wave used (longitudinal, transverse shear, torsional, Lamb, Rayleigh, . . .); further divided according to whether propagation is measured in the liquid, in the liquid's container wall, or in a sensor immersed in the liquid; and lastly divided according to whether the effect of the liquid is primarily an attenuation effect or a transit time or sound speed effect. Some special solutions may not fit this categorization exactly; some of the following remarks may also apply to measuring the level of solids, and to liquids in open channels. Asher (1997) carefully distinguishes between noninvasive and nonintrusive sensor installations, covering most cases of interest. However, some weld-in or hybrid situations challenge the common definitions of those terms, as some degree of invasion may be required temporarily as a "weld-prep," with the temporary invasiveness subsequently filled by weld metal during the completion of the sensor's installation. Methods and sensors may also be categorized according to whether their installation and/or their subsequent calibration interrupts the process or interrupts plant operations. A number of level-sensing concepts are diagrammed in Figures 6–9.

b. Nonintrusive Methods. *Nonintrusive* methods typically couple the transducer to a pipe, tank, or structure by techniques such as clamp-on, strap-on, bond-on, press-on, braze-on, or weld-on. Clamp-on methods have the obvious apparent advantage of minimally interfering with normal operations in order to be used, and being transportable from one site to another so that one sensor can yield many measurements. For example, a fuel truck driver could temporarily attach a magnetically clamped transducer to each customer's steel tank that was to be filled, to avoid overfilling. However, a clamp-on device that is seemingly so simple to use might be misused or abused. Permanent *intrusive* sensors can be mass produced largely independent of the tank wall details. To be installed, they are typically threaded into a standard

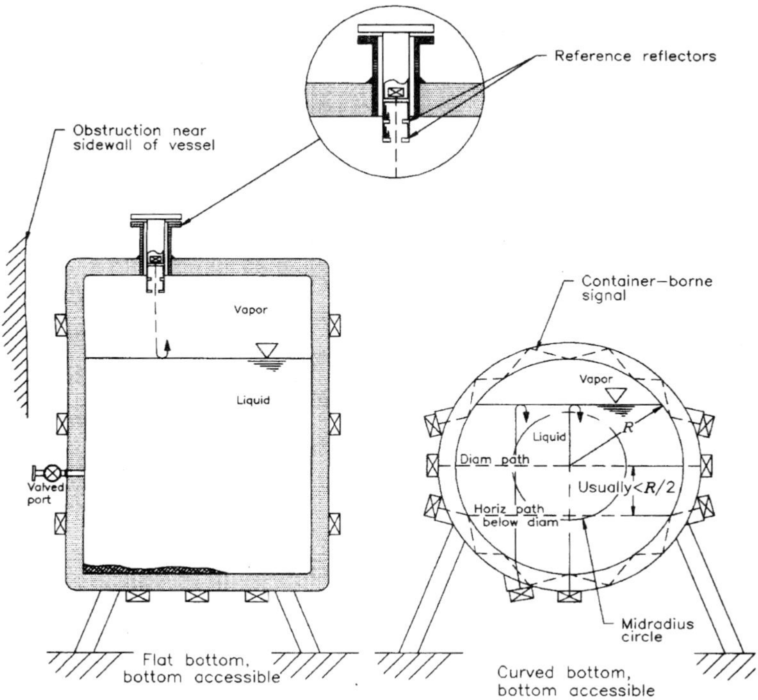

FIG. 6. Schematics showing various interrogation geometries according to what parts of the container are available for use, accessible, and/or penetratable. If the bottom were *not* accessible, but clamp-on is still desired, the methods shown in Figs. 25 or 27 may be considered, in addition to the methods (paths) shown above.

pipe fitting. It is interesting to observe that the clamp-on *flow*meter, compared to, say, an orifice plate, claims the advantage of no pressure drop, which saves money throughout the life of the installation. In the case of a liquid level application that advantage seems absent, when comparing against an intrusive sensor.

The nonintrusive techniques that are well known include liquid presence detecting at a discrete point, based on detecting an echo from the far wall; sensing the change in ring-down when a probe is slid up and down a vessel wall to find the level; sensing change in ring-down at a fixed point when the level passes through that point; sending the sound wave vertically up the liquid column and timing the round-trip. A hybrid case occurs when the nonintrusive transducer is outside the vessel but a stillwell or reflector is placed inside the vessel to improve the measurement.

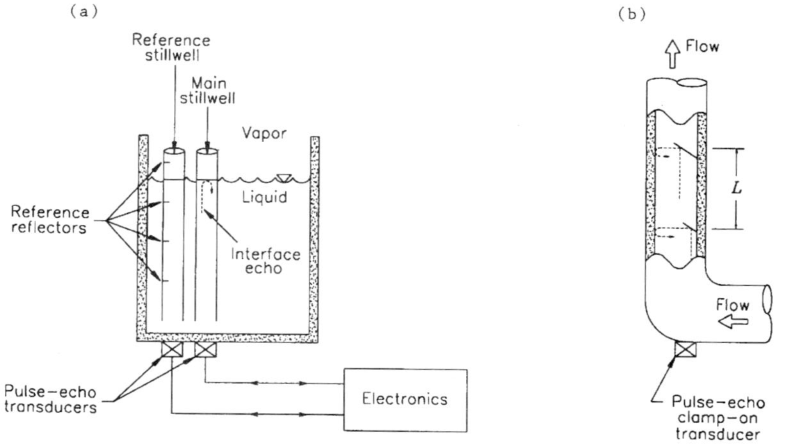

FIG. 7. Reflectors used as (a) reference targets in the reference stillwell of a two-legged liquid level sensor, adapted from a diagram in Frederick (1965), credited to Bogue Electric, and (b) as a means of sensing flow by the *tag* method, using only *one* transducer, adapted from a proposal in Jacobson *et al.* (1988). Somewhat larger reference reflectors, natural or manufactured, are used as landmarks on the sea bottom, shown in Fig. 97, courtesy Sonardyne. Instead of the internal reflectors, one could consider using an array of external thermocouples. One could also consider adding an external concentric "ultrasonic cleaner" transducer to discourage internal residue buildup. These suggestions appear in upward-looking sonar proposals due to Gazis *et al.* (1998).

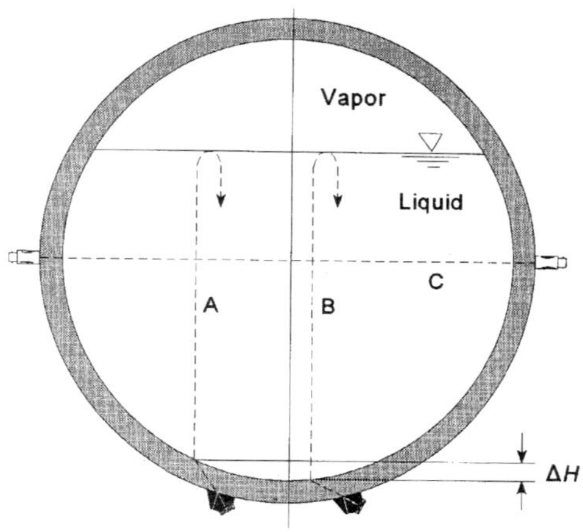

FIG. 8. Different time windows for echoes from *one* interface provide a differential-path way to *measure* sound speed in the path and thereby compute more accurately the distance to the interface. Horizontal path C provides yet another path for measuring sound speed if the tank is at least half full, if the liquid is transmissive, and if there is no attenuative residue or bubble formation on the walls of such a degree as to block transmission.

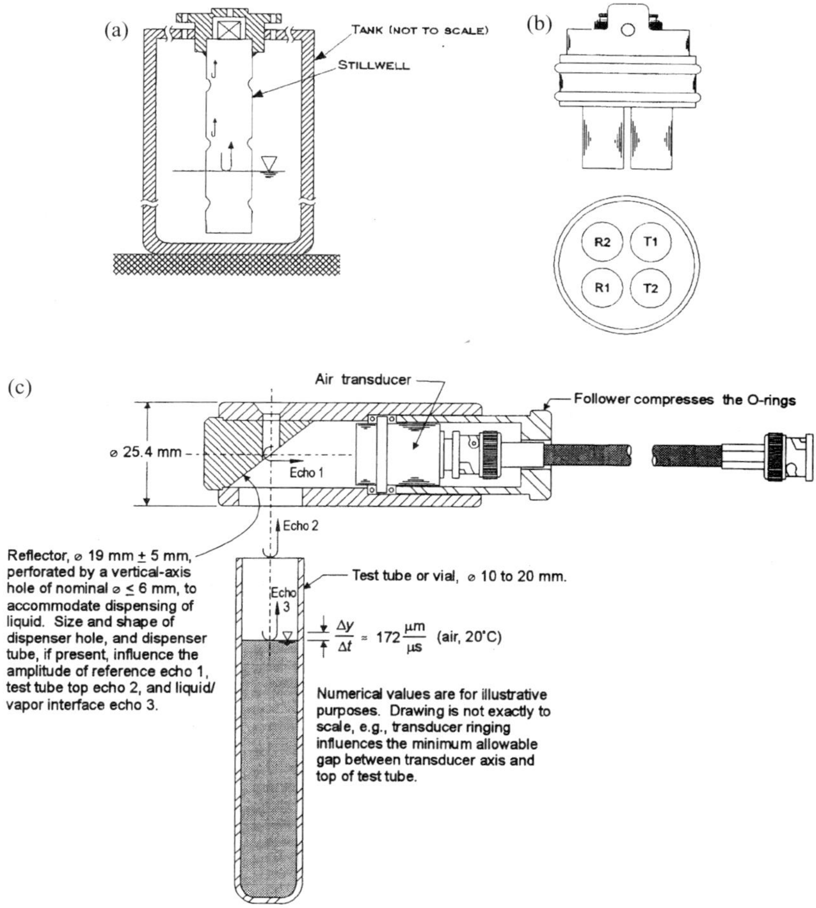

FIG. 9. Downward-looking "air sonars" transmit through the vapor blanket. In (a), the stillwell pipe contains reference reflectors and extends from the transducer down into the liquid, representing a design due to Saab Tank Control. This resembles a stillwell sensor concept due to Tomioka (1968). The paired arrangement in (b) originated as a phased array flowmeter transducer, according to a Panametrics design by Jacobson (1995). Pairs may be operated all at the same frequency or at different frequencies to provide different range/resolution combinations in one given assembly. (c) Schematic shows how a 45° reflector can redirect the interrogation pulse from a horizontal-axis transducer to a vertical path. (Compare with Fig. 21 on pp. 322 and 323.)

Nonintrusive techniques that have been overlooked in numerous surveys of liquid level sensors are those using plate waves and transverse shear waves. One plate wave solution, due to Royer *et al.* (1992), makes use of the leakage of symmetric plate waves. Another plate wave of flexural wave solution makes use of the retarding effect of a liquid adjacent a tank wall (Liu and Lynnworth, 1993, 1995; Sylvia *et al.*, 1994). This latter plate wave method can be applied with the path horizonal or vertical depending on whether discrete or continuous measurements, respectively, are desired. By 1996 a prototype handheld magnetically attachable transducer assembly had been demonstrated on locomotive diesel fuel tanks. Flexural wave transit time responded (increased) when liquid reached the sensor level.

Nonintrusive methods cannot be applied in some situations. For example, sometimes the side wall of a tank is accessible but not available for use because a sensor mounted thereon might pose a safety hazard (e.g., too near a spiral staircase). Or the placement of a sensor might create a clearance problem (e.g., sidewall protrusions from railroad diesel fuel tanks). Double-walled vessels rule out or certainly complicate most or all clamp-on level sensors if physical access to the inner wall is denied.

c. Intrusive Methods. Referring to Figure 9, the top-mounted downward-looking sonars may operate free-field or in a multiple notch stillwell. In either case the transducers do not contact the product. Manufacturers of such equipment include Milltronics and Saab Tank Gaging, respectively. If the product is clean and not residue-bearing and not bubble- or microbubble-bearing, a wetted probe such as a gap probe (one of a number of liquid level products from Cosense) or a torsional waveguide (Panametrics) may be used (Kim *et al.*, 1993). Sometimes a magnetic float rides up and down a magnetostrictive waveguide. The Wiedemann effect is used to achieve accuracy of a fraction of one mm over long spans (MTI). The successful use of this product would seem to contradict the claimed importance of "no moving parts" that is often associated with other applications of ultrasonics, e.g., in flowmeter comparisons with turbines. The torsional sensor, under development for many years for specialized applications such as collapsed liquid level at high pressure and high temperature, may also find applications in cryogenic level sensing, e.g., liquid oxygen, as this is a clean liquid and one into which it is desirable to not introduce electrical energy.

Tuning forks and gap sensors are in wide use, installed by threading or flanging. Liquid dampens the tuning fork vibration, an "attenuation" effect, but liquid increases the gap signal compared to the near-zero ultrasonic signal strength transmitted across the vapor. On the other hand, an airborne signal across a wall-to-wall "gap" will be interrupted (attenuated) by the presence of solids piling up in the path.

The upward-looking transducer runs the risk of being buried in sludge, but if cleaned or if used only in clean liquids, it can provide a reliable echo from the liquid–vapor interface or some liquid–liquid interfaces. A stillwell stabilizes the interface and can contain reference reflectors to compensate for sound speed gradients, the importance of which increases if a temperature gradient exists. A need for *multiple* reflectors is readily appreciated if one recognizes that for water near room temperature, and for many other liquids, sound speed changes by about 1/4% per °C. This means that a small temperature gradient, if not compensated, can lead to an unacceptably large error in computing the liquid level because the c reference is in fact not sufficiently representative of sound speed along the path to the surface. This point should be kept in mind when analyzing systems employing horizontal reference paths to compensate vertical interrogation paths. (A stepped reference reflector with *two* reflecting surfaces was used by Hill and Ruoff (1965) to measure the speed of sound in molten indium between 167 and 345°C.)

Upward- *and* downward-looking sonars were illustrated in Lynnworth (1989, p. 492) rather schematically. Around 1995 U.S. Test began manufacturing *bidirectional* interrogation systems for use in underground fuel storage tanks (such as those at gas stations) to measure the water level at the bottom and the gasoline level according to a method described in Rountree and Berjaoui (1996). The purpose of such measurements is to determine the mass of fuel in the tank and whether the tank is leaking. The mass determination requires a volumetric computation first, followed by compensation for thermal expansion or contraction. The fuel volume measurements are based on obtaining within a stillwell, echoes of the gasoline–water interface and of the gasoline liquid–vapor interface, together with reference echo(es). Note that an accurate measurement of *volume* vs time does not suffice for leak detection because changes in volume due to thermal contraction could easily be misinterpreted as, or mask, an actual leak. In one form of this equipment the relative vertical density (ρ) distribution in the hydrocarbon (gasoline) is obtained from the sound speed (c) distribution, the connection being made by Rao's Rule: $\Delta\rho/\rho \approx \frac{1}{3}\Delta c/c$. (The validity of this approximation is discussed in Rao (1940), Bhatia (1967, p. 24) and Schaaffs (1963).) Instead of placing thermocouples up and down the stillwell, this solution, through an acoustic interrogation from a submerged transducer, yields a measurement of the vertical density distribution near and also remote from the transducer, with very few wires passing through the gasoline.

In the NuSonics analyzer, an RTD (Resistance Temperature Detector) measures the temperature of the liquid whose c is measured between the transducer and the corner reflector. In energy flowmeters T is measured in the hot and cold legs, and flow is measured in at least one of these legs. In the SEI flowmeter, both the temperature near the wall and a local velocity are measured as part of a profile fitting routine. In numerous laboratory experiments and in industrial ultrasonic instruments, one can find further examples of the combined sensing of T and c.

The relationships among density ρ, sound speed c, and temperature T for various fuels (including aviation fuels) are contained in a graph in Lynnworth (1989, p. 426). In general, if the fuel *type* is unknown, or if the fuel is an unknown mixture, measurements of c and T are not sufficient to compute ρ. However, during the course of a brief test such as a one-hour leak rate determination, the fuel type remains constant. If the fuel has the same composition throughout the tank, the link between ρ and c is explained by Rao (1940) and may provide a way to determine the ρ distribution from the c distribution. Rao found the empirical relation $(c^{1/3}) \cdot (\text{molecular volume}) =$ constant independent of the temperature of the liquid. The "constant," however, depends on the particular liquid—e.g., 975 for benzene, 1106 for chlorobenzene—as listed in Rao (1940), and depends somewhat on temperature (see also Schaaffs, 1963).

It is well known that for many organic liquids, $dc/dT \approx -2$ to -5 m/s-°C. If the tolerance is $\pm 10\%$, then dc/dT may be said to round off to approximately -4 m/s-°C for hydrocarbon fuels, based on the data compiled in Lynnworth and Carnevale (1972). As numerical examples, the coefficients for Avgas 100 and JP-4 are about 4.4 and 3.9 m/s-°C, respectively. According to Rao's Rule, $\Delta\rho/\rho \approx \frac{1}{3}\Delta c/c$, an approximation changing only 1.3% for octane from 25 to 45°C. (See also Sections II.C.3. and III.B.2.)

Not necessarily the first, but nevertheless an early use of a *multiplicity* of reference reflectors appears in a one-transducer Rayleigh wave sensor shown in Krautkrämer and Krautkrämer (1966, page 453). This is analogous to the downward-looking design of Tomioka (1968). Temperature profilers using multiple-notched wire waveguides and in sideways-looking or upward-looking liquid level sensing arrangements are shown in Lynnworth and Patch (1970) and in Lynnworth *et al.* (1971). Some single-zone and multiple-zone reference path arrangements are reproduced in Lynnworth (1989, p. 63). The use of a single reflector that is movable through a fluid appears in high-temperature acoustic interferometer designs such as Macedo and Litovitz (1965). The use of a fixed reflector to define the beginning of a high-temperature sensor that is remote from the transducer appears in Frederick (1947 and 1948) for a thick rod, and ten years later in Bell (1957) for a thin rod. The immersed stepped reflector of Hill and Ruoff (1965) might be said to be analogous to these and other stepped solid sensors in providing fixed differential paths for determining c remote from the piezoceramic.

Liquid level may seem "easier" to measure than flow because it lacks the fluidynamic complexity of flow profile as well as other details. Stationary liquids, however, have their own complexities, some due to lack of motion and opportunity for vapor bubbles or solid buildup along the walls to not be swept away as would be the case if liquid were sweeping past the point of sensing. If the solid wall is wet by the liquid, a variety of external sensing options become practical. Most are amplitude- or attenuation-based, e.g., ring-down in the wall (Müller, 1996) (see commercial examples in Figure 25 due to Sunx, Endress + Hauser, and Canongate Technology). Other possibilities include attenuation related to leakage of zigzagging SV (vertically polarized) shear wave (Lynnworth *et al.*, 1982) (its principles are illustrated in Lynnworth (1979, pp. 507–510) and in Figure 6) or of a leaking S_o symmetrical plate wave (Royer *et al.*, 1992) used in products manufactured by Auxitrol. Flexural waves are slowed down by the mass loading of liquid adjacent the wall and so provide a way to sense liquid level continuously, locally continuously, or discretely by timing a low-frequency A_o wave (Liu and Lynnworth, 1993, 1995; Stein *et al.*, 1996; Craster, 1997). A practical implementation in the field is illustrated in Figure 27. The flexural wave time delay is thus analogous to the slowing down of a torsional wave in a noncircular waveguide immersed in a liquid. The flexural wave allows the measurement to be noninvasive, and further allows for external verification by filling a small reservoir adjacent the sensing area with alcohol, water, or other liquid to simulate that the liquid *inside* the tank has reached the level of the reservoir.

Although flexural wave technology was initially developed for sensing the high and high-high levels of clean liquids like gasoline in megagallon ($\sim$10 megaliters) above-ground storage tanks, the foul-resistant behavior of flexural waves is currently being utilized in smaller tanks for gaging sludge level on naval vessels (Stein *et al.*, 1996). A possible future application would be to gage the level of waste liquid in aircraft lavatory storage tanks. Stein found that the use of substantially lower frequency flexural waves avoided the "shielding" effect of buildup that prevents accurate measurements at higher flexural frequencies.

A. Level, Interface, and Ranging

The measurement of the distance between the sensor and an object or the determination of the status of things at a distance is the important task for distance sensors. This comprises presence sensors, distance meters for ranging purposes, and level meters for liquid level in tanks and reservoirs or for different materials in bins or containers. In these applications ultrasonic sensors benefit from important virtues such as ruggedness, contactless operation, and easy installation without impeding the objects to be detected. In comparison to competing noncontact principles (such as optical sensors), the ultrasonic sensors are insensitive to moisture, dirt, and abrasion, and they provide a simpler distance evaluation compared to microwave distance sensors—at least simpler in terms of the time resolution. The radar wave travels through air with no significant speed-dependence on temperature, but at a small reflector distance the echo return time is inconveniently short. Depending on the frequency, ultrasonic pulse-echo ranging in air can operate at distances as short as one or a few cm. (The round-trip transit time in air at 20°C is $t = 2x/c = 100\,\mu\text{s}$ at $x = 1.7$ cm.)

1. Physical Conditions Limiting Distance Range and Resolution in Gases and Liquids

The nominal detection range of an ultrasonic distance is determined by the operating frequency in a twofold way. First, the attenuation coefficient α of ultrasound, increasing with the operating frequency, gives a rather firm limit for the *maximum* distance at which even good reflecting objects are detected. Second, the ring-down or decay time of the transmission signal at the transducer after transmission, being inversely related to the operating frequency, determines a *minimum* distance for objects to be detected. The nominal distance range lies between these two limits, which also depends on the object's reflective properties. To detect relatively weak echoes coming from objects with low reflectivity, one must wait longer until the decaying transmission signal at the transducer has become considerably weaker than the echoes. The local resolution of distance sensors improves by increasing the operating frequency. At similar conditions (constant relative system bandwidth), the local sensor resolution is proportional to the wavelength. This property is convenient, as higher resolution normally is needed when the objects come closer to the sensor. In the case of flow velocity (V) measurements (Section II, B), similar considerations come into play with respect to optimising the operating frequency f. In *gases*, where α is inversely proportional to gas pressure P_{gas}, it is evident that at high pressure one can operate at high f [until limited by turbulence or beam drift] and measure V in small pipes to a high resolution and at a high sampling rate.

Factors favoring the interrogation of air at low vs high frequency are listed in Lynnworth *et al.* (1997a). *Guidelines* specific to air flow measurement are given in Table 2. Actual maximum distances that are practical depend on transducer directivity, sound speed gradients in the path, signal processing, and other factors. For *liquids* the practical range, depending on frequency f, can approach or exceed one km (e.g., Laenen (1984) river flow, 400-m path; Sonardyne positioning system for subsea work (page 458); Baggeroer and Munk (1992) for ocean thermometry *half-way around the globe*).

2. *Ultrasonic Presence Sensors with Time Window Evaluation*

A frequently occurring task is to recognize whether or not an object is at a predetermined position. It has been shown that for this application area ultrasonic *presence sensors* or contactless *proximity switches* are an excellent choice. These sensors use robust ultrasonic transducers (Kleinschmidt and Mágori, 1981) that radiate with high directivity and employ a time window concept to detect objects only if they are within one (or more) predefined distance intervals, ignoring all other objects outside said interval(s). By the use of sensors with different operating frequencies, different application areas—in particular different distance ranges (maximum and minimum distance)—can be covered.

A typical representative of this ultrasonic sensor family was a device with an operating frequency of 200 kHz, introduced to the industrial market by Siemens around 1980 (Mágori and Walker, 1987). Figure 10 shows the 200-kHz Ultrasonic Bero®, which was fabricated for many years in high volume

TABLE 2.
GUIDELINES ON MAXIMUM RECOMMENDED FREQUENCY AND CORRESPONDING WAVELENGTH ($\lambda = c/f$) FOR MEASURING AIR FLOW AT LOW MACH NUMBER AT ROOM TEMPERATURE AND ATMOSPHERIC PRESSURE

Air Distance	f_{max} (kHz)	Wavelength λ (mm)
~1–10 cm	1000	0.343
~10–30 cm	500	0.646
~3 m	100	3.43
~10 m	50	6.46

and, as an OEM (original equipment manufacturer) item, was marketed by other companies too. Its nominal detection range of 0.2–1 m was divided into eight time windows each 10-cm wide, with one of them being selected by a small plug arrangement. All time windows before the selected one can be combined to inhibit the sensor output if an object is in the so-called *blocking range*. This prevents the sensor from being deceived by objects at shorter distances. Otherwise, deception could occur because of multiple transit echoes arriving in later time windows, thereby simulating objects at farther distances. Thus the ultrasonic presence sensor is active as long as an object is in the selected detection range without other objects being at shorter distances. In this way two different operation modes become possible:

1. proximity switch with a presettable detection range, and
2. ultrasonic barrier with a passive reflector at a presettable distance.

Although not fully implemented in this *ultrasonic presence sensor*, the time window concept provides a simple yet efficient signal processing capability. To every time window T_i, a status variable could be associated, having the value H (logic 'high') or L (logic 'low'), depending on whether an echo coincides with the ith time window. The time windows can then be logically combined to form an output statement of higher significance. Figure 11 shows some evaluations of different time windows such as the discrimination of an object by its size (a), the blocking of the output signal for echoes of objects in

FIG. 10. The ultrasonic presence sensor Ultrasonic Bero® with 200-kHz operating frequency. Illustration courtesy of Siemens.

said blocking range (b), and the discrimination of the direction of an object's motion (c). The logical decision can be performed by simple hardware using standard ICs (integrated circuits), as suggested in Figure 11, or in a microcomputer. If the time window status is characterized by a linguistic description, a formal similarity to *fuzzy logic* becomes apparent. For instance, the variables "object too close" or "object too far", initiate a servomotor to correct the distance to the object, unless the object is "too close" and "too far" at the same time (Figure 11, Figure 12). It is left to the reader's imagination to find appropriate linguistic descriptions for the time windows in the simple object recognition setup of Figure 11(d).

The described presence sensor is also available with an "impulse-to-output signal" as an option, rather than with the time window evaluation. This option allows the device to be used as an ultrasonic distance measuring system because the duration of the output pulse is equal to the propagation time of the first ultrasonic echo. Typical applications of ultrasonic presence sensors are compiled in Table 3. As an example, the photograph in Figure 12 shows the automatic adjustment of the tension of a belt as sensed by an ultrasonic presence sensor with time window evaluation. When the tension of the belt is correct, its echo is within the predetermined time window. If the echo occurs in another time window, the tension is accordingly increased or decreased.

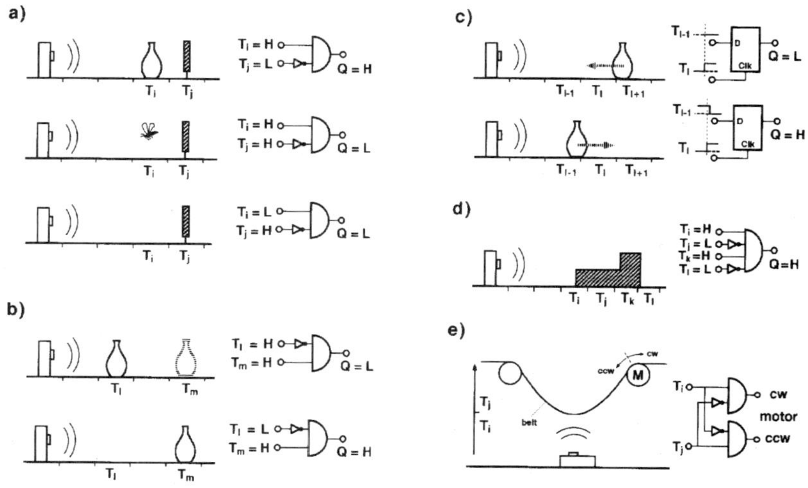

FIG. 11. Evaluation of different time windows.

FIG. 12. Application of an ultrasonic presence sensor: tension adjustment in tape machinery, controlled by an ultrasonic presence sensor. Photo courtesy of Siemens.

3. *Further Developments of Ultrasonic Presence Sensors*

In view of the good results of the presence sensor presented above, its successful use as an *intelligent proximity switch* for many applications, and its acceptance as an innovative product successfully introduced to the market, the further development of a comprehensive family of ultrasonic presence sensors was encouraged. At Siemens a versatile family of ultrasonic distance sensors for a wide variety of applications with different comfort levels at different price classes was developed and improved during the following year (Figure 13). A number of different operating frequencies, whose nominal detection ranges overlap, are in use to cover a wide range of nominal detection distances.

TABLE 3.
TYPICAL APPLICATIONS OF ULTRASONIC PRESENCE SENSORS

Material and machinery control	Object presence detection, sensing, identification and counting; height differentiation; monitoring of spacing; measuring position
Transport control	Monitoring of transport belts and collision control (cranes, ramps, railcars)
Level sensing	Height of material stacks; level in bins and silos; liquid level of industrial fluids; level of water (wells, lakes, rivers, channels)
Detection of persons	Registration of persons; automatic door-openers; safety systems; security

FIG. 13. Ultrasonic presence sensors for different detection ranges. Operating frequency = 40 to 400 kHz. Courtesy Siemens.

This modular system comprises different sensor units ("sensor heads") at different frequencies and using a common evaluation electronic circuitry. The block diagram of this design approach is shown in Figure 14. Additional auxiliary devices can be used as options. As an example, sensor units are shown in Figure 13, which by operating with different frequencies yield overlapping ranges of the following different nominal detection distances (compare with Table 2, which was compiled for *flowmetering* applications).

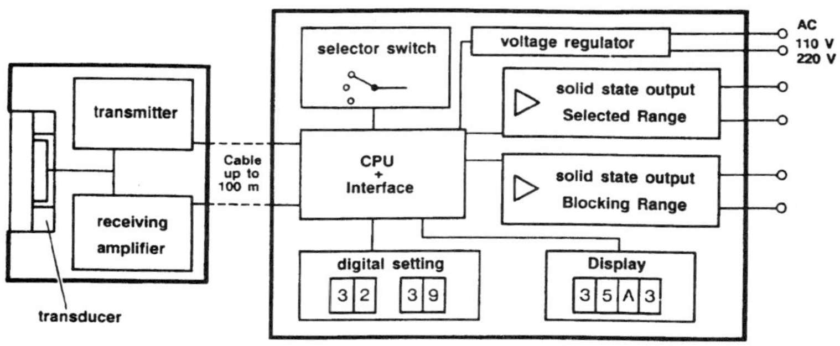

FIG. 14. Block diagram of a modular ultrasonic presence sensor.

1. A short-distance unit with 400-kHz operating frequency for a 6–30 cm nominal detection range
2. A medium-distance unit with 200-kHz operating frequency for a 20–100 cm nominal detection range
3. a long-distance unit with 80-kHz operating frequency for a 60–600 cm nominal detection range

These ultrasonic sensor units comprise the robust, high-directivity ultrasonic transducers of the type RU 400, RU 200, and RU 80 (Kleinschmidt and Mágori, 1981), respectively, and the necessary transmit/receive and echo evaluation circuitry. The direction of the ultrasonic beam can be adjusted easily and accurately, as the transducer housing can be rotated 90° in the azimuthal and 180° in the radial directions and can be fixed in any position.

The evaluation unit contains a microprocessor and can be connected to any one of the sensor units (Figure 14(b)). It supplies the required power, initiates the ultrasonic burst transmission, and processes the echoes by the sensor unit. The upper and lower limits of a time window can be preset, accordingly defining the associated selected detection range. All shorter distances fall automatically into the blocking range, as previously discussed. Optionally, a display is also available to indicate the actual distance to the object being detected. The system comprises an auxiliary multiplexer for the common operation of a number of sensor units with the same evaluation unit. Different output signals are available; e.g., analog, digital, or bus. Using the bus, these distance sensors can be configured for various applications (e.g., to set the width and position of the time windows, the limits for distances, and other operational application-oriented parameters).

As an economical alternative for less demanding applications or as replacement for optical barriers, a range of compact ultrasonic presence sensors is offered, with the same operating frequencies (e.g., 400 kHz, 200 kHz, and 80 kHz). Figure 13 shows such sensors, which in a threaded cylindrical metal housing contain the ultrasonic transducer and the transmit, receive, and evaluation electronics. For low cost applications, the transducer and the electronic circuitry are built into a rectangular plastic housing.

The typical properties of the different ultrasonic sensors are shown in Table 4. All fulfill the appropriate mechanical and electrical specifications for industrial sensors. In particular, the front face of the ultrasonic transducers is not affected by sprinkling water and dirt, which could be a problem for thin foil (membrane) transducers. The applications include detection or monitoring of a wide variety of objects in different application areas. The objects can

TABLE 4.
TYPICAL PROPERTIES OF ULTRASONIC PRESENCE SENSORS

Operating frequency	400 kHz	200 kHz	80 kHz
Nominal sensing range	6–30 cm	20–100 cm	80–600 cm
Activating element (flat area, e.g., minimum area of steel flag)	1 cm × 1 cm	2 cm × 2 cm	10 cm × 10 cm
Switching frequency	8 Hz	4 Hz	1 Hz
Repeatability	0.45 mm	1.5 mm	9 mm
Hysteresis	1 cm	1 cm	6 cm
Maximum length of the sensor (except plug)	140 mm	140 mm	161 mm

be solid or liquid, granular or powder materials. The function of the ultrasonic sensors does not depend on the object's surface quality or color, even if the surface is optically transparent or opaque, polished or dull.

The presence sensors described above are in the market with a high annual number of units, marketed by the manufacturer or by other companies as OEMs. Other firms have developed similar devices with different ultrasonic transducers. Especially after the patents concerning the ultrasonic transducers (Mágori, 1980) used in the described presence sensors have expired, an increasing number of companies will develop their own devices and try to share the market.

Another type of robust air transducer was derived from the continuous emissions monitor (cem) work reported by Matson and Davis (1994) and specifically from transducer arrays designed according to the noise-cancellation method of Jacobson (1995). This type, the Panametrics T7, became commercially available in 1996. It consists of a solid piezoelectric disk that is quarter-wave matched on its front face, supported by a fully potted, bonded, strong backing on its rear face, capable of withstanding hydrostatic test pressures to 1000 psig (70 bar), and sealed within a corrosion-resistant metal case. The thickness of this case (particularly the window thickness in front) depending on the density of the metal used for the case, limits the maximum frequency to about 500 kHz at one bar. The envelope shown in Figure 15 is less than $\phi 25 \times \sim 25$-mm long and is appropriate for T7 transducers operating at 100 and 200 kHz. A longer version operates at 50 kHz. The O-rings acoustically isolate the assembly. This means that either the assembly can be installed inside a standard one-inch pipe plug or pairs can be installed side by side in a two-inch pipe plug as illustrated. In early designs, because of ringing, pulse-echo response was not very good below 500 kHz, but the two-transducer arrangement overcomes some of this limitation. When the T7 is supplied already installed in a pipe plug, the user can usually work with it

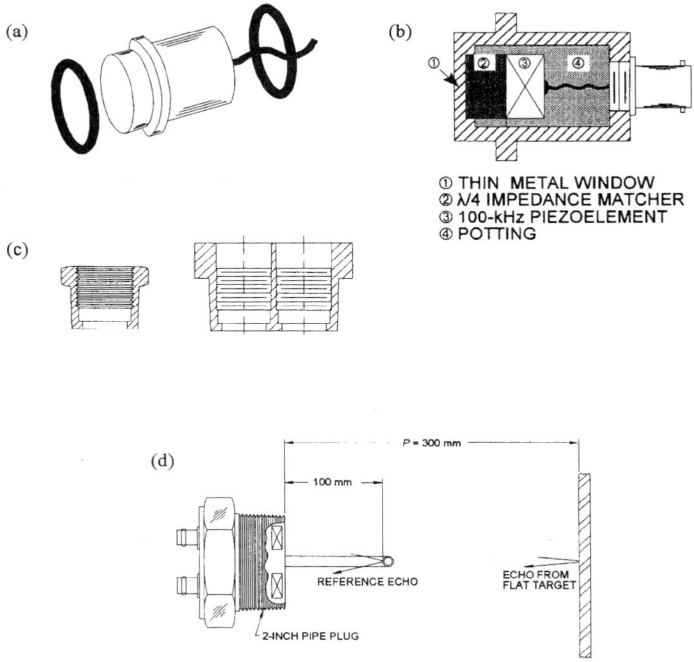

FIG. 15. (a) The 100-kHz transducer's flange is supported (sandwiched) between two soft attenuating O-rings made of, for example, silicone or fluorosilicone. (b) Internal details. (c) Examples of threaded hole patterns for mounting the O-ring flange sandwich transducer in standard 1-inch and 2-inch hex head pipe plugs or flanges. (d) Example of the use of the 2-inch (nominal pipe size) plug with an open-ended tubular reference reflector that could contain conventional (nonultrasonic) temperature and pressure sensors. Test waveforms and other details in Lynnworth *et al.* (1997a).

without having to be an expert in ultrasonic isolation (Lynnworth *et al.*, 1997a; Lynnworth, 1998b).

For applications where the air (or gas) pressure is high and pulsating, the O-ring isolation method may not by itself, provide adequate mechanical stability. In such cases, other designs may be considered. These include: (a) transducer and tube assembly welded to a standard pipe flange; (b) swaged assembly; and (c) a buffered transducer typically held in place by a lap joint flange as in Figure 43, page 365 (Liu *et al.*, 1998).

4. *Ultrasonic Sensors for Traffic Control and Vehicular Technology*

For examples of the state of the art approximately ten years ago, the reader is referred to Lynnworth (1989, pp. 598–601). Ultrasonic presence detectors or distance-measuring devices now find useful applications in automotive technology and for traffic control systems. Due to the rapidly increasing road traffic, these systems and their key sensor technologies can be expected to rise in importance in the future.

So-called *back-sensors*, intended for installation on the rear side of vehicles, facilitate reverse park-in maneuvers. Such sensors are supplied by various manufacturers (e.g., ITT, Bosch). Attempts to reliably measure the distance between cars at normal cruising conditions, however, proved to be not feasible. However, experiments performed with the car of one of the authors demonstrated that it is possible to measure the distance between the car body and the road surface with an ultrasonic distance sensor even at a cruising speed of 200 km/h (approximately 130 miles/h). Such sensors could be mounted in the front bumper to recognize changes in the road surface condition early enough for a potential fast intelligent suspension system to obtain a sufficient reaction.

A further example for a time window evaluation is the parking garage sensor, intended to be placed above (e.g., in the ceiling) parking spaces in a large parking garage to monitor whether the parking place is free or occupied. Figure 16 shows an example of such a sensor operating at ~120 kHz. At a free place, the sensor receives the echo from the pavement. At an occupied place, the pavement or ground echo is blocked by the parked car, an echo from the car's top is received, but with a considerably smaller propagation time compared to the ground echo. In this way, the sensor can discriminate

FIG. 16. Parking garage sensor. Image courtesy of Siemens.

between small objects that may be lying on the ground and a parked car. By a wire network (bus), the information from the individual parking positions are communicated to a central system, which informs the customers about the availability of parking places and guides the driver to a free position. These ultrasonic parking sensors have been built by various manufacturers and have been installed in a number of parking garages. In a similar manner, sensors could be suspended above the lanes of a highway to count the vehicles passing by and to recognize the type of vehicles by their length and the shape of their upper contour. (See also Figure 19(f), board sizing in Taylor-designed apparatus using Massa ultrasonic equipment, and Rosa, 1991.)

5. Ultrasonic Water Tap Control

For noncontact-activated water faucets, a special ultrasonic presence sensor was introduced, using robust, high-directivity ultrasound transducers and a very elaborate evaluation circuitry realized at very low cost. The sensor is not adversely affected by dirt or sprinkle water, and the closed rugged front face of the used transducer can be cleansed easily. The transducer, operating at 200 kHz, is fitted into the tap, connected by a coaxial cable to a control box, and mounted below the hand basin. The control box contains the electronic circuitry (transmitter, receiver, and the echo signal evaluation; Figure 17) together with a power supply. The flow of the water (hot and cold) becomes activated by magnetic valves when hands to be washed are held under the water tap at a position where the hands normally would be held if water were running. The ultrasonic transducer is aimed in a direction such that its sound rays intersect the water jet at an angle at which the echoes will not be directed back to the ultrasonic transducer. A short selected detection range defined by an associated short time interval, which comes at a predetermined delay time

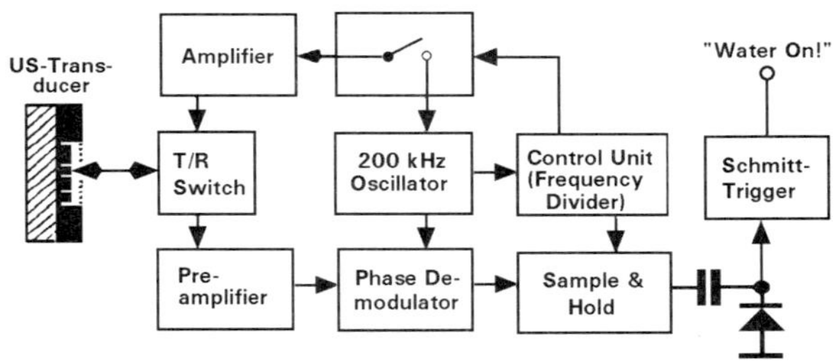

FIG. 17. Block diagram of the water tap control electronic circuitry.

after the transmitted ultrasonic signal, is centered at the intersection between the water jet and the ultrasound beam. This is the intended position for the hands that are to be washed. Within this time window, a Doppler evaluation discriminates between hands (which always have a slight random movement) and inanimate objects (a wet umbrella, for instance) placed in the wash basin.

The transmission signals—short bursts—are coherently gated out from a reference oscillator (about 200 kHz), from which the control signals (repetition rate, time window) are also coherently derived. At every operation cycle (repetition frequency about 100 Hz), the mean phase angle of the echo signal prevailing within the time window becomes stored by sample-and-hold (S&H) circuitry. If slightly moving objects such as hands are in the right position, the output signal of the S&H becomes different in every period, forming an ac signal that activates the water jet. If no echo returns, no signal is stored. Static objects have mainly constant phase angle signals in every period and consequently the S&H output is mainly a dc signal activating nothing. A similar situation is obtained with water droplets separating from the water jet. These droplets cause Doppler-modulated echo signals with sufficient amplitude. The phase angle signal varies in the time window periodically with several periods, in accordance to the relative high speed of the droplets. The S&H, however, stores the mean value of many sinusoidal cycles over the time window. These mean values are mainly constant from period to period and therefore do not activate the water tap.

The water tap control further takes advantage of the ultrasonic transducer used: its high directivity, together with the short time window, allows sensitivity to be confined to a small, well-defined volume. Thus the water flow stops as soon as the hands are removed from this critical volume. The flow of water is neither initiated by the water jet itself nor by other physical interferences such as people passing by nor by acoustic noise interferences such as jingling keys. The use of a relatively high operating frequency of 200 kHz additionally reduces disturbance potential because the high propagation attenuation in air at this high frequency limits the range to the order of 1 m or so (see Table 2).

These automatic water taps are used in lavatories and restrooms in restaurants, hospitals and so on. Figure 18 shows an automatic water tap with continuously controlled water temperature, equipped with the built-in ultrasonic transducer. The ultrasonic water tap control is a good example for the realization of an ambitious concept like a range-gated Doppler moving target sonar at very low cost. An intelligent design approach was realized

FIG. 18. Water tap with temperature adjustment and built-in 200-kHz ultrasonic transducer. Courtesy of Siemens.

using a few standard CMOSs and one radio intermediate frequency integrated circuit together with a robust, high-performance ultrasonic transducer as the key component. A larger number of these sensors have been supplied to a manufacturer of appliances.

6. *Ultrasonic Level Sensors in Air*

The development of transducers for ranging in air and for gas flow measurements opens up other test and measurement possibilities as a part of a progression from intrusive sensing, to clamp-on sensing, to noncontract sensing. Figure 19(a)–(j) shows several examples of air-coupled applications.

Air-coupled ultrasound has been used for many years to test products such as rubber which, although solid, exhibit acoustic impedances more like a liquid. Other solids with Z on the order of $\sim (1\ \mathrm{kg/m^3})\ (1500\ \mathrm{m/s}) = 1.5 \times 10^6\ \mathrm{kg/m^2 s}$ are wood, plastics, and foamy plastics like Styrofoam. Around 1996, several laboratories and one pilot plant began evaluating T7 and other air transducers (Section II.A.3) for applications where it is desirable to obtain information about acoustic propagation *in* the solid without actually

(a)

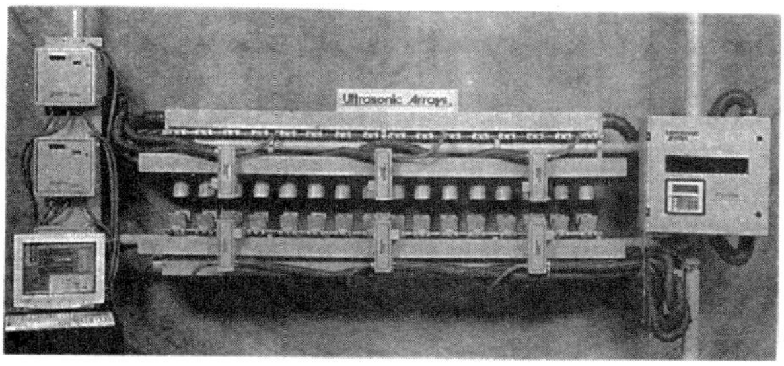

FIG. 19. (a) Example of ultrasonic air-coupled dimensional gaging equipment from Ultrasonic Arrays. One of their air-coupled systems, the BMS-1000 bond measurement system, is designed to monitor the internal integrity of composite products. Compare with *water*-coupled analog twenty-one years earlier, in the 1977 Krautkrämer and Krautkrämer diagram from which Fig. 4(a) was derived. (b)–(j). Further examples of air-coupled applications or transducers, courtesy Airmar, Delta, Hyde Park, Lundahl Instruments, Massa (Board sizing; see Rosa, 1991), Merritt Systems, Migatron, Polaroid, and Senix, respectively. Asher (1997), *Sensors 1998 Buyer's Guide*, and websites of transducer manufacturers may be consulted for additional illustrations.

(b)

(c)

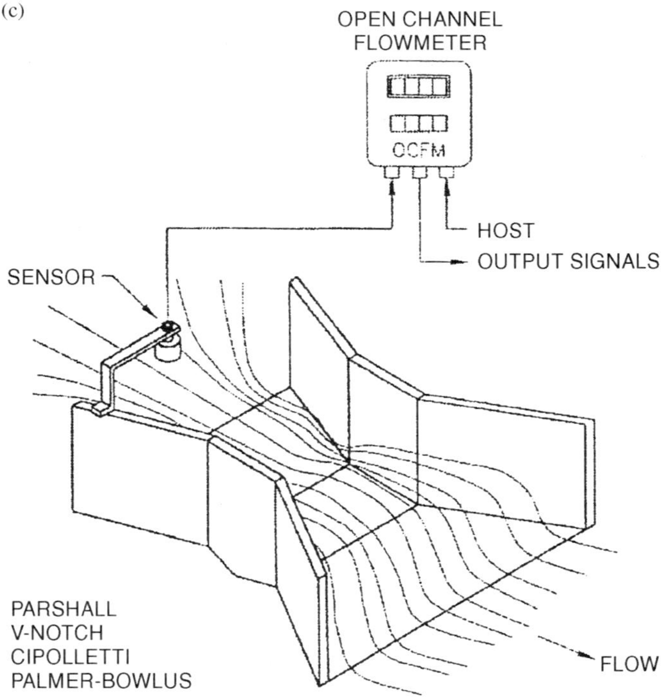

FIG. 19. (*continued*)

touching the solid. Air-coupled inspection of rubber aircraft tires was investigated twenty-five years ago by Van Valkenburg (1973), while noncontact air-coupled measurements of paper webbing in motion were reported twenty-one years ago by Jartti and Luukkala (1977). Examples of recent

(d)

FIG. 19. (*continued*)

papers on air-coupled (or water-coupled) vibration monitoring are Bou Matar *et al.* (1996, 1997).

High-impedance air transducers have an interesting characteristic that allows them to be clamped externally onto plastic walls of a wind tunnel or plastic walls surrounding a process. They can also be installed in a thick-end-wall thermowell-type probe. When surrounded by a small flange that protrudes radially < 3 mm, the transducer can be sandwiched between two

(e)

FIG. 19. *(continued)*

standard soft silicone, fluorosilicone or ethylene propylene O-rings and thereby be isolated acoustically from the surrounding structure, particularly against cross talk between itself and another similar transducer mounted nearby. This means that the transducers can be supplied to users already mounted in a standard one-inch (or larger) pipe plug or flange, for use in do-it-yourself flowcells (see Figure 31), multifrequency arrays, or other applications. Made of solid piezoelectric material, the Panametrics T7 design is rugged, requires no bias, and is operable with standard pulser/receivers, function generators, or flaw detectors that go down to the ~100-kHz range. Thus it shares some of the features mentioned for the Siemens RU rugged ultrasonic transducers that were developed years before the T7. (Some NDT transducers—e.g., immersion and angle beam types—are potentially useful in clamp-on air applications where the medium against which they are clamped is of low impedance, as provided by plastics. For air paths on the order of 1 m or less and where turbulence is not too severe, a frequency of 500 kHz may be appropriate.)

If the "air" is a gas at high temperature, high-temperature air transducers may still be practical. Above some limit, however (usually well below the Curie point), buffers provide a more durable solution. Buffers can be a long air column, a solid waveguide, a platelike structure, one or more thin rods

(f)

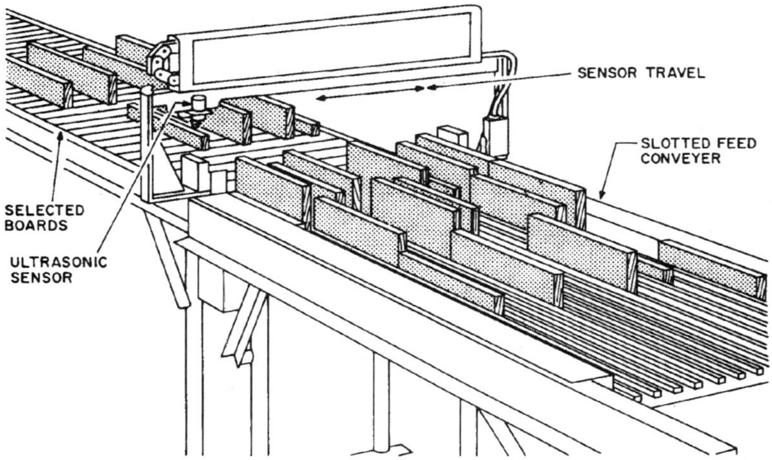

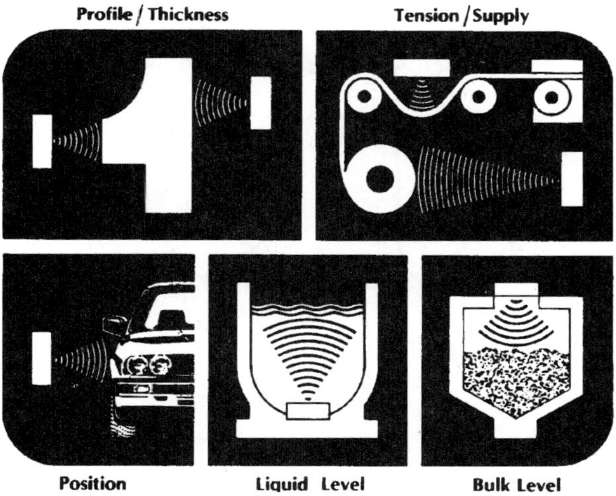

FIG. 19. *(continued)*

(Liu *et al.*, 1998), or a clad structure (Jen and Legoux, 1996). In all these cases except the long air column, the preferred transducer is likely to be a high-acoustic-impedance ceramic. Kleppe (1995a, 1995b, 1996) uses compressed air as a source of low-frequency (~1500 Hz) acoustic noise pulses to measure gas temperature and/or flow at high temperature in boilers and cem (continuous emissions monitor) applications (Section II.C.3; Figures 78–80).

(g)

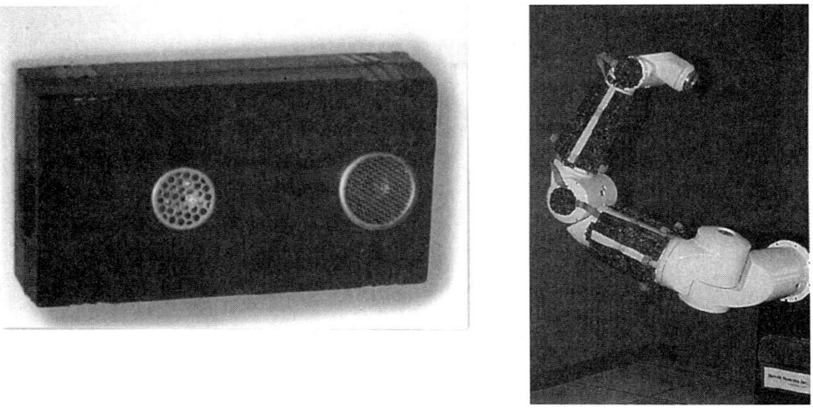

(h)

FIG. 19. (*continued*)

As evident from recent reviews such as Hayward (1997), transducer choices for air-coupled applications include PVDF, electrostatics (Kelly *et al.*, 1996; Ladabaum *et al.*, 1997) composites (Manthey *et al.*, 1992), and the rugged designs discussed in Sections II.A.2 and II.A.3. An O-ring flange sandwich design in a titanium housing (similar to Figure 15) has been proposed for low-pressure steam flowmeters ($T < 150°C$), whereas the buffered design of Figure 43 has been used to measure steam flow at approximately 340°C, 35 bar, and 45 m/s. Special devices (e.g., actuators such as "moonies"

(i)

(j)

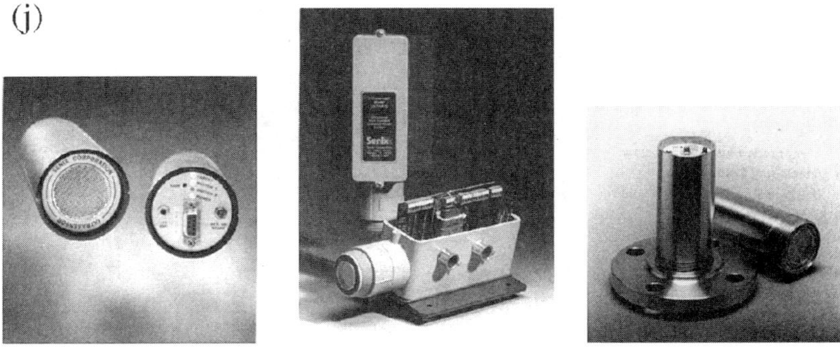

FIG. 19. *(continued)*

(Dogan *et al.*, 1994)) are also of interest. The leaky cone of Dabirikhah and Turner (1994) is another novel and potentially useful air transducer.

For industrial automation a very important task is the continuous measurement of the level of goods in reservoirs or bins or of liquids in tanks. At water supplying and waste water processing facilities, measurements must be made of the level in reservoirs and basins and also in channels to calculate the flow velocity from the liquid level height at known channel cross sections. Similar measurement tasks occur in the mining and building industry, including the height measurement in ore, coal, gravel, sand, and other materials' bins as well as determining the material height on conveyor belts to derive the mass of transported material.

Turning again to technical issues, the diagrams in Figures 6–9, 19, 27, and 30 represent a number of possibilities, e.g., transducer locations and paths for vessels whose axes are vertical, horizontal, or tilted. But certainly not all

(a)

FIG. 20. Examples of Milltronics liquid level equipment. (a) MultiRanger Plus for nasty applications having air paths to 15 m. (b) Slurry, liquid, and bulk solids applications. (c) AiRanger DPL for long-range (≤ 61 m) measurements in one or two vessels, at temperatures up to 150°C. (d) AiRanger XPL to monitor one to ten vessels, over same distance ranges and temperatures as (c). Illustrations courtesy of Milltronics.

(b)

FIG. 20. (*continued*)

(c)

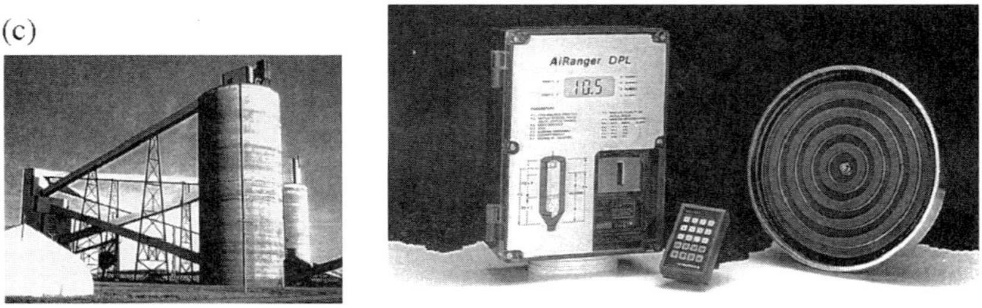

(d)

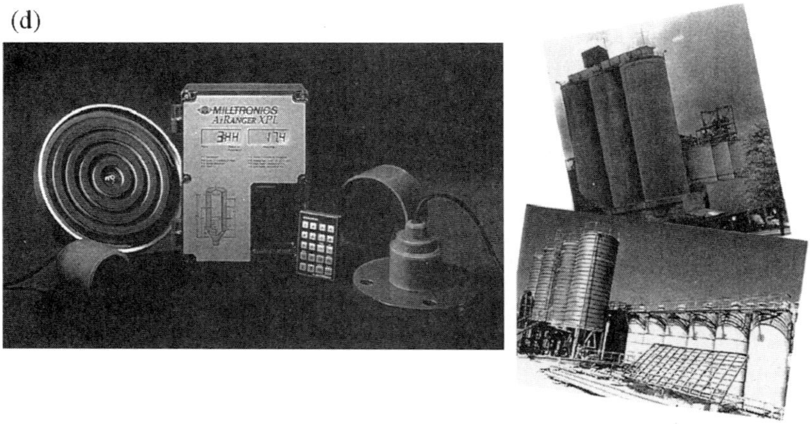

FIG. 20. (*continued*)

possibilities can be included. Some of the issues to consider in analyzing a downward-looking sonar system are discussed by Evans (1997), including noise suppression algorithms used by Milltronics. Examples of Milltronics and Cosense liquid level equipment appear in Figures 20 and 21, respectively. By 1997 there were 800,000 installed Milltronics downward-looking air sonars. These high-power, low-frequency transducers can measure level at ranges to 60 m and temperatures to 150°C (Evans, 1997).

The Cosense ML-101 Micromeasurement System (Figure 21) is designed to accurately measure levels in small vials, test tubes, and microplates. The sensor enables the ML-101 to be used with automatic and robotic samplers and microsampling devices. Micromeasurement in the ML-101 is based on a modular design that accommodates a single-channel to a 24-channel unit. The unit is capable of measuring level to ± 0.1 mm or better. For the microplates application, the ultrasonic transducer transmits a 1-MHz signal through air with a beam angle less than 1° up to 100-mm air distance. Cosense manufactures a variety of ultrasonic level switches and transmitters that have been used from cryogenic temperatures to as high as 375°C.

Although many of the described presence sensors could perform such level measurements, dedicated ultrasonic level sensors have been developed to meet the special requirements in this application area. In some cases relatively large measuring distances are needed, 40 m and more, requiring operating frequencies lower than 30 kHz. Often, the tanks or bins are relatively slim and may have ledges, traverses, and even large pieces of the stored material sticking to the walls. These extraneous objects produce spurious echoes, disturbing the measurement. Sometimes vapors and gases emanating from the material in response to temperature changes and temperature gradient influences can alter the speed of sound significantly. Another interfering variable, easily overlooked, is the *pressure* of the gas. Changing *temperatures* are of concern with respect to accuracy as well as to a safety issue related to the pyroelectric effect, discussed below. In underground coal mining locations, there is the potential for the presence of explosive levels of methane and air. In these applications it is essential to prevent the release of ignition-capable energies. This can be accomplished through flameproof enclosures or by limiting the energy to acceptable levels.

In these applications the piezoceramic transducer is also subject to being impacted. These impacts could generate voltage spikes, but this is considered to be a minor danger compared to the pyroelectric effect. Very high voltages on a piezoceramic transducer, high enough to cause ignition sparks, can be generated by the pyroelectric effect associated with piezoelectricity. By this effect, large electric charges are produced even by minor temperature changes. Electrical charge corresponds to high voltage on the relatively high capacitance of the piezoceramic body. In one method of protecting against ignition due to the release of the electrical energy, three parallel resistors are placed across the piezoceramic body. The value of the resistors is such that one is sufficient to discharge the electric charge produced by the maximum possible temperature change.

In response to the said applications' demands, a variety of airborne ultrasound distance meters have been developed. Figure 22 shows ultrasonic distance meters for mining applications: a 30-kHz device for large material bins and an 80-kHz device for smaller containers and for conveyer belts.

Another example of an ultrasonic proximity sensor is the *Ultra21* series from Cleveland Machine Controls. This device has one analog output and two

(a)

(b)

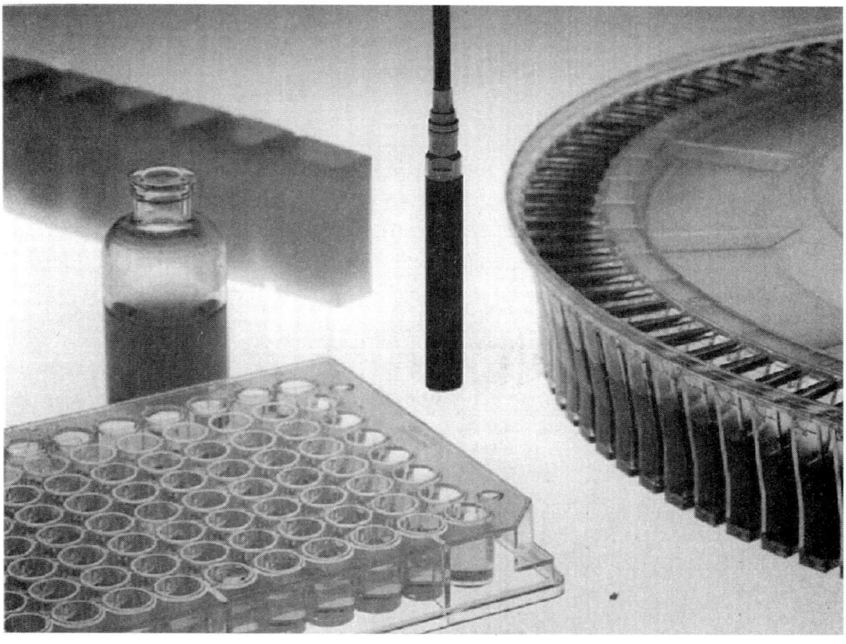

(c)

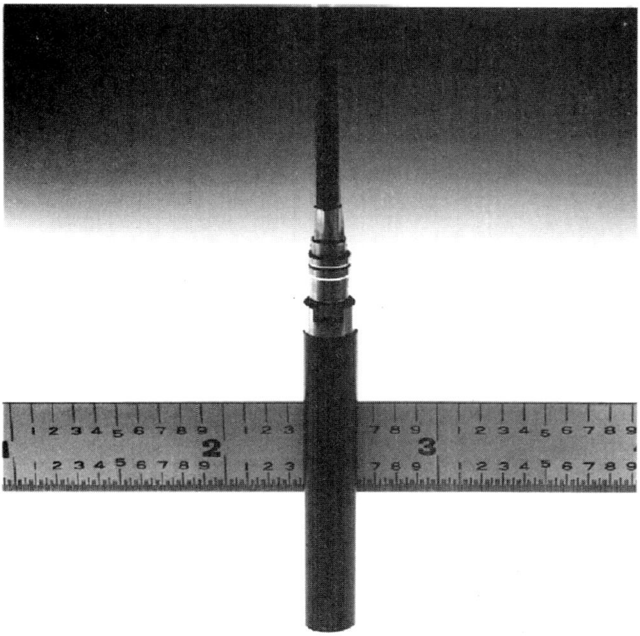

(d)

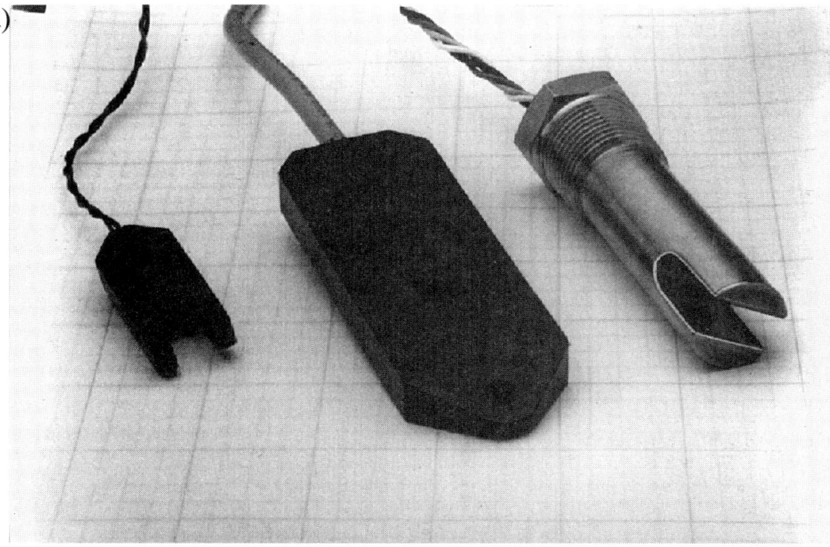

FIG. 21. Example of Cosense liquid level equipment. (a) Sanitary, ultrapure products, discrete level switches. (b) Micromeasurement system, $f = 1$ MHz, ϕ 6.4-mm, air distance $\leq$ 100-mm. (c) Close-up of transducer. (d) Gasoline leak detection sensor for double-wall tank, PD-101 series. Illustrations courtesy of Cosense.

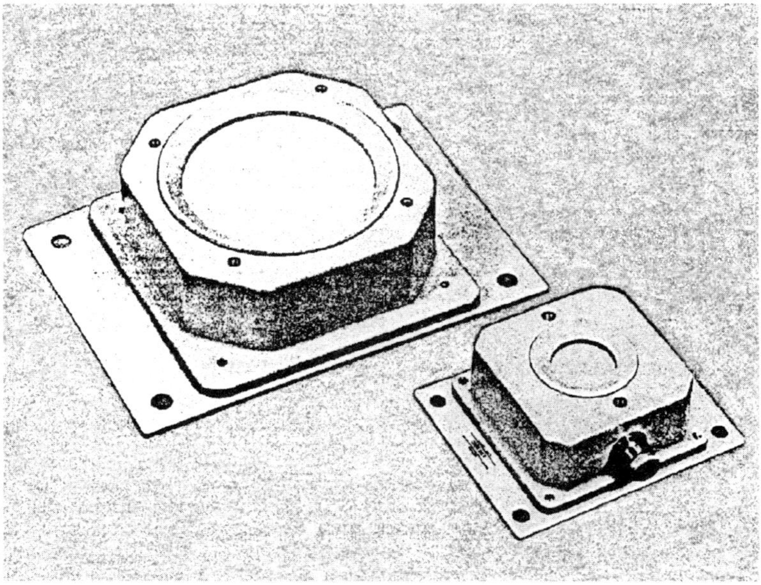

FIG. 22. Ultrasonic level meters for mining applications. 30-kHz operating frequency (bigger device) for up to 30 m for coal bins, 80-kHz operating frequency (smaller device) for up to 8 m for conveyor belts (Siemens).

discrete outputs. The distance range is ~125 mm to ~4 m. Its electronic companion provides self-test diagnostics, push-button programming, and serial communications.

"Air"-ranging depends on the speed of sound c in the gas along the path, which in turn depends on the gas composition (average molecular weight MW), the temperature T, and to a generally lesser extent, the gas pressure P. The influence of MW and T may be illustrated by a graph such as Figure 23. Sensors or reference reflectors close to the transducer obviously provide sound speed compensation in that vicinity but not necessarily elsewhere, where temperature or vapors contribute to a sound speed different from that near the transducer.

A comprehensive range of ultrasonic distance–level meters has been developed for many industrial and similar applications, as shown, for example, in the U.S. publication *Measurements & Control*, April issues in 1995, . . . , 1998. In the products of some manufacturers listed in these April issues, different transducer principles were employed, as for instance high-power bending type transducers for lower frequencies (down to 16 kHz) for

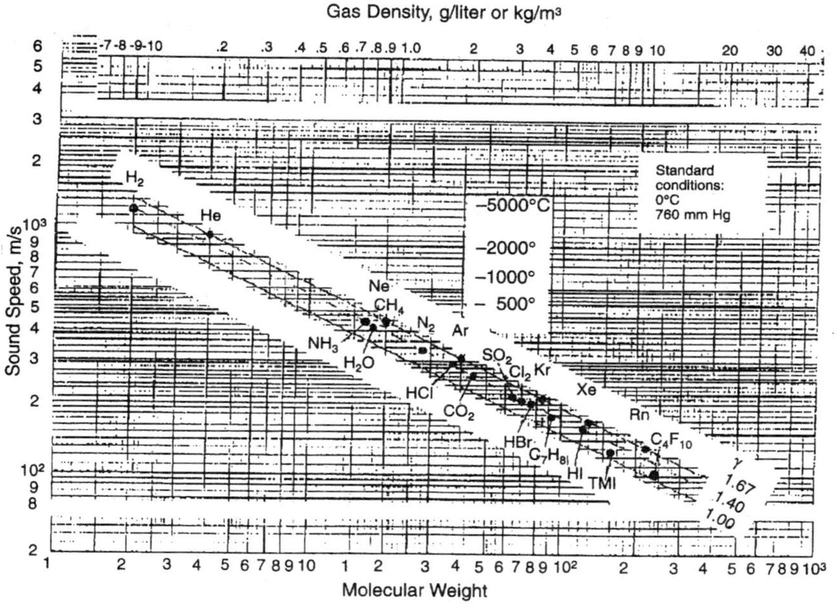

FIG. 23. Graph of sound speed c as a function of gas molecular weight (MW), at standard conditions, after Lynnworth (1995). As illustrated for argon, if temperature T increases, c increases. Sound speed is also a function of pressure, not shown here. The density scale is applicable to standard pressure. For a *mixture* of two gases, the resultant sound speed may be calculated as described in Valdes and Cadet (1991).

maximum distances (up to 60 m). At higher frequencies, piezoceramic transducers with matching layers are used. Aided by finite element simulation, the transducer design of some manufacturers is optimized, ensuring an ultrasound radiation mainly orthogonal to the front face with a narrow main beam and efficiently suppressing side radial radiation. By this design it is possible for transducers to be mounted freely at the front of a (coaxial) housing of smaller diameter, which contains the electronic circuitry. An interesting example of an application-specific sensor is an *intelligent blind flange*, which is intended to be used in tube sections or tanks under pressure.

For efficient and accurate distance meters, much effort has been given to the signal evaluation. For achieving a high accuracy, the amplitude of the received echo of interest should remain constant, independent of the object's distance, orientation, and reflectivity. Therefore, the distance-dependent signal decrease normally becomes equalized by a time-dependent amplification control. The object's reflecting properties are automatically adjusted by a closed-loop control in regard to the echo amplitude (Note the recommendation on

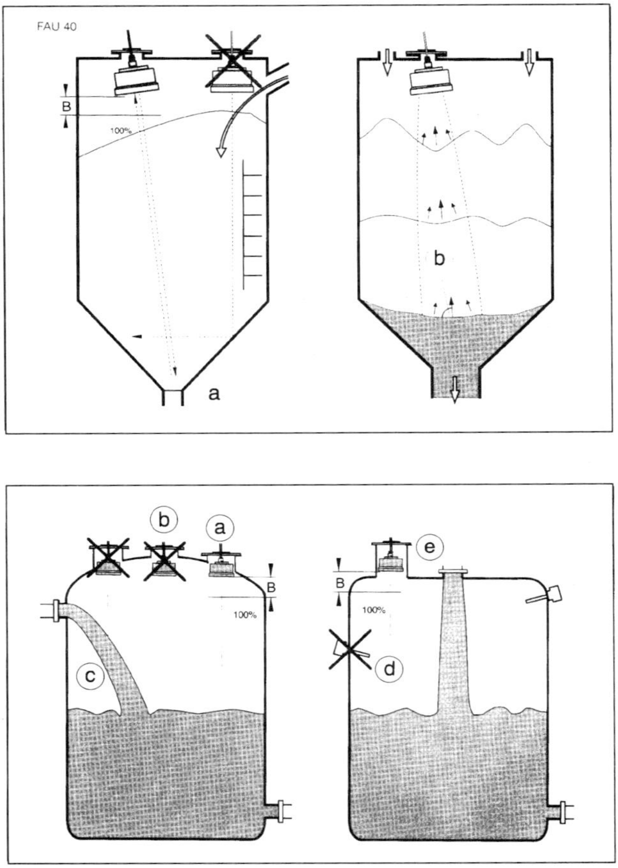

FIG. 24. Guidelines on transducer mounting, illustrated with Endress + Hauser Prosonic P level measuring equipment. Illustration courtesy of Endress + Hauser.

aiming the transducer in Figure 24). The influence of temperature on the speed of sound is compensated in accordance with a (built-in) temperature sensor or, better, to a reference object at known distance. Problems can occur with liquids having a high vapor pressure, e.g., gasoline or organic solvents, whose vapors can cause propagation velocity changes that hardly can be predicted. As mentioned before, the influence of gas molecular weight (*MW*) on sound speed may be illustrated by a graph such as Figure 23. A similar relation is not available for liquids. For liquids the dependence of sound speed on density, temperature and type of liquid has been explored, for example, by Rao (1940), Schaaffs (1963) and others.

Figure 24 shows transducer installation guidelines recommended for ultrasonic level meters such as those manufactured by Endress + Hauser for many years. Other transducer technologies, such as bending-type or electrostatic foil devices, are in use in less demanding environmental conditions (Hayward, 1997). Significant differences can be found in the housing design, evaluation electronics, (especially in the user interface), and the availability of different bus technologies. Some of these differences will be evident in the illustrations comprising Figures 19–21, 25–30.

7. *Ultrasonic Level Sensors in Liquids*

There are many ways of approaching the problems of sensing liquid level. Categories may be constructed based on clamp-on vs wetted; transmitting from the top, side, or bottom; measurement based on amplitude or transit time; utilization of existing structures as reference reflectors or not; vessel axis vertical, horizontal, or inclined; vessel material ferrous, nonferrous, plastic, or other; wall thin or thick, lined or unlined; and so on. These technical issues must be considered along with economic issues.

An important economic consideration is the cost of maintenance, especially for a wetted sensor that becomes fouled with hazardous waste. To appreciate the high cost of replacing a sensor, consider, for example, what would happen if a 300-passenger commercial jet were not allowed to take off on schedule because of a failed sensor on a lavatory waste storage tank. One avionic sensor manufacturer estimated in 1997 that the cost of moving all the passengers and their baggage to another plane is on the order of $250,000, or nearly $1000 per passenger. This estimate probably does not include the inconvenience to the passengers and crew. As another example, consider what would happen if the failure of a water level sensor requires either a nuclear or a fossil-fueled power plant to be shut down for a replacement sensor to be installed. There may be an interrruption of power generation for several days, as plants of this magnitude cannot quickly be turned on again to full power. Considering that these plants generate power at rates >100 MW, the loss of revenue—especially if it should occur during a time when the region's electricity needs were greatest—far outweighs the sensor cost.

In the case of safety devices like high and high-high level sensors installed to prevent overfill or their counterpart low and low-low sensors to prevent running out of liquid, one wants to never have a false alarm yet never miss the real event that must be sensed. False alarm rates are specified in some applications as $<10^{-4}$.

Clearly, when evaluating acoustic/ultrasonic level sensors one must very carefully take into account the influence of unwanted variables such as residue buildup, bubble formation, or other technical details that influence sound propagation and hence the reliable sensing of level. One approach is to take two approaches, i.e., to use sensors based on two different physical principles rather than simply using redundant identical sensors. In other words, it may be possible to eliminate the weak points of one sensor (e.g., nonlinearity, temperature dependence) by a joint model-based signal evaluation that takes into account the results of the other sensor. An example is the combined application of an ultrasonic and a microwave distance sensor (Ruser and Mágori, 1997).

Even if attention is limited to *external* clamp-on, bond-on, or weld-on sensors, coupled temporarily or permanently, there are still many installation variations possible. Perhaps the most obvious clamp-on or externally bonded method is one that requires only a single transducer coupled to the sidewall. The ring-down would be a measure of liquid present or absent on the inside. If there were no interfering variables, this would be a simple solution. At least two companies manufacture point sensors based on this idea: Endress + Hauser (Figure 25(b), and described in an article by Babb (1996)), and Canongate Technology (Figure 25(c)). (See Section II.A.8.) An earlier version of this idea is illustrated in Krautkrämer and Krautkrämer (1966, p. 453). Two transducers were coupled to the outside of the vessel, one above and one below the level of interest. The progress in sideways-looking ring-down from two transducers (ca. 1966) to one transducer (ca. 1996) is also seen in upward- and downward-looking sonar. The early two-stillwell versions include a downward-looking design (Tomioka, 1968) and an upward-looking design (Frederick, 1965). These may be compared with the single- or multiple-path pulse-echo transducer interrogations in Figures 20 and 28.)

The Canongate Technology product line includes a clamp-on version where the reference sound speed is obtained from a horizontal interrogation so that the vertical round-trip transit time can be corrected for sound speed. The method of combining a reference path perpendicular to a measuring path is also utilized in some of the wetted probes of CTI Manufacturing (Figure 26). The clamp-on horizontal path assumes good sound transmission through the liquid, and further assumes that there are no significant vertical gradients in sound speed, i.e., no stratification. To the extent that these assumptions are justified, the continuous level detection equipment in Figures 26 and 28 suffices. Sunx makes a point sensor that looks for echoes from the opposite wall, for tank ϕ 0.3 to 3 m (Figure 25(a)).

To deal with the complexities of buildup, floating roof tanks, floating roof seals of several varieties, and with the problem of having no opportunity to

(a)

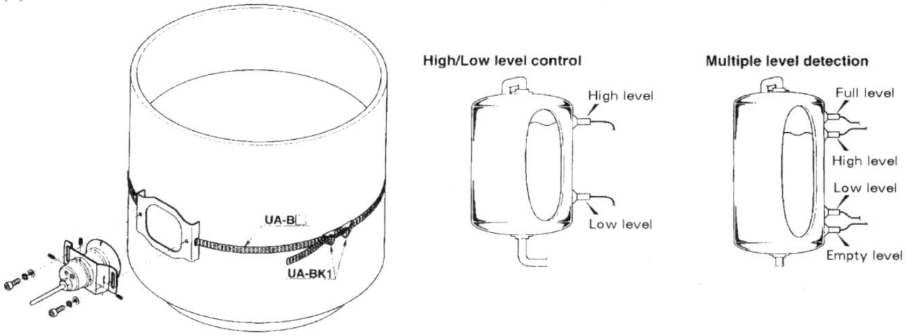

(b)

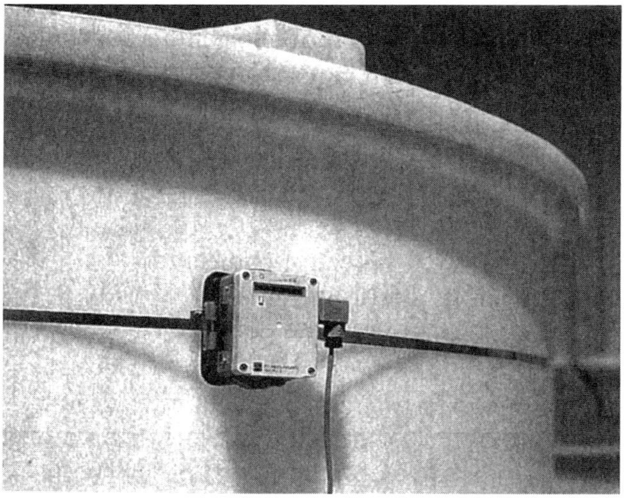

FIG. 25. Examples of external noninvasive sidewall-mounted sensors. (a) Strap-on UA-11 liquid level presence sensor and application examples, courtesy of Sunx. (b) Nivopuls point sensor straps to the outside, requires no holes, and uses resonance in the pipe wall to distinguish liquid inside or not. Application details in Babb (1996). Illustration courtesy Endress + Hauser. (c) SpotCheck from Canongate Technology is bonded to tank surface and claims ±2-mm repeatability, response time user-adjustable, $\frac{1}{2}$ to 8 s, and $T = -40$ to $+110°C$ transducer operating temperature. (Not shown: Stresswave Technology's Pulsar*point* 610 attached to a vessel by means of their optional self-adhesive metal plate.)

actually run the liquid up and down to calibrate the ring-down pattern, and to utilize a *time*-based rather than an *amplitude*-based system, one can consider utilizing guided waves in the tank wall—in particular, the lowest-order asymmetric plate waves (A_o mode). These flexural or bending waves travel at predictable phase and group velocities in the absence of liquid (Viktorov,

(c)

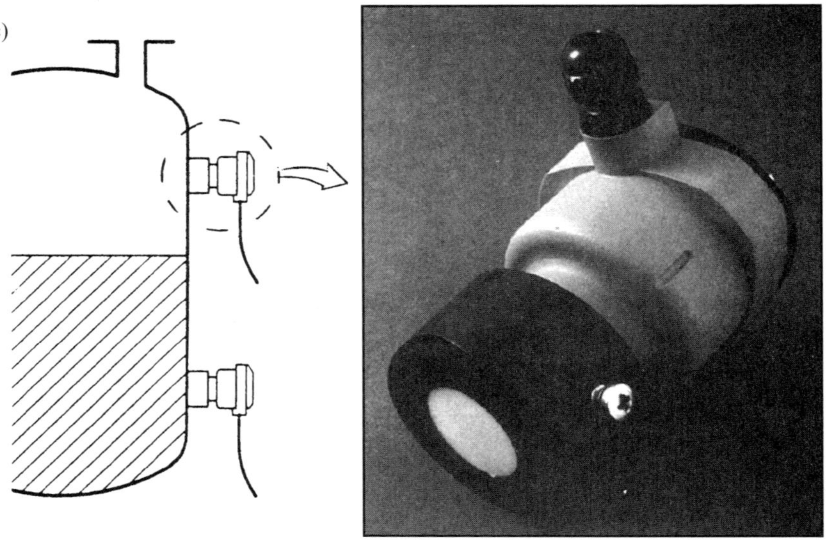

FIG. 25. *(continued)*

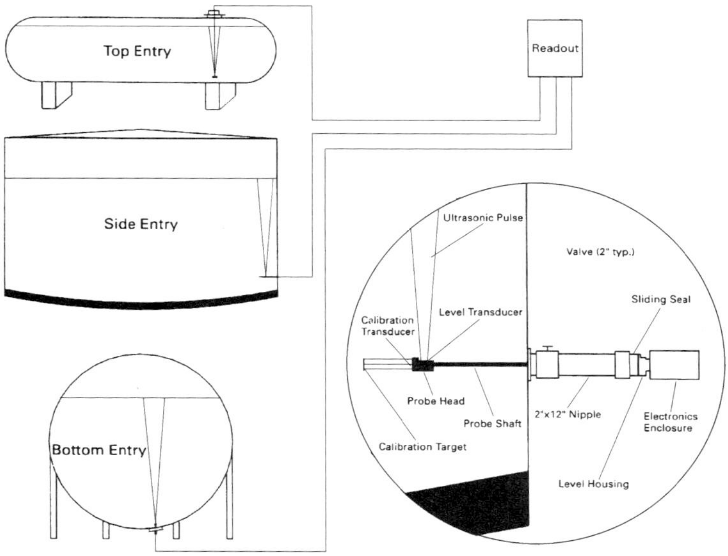

FIG. 26. Example of a wetted probe with the reference path perpendicular to the measuring path. Courtesy of CTI Manufacturing. In contrast with clamp-on versions shown in Fig. 28, to establish a c reference with this probe, transmission all the way across the vessel diameter is *not* required.

1967). Ageeva (1960) found the flexural phase velocity in an aluminum strip to decrease upon immersion. (For a recent analysis of fluid loading, see Craster (1996, 1997).)

Retardation of a flexural wave pulse (tone-burst) is the basis for measuring discrete or continuous levels from outside the tank, in a Panametrics time-of-flight system due to Liu and Lynnworth (1993). This high-level acoustic system (HLAS) is *not* a leaky-wave method and is not limited to *high* levels. If the waves leaked into the adjacent liquid, the resulting attenuation would limit the path. If the system depended on signal attenuation, it would become sensitive to coupling, viscosity, and probably other variables that influence signal strength. Instead, the frequency is kept low enough so that the frequency $\times$ thickness product yields a flexural phase velocity c_{FLEX} < sound speed in the adjacent liquid (e.g., <1000 m/s if the liquid is gasoline). (An *attenuation*-based system (Royer *et al.*, 1992) uses the symmetric or S_o wave, which leaks into the liquid when the liquid is present. Equipment based on leakage and attenuation is marketed by Auxitrol. It is interesting to note that the frequency $\times$ thickness conditions for maximum leakage can also be used for transmission into, or receiving from, an adjacent fluid. This aspect is utilized in the *air* transducers due to Dabirikhah and Turner (1994)).

Regarding the low-speed leak-free flexural waves, it was found that the transit time in a horizontal steel path increased by about 10 μs as gasoline reached the level of the transducers. In this test, (Liu and Lynnworth, 1993) the transducers were spaced about 225 mm apart and the steel was about 6.4-mm thick. The tank diameter was ~50 m, and its height was ~15 m. In another test with transducers ~8 m apart, the flexural transit time was found to increase very nearly in proportion to fuel level *H*, where *H* was measured by a preexisting HTG (hydrostatic tank gage). Precision is ±25 mm for a fixed (discrete) level determination and about 1.5% of span in the continuous measurement just described. HTGs are commonly used for such applications, but they require two or three pressure-sensing penetrations of the vessel to compensate for liquid density and vapor pressure above the liquid. In this application, the customer sought a redundant, independent (e.g., acoustic), external noninvasive system. The HLAS satisfies the independent requirement because it operates on physical principles different from the HTG. Field tests demonstrated that the sensor did not falsely alarm when the floating roof passed by. Sensors, whether installed on an empty or full tank, can be demonstrated to be operational by temporarily filling their (optional) reservoir with alcohol or other liquid of density ρ similar to that of the fuel inside the tank. This increases the flexural transit time by about 10 μs. Verification takes about one minute.

As we go to press, some two dozen HLAS flexural wave systems are operating on tank farms in the northeastern part of the United States. One of these systems, installed on a tank containing lubrication oil, has been operating almost continuously since December 1992 (Figure 27). Its performance is checked monthly using either the external reservoir or an independent manual dip gage (a weighted line is lowered to the very bottom of the tank, and then the liquid residue mark on the tape is read when the line is withdrawn from the tank).

Applications are also being evaluated on tanks that store kitchen wastewater on board large naval vessels such as submarines or aircraft carriers. At the outset of this shipboard R&D study (Stein *et al.*, 1996) it was convenient to try to solve the problem using the same HLAS sensors as were being used on the gasoline and lubricating oil tanks. These worked on clean water, but it was found that greasy deposits isolated the HLAS sensors from the liquid. The solution was to reduce the frequency by a factor of about 20. R&D is currently focused on making a practical version of the lower-frequency HLAS. On the other hand, if the tank wall is thin (~1 mm) and the liquid is inviscid and nonresidue-bearing, a higher frequency (~100 kHz) is satisfactory.

On tanks that store waste on commercial aircraft, the tank size is obviously much smaller than for an aircraft carrier kitchen waste tank or the gasoline tanks on a tank farm. The aircraft waste storage tanks have a capacity of some 500 liters, compared with the several megaliter capacity of the gasoline tanks referred to above and by Liu and Lynnworth, (1993). Even so, the acoustic problems due to buildup are rather similar in the airplane lavatory waste tank and on the aircraft carrier kitchen waste tank. Accordingly, the HLAS flexural wave approach is a candidate solution, provided it operates at a frequency low enough to look beyond the buildup to sense liquid presence. Its purchase price is probably greater than the simpler pulse-echo transmission or ring-down methods but it would appear to be less sensitive to the interfering effects of buildup.

The flexural or bending wave used in the HLAS sensors shown in Figure 27 can also be applied to common pipe sizes to measure either liquid level, density of the fluid therein, or the integrated product of density × liquid level. The method so far seems to be most sensitive when the pipe wall is not too massive. This means if it is metal, it should be thin (Yon, 1991); either that or it should be made of plastic. The effect of viscosity on the measurement of density is not yet determined.

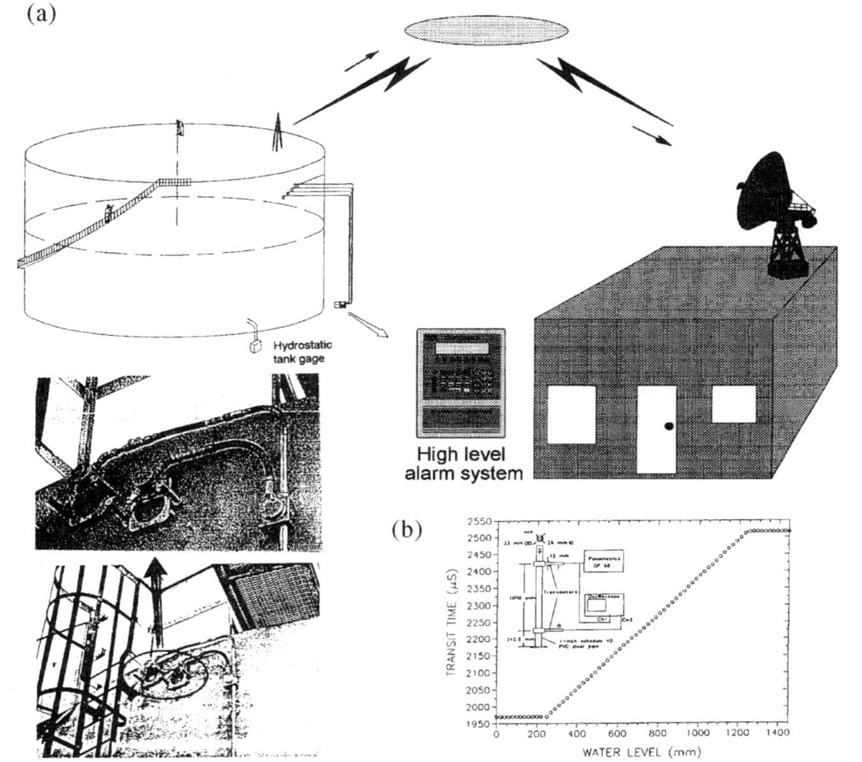

FIG. 27. (a) Schematic and photo of flexural wave HLAS (high-level acoustic sensor) illustrate the principle and the application. One of these short-horizontal-path systems, shown next to the ladder in the closeup photo, was installed on a tank containing lubricating oil. This particular high and high-high pair have been operating almost continuously since December 1992 (> five years at press time). It is normally not allowed, because of safety regulations, to verify the sensor's readiness and accuracy by actually filling the tank to the high or especially the high-high level. Instead, their operation can be checked by using the external reservoir, or during an upset event, by comparing the time of the overfill event with the alarm triggering. Tests can also be conducted safely at midheight, without risking an overfill. The wireless link shown schematically for this Panametrics HLAS sensor, is in fact part of some liquid level sensors available from other manufacturers (example: the air-coupled radio-linked sensor from Arichell Technologies shown in Fig. 96). In (b), the flexural or bending wave idea is applied to noninvasively measuring water height in a one-inch schedule 40 PVC plastic pipe, $\sim \phi 33$ mm. The transducers were clamped 1000 mm apart for this test.

Despite all the potential advantages of the flexural wave system in which transit time is sensed, such systems may be limited to walls <10-mm thick unless audio frequencies below 10 kHz are used. It is well known that MHz shear waves readily penetrate *thick*-walled vessels. Leakage of these waves from the vessel wall into the adjacent liquid can be interpreted in terms of liquid presence or liquid level (Lynnworth *et al.*, 1982). Even easier, of course, are standard wall thicknesses <10-mm. Clamp-on ultrasonic flowmeter geometries utilizing paths tilted in the axial direction, or confined to a plane perpendicular to the axis of the tank or pipe, can also be used to measure liquid presence based on the *amplitude* of the received signal. Such paths are shown in Figures 6 and 30. When measurements are required at high temperature, the buffered transducers used in high-temperature clamp-on flow measurements may be considered (Section II.B.2.b).

Another method of ultrasonic liquid level measurement uses the sound propagation in liquids using an ultrasonic transducer at the bottom of a vessel. The transducer emits ultrasonic signals into the liquid and receives the echoes from the liquid surface. At known speed of sound of the liquid, the level height is determined from the time lapse between signal emission and echo reception. For the determination of the speed of sound, the echo of a reference object at known distance can be evaluated. The ultrasonic transducers can be inserted from the top by mechanical supporting constructions (as "diving" tubes) or through bottom holes. A very favorable possibility is the use of a clamp-on transducer mounted outside and/or beneath the tank, as in equipment made by Sonotec. A necessary condition is an unobstructed ultrasonic path between the transducer and surface, which must be met orthogonally. Problems can occur with soft sediments that may cause a high attenuation, such as, e.g., yeast in beer clearing tanks. The propagation of ultrasound in liquids can also be employed to watch over level limits of liquids in tanks. By the use of an ultrasonic transducer clamped to a (vertical) wall, the echo of a reflector, (for instance, an opposite wall) can only be recognized when the whole propagation path is "flooded," i.e., when the liquid level is high enough. The sidewall-mounted transducer not only measures liquid presence at its own level, but when the liquid is at least that high it can provide a sound speed c reference so that the vertical round-trip time to the liquid surface, obtained from a bottom-mounted, upward-looking transducer, can be properly interpreted. This interpretation assumes that c gradients are absent. In some tanks (Figure 30), it is possible to check that assumption. "Bottom-up" externally-mounted liquid level sensors are available from different firms, e.g., Canongate Technology, Panametrics, and Sonotec, among others. Examples from Canongate Technology and Sonotec are given in Figure 28.

Gazis, Kane, and von Gutfeld (1998) proposed solutions to potential problems associated with sediment or debris at the tank bottom, and echo jitter from a rippled surface. The rippled surface might be present in tanks subject to movement and vibration, or during filling or withdrawal of liquid. The solution of Gazis *et al.* for dislodging debris or preventing its accumulation above the upward-looking clamp-on transducer is to surround that transducer with another one, donut-shaped, excited strongly and independently at a low frequency (e.g., kHz range). The immediate neighborhood *inside* the tank of the externally mounted liquid level sensor is thereby sonically cleaned using a second annular transducer also mounted external to the tank bottom.

A multiplicity of horizontal paths at levels $H_1, H_2, \ldots$ can be imagined that provide resolution of level equal to ΔH, the distance between paths. Often, only paths near the top and near the bottom of a vessel are important. Otherwise, especially in a tall tank, the number of transducers can become too large to be practical with this method. However, in Figure 30(f), many such point level sensors were used (Rod, 1962).

A hybrid case occurs when the nonintrusive transducer is outside the vessel but a stillwell or reflectors are placed inside the vessel to improve the measurement. A stillwell is a tube of relatively small diameter compared to length, so that sloshing and wavy action is smoothed out or stilled. The stillwell sometimes provides the further advantage of guiding the wave. In

(a)

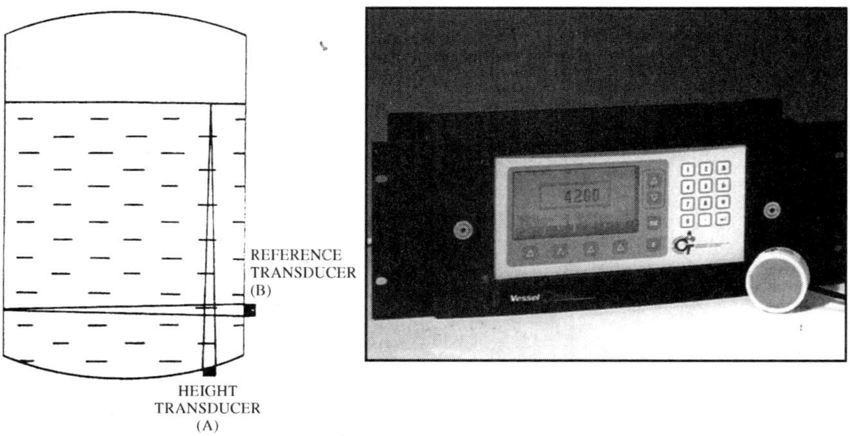

FIG. 28. Side-to-side paths are used alone for liquid presence sensing or to calibrate or cross-check an upward-looking ("bottom up") sonar in liquid level sensors available from firms such as (a) Canongate Technology (VesselCheck) and (b) Sonotec. In (b), the sensor at the lower left operates with a Sonocontrol unit to provide a low-limit control sensor; in the lower middle photo, the sensor with Sonometer measures transit time up to the liquid–vapor interface.

(b)

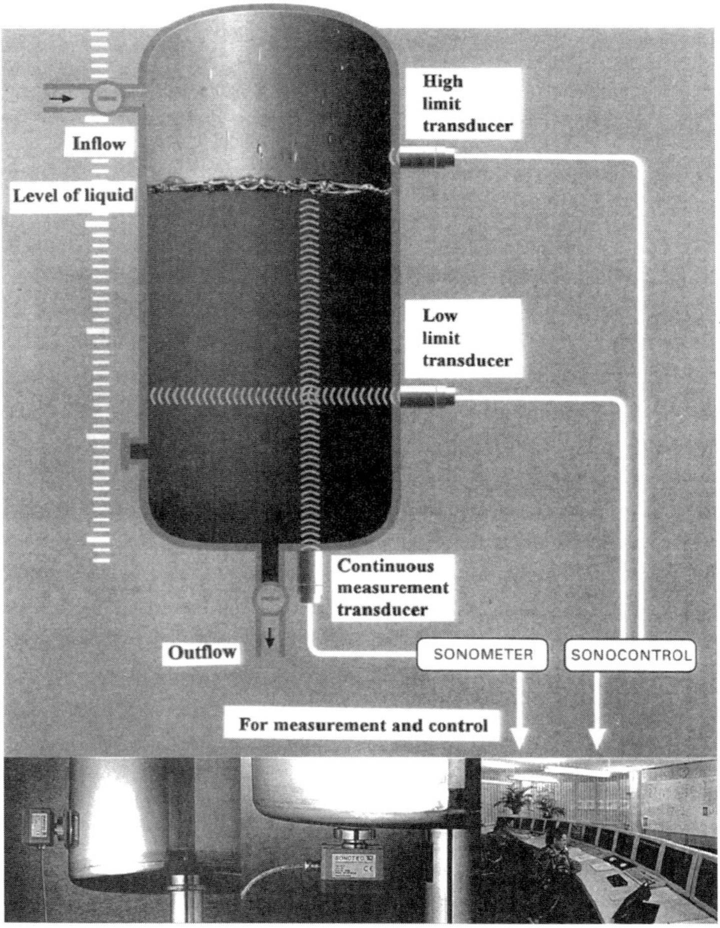

FIG. 28. *(continued)*

other words, the stillwell, depending on the diameter-to-wavelength ratio and the stillwell material or lining, may reduce attenuation that otherwise would be caused by beam spread. The stillwell can avoid spurious echoes from off-axis structural reflectors, yet it can provide a number of reference echoes that are quite useful. A potential disadvantage is that the density of liquid inside the stillwell could end up being different from that outside the stillwell. A slotted or perforated stillwell solves this problem, if the slot or holes remain open.

In the situation shown in Figure 9, page 293, the downward-looking sonars may operate free-field or in a multiple notch stillwell. In either case the transducers do not contact the product, but a hole in the roof is necessary. The hole is sealed by a flange when the sensor or transducer is installed. Recent advances in transducers for use in air, as may be used in downward-looking sonars, are discussed or reviewed by Hutchins and Schindel (1994) and Mágori (1994). If the product beneath the vapor is a settled liquid that is clean and not residue bearing, a wetted probe such as a torsional waveguide may be used (Kim *et al.*, 1993). (Some versions of this waveguide may be used to sense liquid density ρ and viscosity η. Whether this is better than alternative and especially noninvasive solutions based on the complex reflection coefficient (Sheen *et al.*, 1996) or streaming (Hertz *et al.*, 1991; Rudenko *et al.*, 1996; Sarvazyan, 1997) needs to be determined in each case.) Sometimes a magnetic float rides up and down a magnetostrictive waveguide, using the Wiedemann effect to achieve positional or liquid level accuracy of a fraction of one mm over long spans (Figure 29(a)).

Also in wide use are tuning forks and gap sensors installed by threading or flanging. Liquid dampens the tuning fork vibration—an "attenuation" affect—but liquid increases the gap signal compared to the near-zero ultrasonic signal strength transmitted across the vapor. On the other hand, an airborne signal across a wall-to-wall "gap" will be interrupted (attenuated) by the presence of solids piling up in the path. Surface acoustic wave (SAW) and plate wave sensors are often proposed to sense the presence of particular substances (Wenzel *et al.*, 1994; Ballantine *et al.*, 1996).

The upward-looking transducer runs the risk of being buried in sludge, but if cleaned or if used only in clean liquids, it can provide a reliable echo from the liquid–vapor interface. A stillwell stabilizes the interface and can contain reference reflectors to compensate for sound speed gradients, the importance of which increases if a temperature gradient exists. This need for multiple reflectors is readily appreciated if one recognizes that, for water near room temperature, sound speed changes by about 0.2% per °C. For gasoline, the coefficient is around $-4\,\text{m/s}$ per °C, or roughly -0.4% per °C, and for cryogenic liquid nitrogen or oxygen it is around -1% per °C.

A 45° internal (immersed) reflector sometimes allows a sidewall-mounted external transducer to interrogate vertically, according to Figure 30(b). The difficult problem of measuring the level of molten aluminum is discussed by Nygaard and Mylvaganam (1993).

The *tomographic* multichord method depicted in Figures 30(a), (c), was developed in response to a chemical company's request for a clamp-on means

(a)

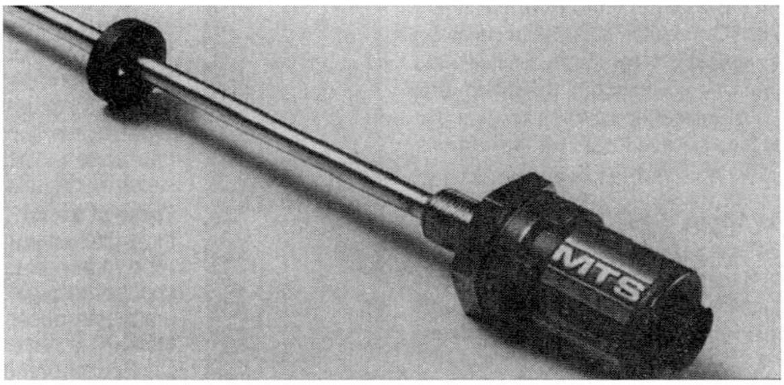

(b)

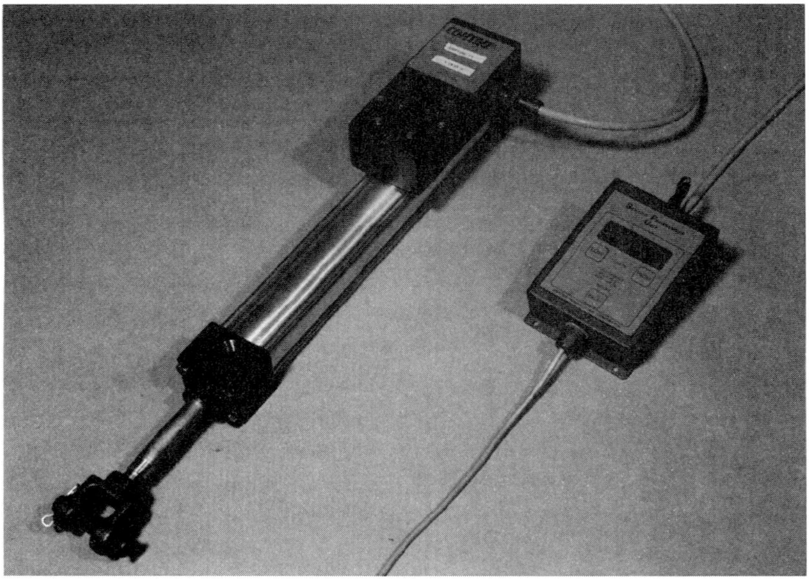

FIG. 29. (a) The Wiedemann effect is used in this floating magnet, magnetostrictive sensor manufactured by MTS Systems Corp. It measures both displacement x and velocity dx/dt. Applications include injection molding machines and underground mine-fault detection. (Talmadge and Appley, 1991; Denver, 1998). (b) Linear displacement transducer from Contaq Technologies.

to determine when "rag layer" was at the same level as one of the nozzles existing on that tank. Here the electronics was a four-channel flowmeter that could sense the presence of rag layer by its attenuating effect and confirm the presence of water below the rag, and hexane above, by their sound speeds. Hexane's c is about one-third slower than that of water at the temperature of this application. This application has so far been demonstrated only at one plant on outdoor tanks, but it uses standard equipment that is used elsewhere in much greater quantities as clamp-on ultrasonic flowmeters. The tanks are cylindrical, SS (stainless steel), axis horizontal, and $\sim$1.5 m in diameter. The clamp-on transducers, in particular, are of a completely standard explosion-proof design. Their plastic wedges are each coupled with a rubbery sheet. SS bands encircle the tanks, maintaining the coupling pressure. Other liquid level sensing paths comprise the rest of Figure 30.

An application that would seem important around the world (wherever gasoline is sold) is the detection of leakage of gasoline from underground storage tanks at automobile and truck service stations. The problem is difficult because (1) the tank itself and the contained liquid expands or contracts as temperature changes; (2) water may be at the bottom; and (3) gasolines vary enough so one cannot assume a particular sound speed derived from a temperature measurement. (According to a graph in Zacharias and Ord (1981), gasolines have specific gravities near 0.71 to 0.73 and sound speeds near 1120 to 1160 m/s.) The leakage rates of interest are so low that a very accurate measure of fuel mass in the tank is required. Recognizing that the *mass* of gasoline is constant if the leak rate is zero and no fuel is being added or subtracted, the ideal solution might take into account the fuel density independent of the expansion coefficient, i.e., independent of temperature. Some of U.S. Test's equipment for this application is based in part on Rao's Rule (an explanation of which appears, for example, in Bhatia (1967, p. 24)). Rao (1940) found that for hydrocarbon liquids, the fractional change in sound speed is very nearly three times greater than the fractional change in (fuel) density. One measuring system that was available in 1997 combined upward- and downward-looking ultrasound generated by one transducer and included a stillwell and reference reflector. (*Bidirectional* interrogation is required to separate water level from total level [water + gasoline].) See also Section II.1.

Some earlier work on liquid level will be found in patent searches around 1960. Hundreds of single-point, multipoint, and continuous level sensors were manufactured based on work by Rod (1962) (see also Rod and Massa (1962) and one version is illustrated in Frederick (1965, p. 213). In the Russian literature, the liquid level work of O. I. Babikov will be of interest, from about 1960 to the present. (See Babikov, 1960; Hamidullin, 1989.)

8. *Acoustic Impedance, Density, and Level Limit Monitoring*

Consider the transmission of an ultrasonic signal between materials having different acoustic impedance values. Only a part of the signal passes; the rest

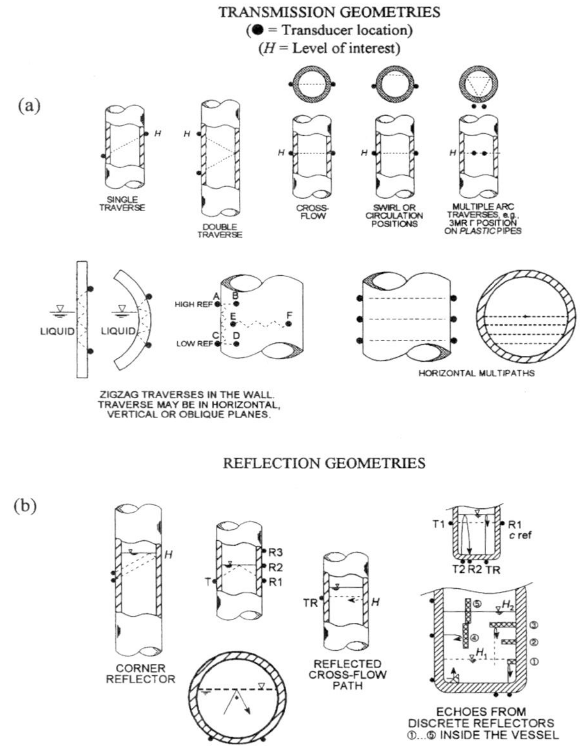

FIG. 30. Examples of adapting standard industrial ultrasonic clamp-on flowmeter equipment or NDT pulse-echo and/or through-transmission standard equipment to external nonintrusive liquid presence and liquid level measurements. (a) Transmission geometries. (b) Reflection geometries. (c) Detail view showing multiple horizontal chords interrogated using external clamp-on transducers to detect rag layer by the attenuating effect of that product. Confirmation by transit time readings above and below the rag is sometimes possible because the rag layer lies between two liquids of different density and different sound speeds. Snell's law limits the chords that may be interrogated from the outside. (d) Pipe axis tilted: upward-looking clamp-on proposed for a pipe inclined at an angle $\theta_{\mathrm{PIPE}} \leq$ refracted angle in the liquid. If two transducers are used as shown, $\Delta H = H_2 - H_1 = S \sin \theta_{\mathrm{PIPE}}$, from which sound speed c_3 in the liquid is obtained as $c_3 = 2\Delta H/\Delta t$, where Δt is the round-trip time difference. (e) Liquid level, axis horizontal, differential clamp-on vertical path and horizontal diameter path allow one to compare the liquid's sound speed at the bottom and across the middle of vessel. (f) Historical flashback to 1962 provides transition from liquid level to the flow part of this chapter. In this 1962 diagram, point sensors arranged one above the other, spacing ΔH, provided an approximation to the propellant volumetric flow rate $Q = A\Delta H/\Delta t$ where A is the cross-sectional area of duct, and Δt is the time between passage of the liquid vapor interface across two liquid presence measuring stations. In a sense, this is a special case of a tag flowmeter. After Rod (1962). Illustration courtesy Robert L. Rod, priv. communication (1996).

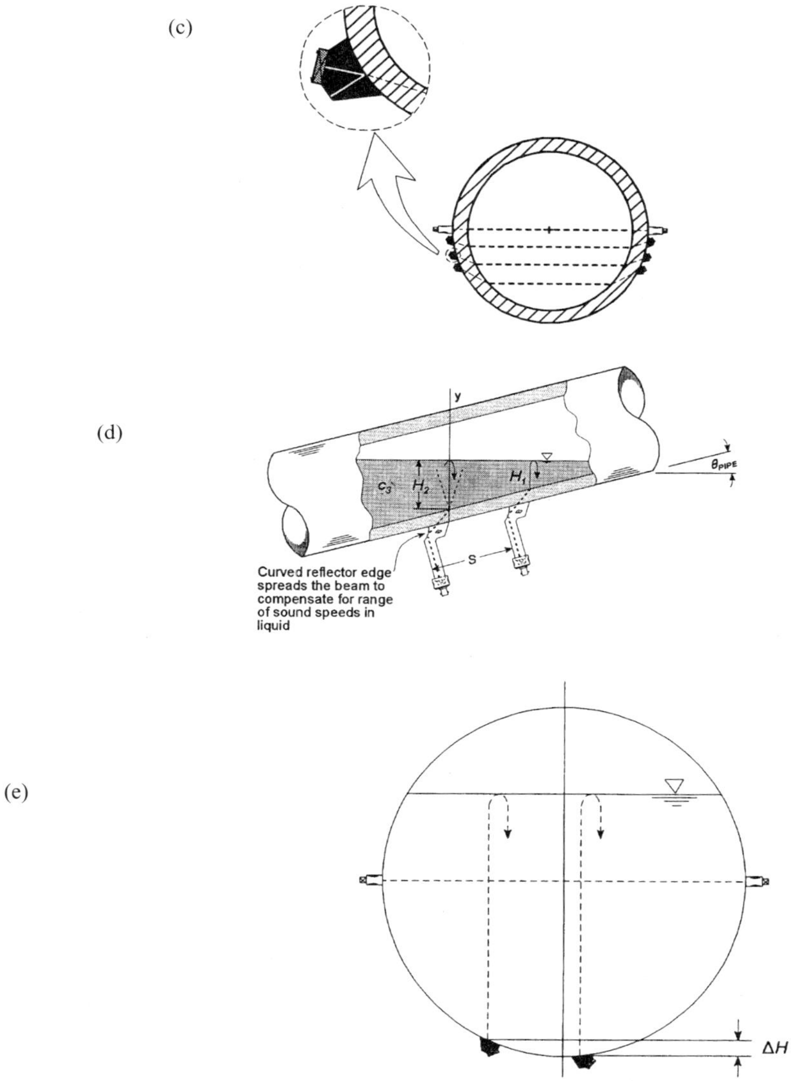

FIG. 30. *(continued)*

(f)

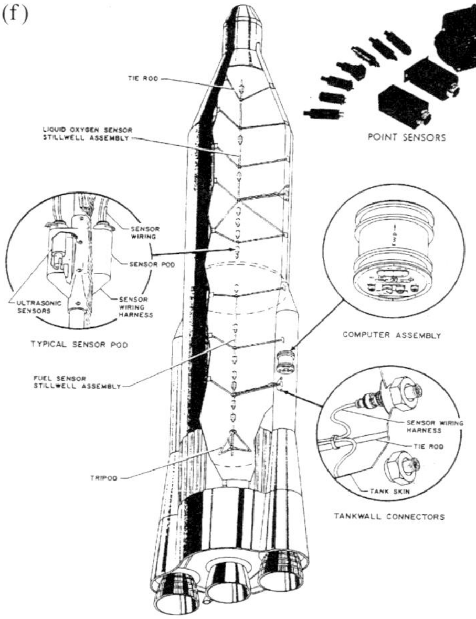

FIG. 30. *(continued)*

is reflected. Thus, from the ratio of the incident signal amplitude and the reflected amplitude and the known impedance of one material, the other material's impedance can be calculated. The characteristic acoustic impedance is the product of density ρ and speed of sound *c*. In principle, the density of liquids could be determined from the acoustic impedance measurement at a known speed of sound, the measurement of which is not difficult. The density determination could be of particular importance for volumetric flowmeters, often blamed for not measuring the mass flow. By the combination of a volume flow sensor with a density sensor, the desired mass flow sensor would be achieved.

Ultrasonic flow sensors have the advantage of delivering the speed of sound without additional efforts. Apart from ultrasonic *flow* measurements, the speed of sound as an indicator of density could be combined with other volume flow sensors, e.g., magnetic-inductive devices. In spite of their high importance and several approaches (Lynnworth and Pedersen (1972): Lynnworth (1979, pp. 507–510); Lynnworth *et al.* (1982); Fischer *et al.* (1995); Puttmer *et al.* (1996); Van Deventer and Delsing (1997); Povey (1997); Adamowski *et al.*, (1998)), no successful commercial liquid impedometer–densimeter (ρmeter) devices are presently known to the authors. (In flare gas flowmeters (Section II.A.2), *c* after compensation for temperature yields molecular weight (*MW*); then *MW* yields ρ. Of course, the conversion from

MW to ρ requires pressure compensation, but this is usually easy in this application as the flare exits to the atmosphere, so *P* usually is pretty close to atmospheric pressure.) One likely explanation for the absence of *Z*-based ρmeters is the necessary high accuracy, which is difficult to obtain for amplitude ratio evaluation, especially considering industrial requirements that the wall material be stainless steel or similar. In these cases, the reflection is almost total. Thus, only a small dependence on the liquid acoustic impedance *Z* exists, and this is very difficult to evaluate with high accuracy in the presence of superimposing spurious signals. Another reason is that the reflection coefficient senses the fluid adjacent the wall, which may not be a good representation of the fluid's average density. Van Deventer and Delsing (1997) and Adamowski *et al.* (1998) used *plastic* instead of stainless steel to increase the *Z* sensitivity of the reflection coefficient measurement in their laboratory studies.

Much easier is the discrimination of whether the acoustic impedance prevailing at the other side of a wall of known material impedance characterizes a gas or a liquid (Figure 25). (See also Krautkrämer and Krautkrämer (1966, p. 423).) Using this principle, Endress + Hauser (E + H) introduced the level limit sensor *Nivopuls* (Figure 25(b)), which is clamped to the outside wall of a tank to recognize whether the liquid level inside has reached the sensor position (Müller, 1996). Rather than measuring the amplitude ratio between the incident and reflected wave, the so-called *tank woodpecker* evaluates the ringing of the wall after a pulse excitation, equivalently determined by the reflective properties of the wall/interior material impedance ratio. The sensor consists of an ultrasonic transducer made from a low-*Q* piezoceramic material and electronic circuitry at the sensor's position. An important advantage of this clamp-on sensor is that no aperture in the wall of the vessel is necessary, and, further, that it does not come in contact with the fluid inside. The sensor can be used for liquids or liquefied gases. Material depositions on the inner wall must be avoided at the sensor's position. The walls can be of metal, enameled metal, glass or plastic materials, or plastic reinforced with glass fiber. The possible wall thickness range is from 2 mm to 12 mm for metal and 1 mm to 10 mm for glass or plastic materials. The sensor operates at temperatures of the liquid to be detected, ranging from −20°C to 100°C. According to E + H, one of the first applications in the United States was where fifty wort tanks at a brewery needed high and high-high alarms. The signal was unaffected by the beer's efferves-cence. In Europe, applications include hydrochloric acid in a glass-lined steel tank; sulfuric acid and other liquids in plastic vessels; and a water-plus-graphite mix in a plastic-lined steel tank. Nivopuls is TÜV-approved for overspill protection.

A difficult development target of this innovative sensor was a long-term stable coupling material that could be inserted between a sensor and a vessel's wall. As auxiliary information enhances the reliability of the sensor, the evaluation of possible echoes from an opposite wall of the vessel (as explained in Section II.A.7) can be implemented.

Canongate Technology's SpotCheck is another external point level sensor, similarly relying on the ring-down to indicate liquid presence on the farside. Factors complicating the application sometimes include a floating roof, whose seals might influence ring-down; and regulations preventing on-site calibration or performance verification by actually raising the liquid to the forbidden high or high-high level.

Let us return now to the general topic of liquid level. The time rate of change of liquid level, dH/dt, compensated for the vessel's cross-sectional area A, not only has the dimensions of volumetric flow velocity, Q, but also has been used to measure flow (Rod, 1962). Also, dH/dt is one of the methods used for calibrating ultrasonic flowmeters. This brings us to the second major measurand in process control, flow.

B. Flow

The measurement of volume flow and mass flow is right alongside the measurement of temperature and pressure as being among the most important measurement tasks in closed-loop process control. This is evident from the multiplicity of different flow measurement methods and their variations that are employed for process control. The current trend in flow measurement is toward nonmechanical meters without moving parts. Highly desired are methods that would provide a minimum length plain measuring tube without crevices or built-in objects, with minimum flow obstruction or pressure loss. The measuring tube should be of the same inner diameter and made from the same material as the tubing to which it is intended to be connected. Further, the measuring tube should be mountable in any tube configuration, including critical ones such as near bends, multiple bends, or other problematic setups. (*Clamp-on ultrasonics* appears to meet all these requirements. Perhaps the only remaining problem is to obtain high accuracy and prove it.)

For entirely electronic flow measurement, the following techniques can be considered: the coriolis method for high-precision measurement of mass-flow, magnetic induction flowmeters for electrically conducting fluids, thermal methods such as hot wire or hot film anemometers (for gases), and ultrasonic flowmeters. The latter technology has many significant advantages, as long as the measurement of *volume* flow is accepted as sufficient.

For some applications, high or extremely high accuracy specifications are essential and therefore high prices are acceptable for such high accuracy meters. By far, however, devices with moderate accuracy generate the major part of sales. Often, a good *repeatability* in a given installation is satisfactory. In addition, the demand for a *mass* flow, rather than volume flow, often seems to be exaggerated. Presuming it were known which kind of liquid is flowing through the tube (the temperature of which is measured by readily available low-cost methods), the mass flow is easily determined by an electronic flowmeter. However, there are atypical situations where several types of hydrocarbons pass through a given pipe section. Is it possible to ascertain what product, and thus what density ρ, is flowing at any moment if the only tools available are means for measuring temperature T and sound speed c? If there are only a small number of possible products, then a solution analogous to that presented in Section III.B.2 may be appropriate. On the other hand, why should it be so important to know the mass flow rather than the volume flow? One answer comes from chemistry. For processing liquid or solid compounds, components are mixed on a mass-balance basis. Sometimes the motivation is an energy extraction (fuel combustion) process that utilizes variable-density, imprecisely known fuels, or fuel mixtures (e.g., military aircraft refueling in adverse conditions). For extracting the density or for recognizing the liquid, the liquid's speed of sound, which is measured simultaneously with flow velocity at no extra expense, sometimes yields an inexpensive solution.

Nevertheless, the development of a consistent ultrasonic density measuring method for upgrading volume flowmeters to mass flowmeters is an ambitious and very necessary task. Perhaps answers will come from sensors employing guided torsional or flexural waves since the phase velocity of these waves slows down as a function of the density of the adjacent fluid (Kim *et al.*, 1993; Liu and Lynnworth, 1993, 1995; Craster, 1996, 1997). Ideally, the pipe wall or the pipe itself would serve as the sensor. However, the interfering effects of viscosity on a density sensor would have to be subtracted, whether the sensor was intrusive or not.

Recall that *gases* mix on a molar basis—say, two volumes of hydrogen plus one volume of oxygen to make one volume of steam, all at the same pressure P and temperature T. Thus it would seem that for *gas* flow, density or mass flow rate is not needed. Wrong conclusion! Gas flowmeters *are* required, in general, to deliver an output Q_s in *standard* cubic units per unit time. This means the product of actual average flow velocity V_{AVG} times area A, which equals the actual volumetric flow rate Q_A, must be compensated for P and T to yield Q_s. When measuring *steam* flow, the desired output is mass per unit of

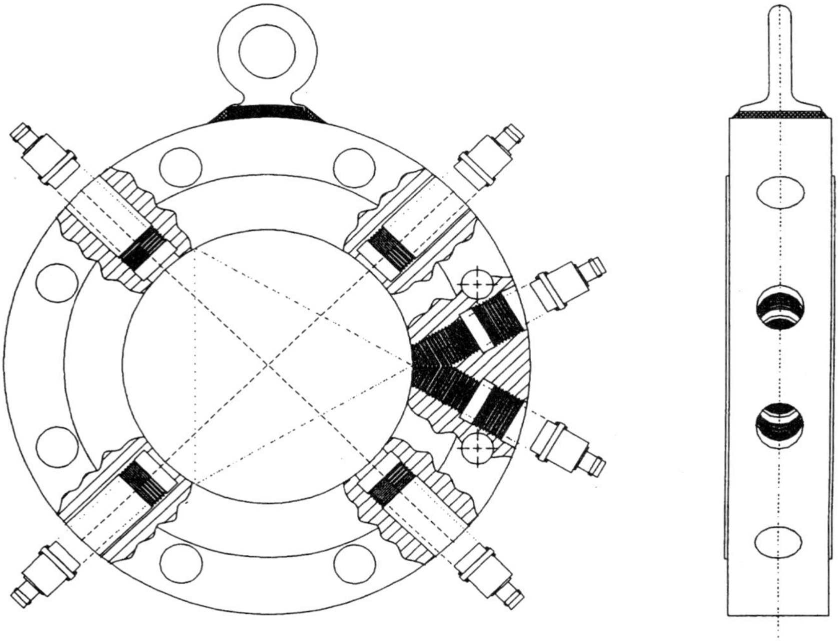

FIG. 31. Wetted-transducer air paths for measuring secondary flows (crossflow, circulation) are obtained in this thick Flangitron® flange design by using spaces between bolt holes. If the flange is to be used at high pressure, e.g., 10 bar air pressure, it must be made extra thick to compensate for the material removed when creating the radial or angled ports. The ports, like those in Fig. 15(c), accommodate the T7 air transducers illustrated in Fig. 15(a). This example was constructed to fit between 6-inch 150-pound standard raised face flanges in a 6-inch nominal pipe size line.

time. If the steam is two-phase (wet steam), then a density or quality measurement is needed. When the purpose is to meter the heating value, as it often is for methane, means using acoustic, thermal, and/or other technologies are sought to determine the calories per kg (or btu/pound). Apart from industrial process control applications, consider that in *aerodynamic* studies the important measurands may be flow velocity V and sound speed c, their ratio being the Mach number. *Lift* depends on circulation γ. The lifting force per unit length of an airfoil, L', equals $\rho V\Gamma$, according to the Kutta-Joukowski theorem, where ρ = gas density, V = upstream velocity, and γ = circulation about the airfoil. This theorem motivates a combination of counterpropagating measurements of V, clockwise and counterclockwise measurements of γ (Schmidt, 1975; Smith, 1994; Smith *et al.*, 1995), *and* a measurement or computation of gas density. Regarding the more common industrial measurements (in gases *or* liquids), measurement of secondary flow (crossflow, circulation) may have a role in providing correction terms to high-accuracy multipath custody transfer flowmeters. Figure 31 shows an example.

1. *Ultrasonic Flow Measurements*

Ultrasonic flow sensors evaluate the influence of the moving liquid or gas on the ultrasonic beam. In this regard, different techniques have been realized. The so-called beam drift or aberration method (Figure 32) and the tag-correlation method (Figure 33) are used in rather special cases. The problem of the classical or simplest ultrasonic Doppler methods is that it is difficult to

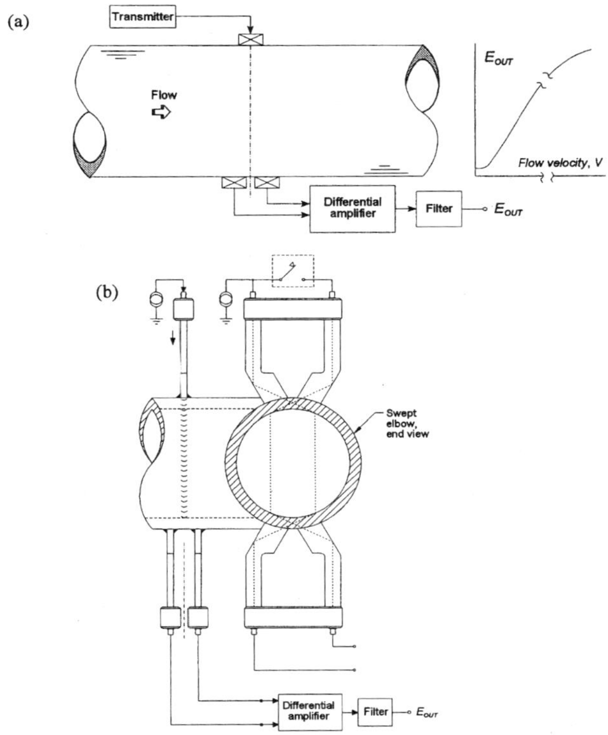

FIG. 32. The beam drift method of measuring flow, illustrated schematically in (a), seems simple in principle. Various designs responding to beam drift, ray rotation, turbulence, and attenuation are illustrated in Lynnworth (1979, pp. 446–449). However, there do not appear to be any industrial flowmeters currently available based on these principles or effects. (b) A 1996 patent application by Lynnworth proposed the measurement of beam drift in *off-diameter chords* using swirl transducers (Fig. 50(d), page 379), as suggested here.

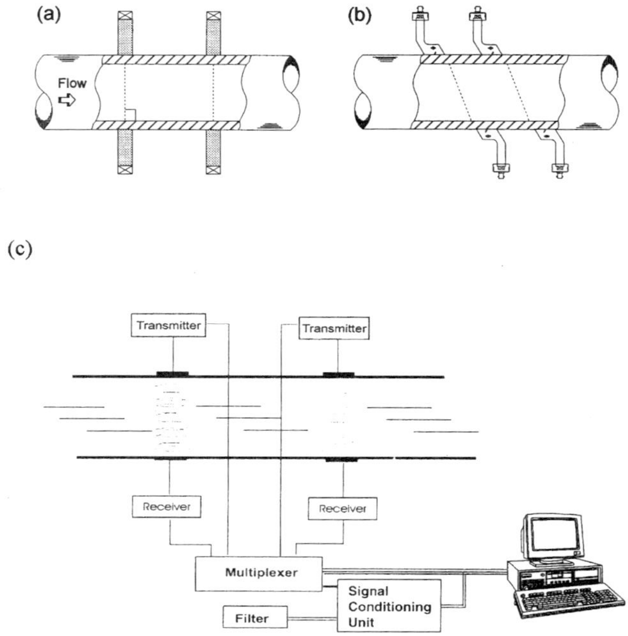

FIG. 33. The tag-correlation method has been used where the fluid is sufficiently inhomogeneous and time-varying in its acoustic propagation properties, so that a sound beam will be modulated significantly when passing through said fluid. (a) Schematic with buffered (Fig. 43) compressional mode piezoceramic elements at normal incidence. Transmission is across the liquid normal to the pipe axis. (b) Proposed use of shear waves at oblique incidence, oblique (refracted) transmission across the fluid, using clamp-on or weld-on hockey stick waveguide buffers of the type shown in Fig. 47(a), or in Lynnworth *et al.* (1996, 1997b). Oblique arrangement would allow transit time (contrapropagation) to be used until transmission against the flow direction is so difficult that only the tag method would remain viable. (c) Block diagram and (d) photograph of transducer clamping fixture, transducers and a commercial tag flowmeter instrument (Crossflow System) manufactured by AMAG (Advanced Measurement & Analysis Group, Subsidiary of Canatom). (e) Reactor coolant feeder flow test data for AMAG system. When the power was held constant, the standard deviation was 0.12%. Illustrations (c)–(e) courtesy of AMAG.

determine in which part of the cross-sectional area of the measuring tube the measurement actually takes place. (See, however, Section II.B.2.i.) Therefore, they are relatively inaccurate and in principle depend on the existence of scattering particles in the moving fluid, normally not present in pure liquids. High accuracy, however, can be expected from transit-time and phase-difference methods based on the "carry-on" effect, $c \pm V$ (Figure 34).

In the arrangement shown in Figure 34, two transducers—one of which is located upstream (transducer 1) and another downstream (transducer 2)—

(d)

(e)

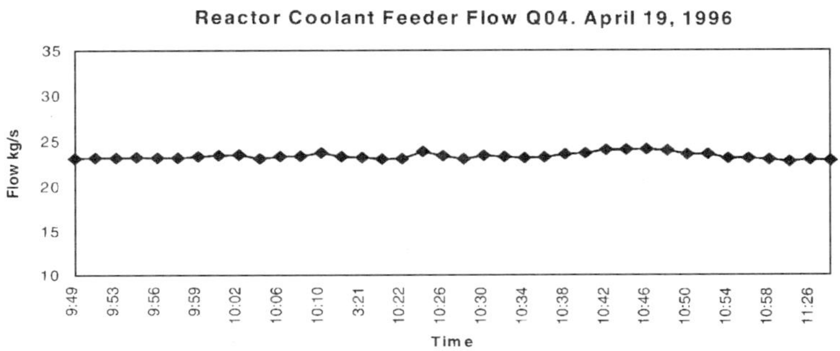

FIG. 33. *(continued)*

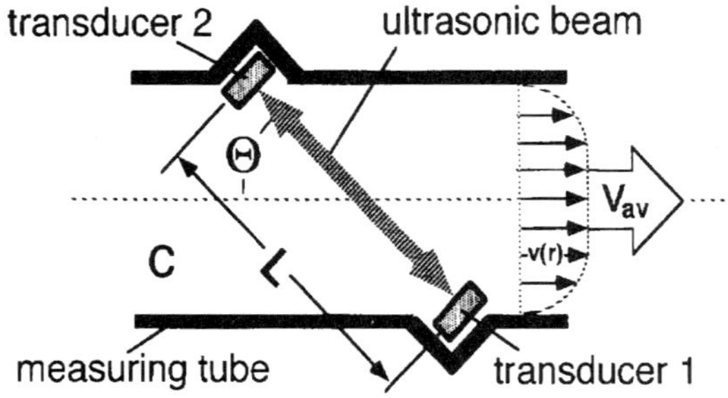

FIG. 34. Basic arrangement of an ultrasonic flowmeter. V_{av} = average (volumetric) flow velocity; v(r) = flow velocity (radial) distribution.

both transmit and receive ultrasonic signals. Thus, either one can be utilized as transmitter and receiver. When the fluid is not in motion, the transit time of the ultrasonic wave depends on the path length and the sound speed c, but is independent of whether transducer 1 is sending and 2 receiving or vice versa (Helmholtz's theorem of reciprocity). With the fluid in motion, the transit time depends on the direction of the sonic wave relative to the direction of the fluid flow. Depending on this direction, the wave will either be carried with the flow or it will run against it. According to this, the transit time needed for the sonic wave to cover the measuring distance L is t_1 in the direction of the flow and t_2 in the opposite direction. Accordingly, t_1 is shorter and t_2 longer than the time needed with the fluid at a standstill. The transit time difference is directly proportional to the velocity of the flow, V, to the extent that $V^2 \ll c^2$. (Nonlinearity between Δt and V becomes noticeable when the Mach number V/c exceeds about 0.1.) The measured flow velocity value is an average along the chosen measurement path.

Arrangements such as in Figure 34 are preferred where the path is laid at an angle θ across the measuring tube. Then the following equations hold:

$$V = \frac{L}{2\cos\theta} \cdot \left(\frac{1}{t_2} - \frac{1}{t_1}\right) \tag{1a}$$

$$V = \frac{L}{2\cos\theta} \cdot \left(\frac{t_1 - t_2}{t_1 t_2}\right) \tag{1b}$$

For the evaluation of the average flow velocity, in addition to the known constants of the measuring tube, the length of the ultrasonic propagation path L and its inclination angle with respect to the measuring tube axis θ, one needs the difference of the two transit times—i.e. the transit time in the upstream and in the downstream direction. The sensitivity—i.e., the output value vs flow velocity—of this ultrasonic flowmeter design is, apart from flow profile influences, given by the geometric properties of the measuring tube. This advantage makes individual calibration unnecessary in many cases and results in an excellent long-term stability.

It can be an advantage for the measurement of the flow to get the reciprocals of the two transit times since the difference of these reciprocals is directly proportional to the flow velocity, regardless of the fluid's speed of sound. This was the reason for the development of methods for measuring the reciprocals of the transit times directly, for instance, by the so-called sing-around method. In this case, the reception of a sonic pulse by one of the transducers triggers the emission of another sonic pulse in the same direction. A repetition rate of $f_1 = 1/t_1$ or of $f_2 = 1/t_2$ is thereby generated according to the directions. In one advanced setup, the repetition rate was stabilized by an *electronic flywheel*, which triggered the emission even when the received signal failed to arrive. This is realized by an oscillator circuit, the frequency of which is coupled to the repetition rate by a phase locked loop (PLL). Similarly, for a continuous wave transmission, the frequency can be stabilized to a value, which gives a certain predetermined number of wavelengths λ in the ultrasonic path by means of a PLL. This condition can be permanently checked by a phase discriminator comparing the received signal and the transmitted signal with high resolution. So for both ultrasonic propagation directions an individual frequency f_1 and f_2 (typically in the megahertz range) is achieved whose difference yields the flow velocity with a high resolution and short response time, independent of the speed of sound in the fluid. This method, virtually an "overtone flywheel", is the so-called lambda locked loop (LLL) and is an appropriate solution for small-diameter measuring tubes. In the past, ultrasonic flowmeters using sing-around methods were manufactured and brought to market.

Nowadays, these generation and explicit evaluations of the reciprocal time are no longer regarded as up to date. Today one can measure the transit times directly and calculate the flow velocity according to Eq. (1b) with the aid of a microprocessor built into the evaluating electronics.

a. Advantages of the Ultrasonic Flow Measurement. The actual sensing element in an ultrasonic sensor is the ultrasonic signal, which is generated anew for every measurement cycle. The ultrasonic wave is highly sensitive to a number of physical and chemical quantities, and this "coding" is performed with high reproducibility. It is the task of the specific sensor signal processor to decode this information and to transmit the derived measured values to the data logger or process controller in a format conforming to the system. As *intelligent sensors*, ultrasonic flow sensors use available or automatically *learned* information about the coding of the sonic signal that have a bearing on the process and its parameters. This leads to a number of advantageous characteristics of the ultrasonic flow measurement method, such as:

- High accuracy over a wide flow velocity range, high reproducibility
- High linearity, correct determination of the direction of the flow
- High resolution, short response time
- Compact, straight measuring tube, low pressure drop
- High long-term stability
- Insensitivity to abrasion and dirt
- Low energy consumption
- Ability to obtain an accurate measure of the average flow velocity even for abnormal flow profiles, by interrogating along appropriate paths
- Applicability to a large number of fluids and gases, no electrical conductivity required
- Ability to carry out the measurement through the measuring tube walls with transducers attached from the outside (clamp-on), when the fluid is a single-phase or multi-phase liquid (but not for all liquids), and probably in the near future, for gases if the pressure is high.

The span of measuring-tube cross sections that have been realized with industrial flowmeters ranges from approximately 1 mm to several meters in diameter. At one extreme, a laboratory model of an ultrasonic microflowmeter with 2-mm inner diameter was built up for a maximal flow of 50 l/h. This sensor could be used for the on-line measurement of gasoline consumption of automotive engines: By the high repetition rate of about 50 kHz an extremely high time resolution was achieved, resolving easily the velocity oscillations in the gasoline tube cause by the opening and closing of the injection valves. For medical, biological, and clinical applications, Transonic Systems has pioneered clamp-on flow measurements for vessels or soft tubes down to sub-mm diameters. See their catalog or Figure 50(j), pages 385–386 for examples, the smallest ID being 25 μm. See also Smith *et al.*, 1994; Heimisch, 1996.) At the other extreme even the flow velocity of rivers hundreds of meters wide can be measured by ultrasound (Laenen, 1984) (Figure 35). Several such stations have been installed and the measured flow values are automatically communicated to a central station.

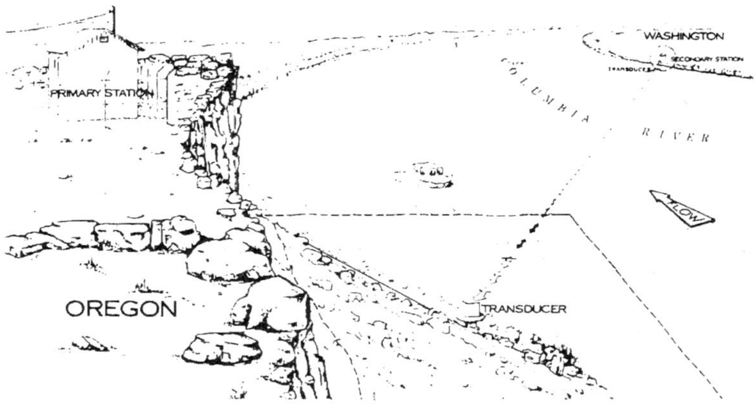

FIG. 35. Ultrasonic flow velocity measurement at a river at The Dalles, Oregon. After Laenen (1984). Path length ~ 400 m.

b. Problems of Ultrasonic Flow Measurement. Despite its many advantages, ultrasonic flow measurement is often accused of being unreliable and imprecise. This is because the different ultrasonic methods are often not distinguished from each other and e.g., the difficulties of the Doppler method are attributed to other methods equally across the board. Furthermore, a lack of knowledge of the fundamental physical effects that participate in ultrasonic flow measurement has led to the construction of measuring arrangements that indeed display serious shortcomings.

A typical problem is the inadequate information about the cross section of the measuring tube. The quantity to be measured is the volume flow, Q, i.e., the volume crossing the cross section per unit of time. In a coordinate system with its x-axis parallel to that of the tube and its y-axis in the plane perpendicular to the tube axis in the direction of the projection of the sonic beam onto that plane, one can write the volume flow in the following form:

$$Q = \int\int V_x(y, z)dy\, dz \tag{2}$$

The quantity $V_x(y, z)$ represents the component of the flow velocity in the direction of the tube's axis. The integral is extended over the entire cross section. In contrast to this, an ultrasonic ray having negligible cross section and located at a coordinate z_0 measures only a one-dimensional integral, which must still be multiplied by a correction factor K given by

$$D = K \int V_x(y, z_0)dy \tag{3}$$

In the range of high flow velocities, one obtains turbulent flow conditions in which the velocity distribution $V_x(y, z)$ inside the tube is almost constant. It seems advisable to choose the factor K in such a way that the value D obtained from the ultrasonic measurement agrees with the exact volume flow Q in the case of a stepwise velocity distribution.

In the range of low flow velocities, the flow profile changes to more laminar patterns. If one continues to employ the correction factor K valid for turbulent flows, the value D can significantly differ in the case of slow flows from the true value Q. An example is a circular tube in which the flow profile yields a parabolic distribution with its maximum on the tube axis. The scanning of the flow by an ultrasonic ray along the tilted diameter leads to a value that is too high by a factor of 4/3. (Profile problems were recognized by 1955. Remedies evolved such as computing K as a function of the Reynolds number Re, finding a path relatively insensitive to profile, and integrating along special paths. These remedies are reviewed in Lynnworth (1979 and 1989). A special helical path provides one of the newer solutions to this old problem, as will be described in Section II.B.2.d.)

Further reasons for the distortion of the linearity of an ultrasonic flowmeter are the diffractive effects of flow velocity or sound speed (thermal) gradients. Following the basic physical principle that the waves traveling between two points A and B seek the path of minimal transit time (Fermat's principle) and considering boundary conditions such as reflecting walls and the like, the upstream and downstream ultrasonic path are not exactly the same. It is obvious that in a configuration like Figure 34 the downstream ultrasonic signal must try to take advantage of high flow velocity parts of the flow profile, whereas the upstream signals tend to avoid them (Bruner, 1977).

For example, the focusing effect of the flow on the propagation of ultrasound and the waveguiding effect of the measuring tube, can lead to especially unsatisfactory results if the sonic beam is launched coaxially to the tube, i.e., straight down the tube. Figure 36 shows the results of a simulation. In this simulation the axial-path sonic pressure distribution was computed in a tube as a function of the flow. For better visibility of the effects, the flow velocity was strongly overemphasized. But due to the high sensitivity of the ultrasonic method for transit time differences in upstream and downstream directions, these effects can play a significant role also at very low flow velocities.

Another effect that reduces accuracy is the so-called zero-flow error, which erroneously indicates a finite flow caused by minor differences in the transit times t_1 and t_2 even when the flow is at a standstill. This error is particularly bothersome at low flow velocities, when very small time differences must be measured with high accuracy. The typical transit-time differences at low flows

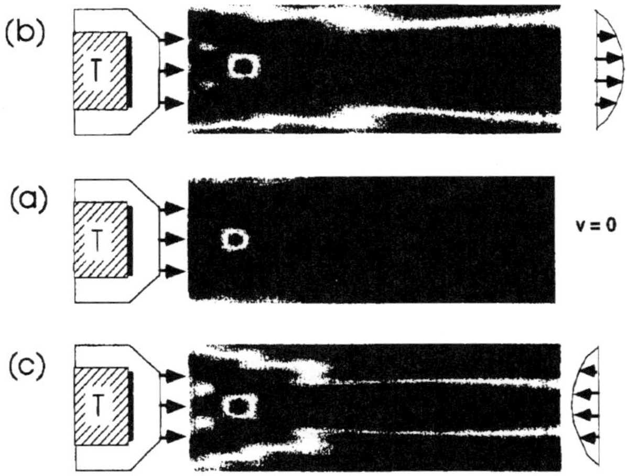

FIG. 36. Diffraction effects in a tube on ultrasound at a coaxially insonified tube by flow velocity gradients. T ultrasonic transducer (a) at rest; (b) sound propagation in flow direction; (c) sound propagation antiflow direction. Courtesy of Siemens.

amount to about 10 to 100 ns, and these have to be measured with an accuracy in the subnanosecond range (50 ps to 1 ns) if the desired accuracy of the flow measurement of 1% is to be attained. This represents a very demanding task, considering the typical range of operational frequencies of about 100 kHz (gases) or 2 MHz (liquids), with corresponding time periods of 10 to 0.5 μs. The zero-flow errors are actual deviations in the transit times t_1 and t_2 caused by asymmetries in the characteristics of the transducers and in the emitter and receiver circuits. Strict requirements for the uniformity of transducer characteristics drive the costs high and cannot be satisfied over a long period of time anyway because of differences in the aging characteristics and changes caused by dirt. In fact, the symmetry of the transducers would be of reduced significance if one could succeed in setting up the arrangement in such a way that the principle of reciprocity (Helmholtz, 1860) were satisfied (Figure 37). It is difficult, however, to accomplish this, and small deviations from linearity disturb the principle of reciprocity even at zero flow. A further experimental difficulty stems from the thermally driven currents having a component of flow along the acoustic path. This is more evident in gases than in liquids.

Further causes of unsatisfactory results are unsuitable ultrasonic transducers and electronic circuits that are based on false concepts or do not function to achieve the desired accuracy. In particular, acoustic cross talk (short circuit or body sound), which is transported from one transducer to the

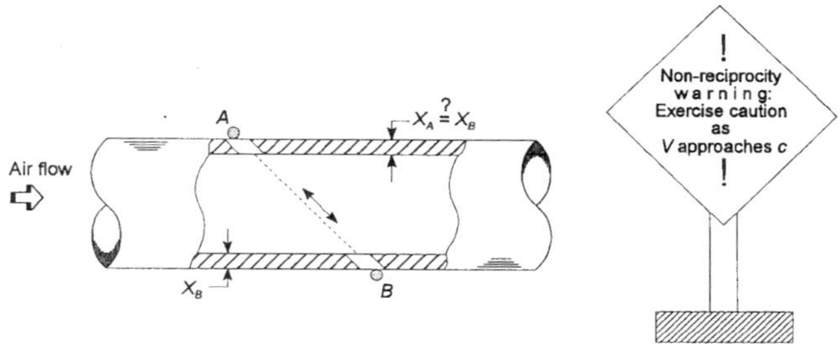

FIG. 37. According to a translation of Helmholz's reciprocity theorem appearing in Rschevkin (1963): "In an air-filled region having bounding surfaces S_1, S_2, $S_3, \ldots$, the region containing only one point source at *A*, the velocity potential of sound waves is the same in magnitude and phase at another point *B* as it would have been at *A* had the source been located at *B*." Flowmeter designers need to remember that this theorem applies to a medium at rest with respect to the transducers. Some further details concerning the boundary of the air, especially near the transducer(s), and *simple* vs *double* sources, are spelled out in Rayleigh (1896, 2, pp. 145–146). See also, Pierce (1981), pp. 195–203.

other by the measuring tube body becomes superimposed on the received signal and spoils the measurement accuracy. It must be avoided by all means. The application of ultrasonic flow measurement is also limited by highly attenuating media, such as pastes or suspensions transported in pipelines, or coal or ore slurries, that prevent the propagation of the ultrasonic beam through the tube. The measurement of multiphase flow is in many cases impossible or at least very difficult because consistent ultrasonic paths hardly exist.

One Japanese company's method of dealing with flow profiles and also rapidly changing (pulsating) flows is represented in Figure 38. PZT transducers interrogate the entire cross section of a symmetrical flow tube that is made of stainless steel or PVC. The manufacturer cautions against using this device with slurries and very viscous liquids. Response time is selectable down to 0.25 s, according to a 1996 brochure. Flowmeter equipment made by Tokyo Keiki (now Tokimec) is shown in Lynnworth (1989, pp. 33 and 256). Fuji and Keijo are two of the other ultrasonic flowmeter suppliers well known in Japan and elsewhere.

Instead of integrating the flow profile by a special helical beam, by multipaths such as the midradius chords or the GC (Gauss-Chebyshev) quadrature chords, one can intentionally interrogate at a point or along a short segment and derive a meter factor *K* for that sampled path (Lynnworth and Lynnworth, 1985). The bias-90 path shown in Smalling *et al.* (1984) is one such segment that is used in hundreds of flare gas applications, especially where access is limited to one side of a pipe and echoes from the opposite wall cannot be guaranteed to be detectable.

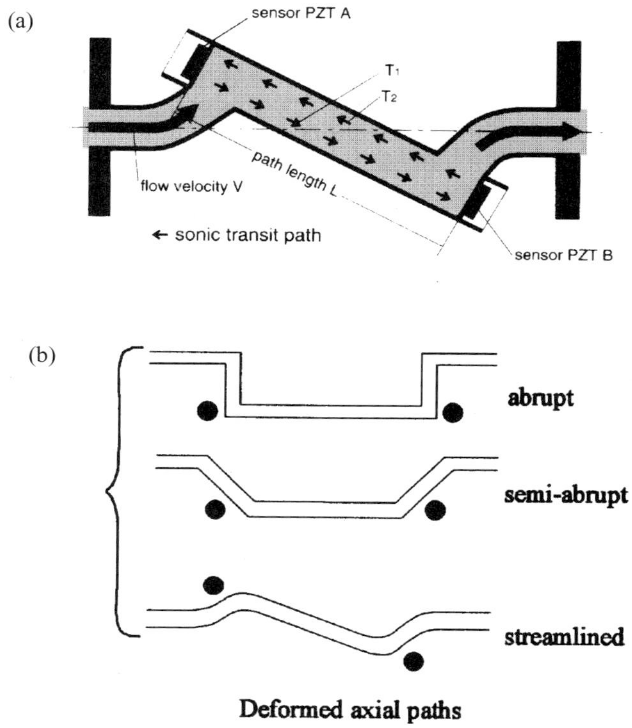

FIG. 38. (a) Model SF-1000 from Kay Instruments Co., Ltd interrogates the full cross section in a symmetrical construction. This construction may be compared with the several axial-interrogation concepts shown in (b), in which varying degrees of abruptness are present at the flow inlet or outlet.

Figure 39 shows a wetted *reflection*-mode (R-mode) flowcell concept, proposed here as an R transform of the T version of Lake (1962), with some additional paths for sensing c, ρ, and η. Pfau (1970) introduced a *nonintrusive* Doppler version of sensing in a defined region near the wall. The insonifying beam and the receiving beam intersect over a preferred region near the wall of a pipe to obtain a reading relatively immune to changes in profile. If one has an equation representing the flow profile, one can find two points A and B in the fluid between which the integrated profile changes very little as a function of Reynolds number Re. At least one of these points can be at the wall. In principle, this provides a reasonable starting point for designing a flow sampling system (Lynnworth, 1989, p. 249).

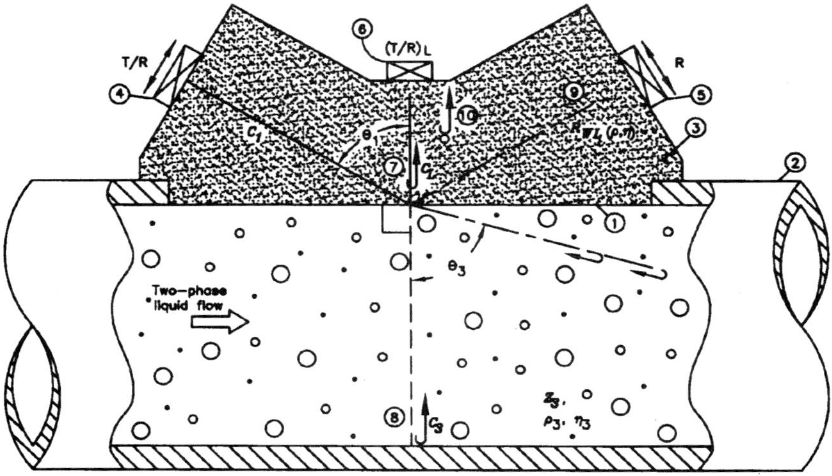

FIG. 39. Concept of *wetted* flush-mounted reflection-mode flow sensor that requires but *one* hole ① in the pipe ②. Plastic wedge ③ is nonintrusive. Transducers ④ and ⑤ operate in *shear* mode so that incident velocity c_1 is lower than longitudinal velocity c_3 in the two-phase liquid. This makes the refracted angle θ_3 larger than the angle of incidence θ_1. (In this respect, the R-mode design resembles the T mode of Lake (1962) shown in Lynnworth (1979), p. 422, Fig. 5(a).] Pulse-echo transducer ⑥ operates as a longitudinal (thickness) mode device, yielding a reference echo from reflector ⑩ and an interface echo ⑦ from which c_1, the *longitudinal* velocity in the wedge, is obtained. Transducer ⑤ detects an oblique reflected shear wave ⑨ from which the density- and viscosity-responsive reflection coefficient R_{WL} (ρ, η) at the wedge–liquid interface can be determined. Echo ⑦ in principle also yields the liquid's impedance Z_3 from which the liquid density ρ_3 can be determined. This normal-incidence method of measuring *density* in flowcells proposed for mass flowmeters is discussed in Lynnworth and Pedersen (1972) and more recently in Van Deventer and Delsing (1997), but is not yet available commercially. The *viscosity*-sensing aspects of utilizing a measurement of reflection coefficient is discussed in various papers, a recent example being Sheen *et al.* (1996).

2. *Solution of the Problems: Innovative Ultrasonic Flowmeters as Results*

To develop competitive ultrasonic flowmeters, one must have a thorough understanding of the physical relations participating in ultrasonic flow measurement and be able to analyze the problematic cases and find creative solutions to eliminate the problems. As complete an understanding as possible of the physics of sound propagation and of the flow is needed to conceive and realize specific ultrasonic transducers for sensors. Also required is knowledge of electronic measurement technology and of typical cases of application. Obtaining solutions to these problems is a rewarding task because it opens up a valuable application potential for ultrasonic flowmeters in process control.

The following examples of successful realizations of ultrasonic flow sensors demonstrate how principal advantages of ultrasonic flow measurement were successfully used, how principal problems were solved, and what basic ideas stood behind each design approach.

a. Automotive Intake Air Sensor with High Time Resolution. The rapid response time and the flow direction recognition of ultrasonic flow measurement methods are important for *air mass meters* (von Jena and Mágori, 1992), which measure the intake air flow of internal combustion engines (Figure 40) used in cars. On the test stand, this sensor displays a linear flow vs output signal characteristic, independent of temperature over the whole temperature range. As the sensor is intended to be installed in a series of vehicles with the same intake duct configuration, it would be possible even to equalize linearity deviations, if they occurred.

In pulsating and reversible flows, ultrasonic sensors prove to have a faster response time than comparable hot wire and hot film sensors. Figure 41 demonstrates this by comparing the results obtained with an automotive hot wire sensor, a laboratory-type fast hot wire sensor (too delicate to be used permanently on engines), and the ultrasonic sensor. Both the laboratory-type hot wire and the ultrasonic sensor show a considerably faster response. In addition, the ultrasonic sensor correctly evaluates the reversed flow, whereas the two thermal sensors respond only to the absolute value of the flow. The rise time of the widely used hot film sensors is even about one order of magnitude slower than the automotive hot wire sensor. Together with the nonlinearity of the thermal principle, this slow response can cause measuring errors up to 100% in the case of strong pulsations typical for four-cylinder

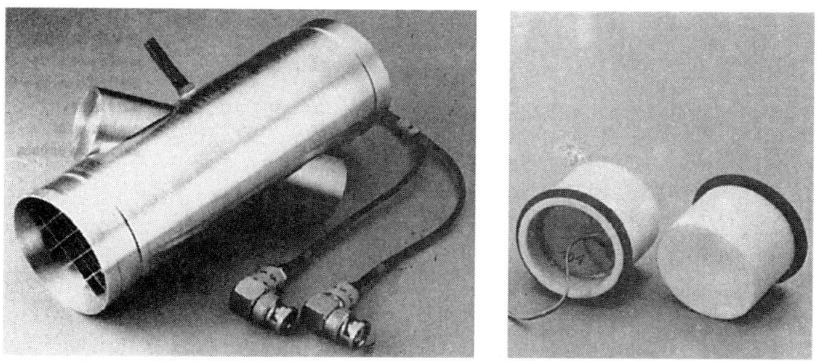

FIG. 40. Ultrasonic automotive intake air flow sensor that used ultrasonic transducers operating at $f_0 \approx 150$ kHz. Images courtesy of Siemens.

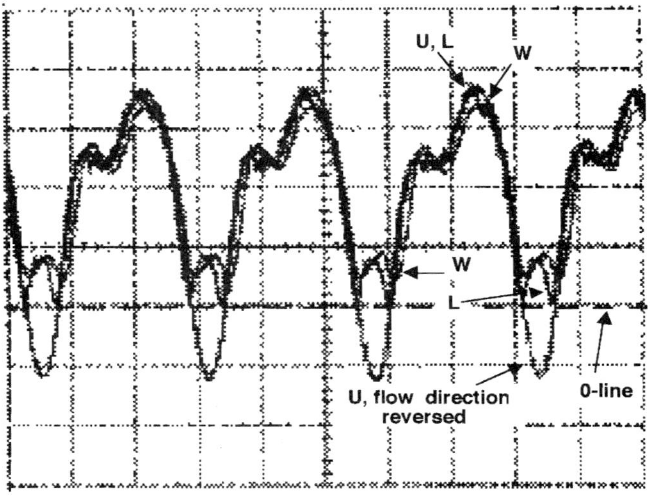

FIG. 41. Function of the ultrasonic intake air flow sensor (U) in a pulsating air flow compared to a laboratory-type hot wire anemometer (L) and an automotive hot wire air mass sensor (W).

engines at partly loaded conditions. These advantages of the ultrasonic intake air sensors could open a new technical potential for improved fuel injection systems that can manage individual cylinders and transient loads. By the correct and fast response measurement of the aspirated air flow eventually combined with fast injection valves, many improvements of the air–gasoline mixture may become feasible, resulting in lower gasoline consumption and cleaner exhaust.

The measured volume air flow becomes equivalent to mass air flow by multiplication with the density, which, in the case of pure air at known pressure P could be derived from the speed of sound that is easily available in ultrasonic flowmeters. As an alternative, the temperature and the pressure could be measured with additional pressure and temperature sensors, and the density could be derived from these measured values. However, as the atmospheric pressure varies very slowly, in an engine management system the air density can be estimated from the atmospheric pressure based on a large time constant evaluation of the exhaust sensor results. An advantage is that this method will work also with changing gasoline quality. Even the concentration of gasoline vapor and rapid pressure changes could be measured easily on-line, as discussed later (Section II.B.3.b). (See also Section II.D; the possibility considered there is that of obtaining P from the amplitude of the received ultrasonic signal.)

Ultrasonic air flowmeters intrinsically have excellent long-term stability, are inexpensive to build, and do not require an individual calibration. In spite of their significant advantages (which mean significant improvement potential of the system performance of engine controls, in particular in highly dynamic operation [working cycle resolving and cylinder-specific determination of the aspirated air] shown in test stands and on the road by test vehicles) and in spite of having a competitive cost potential, only a minor number of samples of the ultrasonic air flow sensor have been fabricated. (See also Gutterman, 1985.)

In designing ultrasonic flowcells using wetted transducers, one tries to minimize flow disturbances. So too in the design of sonic anemometers (Figure 42), an important design objective is to not disturb the flow by the transducers or their supporting structure. Some hints from a related technology, due to Allen (1995), are summarized in *NASA Tech Briefs*, pp. 44–45 (Oct. 1996).

b. Noninvasive Clamp-On Measurement of Liquids. One of the early *analyses* of clamp-on for thick-wall metal pipe is due to Del Grosso and Spurlock (1957). One of the suggestions in that analysis was to modify the wall, i.e., introduce cavities or wedges to avoid refraction and thereby better control the *tilted-diameter beam* (A variation of this suggestion has been recently used by Lynnworth *et al.* (1996, 1997b) to obtain, from outside a pipe, tilted vee paths along *off-diameter* Gauss-Chebyshev, midradius or other chords that would be unreachable in an unperturbed steel pipe. Material is removed from inside and outside the pipe wall, but the wall is not totally penetrated (Figure 50(g)). The process leaves leakproof parallel-faced webs, or *websites*. A transducer is mounted (coupled) against the outer surface of each website. In principle, material could be added inside and out, so the wall would not be weakened at the transducer sites.) In Figure 50(h) several hybrids are shown that use a section of plastic as a window between the external transducer and the air (or other gas) inside the vessel. In Figure 50(f) other hybrids are shown, including a clamp-on adaptation of Chernyshev's idea (1994), where an intrusive reflector is placed at a distance from the wall such that a representative "average" flow is sensed. By installing not just one reflector but a square tube within the pipe, flow symmetry, and perhaps other parameters of interest can be checked.

Clamp-on *measurements* of flow of water apparently had its first large-scale success in Japan. As reported in Vol. 14 (1979) in this *Physical Acoustics* series, by 1964 flow had been measured in Japan in large metal pipes from outside the pipe without modifying the pipe. Today such measurements would

(a)

(b)

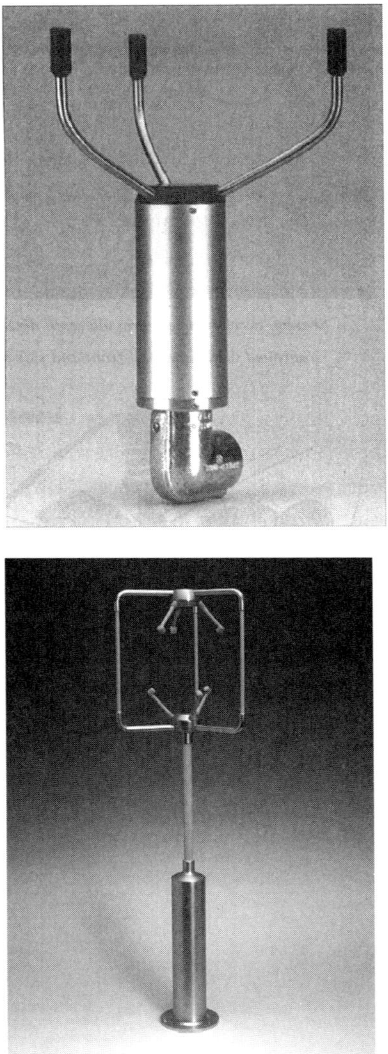

FIG. 42. Sonic anemometers provide examples of three-path ultrasonic flowmeters. Three-path sonic anemometers are available from several manufacturers, e.g., Gill, Handar, IN USA, R. M. Young, and Focal. The Handar Model 425A in (a) contains three paths in a horizontal plane; details appear in Lockyer (1996). The IN USA three-axis anemometer shown in (b) measures turbulence as well as flow and is not limited to a horizontal plane. Similar in outward appearance is the design in (c), which is another typical commercial sonic anemometer for wind measurement in three directions; illustration courtesy of R. M. Young Company. Focal's VDV-1 flow sensor head (d) contains eight miniature transducers mounted in two planes defined by a cage structure. This provides 3D velocity and simultaneous mass flow rate, heat flux, and 2D vorticity measurements.

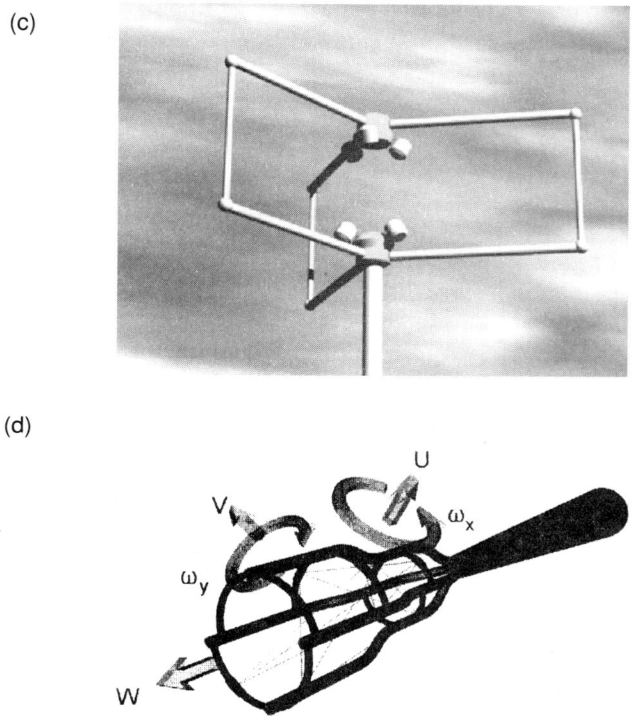

FIG. 42. *(continued)*

be considered routine, provided there were no complications such as a disbonded liner, very complicated flow conditions, aeration, or some other unusual interference.

Clamp-on flow measurements have now been accomplished in probably 10^5 to 10^7 applications. An exact count is not available, as one clamp-on instrument might be dedicated to one site, or it might be used on hundreds of pipes. Examples of companies where clamp-on flowmeters currently comprise a large or major part of their ultrasonic flowmeter business are Controlotron, Panametrics, Peek, and Tokyo Keiki (now Tokimec). Each of these firms has been in the clamp-on business since the 1960s or the 1970s. Applications now include cryogenic flows, where the couplant may be indium, to superheated water (260 to 300°C), where the couplant may be pressure-coupled soft metal foil. Among the metal couplants used at high temperature are gold, nickel, aluminum, tin and zinc foil. (O. I. Babikov, private communication via V. K.

Hamidullin, suggests red Cu as a *cryo* couplant. Another possible metal couplant cryogenic solution is aluminum.) For each couplant, one must check its compatibility with the pipe, environmental and safety concerns, and perhaps other issues.

At ordinary temperatures, the couplant can be a silicone rubber, propylene glycol, gel, grease, or a bonding agent (e.g., urethane adhesive). Sometimes "five minute epoxy" suffices. Generally speaking, transit time (contrapropagation) is used not only on clean liquids, but also on two-phase fluids that are transmissive. Reflection methods (Doppler or stroboscopic scattering) are preferred if, because of scatterers, the fluid is *not* transmissive. (See, for example, Takeda 1995.) *Tag* cross-correlation is used too, but not nearly as often (Tornberg *et al.*, 1983; Tornberg, 1986). See, however, Figure 45(a).

There are also *hybrids* containing an intrusive vortex shedding strut inside the pipe, with noninvasive sensing transducers completely outside the pipe. A more common form of hybrid is where the spoolpiece is introduced during a scheduled maintenance shutdown or, in a new plant, before startup. The spoolpiece contains the transducers, which are *outside* the pressure boundary. In this case, the spoolpiece typically contains plugged ports. The clamp-on transducers are removably coupled to the outside of these plugs, or, alternatively, to the outside of an ordinary section of pipe, or perhaps to a section of specially prepared pipe comprising a precision and calibratable spoolpiece.

The explosionproof design transducer removably coupled to the compressional buffer (Figure 43) is an example of this hybrid concept adapted to a high-temperature situation in a refinery (Liu *et al.*, 1998). Some instruments are multimode, T and R (Figure 48). Some clamp-on flowmeters are passive, where the noise due to flow is analyzed and interpreted either as a leak or as flow above some nonzero threshold (flow switch), or is translated into a flow reading. Figure 44 illustrates the leak detection and the flow measurement versions of *passive* clamp-on flowmeters made by the Norwegian firm ClampOn. The U.S. firm Controlotron (Figure 45(c)) manufactures several types of clamp-on flowmeters, including an active differential flowmeter for leak detection. Badger Meter (Figure 45(b)) combines liquid level with flow in a compound or area velocity meter. Some Controlotron and Badger Meter equipment from about ten years ago appeared in Lynnworth (1989, p. 33). The UE Systems equipment shown in Figure 46 remotely senses leaks; this is another example of remote sensing, here using airborne ultrasound. Contact testing is also used.

Examples of the passive listening method of leak detection are given in Mohr (1995), and a related illustration appears in Figure 46. The Norwegian

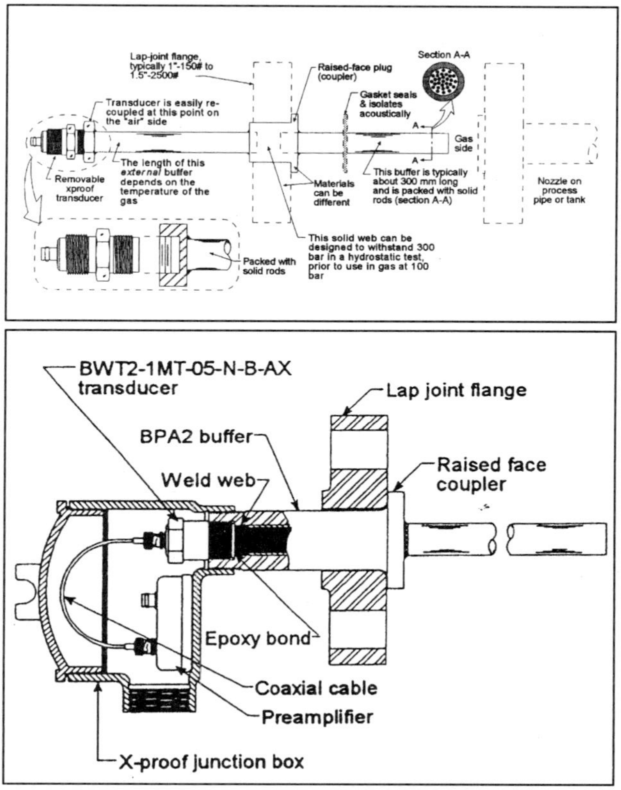

FIG. 43. The explosion-proof design transducer removably coupled to the compressional buffer is an example of the hybrid concept adapted to a high-temperature situation in a refinery (Liu *et al.*, 1998).

company ClampOn uses passive acoustic methods to detect particles in process flows, e.g., in their particle monitors, pig monitors, and corrosion-erosion monitors (Figure 44). ClampOn offers these three functions combined in one instrument. These sensors are intrinsically safe (2000 Series), runs under Windows, detects particles down to 15 μm, and have IP56 enclosures made of 316SS.

The accuracy of a clamp-on is difficult to verify in a typical field installation unless another meter of known accuracy is in the line. This is especially difficult at high temperature. For this reason, a multipath spool-piece has been designed to be the "reference" for clamp-on transducers of the type in Figure 47. However, in most process control applications, repeatability

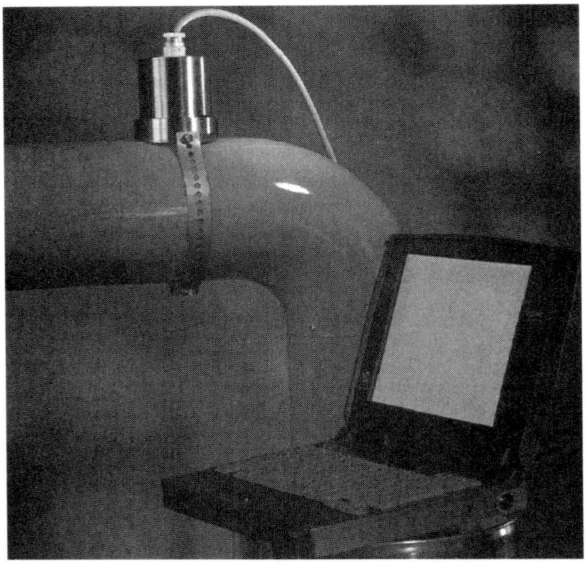

FIG. 44. Passive system for detecting the noise created by passage of pigs in pipeline, or flow of sand, without being confused by mechanical/structural noise or other noise sources. The transducer is mounted in a stainless steel housing and operates against pipe walls at temperatures from -40 to $+180^{\circ}$C. Illustration courtesy of ClampOn.

is more important than accuracy. Because the flow profile influences not only the accuracy but even the repeatability of a flowmeter, designers seek ways of either estimating the flow profile or eliminating its influence. To do this, it may be necessary to measure secondary flow components such as crossflow and swirl, or to range gate, or to measure along multiple diameters and multiple chords, or some combination of the above. In today's clamp-on flowmeters, typical specifications claim accuracy around 1% of reading at ordinary industrial flow rates. NIST, under EPRI support, is currently evaluating several brands of clamp-on ultrasonic flowmeters, and results for ten-inch pipe may be expected soon.

Electronics from Panametrics for use with clamp-on (or wetted) flow transducers are shown in Figure 48(a)–(c). The portable instrument in Figure 48(a) (depending on options selected at the time of purchase) can operate in T or R modes, includes a built-in pipe wall ultrasonic thickness gage, and can accept temperature inputs for heat meter applications. One of the transducer clamping fixtures with which it operates is the quick-on nearly-

(a)

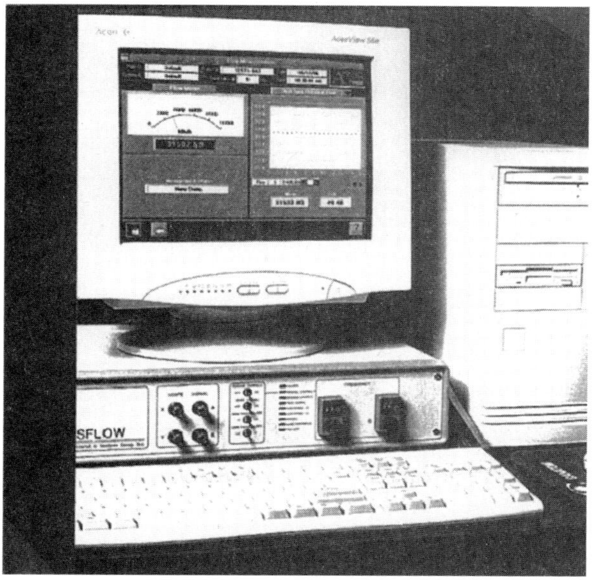

FIG. 45. (a) Amag tag cross-correlation electronics unit (Crossflow); see also, Fig. 33 (c)–(e) for transducer and clamp details. (b) Badger Meter flowmeters include liquid level sensing for measuring flows in partly full pipes (compound meter) and for open-channel flow where measuring level suffices for a flow measurement at a properly designed weir. (c) Controlotron clamp-on 1012TPB Mounting Frame, and leak detection system based on differential volumetric flow measurements, corrected for temperature and pressure. Controlotron also manufactures spoolpieces using nonwetted transducers. Applications for their equipment include volumetric flow rate Q, mass flow rate in cases where sound speed c can be related to fluid density ρ, energy flow rate where hot and cold leg temperatures are measured in addition to flow rate, and leak detection (see block diagram, after Baumoel, 1995), where flow is measured at two or more points in a pipeline, and flow rate difference can be attributed to a leak. Illustrations (a), (b), (c) courtesy of Amag, Badger Meter, and Controlotron, respectively.

as-quick-release device shown in Figure 49. This is but one of a large number of clamp-on arrangements. Others, and numerous hybrids, comprise Figure 50. Not all hybrids are expected to achieve commercial success. Even the unsuccessful versions, however, may satisfy niche markets or solve special problems in an R&D environment.

Long-range goals of the clamp-on flowmeter designer include extending its range to gases (including steam) and measuring not just flow velocity but also mass flow rate. Gases such as air flowing at atmospheric pressure already can be measured in plastic pipes such as two- to ten-inch diameter PVC, using

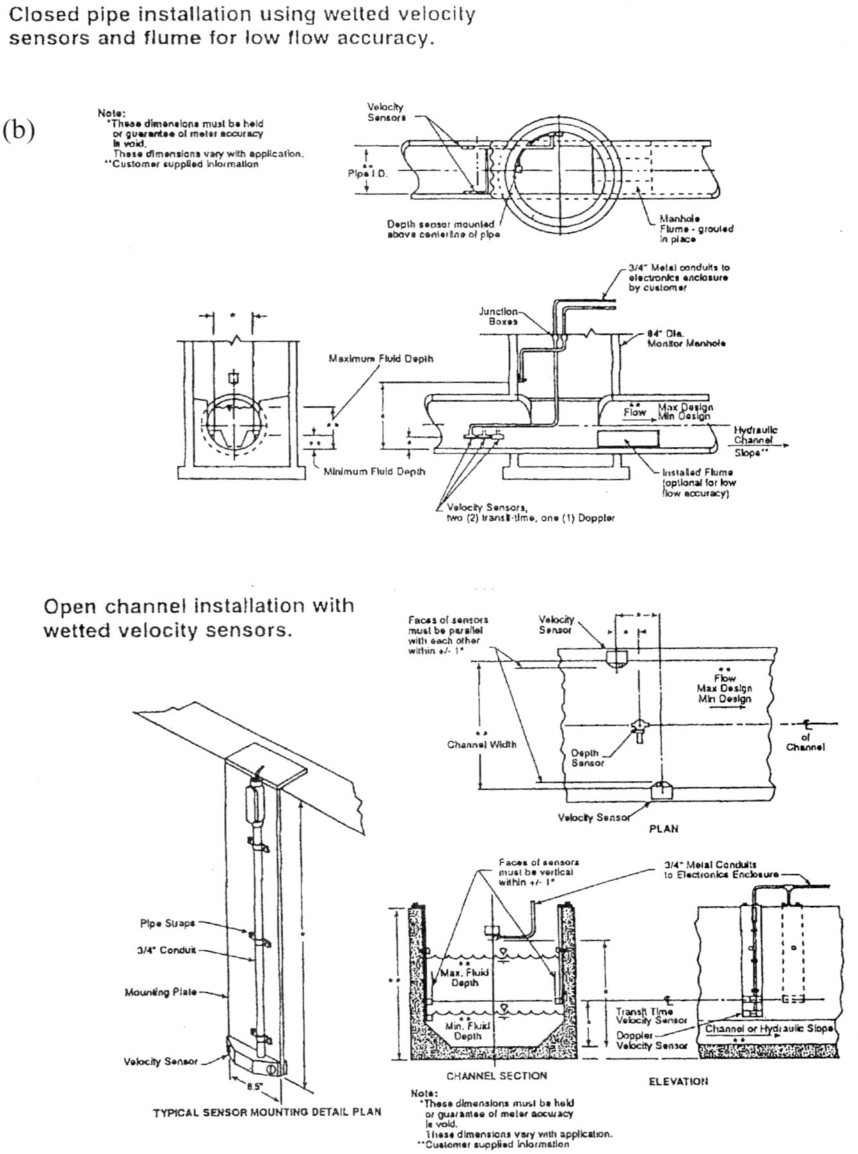

FIG. 45. (*continued*)

(b) (*cont.*)

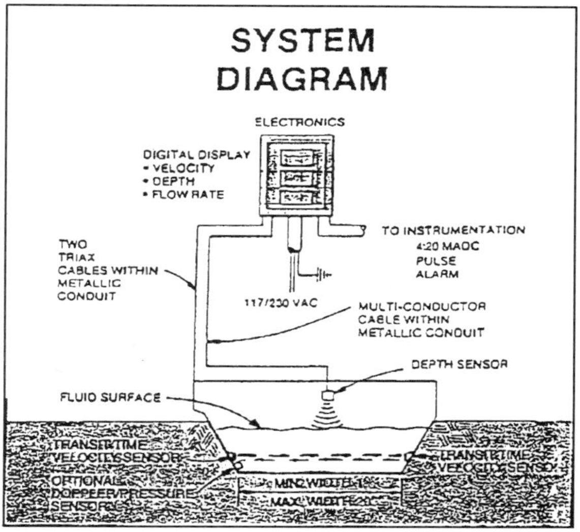

FIG. 45. (*continued*)

(c)

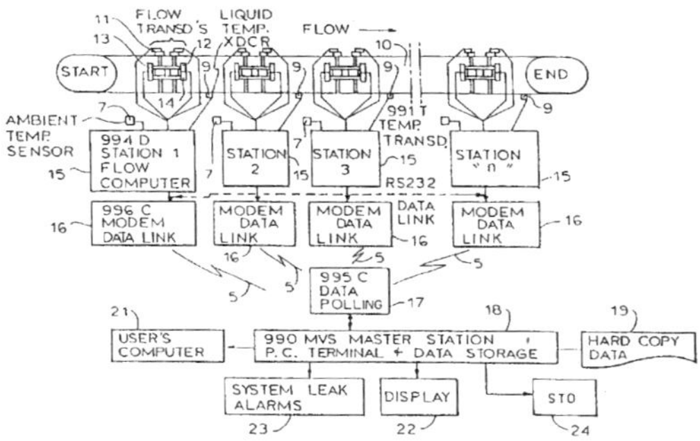

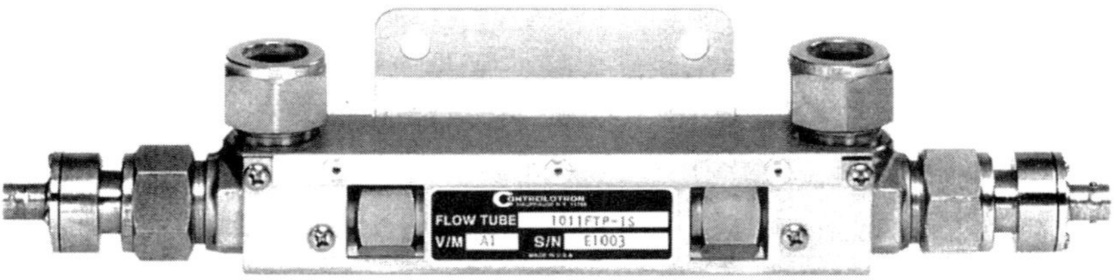

FIG. 45. (*Continued*)

so-called liquid clamp-on flowmeter equipment. In steel pipe of standard or heavy wall thickness, the task is much more difficult because of the acoustic cross talk around the pipe, from transmitter to receiver transducer. Here the flow of gases can still be measured by clamp-on provided the gas pressure is high enough, say, 50 or 100 bar, depending on the pipe, the accuracy/repeatability sought, Mach number, Reynolds number, absorptive or electronic means available to reduce cross talk (acoustic short circuit around the pipe), response time, and other factors.

For processes that are virtually impossible to interrupt in order to install a flowmeter, the clamp-on becomes the obvious choice. Such uninterruptable processes occur in electricity generating plants, in refineries, and in numerous pharmaceutical and semiconductor cases where the sealed system cannot be broken. The illustrations in Figure 47 include applications at elevated temperature where the buffer isolates the shear wave piezoelement from exposure to the hot process (Lynnworth *et al.*, 1996, 1997a; Liu *et al.*, 1998). As piezoelectric materials become available with higher Curie points and higher oxygen-retention temperatures, accompanied by reliable methods of encapsulating them in housings, the need for buffers may diminish. The design trade-off will be made with respect to the cost and performance of special piezoelectrics vs special buffers. The latter, however, can often be made of standard engineering materials, e.g., ordinary steel or common stainless steel alloys like 304SS or 316SS. There are also emat and laser transduction processes that do not require contact with the pipe, and some day these might yield a better solution than clamp-on at the highest temperatures of interest, or in other cases where contactless or remote sensing is required. So far, however, buffers have yielded the practical solutions to high-temperature flowmeter problems, whether the transducers touch the hot fluid or touch the hot pipe.

The density and mass flow rate part of this clamp-on problem might be addressed using flexural waves, the speed of which depends in part on the density of the fluid within the pipe, according to the same principles referred to earlier for noninvasive liquid level measurements in large storage tanks. The mass loading by the fluid slows down the flexural (bending) waves. This explanation may not be complete, but the observation is simple: As the pipe is filled, the flexural wave transit time increases. A more complete explanation considers the interfacial, leaky, and evanescent waves that may be present as well as the fluid's viscosity. See, for example, Craster (1996, 1997). (See remarks in Section IV on mass loading and other sensor effects; see also, Ballantine *et al.*, 1996.)

(a)

(b)

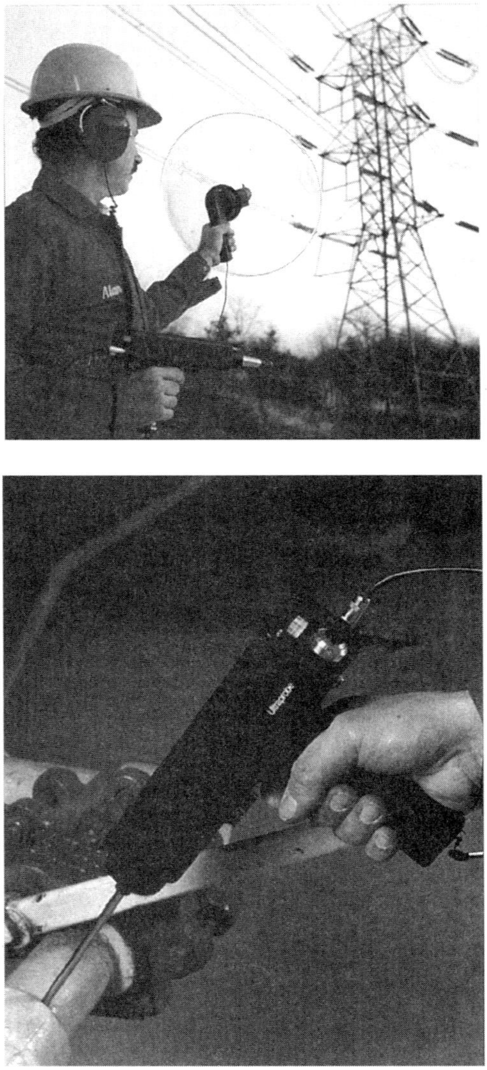

FIG. 46. Applications of UE Systems leak detection equipment. (a) Sensing via air-coupled ultrasound, corona, or other electrical discharge. (b) Recognizing a leaky valve using contact, like a stethoscope. Noise level is compared downstream vs upstream of a valve. Contact test can be applied to frame of rotating machinery to sense bearing wear. (c) Source of a warble tone that floods a heat exchanger or other system. If the system has a leaky seal, ultrasound leaks out and can be detected. Ultrasound is heterodyned down so an operator can listen and interpret audible sound. Illustrations courtesy of UE Systems.

(c)

FIG. 46. *(continued)*

c. ***Portable Clamp-On Flowmeters.*** By about 1990, the technology existed to combine several ultrasonic flowmeter features and related experience and existing mass-produced parts into a new product, the *portable* clamp-on ultrasonic flowmeter. This type of flowmeter offered several advantages:

- Low power consumption (≤ 2 W), as demonstrated in a downhole flow tool design of Jacobson and colleagues, reported in Lynnworth *et al.* (1993).
- Easy-to-use correlation-based receiver and timing (Jacobson *et al.*, 1988), including architecture adaptable to multichannel, multipath, and multimode operation: T1, T2, ..., R1, R2, ..., tag and noise. (T = transmission; R = reflection.)
- Portable digital *flaw* detector and thickness-gage experience, case, keypad, and other components, mainly the work of Elfbaum and colleagues, exemplified by the Panametrics EPOCH and the 26DL PLUS corrosion gage, of which several thousand were in use by 1992.

This development led to the PT868 (Figure 48a), which was introduced in 1992 and won a Vaaler Award in 1993 for being a significant new product for the chemical process industries.

Over one thousand PT868s were sold in its first year on the market. The stroboscopic scattering Transflection® mode was added to a later version, extending its use to many two-phase applications where the fluid was too attenuating to be measured by the contrapropagation transmission mode. (This reflection mode is sometimes called echo tracking, sometimes time

(a)

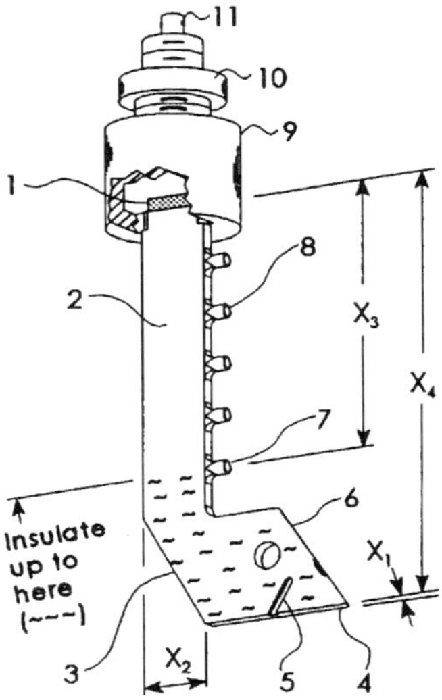

(b)

FIG. 47. High-temperature "hockey stick" clamp-on transducers. (a) Schematic of wave-guide buffer. Principal items: ① Shear wave piezoelement. ② Standoff. ③ Reflecting, *non*-mode converting edge. ④ Radiating edge. ⑤ Reference reflector. (b) Yokes welded to heavy-duty two-piece collar and pipe riser clamp. (c) Temperature along the hockey stick transducer, measured in an electricity generating plant in 1996. Ambient temperature: 30°C. (d) End view of an experimental spoolpiece designed for simultaneously testing the hockey stick clamp-on and wetted transducers at high temperature, including: ① Wetted transducer. ② Flanged nozzle. ③ Yoke. ④ Clamp-on transducer. ⑤ Screw to apply coupling pressure.

(c)

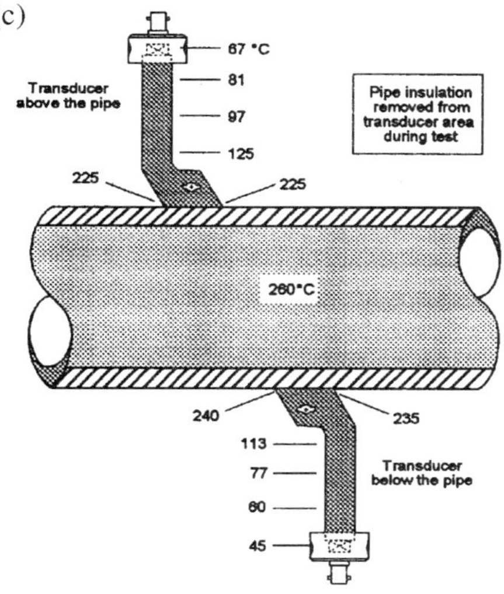

(d)

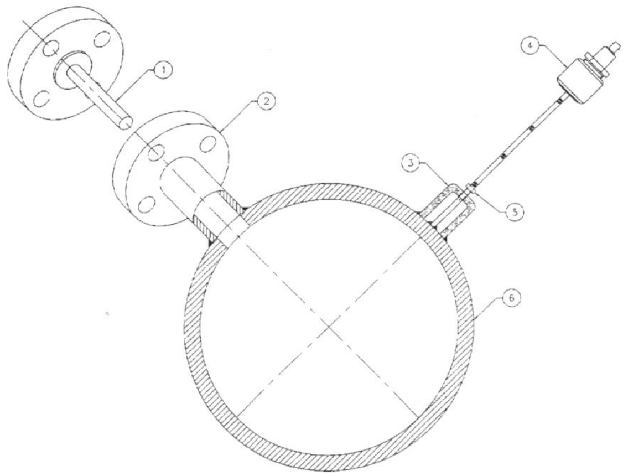

FIG. 47. *(continued)*

domain correlation. Whatever its name, this R mode is finding increasing use in blood flow and vessel motion studies. For example, this *mode* [*not* the PT868] has been used to measure arterial wall motion as well as flow (Bonnefous, 1994). See also Embree and O'Brien, 1985; Trahey *et al.*, 1987; Wilhjem and Pedersen, 1993.) The original, and subsequent PT868 versions, included an optional thickness mode to sense pipe wall thickness (a manual setup procedure). With inputs for temperature data, the PT868 became a portable clamp-on *energy* meter.

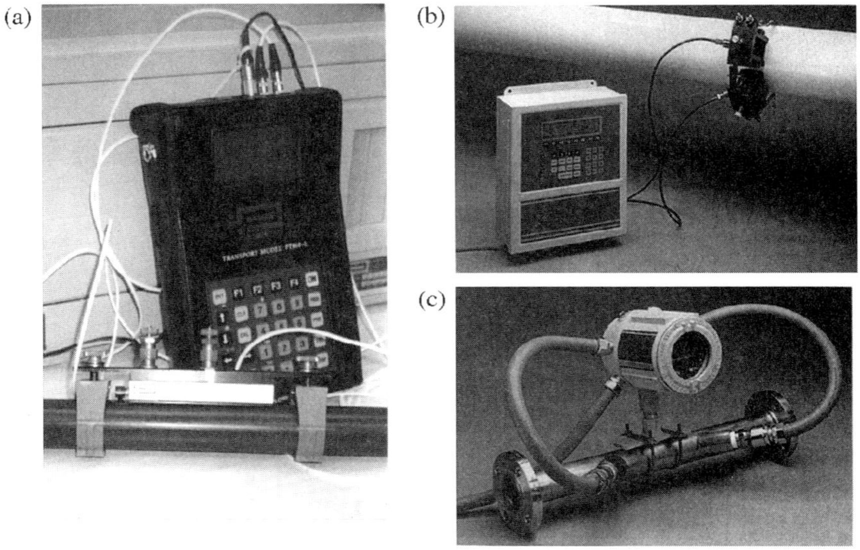

FIG. 48. Comparison of portable vs dedicated ultrasonic flowmeters. (a) PT868 portable ultrasonic flowmeter, introduced in 1992. Its primary use has been as a transit-time flowmeter, T mode, for single-phase and for many two-phase liquids. But in addition, if designated PT868R, it responds to scatterers in its "Transflection" (reflection, or R mode) mode. It is also configurable in a heat flow mode, responsive to hot and cold leg temperatures in addition to flow velocity. T and R modes utilize coded transmission, cross-correlation detection, and clamp-on transducers (Scelzo and Jacobson, 1994). Wetted transducers can be used too, e.g., in measuring air flow in small ducts, $\phi \sim 25$ mm. A pitch-catch thickness mode is included to measure pipe wall thickness. (b) Dedicated DF868 for fixed installations. (c) XMT868 flow transmitter, designed for fixed installations, lower cost per point, explosionproof design housing.

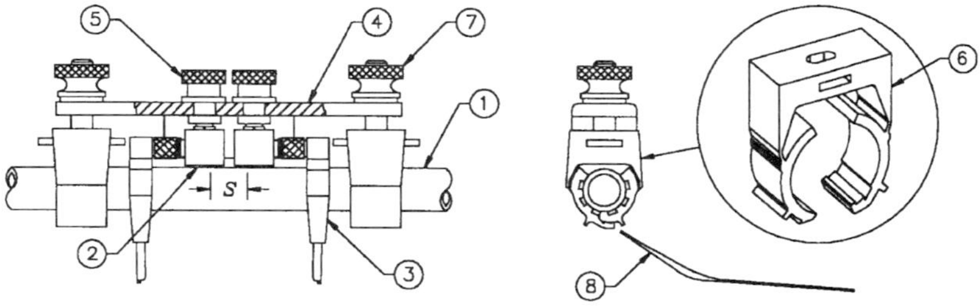

FIG. 49. A snap-on snap-off flowmeter transducer arrangement. ① Pipe/tube. ② Angle beam transducer. ③ Transducer cable. ④ Aluminum channel. ⑤ Transducer hardware for adjusting coupling pressure. ⑥ Pipe/tube clamp ⑦ Clamp hardware. ⑧ Clamp opener.

(a)

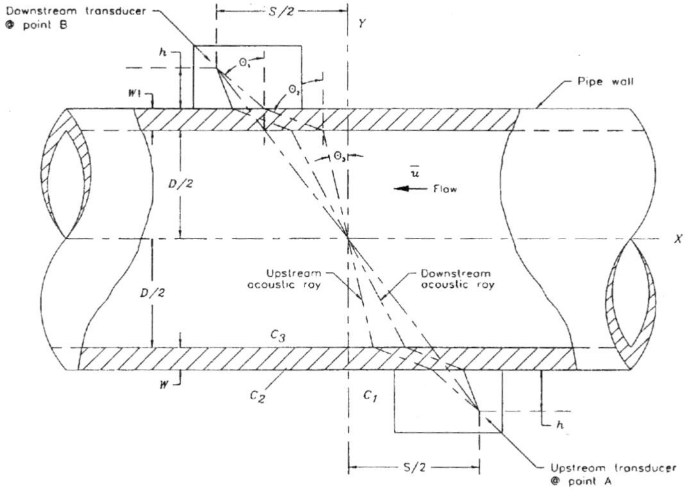

FIG. 50. Because of the importance of clamp-on in the field of ultrasonic flowmetering, this figure deals with a number of clamp-on aspects. (a) Result of a 1977 analysis by Bruner, in which he showed that a consideration of Fermat's Principle of Least Action proves that the upstream and downstream paths cannot be coincident. (Actual paths are curved, curvature depending on flow profile.) Bruner (1977) also pointed out that the clamp-on flowmeter should obey time reversal, a principle finding other ultrasonic uses in recent years. See Tanter *et al.*, 1997; Fink, 1992; Roux and Fink (1996).

Consider again the snap-on/snap-off flowmeter transducer arrangement shown in Figure 49. This arrangement illustrates the achievement of one of the clamp-on designer's objectives: Make it easier to use than any other flowmeter. However, no design is perfect, and no design ever achieves all the objectives one might seek. For example, the illustrated design is temperature-limited by the plastic spring-like retaining structure. For high temperature, the plastic clamp is replaced by a more expensive metal clamp involving mechanical fasteners, e.g., screws or bolts and nuts. Also, as one strives for use at ever-increasing temperatures, the wedge, the couplant, or the piezoelement limits will be reached.

One remedy to avoid T limits is to buffer the piezoelement. A clamp-on buffer consisting of a solid steel or stainless steel waveguide in the form of a thin hockey stick was introduced into evaluation programs in 1996 (see Figure 47). The first pair of such high-temperature transducers was reported to have operated satisfactorily with no maintenance and no degradation for eight months at 260°C. At that time they were moved to another measuring site. The buffers kept the piezoelement temperature below 100°C; passive cooling

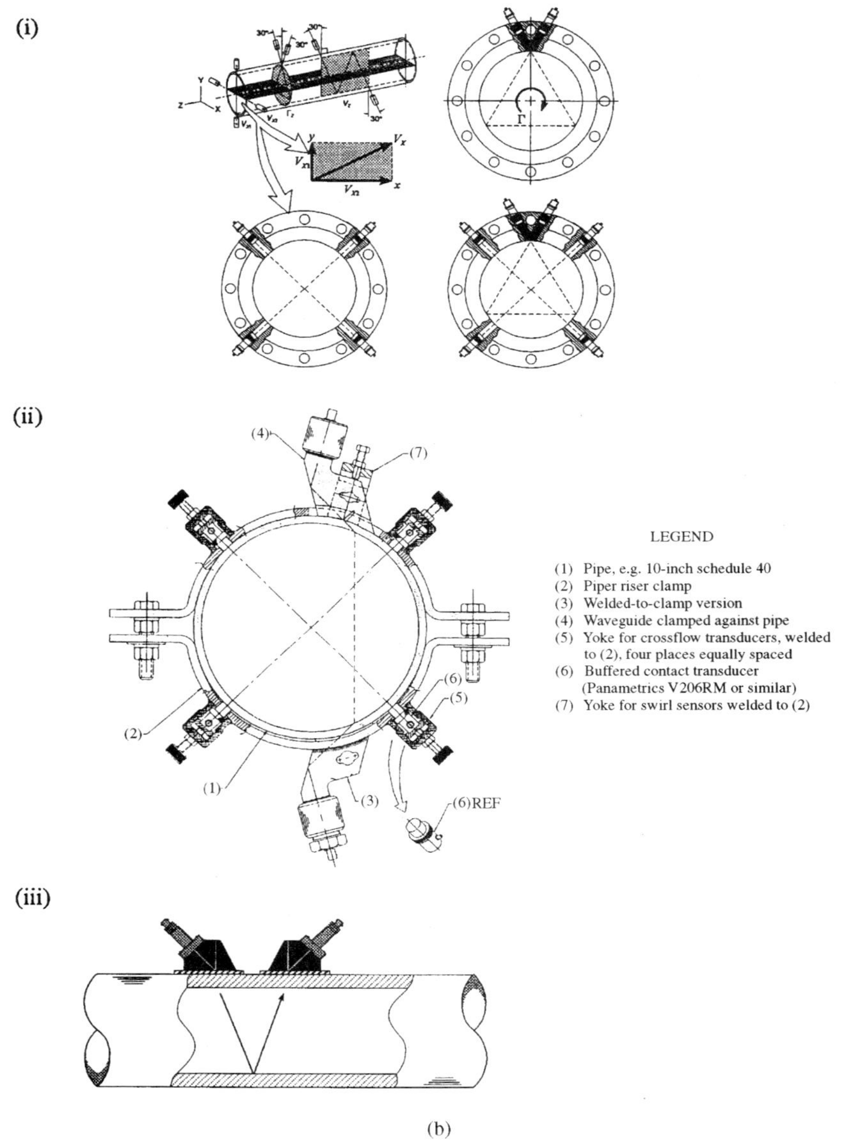

FIG. 50. *(continued)* (b) Concept for clamp-on flowmeter that measures secondary flows as well as the axial flow. This concept for *clamp-on multipath* is derived from *wetted* forerunner (i) shown at top, after Lynnworth *et al.* (1997a). Bottom views show crossflow and swirl clamp-on sensors (ii) to be combined with conventional clamp-on transducers (iii) to measure axial flow component V_Z.

(c)

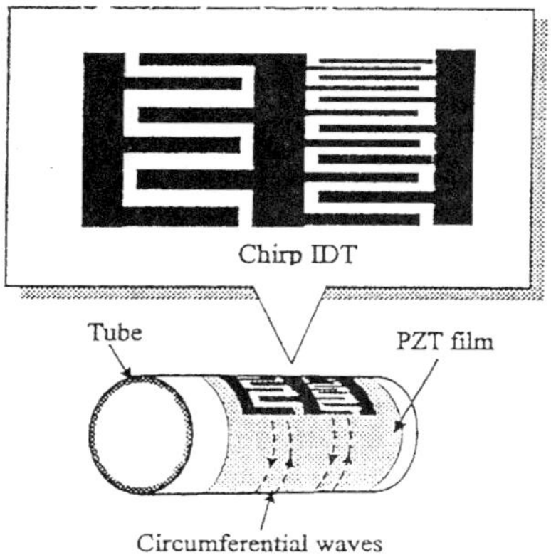

(d)

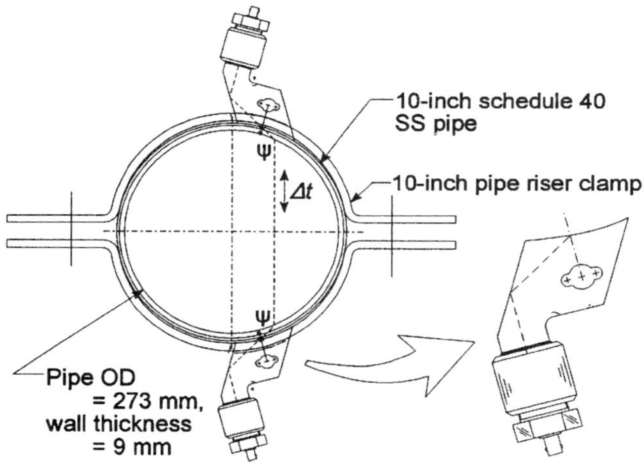

FIG. 50. *(continued)* (c) In contrast to current commercial clamp-on practice of discrete transducers coupled to a small number of regions on the pipe, e.g., Fig. 45, 48, or 49, transducers can be distributed over the surface of the pipe, as in a sol-gel PZT deposition laboratory study by Li *et al.* (1997), © 1997 IEEE. (d) Curved clamp-on transducers for measuring swirl or circulatory components in plane ⊥ pipe axis. These transducers correspond to item ③ in part (b) (ii).

by ambient air was sufficient for this purpose. Details on this waveguide buffer appear in Lynnworth *et al.* (1996, 1997b) and Liu *et al.* (1998).

This buffer might solve the high-temperature problem with respect to the piezoelement, but it introduces other problems: sound speed gradients exist in the buffer; the equipment must be able to be set up at room temperature yet operate all the way up to 300°C; good signals must be provided despite the

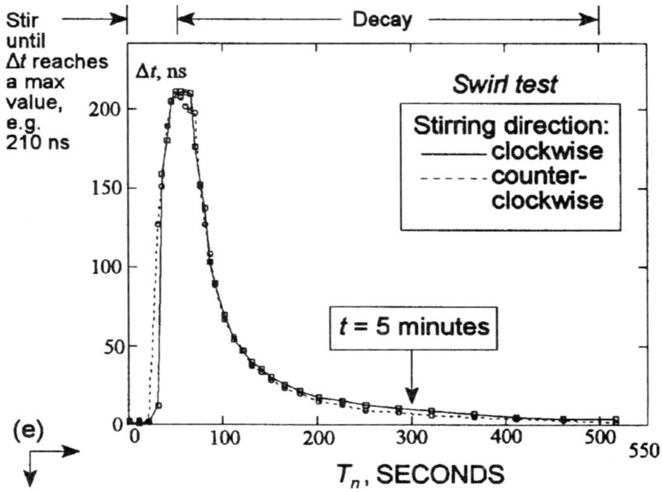

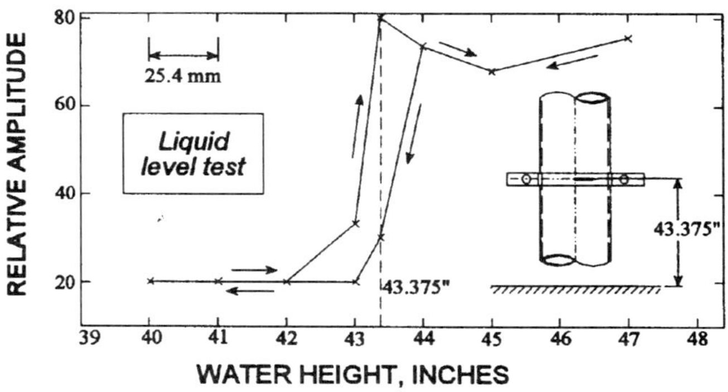

FIG. 50. (*continued*) (e) Results of laboratory experiments corresponding to part (d). Top: Swirl test. Bottom: Liquid level test. (f) Various forms of hybrids, namely, intrusive plugs or reflectors, e.g., solitary intrusive reflector near a smooth wall to sample a representative average velocity between the wall and y/R, as used in a *wetted* transducer probe assembly due to Chernyshev (1994) (and thereby analogous to the *area-averaging Doppler* measurement near the wall, analyzed by Pfau, 1970) but shown here as a hybrid with *clamp-on* transducers; intrusive vortex shedder combined with external, possibly removable clamp-on transducers (e.g., Menz and Dittes, 1997); and an invasive cavity oscillator (Kim, 1992) with pressure sensors located according to Kim (1998, private communication). Hybrids include a multipath concept that provides a four-quadrant symmetry test as well as path averages from the wall to the square-tube insert having four outer and four inner reflecting surfaces. The side view shows some of the paths. Section A-A shows the transducer locations symbolically in the end view schematic.

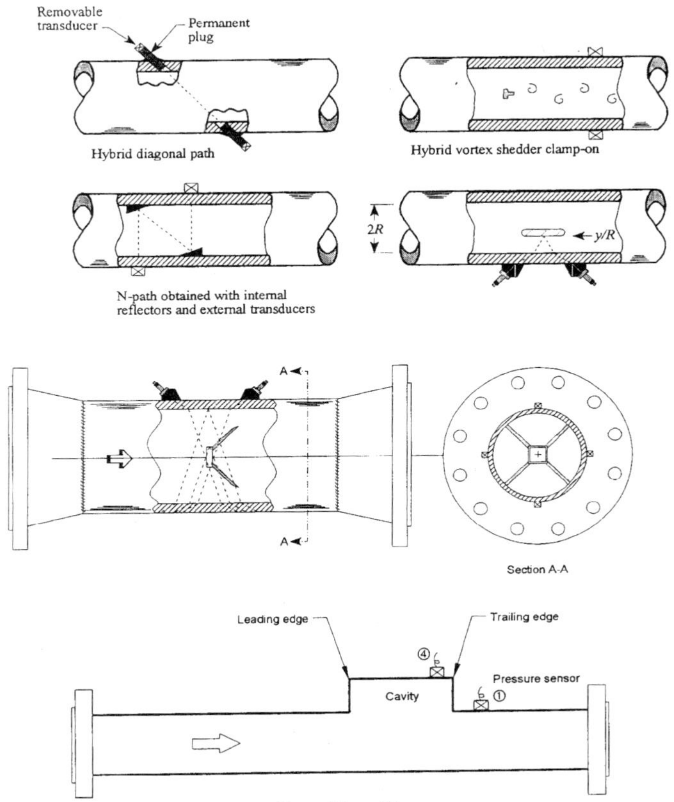

FIG. 50. (f)

changing refracted angle as the sound speed in the liquid first increases above 1500 m/s (in the case of water) and then decreases to 1000 m/s. Other high-temperature problems include the wide range of viscosity and attenuation for "liquids" such as asphalt; the increasing sensitivity to cross flow if the number of diagonal traverses is odd; and possible relaxation of the coupling pressure due to thermal expansion differences or creep. Also, the long steel buffer may be expected to ring a lot longer than a small plastic wedge; this can complicate the acoustic cross talk problem.

Despite these potential or real problems, several forms of such "hockey stick" transducers were requested by beta site customers in the 1996–1997 period. By press time, these clamp-on buffers had measured the flow of superheated water and hot hydrocarbon liquids at some twenty different sites. Temperatures were 260°C to 300°C. Pipe sizes were 3 inch to 16 inch.

(g)

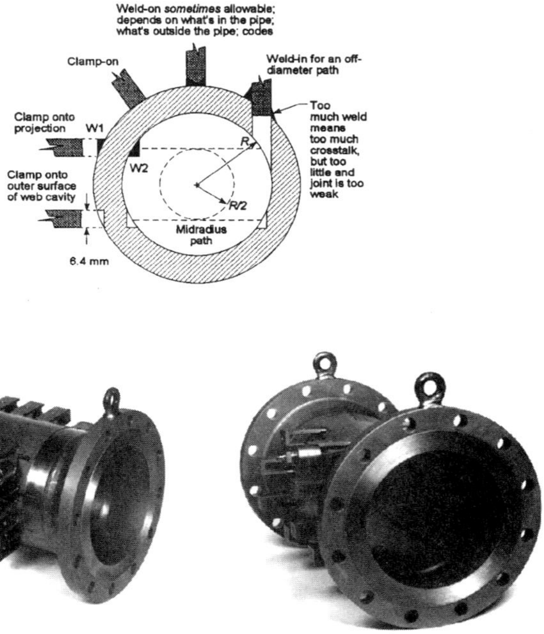

FIG. 50. (*continued*) (g) Hybrid path controllers for off-diameter interrogation as in a GC (Gauss-Chebyshev) flowcell.

Examples of such equipment are presented in Figure 47. In some designs the waveguide is welded to the clamp for easy installation. In other cases a yoke is used. The yoke is more flexible with respect to changing transducers and reduces cross talk around the pipe, but it takes longer to install than the welded-to-clamp version. Pressure-coupled metal foil is the typical coupling solution. Zinc is an example of a relatively economical choice, where allowed; inert gold, although more expensive, is preferred in other cases. Other solutions are possible, e.g., soft Ni, or Al. For permanent coupling, the waveguide has been welded to the pipe or to the spoolpiece. For demonstrations or for conducting a quick measurement (duration of only a few minutes to a few hours) high-temperature greases are satisfactory as couplants.

Note that thin-bladed hockey stick geometries shown previously for liquid level, i.e., measurements in a plane perpendicular to the pipe axis, are appropriate for sensing swirl, based on clockwise and counterclockwise measurements of transit time. If we combine this with cross flow measurements, obtainable using the bundle buffer (Figure 43) or other compressional wave normal incidence buffers, we can obtain data on secondary flow and

(h)

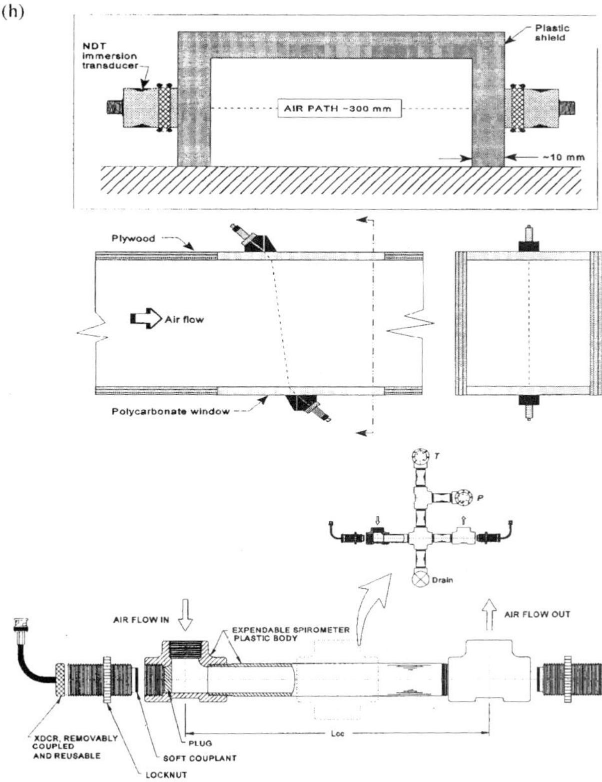

FIG. 50. (*continued*) (h) Top, middle: Straight and angle beam transducers coupled to a flat *plastic* wall of a small wind tunnel. Bottom: Hybrid air flowcell includes low-cost expendable plastic spirometer body with cast-in or molded-in PanAdapta® plugs against which the reusable transducers are removably coupled with gel, soft rubber, or urethane. The dashed pipe-cross at the center of the spirometer represents options for measuring temperature T and pressure P, and for draining.

thereby reduce errors due to uncertainties in such components of flow. Thus we have multipath clamp-on, somewhat analogous to the multipath wetted-transducer arrangements of Figure 53(a)–(d). Further examples of clamp-on multipaths, including paths off the diameter, appear in Figure 53(f, g).

Refraction (Snell's law) limits that are associated with clamp-on prevent one, in general, from interrogating over those paths that would provide the most valuable information on circulation. In other words, the midradius chords are generally inaccessible with today's technology. An exception occurs if the pipe is plastic or, if metal, very thin compared to wavelength. Then the midradius chord or other off-diameter chords may be accessible for

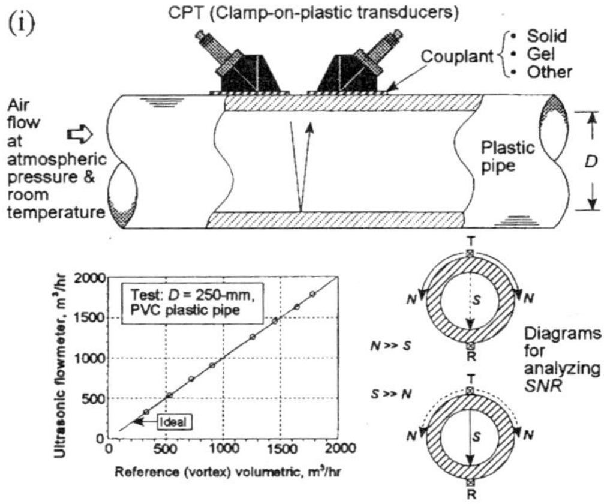

FIG. 50. (*continued*) (i) Angle beam NDT transducer coupled to outside of plastic pipe, ϕ 10-inch (ID ~254 mm) for clamp-on flow measurement of air at ordinary temperature and pressure.

multi-axially inclined paths analogous to the GC (Gauss-Chebyshev) paths and to the triple midradius for circulation measurement. In the simplest case, the circulation $\Gamma = 0.605c^2\Delta t$, where Δt is the cw-ccw time difference measured along the triple midradius path (Smith, 1994). Hybrid spoolpieces with special cavities or built-up internal wedges provide the sought off-diameter paths in designs like those in Figures 50(g) and 51(b), (c).

Clamp-on buffered measurements of crossflow (or swirl) may be compared at axially displaced locations, where the signals, modulated by the fluid eddies or sound speed fluctuations, can be cross-correlated in the *tag method* of flow measurement (Coulthard, 1973; Mayranen *et al.*, 1996). Such measurements have a potential advantage over contrapropagation, in that propagation *against* the flow is not required. Sometimes propagation against the flow is attenuated too much to yield detectable signals; or the signals may jitter so much that timing them becomes an exceedingly difficult task. (These difficulties are encountered more often with gases than with liquids.) The tag method, however, responds to turbulence in the flow profile in a more complicated way than do transit-time flowmeters. Tag is inappropriate for laminar flow unless inhomogeneities are present or can be introduced. Although the tag method might be an interesting complement to transit-time flowmeters, it is unlikely to replace them because it is not as general a solution and it is unlikely to match transit-time accuracy or response time where both can work optimally. Tag principles have been known for over

(j)

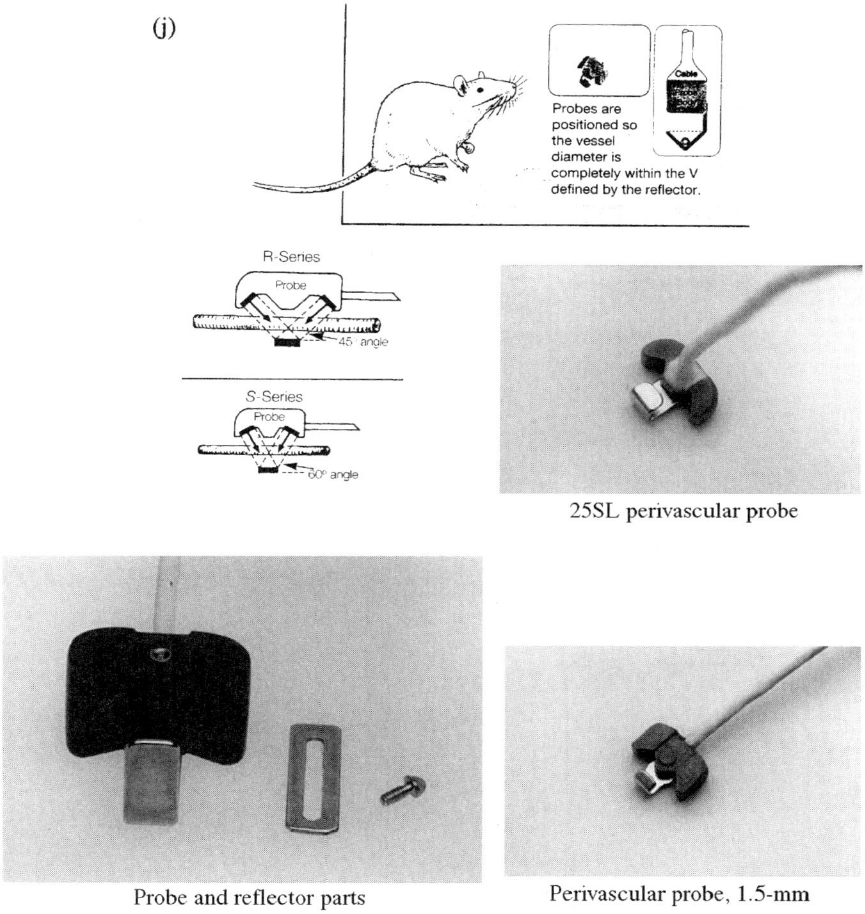

FIG. 50. (*continued*) (j) Small-diameter transit-time clamp-on flowmeter for biological applications, developed by Transonic Systems. Direct volume flow measurement in rodent (or other) vessels as small as 250 μm is possible with their V-reflector probes, used with their clinical and research flowmeters. Illustrations courtesy of Transonic Systems.

twenty-five years, yet apparently only one company manufactures ultrasonic tag flowmeters, in contrast to the many companies that manufacture transit-time flowmeters.

d. Measurement Independent of Flow Profile. To eliminate the dependence on flow profiles, one can lay out several sonic paths in tubes of larger nominal diameters in a targeted way so that flow profiles or disturbances in

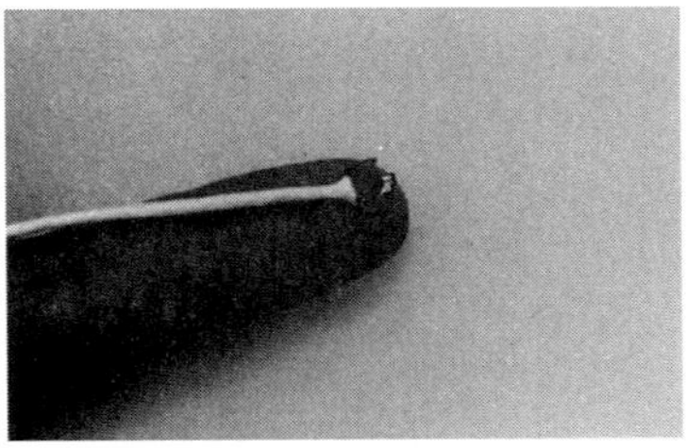

Perivascular 1R probe
on fingertip

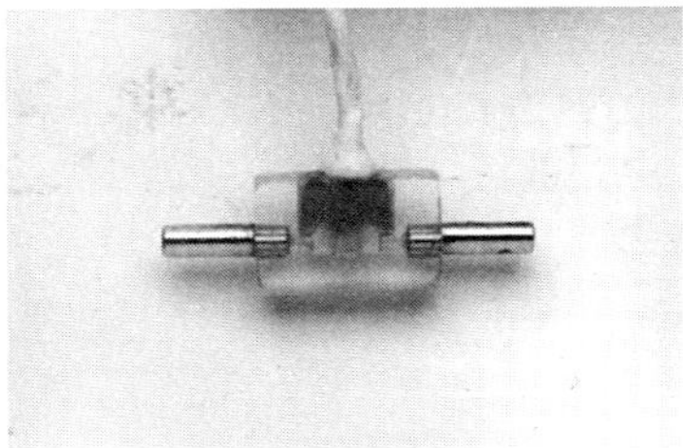

Extracorporeal probe
1N inline

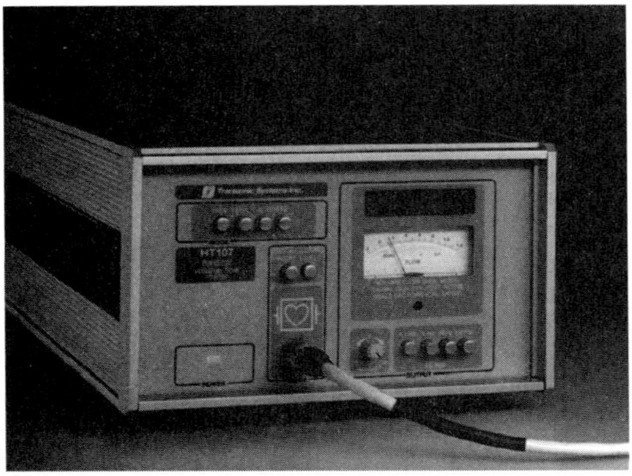

HT107 single channel
flowmeter

FIG. 50. (j) (*continued*)

(k)

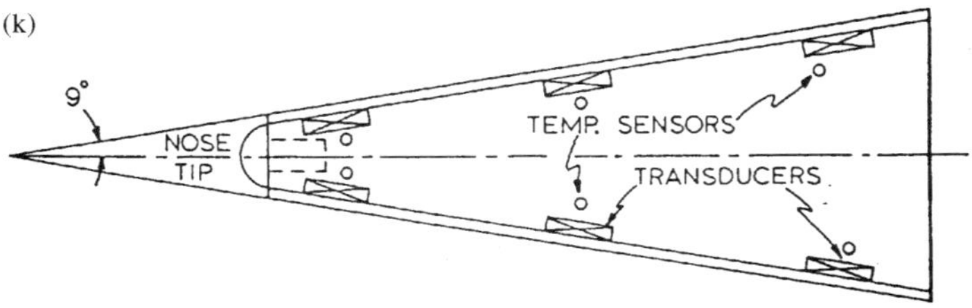

FIG. 50. (*continued*) (k) Clamp-*in*, looking *out*: boundary layer acoustic monitor, manufactured by Kaman Scientific. After Sachs *et al.*, 1977.

flow become suppressed to a large extent. In doing so, one can utilize the fact described above that sonic rays emphasizing the center read too high for laminar flows, while rays that project onto the tube periphery measure too low. The sensitivity of sonic paths whose distance from the center lies between the two extremes lies also between the two cases. A sonic beam positioned at

(l)

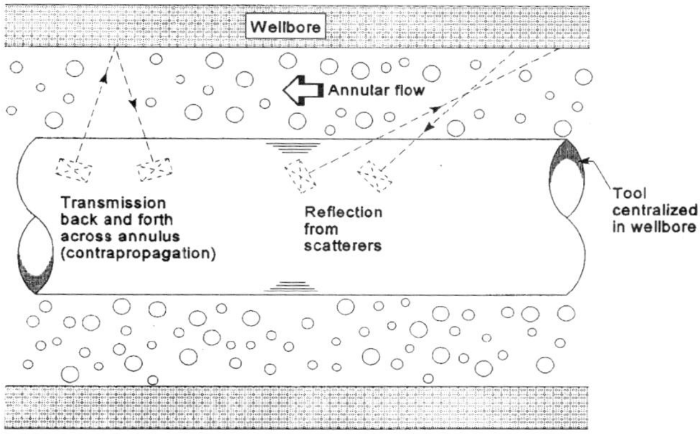

FIG. 50. (*continued*) (l) downhole flow tool *prototype* developed 1988–1992 at Panametrics (Lynnworth, 1988; Lynnworth *et al*., 1993). Its low-power (<2 W) downhole electronics were later adapted to the battery-powered portable clamp-on flow meter models PT868 and PT868R. (Suffix R denotes reflection mode).

about the midradius achieves a by-and-large constant measurement sensitivity for laminar and turbulent flows and, according to some reports, even for the transition region in between (Figure 51). However, such an off-center sonic path will still be disturbed by rotational flows and asymmetrical velocity distributions. It is therefore advisable to add a second beam to the first, symmetrically positioned with respect to it. High accuracies (better than 1% of the measured value) can be achieved with a larger number of cleverly positioned sonic paths, which is possible with large nominal diameters (e.g., >100 mm). Note that the independence of the measured value of a flowmeter from the conditions of the flowing medium is of the highest importance in characterizing the performance. In particular, the linearity of the "measuring curve," i.e., the measured volume flow velocity vs the real flow velocity as measured by a high accuracy standard, is essential.

Such flowmeters are made by various manufacturers in Europe and in the United States. For example, the Danish company Danfoss offers a wide range of ultrasonic flowmeters (Figure 52), ranging from 25 mm to 1200 mm (1″ to 48″). For these measuring tubes, transducer arrangements with one, two, or four acoustic paths are employed, with the two-path configuration being the standard. With mounting kits that include a set of transducers to be applied to the customers' existing tubes, the ultrasonic flow measurement principle is extended to large diameters, up to 4000 mm. The ultrasonic flowmeter consists of the measuring tube and an electronic evaluation unit, which in a compact series is combined with the measuring tube, or, particularly for larger

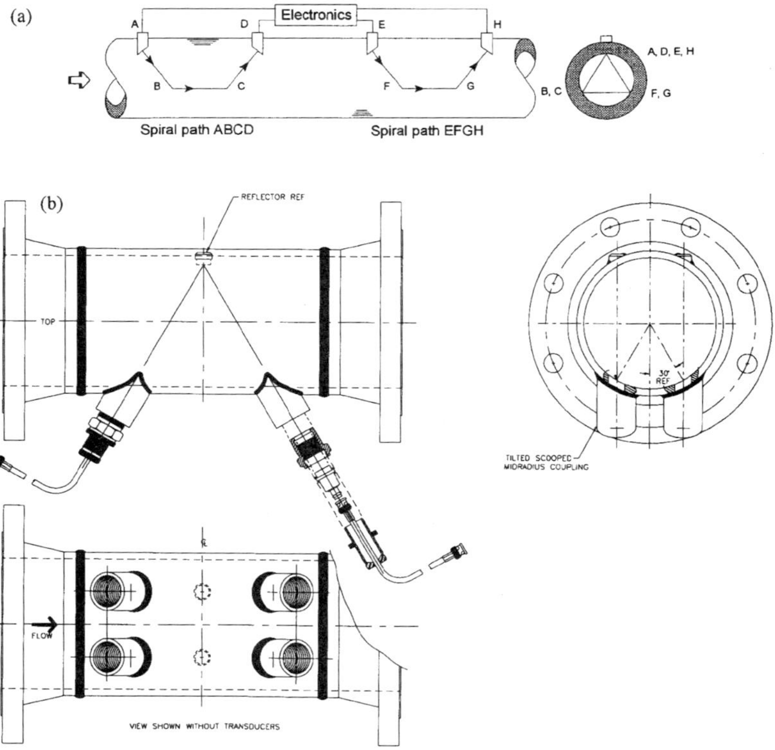

FIG. 51. A sonic beam positioned at about the midradius achieves a substantially constant meter factor K for laminar and turbulent flows and, according to some reports, even for the transition region in between. Midradius spoolpieces have been available from Krohne, Stork (now from Instromet) and Panametrics. (a) Triple traverse, according to Drenthen (1989). Paths can spiral cw or ccw; typically both ways to eliminate swirl effects. Other midradius illustrations appear in Lynnworth (1979, p. 431; 1989, p. 285). (b) Pair of midradius vee paths, each utilizing a flat reflector welded into wall opposite the transducers. The vee path suppresses the influence of crossflow. Two paths allow the meter to suppress swirl effects. This design from Panametrics accommodates their T7 air or gas transducers (Fig. 15). One transducer is shown in exploded view. (c) In this Panametrics design the reflectors 110 and 116 are welded in the flange regions, such that three of the five path segments in the fluid are midradius chords (Lynnworth, 1978). (d) Two single-traverse midradius paths, after Baker and Thompson (1978).

diameters, is connected to the transducers by special coaxial cables. The electronic evaluation unit is either mounted on the measuring tube or, especially for high-temperature devices, can be placed separately.

For different application standards, different tube materials (e.g., stainless steel), and different operational conditions (e.g., explosionproof), different transducer installation methods are applied. The specified temperature range

(c)

(d)

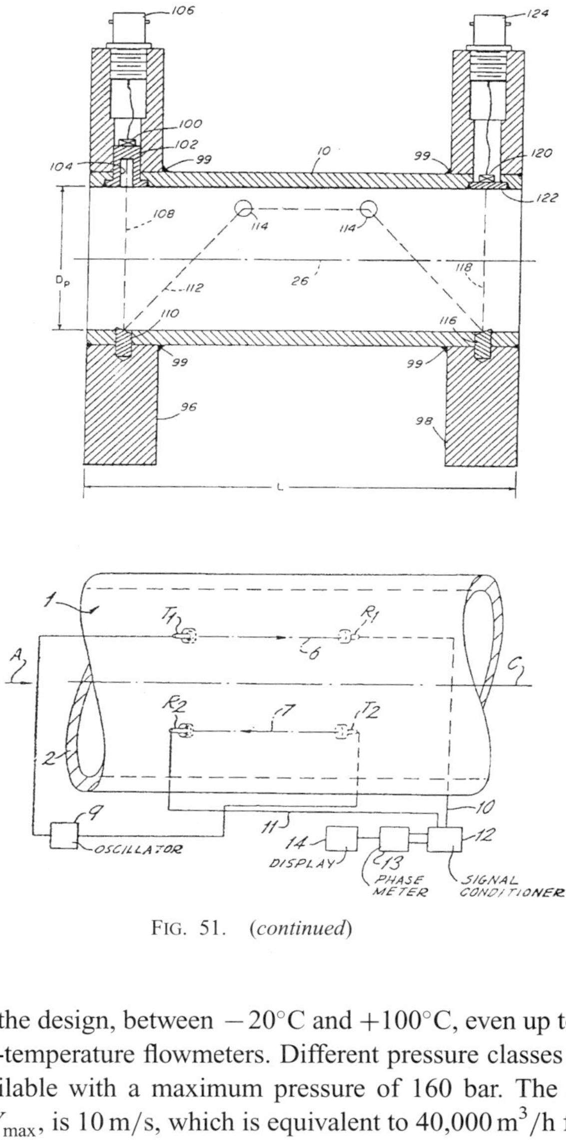

FIG. 51. *(continued)*

is, according to the design, between −20°C and +100°C, even up to +200°C for special high-temperature flowmeters. Different pressure classes (PN 20 to PN 50) are available with a maximum pressure of 160 bar. The maximum specified flow, V_{max}, is 10 m/s, which is equivalent to 40,000 m^3/h for ϕ1.2 m (500,000 m^3/h for ϕ40 m) pipes. Both devices with completely sealed transducers are available for measuring tubes, in which the transducers can be removed and replaced during the process without interruption by the use of a special lock tool. Under standard conditions (water at 20°C ± 5°C, straight inlet length of 20 times the tube inner diameter), an error below ± 0.5% of the

SONOKIT ™
Danfoss

measured value within a 20:1 dynamic range of $0.05^* V_{max}$ to V_{max} is specified. In the same dynamic range, a repeatability better than $\pm 0.25\%$ has been reported. The linearity deviation limits are 0.4% at $Re > 5000$ and better than 1% at $Re > 1000$.

The electronic evaluation unit contains the transmit/receive circuitry, digital signal processing, and a wide variety of user interface capabilities such as optional display and different output or bus configurations. Easy installation and operation is maintained, for instance by a small plugable memory device comprising the important operation and calibration parameter values.

The ultrasonic flow sensor program of Krohne comprises flowmeters for liquids with single or dual ultrasonic paths. Tube diameters ranging from 25 mm to >3 m (1 to 120 in.) cover maximum flow velocities from 1 to 450,000 m^3/h. Dual-path ultrasonic gas measuring tubes range from an inner diameter of 50 to 600 mm. Clamp-on devices and meters using transducer kits to be welded in the wall of customers' existing pipes complete the ALTOSONIC® ultrasonic flowmeter range. The wide application variety includes fluids such as high-purity water, sewerage, gasoline, ammonia, natural gas, air, nitrogen, acids, crude oil, and water–oil mixtures.

For dual-path liquid flowmeters with inner diameters ≥50 mm (≥2″) a typical accuracy of ± 0.5% of the measured value over a dynamic range of 0.5 to 18 m/s, and a reproducibility of 0.2% was reported by the manufacturer. The respective measurement conditions were water at 10°C to 630°C and an undisturbed inlet length of at least 10 times the inner diameter. For smaller devices, single-path measuring tubes, and other liquids, less accuracy must be accepted at stricter inlet configuration conditions.

At ACHEMA 1997, an important European fair concerning the process industry, a high-accuracy multipath ultrasonic flowmeter was presented by Krohne. Developed for maintenance-free custody transfer measurements (Figure 53(a)), this meter can be substituted for high-accuracy turbine flowmeters. Compared to the latter, the ultrasonic device has important advantages, such as minimum pressure loss and almost no viscosity dependence. Due to its excellent long-term stability, a regular recalibration is unnecessary. Employing a multipath arrangement in the spoolpiece, high accuracy is

---

FIG. 52. The Danish company Danfoss offers a wide range of ultrasonic flowmeters, including spoolpieces as well as hot-tap versions. Shown here are three versions of their Sonoflow® ultrasonic flowmeters (top) and their Sonokit™ instrumentation (bottom) to simplify installations in the field. Illustrations courtesy of *Danfoss*.

(a)

(b)

FIG. 53. High-accuracy multipath ultrasonic flowmeters from the European manufacturers (a) Krohne, and (b) Fluenta. Flowmeters like these have been developed for maintenance-free custody transfer measurements and can often replace high-accuracy turbine flowmeters. Illustrations in (a) and (b) courtesy of *Krohne* and *Fluenta*, respectively.

(c)

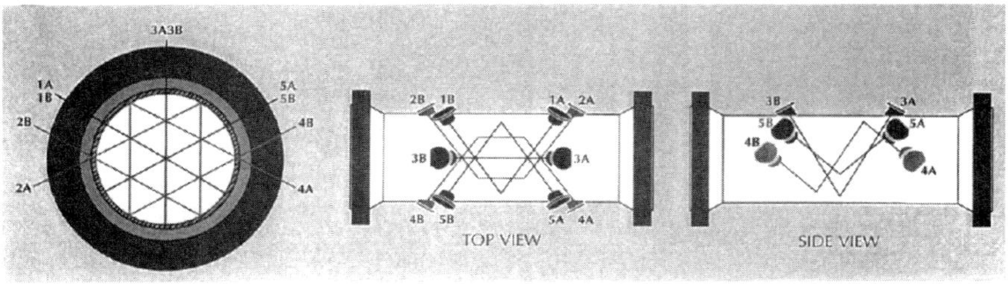

FIG. 53 (c) Multipath *Q.sonic* ultrasonic flowmeter for gases combines cw and ccw spiraling triple midradius paths and tilted-diameter vee paths in planes parallel to each leg of the spiral path. Illustrations courtesy of *Instromet Ultrasonic Technologies*.

FIG. 53 (c) *(continued)*

(d)

FIG. 53 (d) Multipath *SeniorSonic* ultrasonic gas flowmeter utilizes Gauss-Chebyshev paths interlaced as indicated. Transducers are arranged to form an "X" in the plan view. Illustrations courtesy of *Daniel Measurement and Control*.

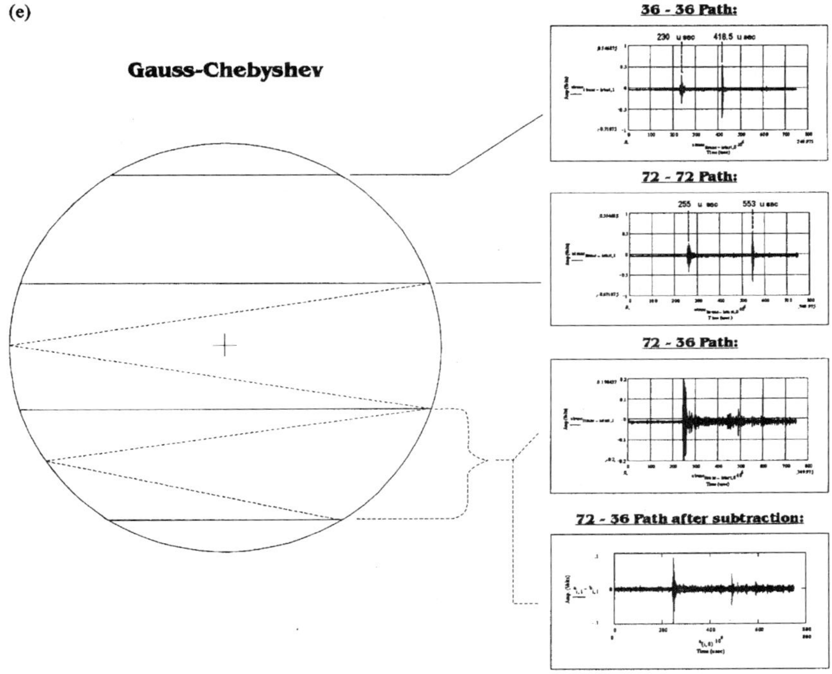

FIG. 53 (e) Multipath *experimental* flow cell utilizes *lateral* beam spreading to yield *skewed* paths that sample the flow between traditional inboard and outboard GC (Gauss-Chebyshev) planes, or between diameter and midradius planes. The total number of paths, including in-plane and skewed paths, exceeds the number of transducer pairs. The use of one flow-sensing transducer to participate in interrogations over more than one path is suggested in Johnson *et al.*, 1975, 1977. [This aspect of tomographic scanning (>1 path per transducer) was also used in acoustic thermometry by Green, 1985 and Kleppe, 1989, 1995a, 1996. See also, Figs. 4(c) and Fig. 78(b).] Waveforms shown alongside their respective paths demonstrate the signal to noise ratio obtainable at $f = 2$ MHz. Examples are chosen from some of the vee paths between in-plane GC transducers and across the inboard and outboard planes. For these tests the fluid was water at room temperature and there was no flow. The spoolpiece was the steel one shown in the photographs in Fig. 50(g). Its nominal pipe size is 10 inches ($ID = 254$ mm) and it has parallel internal and external axially-extended notches to create reflection sites and transducer websites that accommodate, under the welded-on yokes, the clamp-on 6.4-mm wide hockey stick transducer of Fig. 47. The signal from the vee path between the inboard and outboard GC planes is shown before and after short-circuit subtraction. The received signals were acquired using a *Gage Applied Sciences* Compuscope 225 card. The complete received signal was acquired and then the received signal was acquired again after blocking the transmitted signal in the water. Subtraction of these acquired signals resulted in the cleaner signal which is shown for the 72°–36° skewed path. The arrival near 250 μs is undesired crosstalk. The arrival near 490 μs is the waterborne signal. These waveforms were obtained by Brita Dean, 1998, unpubl. Illustration courtesy of *Panametrics*.

(f)

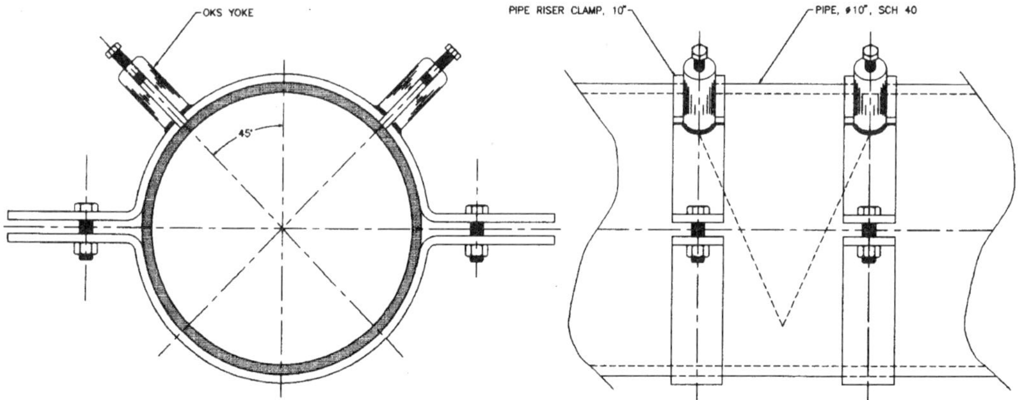

FIG. 53 (f) Multipath clamp-on adapts the transducer and clamping fixture of Fig. 47(a,b) to liquid flow measurements in orthogonal planes. In some respects this arrangement resembles the multipath clamp-on design illustrated in Lynnworth, 1989, page 275. In a multipath flowmeter spoolpiece such as this one, the path geometries, including single *and* double traverse, provide differential path lengths in the fluid. In principle, the *difference* in fluid path lengths can be utilized to measure the sound speed c and/or the attenuation coefficient α, at the frequency or frequencies available from the transducers. In some cases the distribution of paths allows one to measure the distribution of c. The c distribution may be interpreted in terms of the distribution of fluid temperature T and fluid density ρ. The α distribution might be interpretable in terms of scattering or attenuative properties such as absolute viscosity η. If η and ρ have been determined, then the Reynolds number Re can be computed as $Re = \rho VD/\eta$ where D = pipe inside diameter and V is the area-averaged flow velocity.

(g)

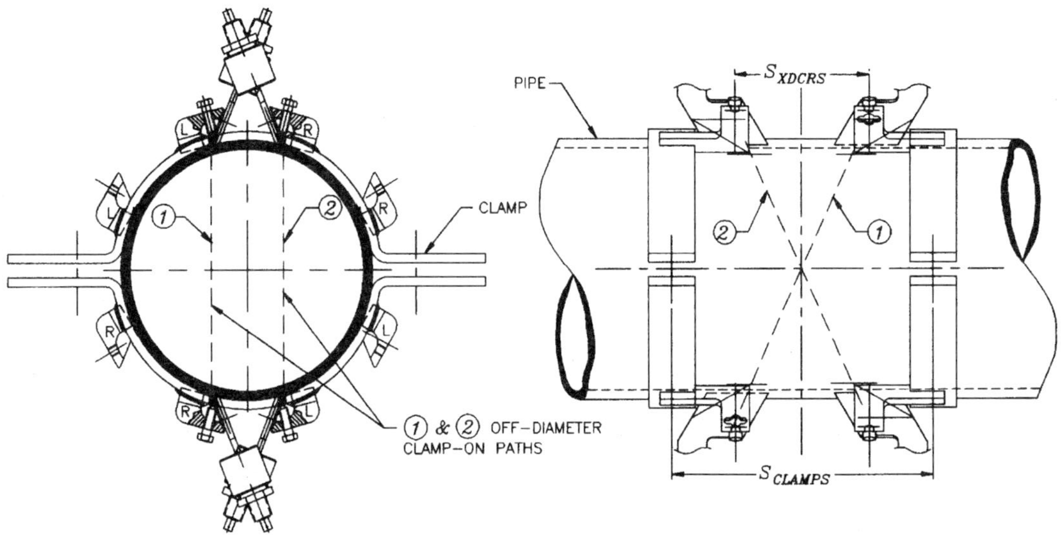

FIG. 53 (g) Clamp-on off-diameter paths obtained with compound angle transducers. Four transducers are shown in this schematic. The other (unused) hold-down clamps are for other transducer locations and other paths.

obtained substantially independent of the flow profile. Flow disturbances are reduced by a reducer at the inlet and a diffuser at the outlet of the spoolpiece.

Krohne's Multisonic V volume flowmeter consists of the measurement tube with the ultrasonic transducers, a field-mounted measuring converter operating the ultrasonic paths, and a "control room converter," which is described by its manufacturer as a user-friendly digital evaluation unit including a keyboard and a display screen. The evaluation unit, to be placed in a nonhazardous area, obtains the multichannel sensor information by a serial bus and, according to the manufacturer, uses intelligent algorithms to check the plausibility of the individual signals, determine the volumetric flow by weighting the multichannel information, and calculate the total volume flow. Furthermore, the sonic velocity of the liquid, c, the Reynolds number, Re, and the kinematic viscosity of the liquid, ν, are determined. (Readers will understand that in general, temperature influences sound speed and viscosity, and ν influences profile and attenuation. In principle, viscosity might be deduced from such relationships.) At press time, however, information on how viscosity is provided in this particular flowmeter was not available to the authors.

The available measuring tube diameters DN are 100, 200, and 250 mm (4, 8, and 10 inch). The specified linearity of $\pm 0.15\%$ of the actual value in a 2:1 dynamic range of $0.5V_{\max}$ to $V_{\max}$ and of $\pm 0.25\%$ for $0.1V_{\max}$ to $0.5V_{\max}$, as well as a reproducibility of $\pm 0.05\%$, was proven with different liquids of different viscosity and at severe flow-disturbing conditions. All measurements were performed using an inlet section of 10–20 DN (20 DN for the disturbed conditions) and an outlet section of 3 DN. The specified temperature range is $-40°C$ to $+80°C$, the operating pressure range is up to 160 bar, and the design and test pressure up to 250 bar. These meters comply with the respective directives of the Norwegian Petroleum Directorate (NPD).

In Europe, other manufacturers of multipath flowmeters are Fluenta (Figure 53(b)), Instromet (Figure 53(c)) and RMG.

In the United States, multipath ultrasonic flowmeters have been manufactured by Caldon, Daniel (Figure 53(d)), ORE, Panametrics and others.

e. Independence of Flow Profiles at Small Tube Cross Section. With small tube diameters it is barely possible to locate more sonic paths side by side because the space is insufficient to position the required number of transducers and, above all, the reflection from walls and waveguide effects prevent the formation of raylike beams. With small nominal diameters, if one can produce a broad oblique beam that fills the entire measuring tube cross-section, all parts of the flow profile contribute to the result of the measurement according to their respective cross-sectional weight. In an ideal case, the whole flow profile will be integrated and the ultrasonic flow measurement becomes independent of flow profile and flow disturbances. The first use of this principle is due to Swengel ca. 1947–1955. His pioneering work in *large* conduits is illustrated in *Physical Acoustics* **14**, Ch. 5. p. 500 (1979). The principle was subsequently and independently adapted by one of the present authors to small conduits (Lynnworth and Pedersen, 1972 [in the first of several R&D programs and a few commercial applications in the United States]) and by the other author (Mágori, 1985) to large-scale high-quality industrial use as a European heat meter. Drost, 1980, has applied this principle to measuring blood flow in small vessels $\lesssim$1-mm diameter (Figure 50 (j)).

The ultrasonic beam configuration for the heat meter as shown in Figure 54, was possible by the use of a special kind of ultrasonic transducer—thin piezoelectric plates with interdigital electrodes. These transducers radiated not normal to their surface plane but at an angle determined by the interdigital electrodes' periodicity and the wavelength in the liquid. By this, the transducers could be mounted parallel to the walls, effecting no flow disturbance or pressure loss. The electronic operation mode was the so-called lambda locked loop, which held the wavelength constant, independent of the sonic velocity in the liquid by a closed-loop control. By this the operation frequency was automatically altered in such a way that the measured phase angle at the receiver input was held constant with respect to the transmitted signal. At flow in both the upstream and downstream direction, different operating frequencies resulted by this control; the frequencies' difference, independent of the actual fluid's sonic velocity, was proportional to the flow velocity. Furthermore, the radiation direction was maintained constant by the achieved constant wavelength (Mágori, 1985).

In work by Mágori, the transmission of ultrasonic waves through a rectangular tube cross section, which covers the whole cross-sectional area, was realized in an ultrasonic flowmeter that was intended to serve as a heat quantity meter in which the width of the ultrasonic transducer was chosen to be slightly larger than that of the measuring tube. Furthermore, the sonic beam was reflected three times from the lower and upper walls of the measuring tube. This resulted in a long, W-shaped measuring path with a high measuring sensitivity and a reduced sensitivity to crossflow. The effectiveness of these measures was demonstrated by an experiment in which a strongly disturbed flow was generated in the measuring tube by a half-closed valve placed directly at the input of the tube. While one could barely notice a deviation from the undisturbed operation with properly

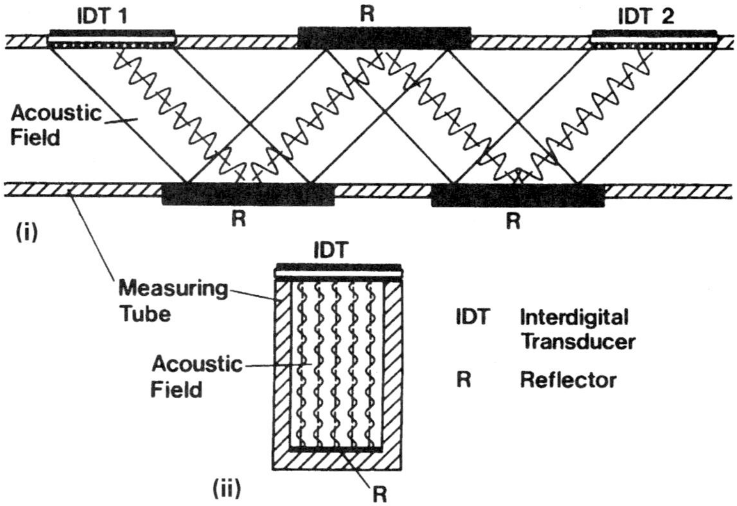

FIG. 54 (a) Ultrasonic beam configuration for an ultrasonic heat meter, using lambda locked loop and interdigital transducers. (i) Longitudinal section. (ii) Cross-section. Illustration courtesy of Siemens.

mounted transducers, transducers whose emitting surface was partly shadowed displayed significant errors (Figure 55).

The described small-diameter ultrasonic volume flow measuring tube serves as the essential part of a domestic heat meter (Figure 56), measuring the heat consumption of hot water heating systems in small houses, flats, and the like. In this device the volume flow of the heating water is measured with the ultrasonic flowmeter and becomes multiplied by the difference of the water temperature at the inlet and at the outlet of the heating system. Due to the low energy consumption of the ultrasonic measuring principle, operation on a built-in battery over more than five years is possible, giving a high flexibility for the installation of the heat meter.

Since ultrasonic heat meters were first introduced to the market several years ago, alterations of the transducer and of the measuring path have been made to reduce manufacturing costs. Ultrasonic heat meters for different maximum flow values also have been developed. Different design philosophies have been successfully realized, for instance, by Landis & Gyr, who manufactured an ultrasonic heat meter with a carefully designed coaxial (axial) measuring path.

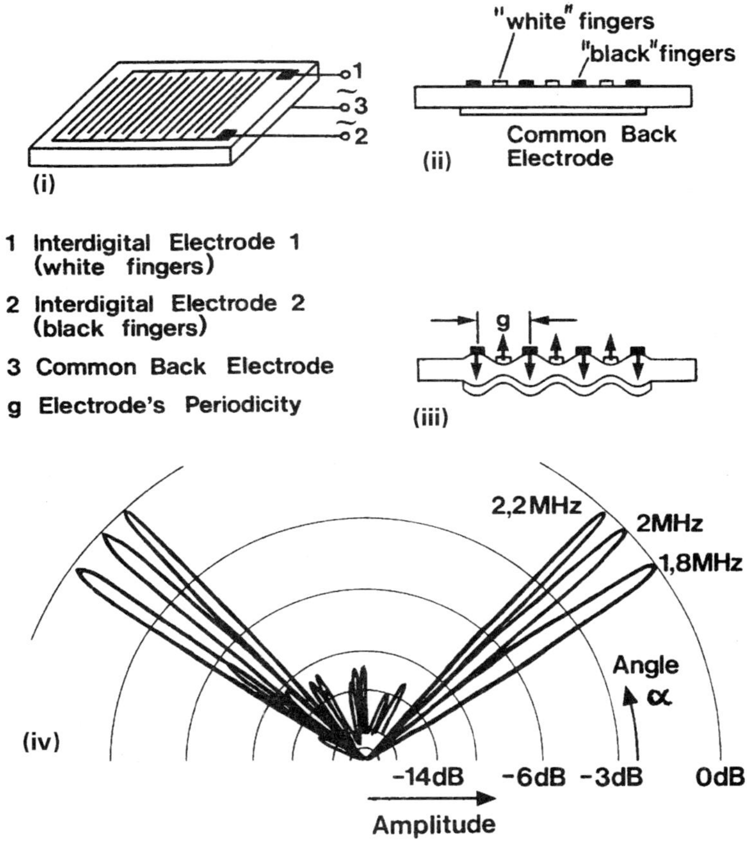

FIG. 54 (b) Interdigital transducers for liquid flow measurement. (i) Overall view. (ii) Cross-sectional view. (iii) Vibration principle. (iv) Polar diagram of radiation in water.

f. High-Accuracy Measurement Over a Wide Velocity Range. The development of an ultrasonic gas meter to be used in residences for the measurement of gas usage was initiated by a competition issued by British Gas, the world's largest gas distributing company with about 18 million households on their net. British Gas was seeking a nonmechanical, self-contained gas meter that would not need an external electric energy supply and would have the typical accuracy, minimum pressure loss, and temperature compensation to replace

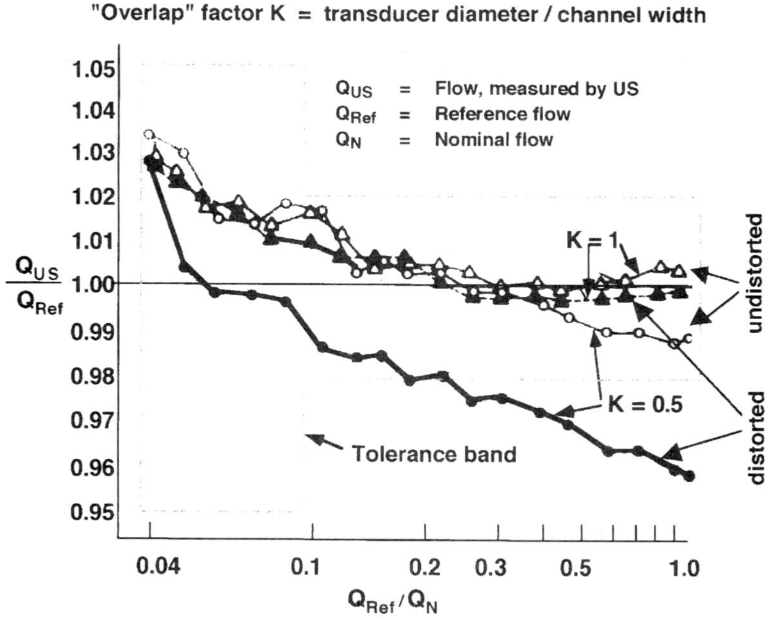

FIG. 55. Experiments with strongly disturbed flow.

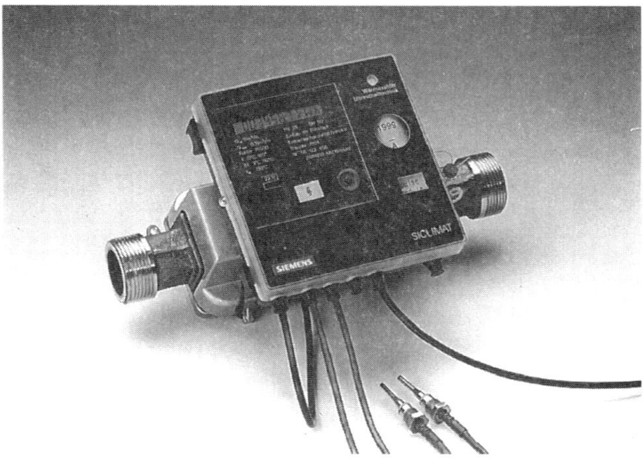

FIG. 56. Ultrasonic Heat Meter (Siemens).

the belly-type conventional gas meters. A large number of companies submitted proposals, and an ultrasonic flowmeter finally won the competition.

During the development of the ultrasonic gas meter, a coaxial arrangement was first investigated employing a circular tube. Large deviations from linearity were observed due to the waveguide properties and the refractive nature of the flow profile, which could not be corrected because of the nature of the gases to be measured and because of their dependence on temperature. Very good results, however, could be attained with a W-shaped sonic path in a measuring tube having a narrow rectangular cross section (Figure 57). The diameter of the transducer exceeds the width of the rectangle such that the sonic beam properly fills the cross section.

A high uniformity of the sonic path employed was achieved by short sonic pulses with spurious paths suppressed by the special design of the measuring tube. By employing high-quality transducers with reproducible quality and by the consequent application of the reciprocity principle in the electronic circuitry as well, the zero-flow error could be kept small. By evaluating the first rise of the received signal, all parasitic propagation modes coming later need not be considered. The problem with a possible V-shaped propagation path, which due to shorter propagation time produced spurious signals interfering with the measuring signals, was eliminated by a *Rayleigh wave trap* at the reflecting position.

The measures described above yielded excellent results. Figure 58 shows the uncorrected and corrected characteristics of a carefully constructed and calibrated laboratory model, which were measured with nitrogen using a precision flow test (Aerotrack) as reference. The correction has been proved to be consistent. A multitude of experiments—with nitrogen, air, and methane—displayed the same typical trend. Over a range from approximately 20 to 20,000 l/hr, the accuracy of the measuring tube stays within $\pm 0.5\%$ of the actual measured value. The pressure drop amounts to less than 2 mbar at

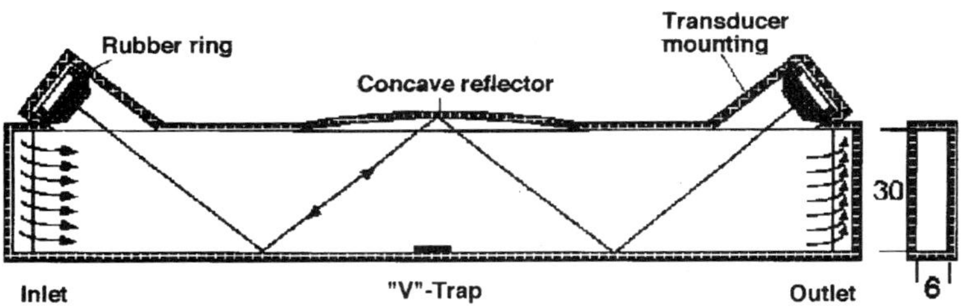

FIG. 57. W-shaped path in gas flow measurement. Courtesy of Siemens.

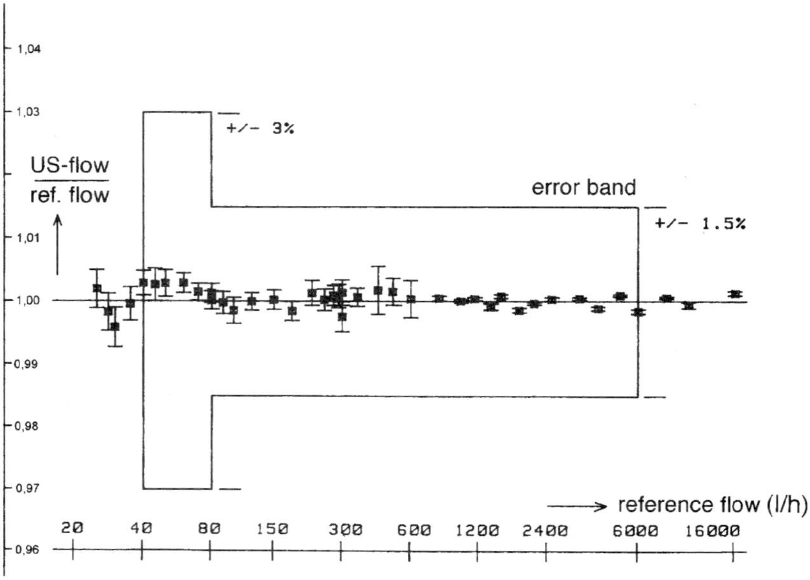

FIG. 58. Calibration curves obtained with the setup of Fig. 57.

the upper limit of the specified measuring range (6000 l/hr). Figure 59 shows the measuring characteristics for six different combustible gases that are to some extent different from each other with respect to their sound velocity and sound absorption.

The ultrasonic gas meter shown in Figure 60 was developed based on the described principles (von Jena *et al.*, 1993) to replace conventional gas meters. The ultrasonic gas meter is supplied from a single lithium battery in the size of a D-cell. With this battery, operation over 10 years is guaranteed, which testifies in an impressive way to the low energy consumption of the ultrasonic flow measurement principle. In the United Kingdom, the application of ultrasonic gas meter is officially certified and, as of 1997, some hundred thousand units have been produced. Their selling price is less than \$100 each when purchased in sufficiently large quantities.

g. High Measurement Accuracy Over a Wide Range of Liquids. The successful introduction of the ultrasonic heat meter and the British Gas meter encouraged its developers to apply the principle of the full cross-sectional insonification with inclined sound fields to flowmeters for industrial applications, e.g., in the chemical or petrochemical process industry. In this wide

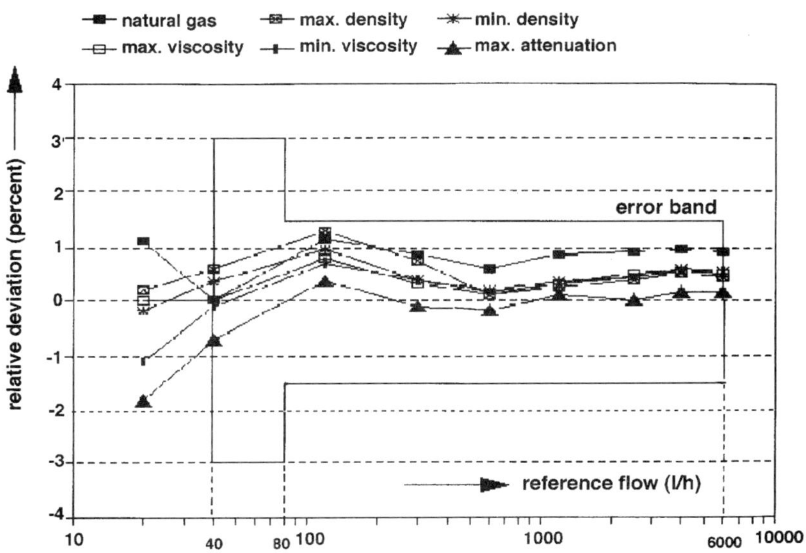

FIG. 59. Calibration curves for six different combustible gases, setup as Fig. 57.

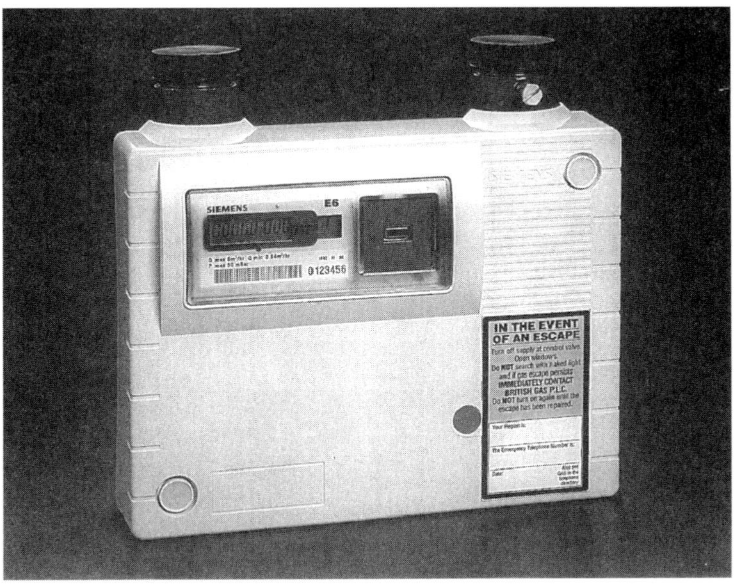

FIG. 60. Domestic-use ultrasonic gas meter. Courtesy of Siemens.

field the competing technology for an electronic flowmeter is the magnetic inductive flowmeter (magmeter), which presently is in wide use. The main advantage of the ultrasonic flowmeter principle is its independence from the electrical conductivity of the fluid, which is a crucial point for the magmeter, in particular concerning organic fluids in the petrochemical industry. Other advantages of the ultrasonic flowmeter are the low power consumption, which facilitates applications in explosive areas, and lower cost for large tube diameters, where the magnets of magmeters generally tend to become unwieldy and expensive.

For the industrial applications, the design approach used in the heat and gas meters, consisting of transducers overlapping the width of a slim rectangular tube cross section, was not feasible. This cross-sectional form, on the one hand, would be rather unattractive for the intended industrial application. On the other hand, bigger tube diameters also need to be considered: To hold the design approach would mean transducers with rather wide diameter. Apart from costs increasing with the tube's size, such transducers are difficult to design because of the high number of vibrational modes to be considered. As an additional drawback, a specific dedicated transducer design would be required for every tube width.

To overcome these difficulties, a different design principle was used. In this design, a narrow aperture transducer is mounted in the wall of a square tube (e.g., in the top wall) to radiate the ultrasound orthogonal to the wall and to the tube's axis. Opposite the transducer an acoustic diverting mirror is inserted into the bottom wall, reflecting the ultrasound at an angle to the tube's axis. In its path from the transducer to the diverting mirror, the originally narrow sound beam diverges, but is marginally affected by the flow oriented mainly orthogonal to the sound propagation between transducer and diverting mirror. After the diverting mirror, the sound field is broadened over the cross-sectional width, due to the necessary interaction with the flow. After being reflected once or more by the top and bottom wall, the sound rays come to a second diverting mirror, which reflects the sound to the other transducer configuration, placed symmetrically, using a path orthogonal to the flow direction.

An important advantage of this configuration is that the same transducer design could be used for different tube sizes. For bigger tubes, the distance between the transducer and the mirror increases, and the sound field diverges wider. For the transducers, due to their orthogonal radiation, no refractive influences need to be considered. A further advantage, in comparison to the use of inclined wet transducers, is that the inclination of the diverting mirrors is only half the value that would be necessary for inclined wet transducers, resulting in much less obstruction and disturbance to the flow in the tube. (This advantage of a relatively small disturbance at the bounding wall is also achieved in N-path designs such as in the clamp-on hybrid in Figure 50(f), or the wetted design of Figure 51(c) for *round* conduits.) Prototype measuring tubes were designed, built, and tested (Vontz and Mágori, 1996) according to this principle. Figure 62 shows such a measuring tube arrangement. The tests, however, showed severe deviations from linearity, increasing with decreasing flow velocity (see Figure 63). The reason for this function deficiency was that

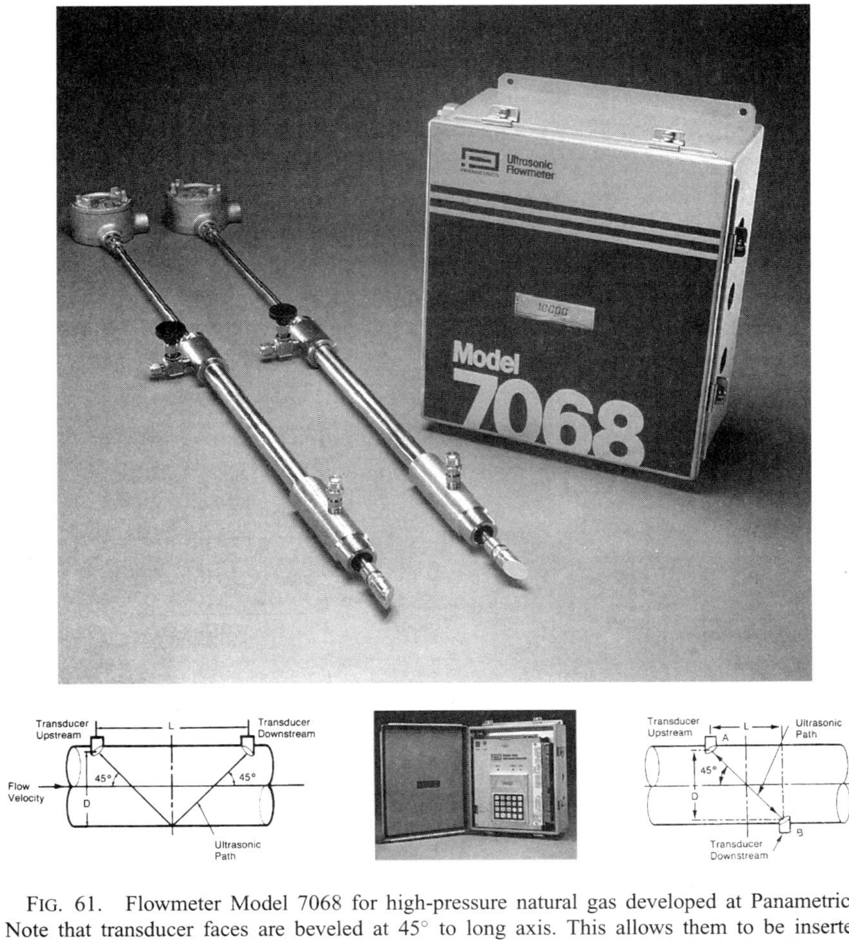

FIG. 61. Flowmeter Model 7068 for high-pressure natural gas developed at Panametrics. Note that transducer faces are beveled at 45° to long axis. This allows them to be inserted through nozzles that are welded perpendicular to the pipe axis. (Scelzo and Munk, 1987.)

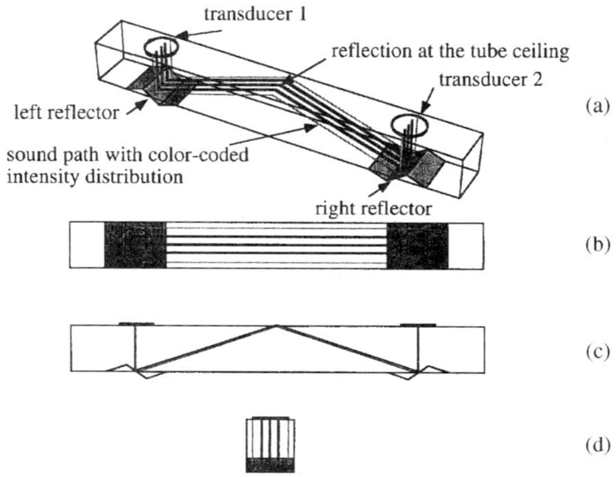

FIG. 62. Measuring tube design, insonification parallel to the side walls.

the equality of the insonification was not reached, maybe caused by the rotational symmetry of the ultrasonic transducers.

The suggestion that the sound distribution density was higher at central parts of the tube (Figure 62(d)) was proved by computer simulations using ray tracing techniques, which showed a great congruence with the measured results. In this simulation (see Figure 64), the flow profile in the rectangular measuring tube was assumed as being of the form $(1 - |x|^n)(1 - |y|^n)$. The

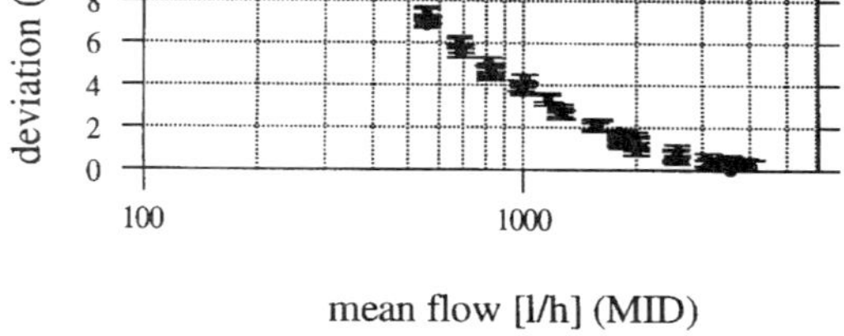

FIG. 63. Measured error curve for water, for the configuration of Fig. 62. Reference: magnetic inductive meter.

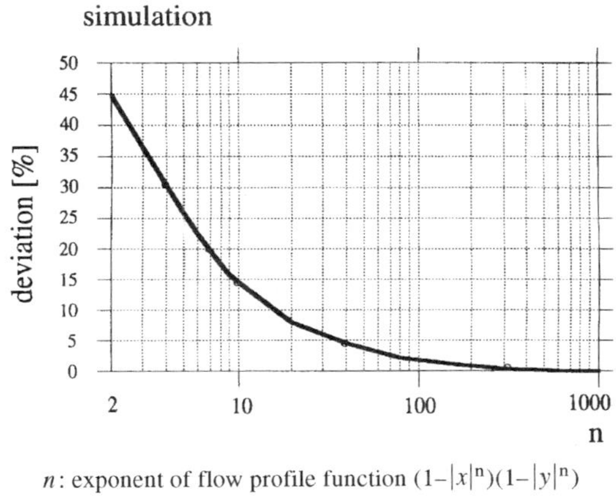

FIG. 64. Simulated error curve for water, for the configuration of Fig. 62. Flow profile approximated by $(1 - |x|^n)(1 - |y|^n)$, $n = n(v)$.

parameter n depends on the flow velocity varying from $n = 2$ ("laminar" parabolic shape) to $n = 1000$ ("box"-type high turbulence shape) at high flow velocity. The calculated flow profiles are in a good approximation to laser Doppler anemometer results, and the exponent n as a function of the flow velocity was fitted accordingly. Figure 65 shows some simulated and measured flow profiles using different n.

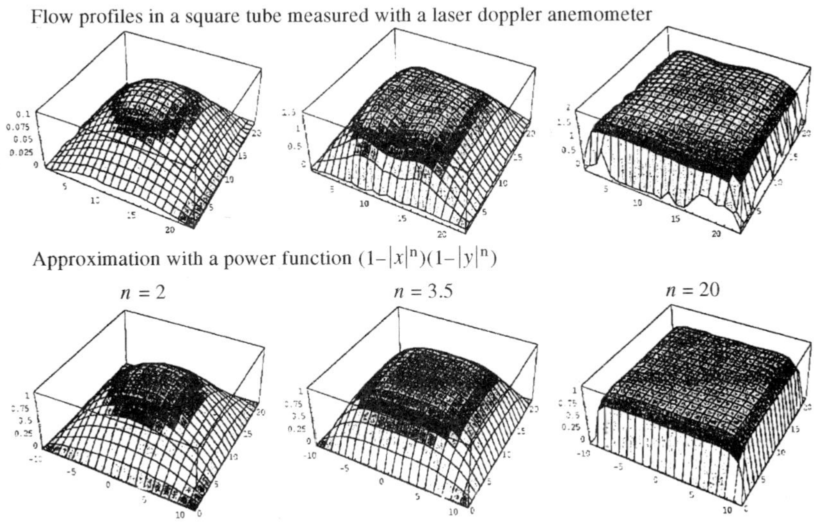

FIG. 65. Simulated flow profiles of the form $v_z = (1 - |x|^n)(1 - |y|^n)$; $n = n(v)$.

To get a better performance, another configuration was introduced. In this configuration, the acoustic diverting mirrors in the wall opposite the transducers were inclined with respect to the tube axis in such a way that the sonic rays were reflected to the side walls rather than keeping them parallel to the side walls. After the diverting mirror in the bottom wall, the sound becomes successively reflected by the first side wall, the top wall, the other side wall, and by the other diverting mirror to the other transducer, forming a screw line or helical path. Figure 66 shows this configuration in different views. In the cross-sectional view (Figure 66(d)) (compared Figure 62(d)), an excellent uniformity of the sound distribution within the measuring tube's interior space becomes evident. As a result of this uniform insonification, an excellent linearity was achieved over the whole flow range, as shown by measurements on the test stand (Figure 67). The simulation, as discussed before, proved this result as being due to the altered insonification (Figure 68).

h. Range of Ultrasonic Flowmeters for the Process Industry. According to the described helical sound path design principle, a wide range of ultrasonic flowmeters was developed (by Siemens), comprising the equivalent nominal tube widths of 25 mm, 50 mm, 80 mm, and 100 mm (the latter two being of circular cross section). The accuracy, depending on the dynamic

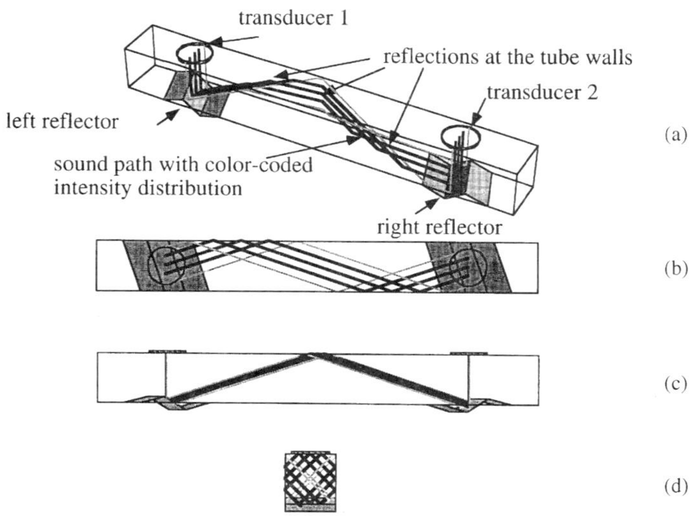

FIG. 66. Measuring tube design, insonification using a helical path.

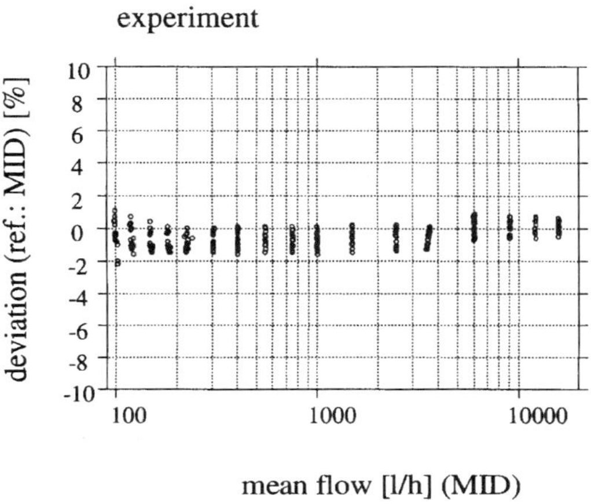

FIG. 67. Measured error curve for water, for the configuration of Fig. 66. Flow profile approach as in Fig. 65.

range, is specified as 0.5% for 25 : 1 turndown and 1% for 100 : 1 turndown. The compiled geometries are given in Table 5.

The ultrasonic flowmeter can be remotely operated by a personal computer or by a dedicated handheld unit using the HART protocol. In situ, control is possible by touching four button positions on a glass window. These are sensed optically with feedback to the operator by an LCD display. The

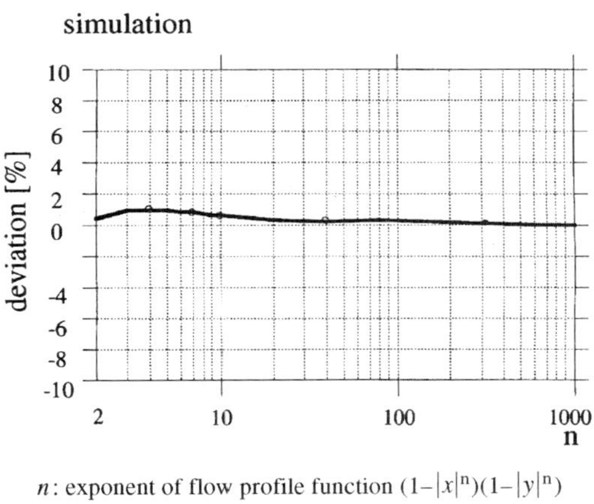

FIG. 68. Simulated error curve for water, for the configuration of Fig. 66. Flow profile approach as in Fig. 65.

TABLE 5.
PROPERTIES OF SITRANS F® ULTRASONIC FLOWMETER RANGE (SIEMENS)

Nominal bores	*Nominal pressure*	*Measured value deviation*
25 mm 50 mm	40 bar (bores 25 to 80 mm)	$\leq \pm 0.5\%$ at 25 : 1 turndown
80 mm 100 mm	16 bar (bores 80 to 100 mm)	$\leq \pm 1.0\%$ at 100 : 1 turndown
Process temperature	*Ambient temperature*	*Power consumption*
−20°C to +180°C	−20°C to +85°C	8 W (120 to 240 V)
Analog output	*Digital output 1*	*Digital output 2*
4 to 20 mA, HART-bus, failure signal configurable	configurable as pulse, frequency, or status	relay output, configurable, e.g., for limit values

parameters that can be selected and changed are the measuring range, units of measurement, unit status, analog or digital output, limit values, display parameters, density, no-flow cutoff, flow direction, and many more.

In operation, the display can show two measured values by choice from the following options: flow volume, flow rate, temperature, ultrasound velocity, and signal intensity. The latter two values provide a simple check for the quality of the liquid during the measuring process. A wide range of diagnostic functions validate an optimum function of the meter. The range of ultrasonic flowmeters denoted Sitrans F® (Figure 69) was presented for the first time in June 1997 at *ACHEMA*, one of the most important fairs for process industry instrumentation.

i. Special Considerations at Temperature Extremes. The ultrasonic measurement of flow, liquid level, and perhaps other process measurands at temperature extremes, particularly high temperature, is often limited by one or more of the following: Curie point of the piezoelement; loss of oxygen from the piezoelement; differential thermal expansion between the piezoelement and adjacent structural or damping media; and couplant degradation. These limits can be avoided if a suitable buffer can be interposed between the piezoelement and the process fluid, or between the piezoelement and the wall containing that fluid. Buffers have been developed to guide either compressional or shear waves to solve high-temperature ultrasonic flowmeter problems encountered in petrochemical refineries and in electricity generating power plants, respectively.

At *ordinary* temperatures, the measurement of flow and liquid level has become routine for ultrasonic equipment such as that described in previous sections. The design of *high*-temperature transducers, including buffers, has been an interesting engineering challenge for the past fifty years or so, and numerous solutions have emerged during this period. Examples of solutions

FIG. 69. Industrial Ultrasonic Flowmeter Sitrans F® in typical application environment (Siemens).

known prior to 1989 are shown or referred to in Lynnworth (1989). Two interesting solutions that evolved in the 1990s are the clad rod described by Jen and Legoux (1996) and the differentially threaded rod described by Nygaard and Mylvaganam (1993). An earlier use of the differentially threaded rod is due to Sather (1968). In the mid-1990s in the laboratory of one of the authors, two new buffers were developed. A design resembling a small hockey stick, denoted OKS and shown in Figure 47, provides a clamp-on solution (Lynnworth *et al.*, 1996, 1997b). A bundle of thin wires, rigidly encapsulated, denoted the BWT (Bundled Waveguide Technology) transducer and shown in Figure 43, provides a wetted solution (Liu *et al.*, 1998).

The compressional buffers are mainly used as *wetted* devices, meaning they contact the process fluid (gas or liquid). Flanged openings in the pipe, called nozzles, are typically provided for wetted transducers, as they are for other standard sensors like thermocouples or other types of intrusive flowmeters.

(The compressional buffers could, in principle, be pressure-coupled at normal or oblique incidence to the outside of a pipe or tank.) So far the main use of these buffers has been to measure the flow of *gas*.

In the Rotterdam-area refinery where the BWT transducers were first used, the gas composition and pressure were not constant. To a considerable extent, although analyzed and tested for safety issues and flow-tested in simulations, this device was still an unreleased product, not yet field-proven at the time of its installation. At the early stage of starting up the refinery in the Netherlands in 1997, for safety reasons the gas initially was nitrogen, first at low pressure, then gradually building up to high pressure. Later, nitrogen was replaced by hydrogen or a hydrogen-rich low-molecular-weight mixture, again first at low pressure, eventually reaching some 200 bar. Flow is to be measured during as much of this procedure as possible. Compressional buffers like those in Figure 43 were able to measure nitrogen gas flowing at pressures above about 10 bar and low-*MW* (molecular weight) gases at pressures above about 30 bar (MW $\sim$ 4). For such gas combinations, the *MW* range is about 7 to 1. As a result of their performance during plant start-up and at normal operating conditions, additional BWT transducers were installed as replacements for "standard" transducers that had been installed previously but where improved performance was sought, along with the easy removability option outside the pressure boundary. Similar BWT transducers were later used to measure steam flow at approximately 340°C, 35 bar, and 45 m/s at another company's test facilities elsewhere in the Netherlands (Urata *et al.*, 1998).

The *shear* wave buffers, on the other hand, were expressly designed for oblique-incidence *clamp-on*, to measure *liquid* flow or *liquid* level without requiring any penetration of the vessel. This requirement came from numerous sources. Early applications were superheated water, hot asphalt and other hot hydrocarbon liquids, $T = 260$ to 300°C. Pipe sizes ranged from nominally 3 inches ($\sim$75-mm ID) to 0.4 m ID (Lynnworth *et al.*, 1996, 1997b).

j. Advanced Doppler Flow Sensors. Though in general not as accurate as the ultrasonic transit time (contrapropagation) flowmeters described above, the ultrasonic Doppler flow sensor can be a good choice if a lower accuracy is acceptable. (Users report uncertainties of several percent of reading, e.g., $\pm 5\%$. Note that Doppler manufacturers' brochures indicate substantially better accuracies. In practical cases, Doppler *precision* may be adequate.) In two-phase and multiphase applications (e.g., waste water), the Doppler sensor (due to its simpler design principle) provides a good performance-to-cost ratio. Necessary for the function of these sensors are particles or microparticles such as dust, gas bubbles, proteins, red blood cells, small liquid droplets, etc. suspended in the fluid. In liquids with a high ultrasonic attenuation, where other ultrasonic meters could fail due to the ultrasonic path interruption, the robust ultrasonic Doppler flow sensor still works. Various suppliers market these ultrasonic flowmeters: Dynasonics, Peek Measurement (Figure 70), and others. (See the U.S. publication *Measurements & Control*, October 1995, 1996, 1997.)

The reason for the low accuracy in the *simplest* Doppler flowmeters is the lack of information about the scattering particles' position and their distribution. In other words, if the scattering volume extends all the way across the pipe, rather than being restricted to a well-defined region, there is no information telling us from which part of the flow profile the individual contributions to the integral Doppler signal come. Further, a weighting of the different signal contributions is given by the liquid's ultrasonic attenuation, which often is not well known and which could vary in a way that is hardly predictable. (*Tag* has some similar problems to overcome.) These drawbacks can be eliminated by more complex methods, which obtain information about the velocity of the scattering particles and about their position as well, for instance by beams intersecting in a particular region near the wall (Pfau, 1970) or near the center or by range gate methods. (See, for example, Brandestini (1978); Brandestini and Forster (1978); Remenieras *et al.* (1996); Anon. (1997).

Range gate methods, motivated initially by and/or proven in expensive medical blood flow measurement systems, increasingly become used for industrial sensor purposes. The ultrasonic water tap control described in II.A.5 demonstrated, however, the possibility of an *inexpensive* realization of the principle. Instead of a continuous ultrasonic wave, short bursts are transmitted, which are cut out (gated) from a continuous reference signal. The received echoes from the moving scattering particles contain the distance information in the time elapsed between the transmission and reception. The velocity information is given by the frequency shift due to the Doppler effect. By measuring the Doppler phase shift at different times after the transmission of the burst, the velocity profile along the ultrasonic beam is obtained after a number of ultrasonic transmissions. The accuracy of the volumetric flow velocity measurement is enhanced substantially by the integration of this velocity distribution, in comparison to a simple Doppler frequency evaluation. With multiple transducer arrangements, forming multiple ultrasonic paths, a two-dimensional flow profile also can be obtained, whose integration gives still more accurate values of the volumetric flow velocity. (See, for example,

(a)

(b)

FIG. 70. (a), (b), (c): PEEK Measurement's DDF portable digital Doppler flowmeter used with transducers strapped onto vertical and horizontal pipe, and close-up of this Doppler instrument being operated under nonideal conditions. Illustrations (a)–(c) courtesy of Peek Measurement.

(c)

FIG. 70. *(continued)*

Takeda (1995).) Such measuring systems are manufactured, for instance, by two different Swiss companies, Signal Processing S.A. and MET-FLOW S.A. Both claim submillimeter spatial resolution along the measuring path, divided in 128 segments.

In Figure 50(f), the solitary intrusive reflector at y/R provides a hybrid clamp-on *transmission* measurement that can be the analog of the area-averaging Doppler sample near the wall, obtained *non*intrusively (Pfau, 1970). The single intrusive reflector idea is due to Chernyshev (1994); the clamp-on transducers and the square tubular reflector were subsequent suggestions. (Multiple reflector *V*-paths also appear in Liu and Lynnworth, 1995, Fig. 6*C*.)

Doppler flowmeters are usually considered the only choice when the fluid is scattering. But there are at present at least two other acoustic measuring principles that can be utilized: timing successive reflections from the scatterers and timing how long it takes a tag to pass from one beam to another. The first should be obvious to NDE practitioners, as it involves a repetitive pitch-catch interrogation of the ensemble or cloud of scatterers.

Sold by Panametrics under the trade name Transflection®, it is based in part on the time domain correlation work of Embree and O'Brien (1985). (See also McGunigle (1974), Bonnefous (1994); Bonnefous and Pesqué (1986); Trahey *et al.* (1987); and Jensen (1996)). Various investigators choose different generic names for it: time domain correlation, time shift, speckle tracking, stroboscopic scattering. Stroboscopic scattering was combined with contra-propagation transit time by Panametrics in a downhole flow tool similar to that described in Lynnworth (1988) and Lynnworth *et al.* (1993). A prototype downhole flow tool that toggled between transmission (T) and reflection (R) modes was tested in an "oil patch," i.e., in the field, at a depth of ~1.6 km. Some of the results are appended to the 1993 Lynnworth *et al.* reference. Another prototype downhole flow tool, designed for geothermal reservoir assessment and also containing the two modes, was subsequently used in Stanford University's department of petroleum engineering (R. Horne, priv. comm., 1997), but has not yet become a commercial product. The T&R two-mode technology first demonstrated in it, however, was reconfigured into the Panametrics portable flowmeter PT868 and related models, of which several thousand are in use (Scelzo and Jacobson, 1994).

Because it is a correlation-based measurement, stroboscopic scattering can utilize most of the same circuitry already inside a portable ultrasonic flowmeter such as the PT868 (Figure 71) and other correlation transit-time flowmeters for water (Lange, 1994), propellant liquids (Werlink *et al.*, 1996), or gases (Anon., 1990). The stroboscopic scattering mode gives the operator a second chance to measure flow when the transmission all the way across the fluid is too attenuated to yield a reliable signal. One may wonder, if the fluid is so attenuating, how can a reflection technique work? The answer is due in part to the circuit's ability to discriminate between stationary noise and a signal that varies in time in response to the moving scatterers. (In the case of *tag* mode, suppression of the stationary noise may utilize the quadrature demodulation method described by Jacobson *et al.* (1985).) The Jacobson *et al.* (1988) patent underlying correlation-based flowmeters (or intervalometers, in general) included four measuring modes: transmission, reflection, tag, and noise. The instruments in Figure 48 contain the first two of these (T and R).

Consider a fluid that can vary from a dense liquid as the continuous phase, at one extreme, to a gas at the other extreme, with two-phase mixtures between these limits. Figure 71(a) shows schematically how the transmission and reflection modes might cover single-phase fluids to which an increasing

(a)

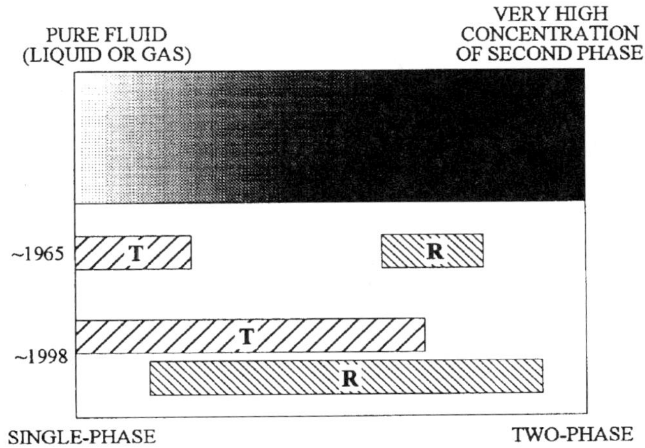

(b)

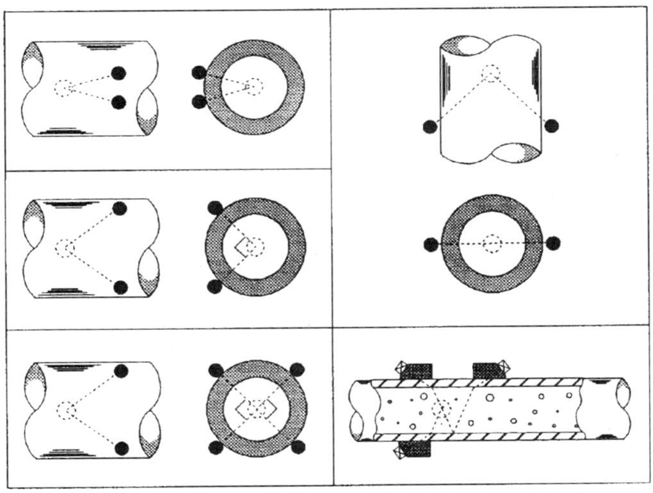

FIG. 71. (a) Schematic indicates T&R (transmission and reflection) progress in the past 33 years. The range of fluids covered by T&R has expanded significantly. Diagram does not indicate which mode of interrogation is more accurate at any given concentration of two phases. (b) Transducer positions for R method. Depending on pipe size and scattering (attenuation) characteristics of fluids, transducers may be placed near one another or at 90° or 180° around the pipe. "Near one another" includes the default situation where one transducer serves as both transmitter and receiver (pulse-echo mode). Small dashed circles represent intersection (scattering) volume. (c) The three arrows near the axis represent the successive strobing of a moving target.

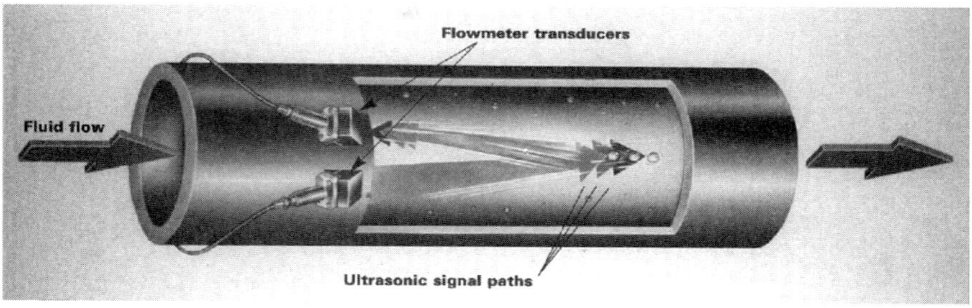

(c)

FIG. 71. *(continued)*

concentration of the second phase may be added. Application examples are given by J. E. Matson in Anon. (1997).

Both Doppler and the stroboscopic scattering technique share an unwanted sensitivity to crossflow. In the T mode, a vee path substantially eliminates the unwanted sensitivity to crossflow. In the R mode, another solution must be found, e.g., interrogation from diametrically opposed points. Figure 71(b) shows several R-mode geometries, some of which allow separation of the crossflow component from axial flow. The three arrowheads near the axis in the bottom part of Figure 71(c) represent three successive interrogations reflecting off an ensemble of moving targets.

The second alternative to Doppler when a fluid is highly scattering is the *tag* method. So far this method has seen very little commercial exploitation. Early studies include those of Coulthard (1973), Tornberg *et al.* (1983), and Tornberg (1986). One of the ideas explored in Tornberg's studies was to cross-correlate Doppler returns to measure tag velocity down the pipe. Another Doppler combination might utilize the frequency shift within the stroboscopic return, instead of just using the change in transit time between successive

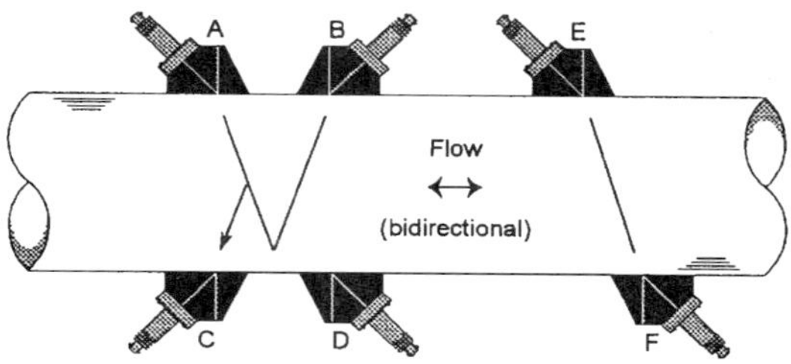

Mode	Transducers/Paths
T	AB, AD, BC
R	AC
Tag	AD, EF
Noise	A-F

FIG. 72. Generalizing from two modes to four modes—T, R, tag, and noise—adapted from four-mode flowmeter concept in Jacobson *et al.* (1988).

interrogations. A multimode flowmeter concept is depicted in Figure 72; the modes are T, R, tag, and noise.

Two-phase and multiphase fluids, however, exist in much more complex forms than is suggested by the foregoing text. Flow patterns for gas–liquid two-phase systems, ranging between annular flow and bubbly flow, are depicted for a *horizontal* pipe in Figure 73, according to Jepson *et al.* (1993). An early example of range-gated Doppler, with 128 gates, is provided by the work of Brandestini (1978). Results obtained with a more modern Doppler flow profiler are reported by Takeda (1995). This equipment was applied subsequently to two-phase bubbly flows in work by Takeda *et al.* (1996).

3. *Ultrasonic Flowmeters as Multiple-Sensor Systems*

a. Further Important Parameters. Other quantities besides the flow velocity can be measured by an ultrasonic flowmeter. One important quantity, for example, is the sound velocity, *c*, in a liquid or in a gas. This quantity can be

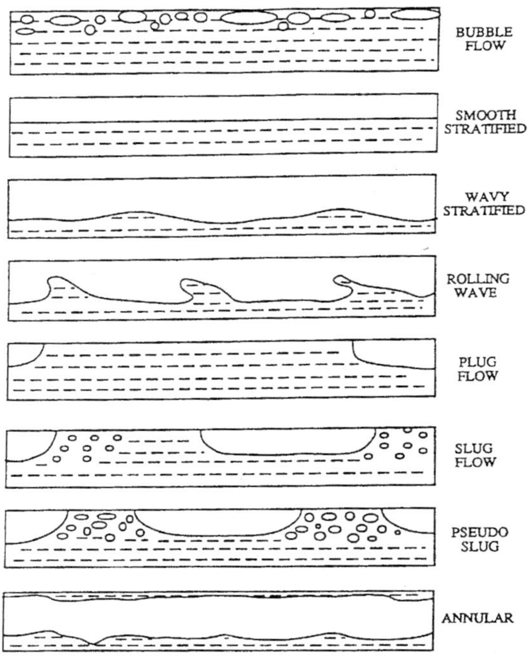

FIG. 73. Flow patterns for gas–liquid two-phase flow system in horizontal pipes, after Jepson *et al.* (1993). Illustration courtesy of Corrosion in Multiphase Systems Center, Ohio University, Athens, OH.

determined with high accuracy simultaneously with the ultrasonic flow measurement from the average of the measured transit time values upstream, t_1 and downstream, t_2, if the acoustic path is known. Another way to measure c is to determine the propagation time for a known path length, e.g., provided by the pipe wall opposite the transducer site, path $\perp$ pipe axis, or given by a reference reflector intentionally placed at a known position. This is a straightforward ultrasonic measurement task. Applications for the c component of a "*V*-meter" or flowmeter include *MW* (molecular weight) determination in gases (suggested by Figure 23) and specifically in a flare gas flowmeter (Section III.A.2; Smalling *et al.*, 1984, 1988) and a mass flowmeter for those liquids in which density ρ and sound speed c are related uniquely (e.g., see NuSonics (Section III.B, Figure 86) pipeline interface detectors, based on data in Zacharias (1970) or Controlotron models based on Baumoel (1995)). In addition, for two-phase steam (provided the fluid satisfies the assumptions of homogeneity, thermal equilibrium between phases, and no slip), steam quality might be obtainable according to the analysis of Michaelides and Zissis (1983) and a proposed implementation in a flowmeter due to Shen (1992).

More difficult is the accurate measurement of the propagation attenuation and the scattering characteristics of the medium. The amplitude of the received signal depends not only on the attenuation, but also on other parameters of the measuring arrangement. The acoustic impedance Z of the medium affects the efficiency of the transducers. The sound velocity c affects their angular radiation properties, and by this the "antenna" gain. Other influences are scattering of particles suspended in the fluid and diffraction at acoustic impedance gradients. Nevertheless, the amplitude of the received signal of a fixed measurement configuration gives reproducible information about attenuation trends, which may be sufficient for many applications.

If the fluid is a gas, the amplitude of the received signal depends on pressure (Section II.D). The simple relationship, linear over limited ranges of pressure and path length, is not easily implemented because turbulence, residue, or other factors also influence the received amplitude.

The sound velocity and the attenuation characterize the medium and its state. Simple questions such as whether the right medium with the proper quality is inside the tube or tank can be efficiently answered. Furthermore, in actual processes the in-line measurement of the velocity and attenuation can be used in closed-loop control, maintaining constant product quality. This was successfully demonstrated in a dairy plant, where a continuous milk homogenization was performed, separating only as much cream as necessary to obtain the specified fat content of the milk (Henning and Hauptmann, 1997).

With substances having different sound velocities (e.g., binary mixtures of gases), the actual mixing ratio can be determined from the measured sound velocity (Valdez and Cadet, 1991; Stagg *et al.*, 1992). This method becomes increasingly sensitive as the difference of sound velocity of the constituents increases. An example of this method extended to a mixture of *many* gases is the determination of the average molecular weight of flare gases occurring in petroleum production (Section II.A.2). On more than one occasion, a very high c was observed, corresponding to an average molecular weight just slightly above 2. This alerted plant operators to a major leak of hydrogen into the flare system. Because hydrogen burns nearly invisibly, its presence could not be detected visually, nor by any other on-line means available at the time (Smalling *et al.*, 1984). Over a thousand ultrasonic flare meters were in use by 1997, designed basically along the lines of the V and c method reported in 1984. Later versions, however, particularly since about 1990, utilized cross-correlation detection.

Another binary gas example is the ultrasonic intake air sensor, as described previously (Section II.B.2.a). Here the gasoline vapor content of the intake air was measured with ultrasound. This can be important when the active carbon filter used to prevent the gasoline vapor from escaping to the environment is flushed by the intake air of the engine.

An ultrasonic concentration sensor has been successfully employed as an important component in a high-speed professional laser printer, the ND 3, manufactured by Siemens. This laser printer is equipped with a novel technology for fixing the toner particles to the paper. The paper is led through a chamber filled with a mixture of air and the vapor of an organic solvent. During the paper's pass through this chamber, the vapor commences to dissolve the toner particles being put loosely onto the paper surface, making the particles sticky to fix them to the paper. Critical is the vapor concentration: too little vapor gives an insufficient fixing; too much gives results like writing with ink on blotting paper. Because no other concentration sensor with satisfactory long-term stability could be found, the ultrasonic concentration sensor was the essential key component for the success of the laser printer, which was marketed at a high sales volume.

In media with a known relationship between the speed of sound and temperature, particularly in gases or gas mixtures with known composition, the determination of the actual speed of sound makes a precise and fast temperature measurement possible (typical risetime is 1 ms or less). Thus, even at a distance of about 10 cm from the outlet valve, the temperature variations of hot exhaust gases (mainly nitrogen) of an internal combustion engine were measured with high time resolution. However, for the intake airflow sensor and other gas measuring devices, it can be useful to employ an additional low-cost temperature sensor in the measuring tube. This can increase the accuracy of the aforementioned estimation of the type of medium or the composition of mixtures. Given the knowledge of the type of medium and its thermal capacity, it is possible to determine the heat transfer, which is associated with the volume flow. By deploying an additional temperature sensor at a different position in the same tube system, the heat loss between the two temperature sensors can be determined (for example, in the heat meter described in Section II.B.2.e).

b. Potential of an Intelligent Multisensor Flowmeter. Evidence has shown that ultrasonic flowmeters can derive additional information about the measuring path besides just the flow velocity. By sensing several parameters by means of a "sensor-fusion," important state quantities relevant for the process can be extracted in a decentralized manner at the location of the sensor (Figure 74). Other auxiliary sensors may be integrated to the measuring tube to give additional sensor information. The computing power of a microprocessor or a digital signal processor (DSP) in the electronics of the flowmeter facilitates the flexible and efficient evaluation of those auxiliary sensors and the extensive utilization of the additional sensor information.

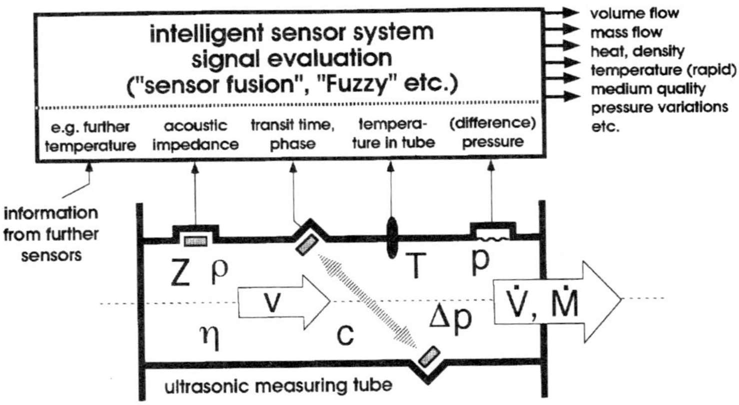

FIG. 74. Ultrasonic flowmeter as a decentralized multisensor system.

Fuzzy algorithms, among others that are especially suitable for this purpose, are processed rapidly and elegantly by microcontrollers employing a fuzzy coprocessor (Eichfeld *et al.*, 1993) or by the implementation of fuzzy algorithms on dedicated digital signal processors, which become available at affordable costs. Also, self-learning strategies could be employed, e.g., Neural Networks, which gain sensor system experience during reference (training) operation cycles or even during normal operation. Thus additional parameters are available for the aforementioned considerations. In addition, evaluation of signal amplitudes plays an important role in the self-checking of the measuring system.

4. *Clamp-On for Gases*

Before considering clamp-on flow measurement of *gases*, a brief review about clamp-on for *liquids* may be in order (Section II.B.2.b). Regarding *liquids* (e.g., water) in plastic pipe, it has been known for a long time that plastic pipe's low sound speed c_2 compared to the sound speed c_3 in water makes it easy to obtain a long axial interaction length L. (Lake, 1962; Lynnworth, 1967, p. 275 or 1979, p. 473.) The low c_2 also makes it easy to utilize an off-diameter path (Lynnworth, 1967, Figure 15; 1979, p. 434). Also, the transmission of acoustic energy between the liquid and the pipe wall is relatively efficient, compared to liquid in a *steel* pipe. This is because the acoustic impedance Z_2 of the *plastic* pipe is usually within a factor of three of the acoustic impedance in the liquid, Z_3. In contrast, for water in steel pipe, $Z_2/Z_3 \approx 30$. Despite this unfavorable impedance ratio, clamp-on flowmeter manufacturers (see Figures 45 and 48) are able to measure flow of liquids in steel pipes and tubes down to the order of 1 cm diameter, and up to several meters in diameter, typically to an accuracy of one or a few percent of reading over normal flow rates. For plastic (low-impedance) pipe, this means one can expect a very high *SNR* (signal-to-noise ratio) and a very high refracted angle θ_3 for ultrasonic measurements of *liquid* flow in plastic pipe.

Can one sacrifice some of the desirable but unnecessarily high *SNR* and high θ_3 and address a fluid of low c and low Z (e.g., air) flowing inside a plastic pipe? In other words, assuming a standard plastic pipe for the conduit, if *water* were replaced by *air* at atmospheric pressure and if no other major changes were made in the clamp-on flow measuring equipment, would it be possible to measure the air flow? The answer depends on the type of plastic, transducer frequency, flow velocity, and fluidynamic conditions. Using clamp-on transducers in the laboratory, it has been possible to measure atmospheric air flow at 0.5 MHz in PVC plastic pipe sizes such as 2-, 4-, 10-, and 20-inch nominal diameter. Results for ϕ10-inch were given in Figure 50(i). The answer to the question, then, appears to be a qualified yes—i.e., yes for some pipes, if the resulting accuracy is adequate. Referring to the two *SNR* diagrams in Figure 50(i), one can interpret the upper diagram to represent air at one bar in a *steel* pipe (poor *SNR*), while the lower diagram could represent one-bar air in a *plastic* pipe ($SNR > 1$ or $\gg 1$). The plastic attenuates the parasitic noise propagation. If air (or gas) pressure increases, impedance mismatch and attenuation coefficient α both decrease, so the signal increases and so does the *SNR*. If pressure is high enough, $S > N$, even if the conduit were steel. (Plastic might not be strong enough to contain the high pressure and would have to be replaced by a stronger material, say, steel.) Bear in mind that high *SNR* does not guarantee success. The small θ_3 obtained with air in steel pipe means it may still be too difficult to achieve by clamp-on the resolution or accuracy sought. In that case, and assuming flow was to be determined by the contrapropagation method, the answer to the question posed above would be no. Recourse to wetted transducers, Figure 61, may then be advisable.

C. TEMPERATURE

This section covers average temperature, temperature profile, and tomographic reconstruction. The *wireless* measurement of temperature at a remote point, using a SAW sensor, is covered in Section IV.B. In most cases, the temperature is derived from a measurement of sound speed (Figure 75).

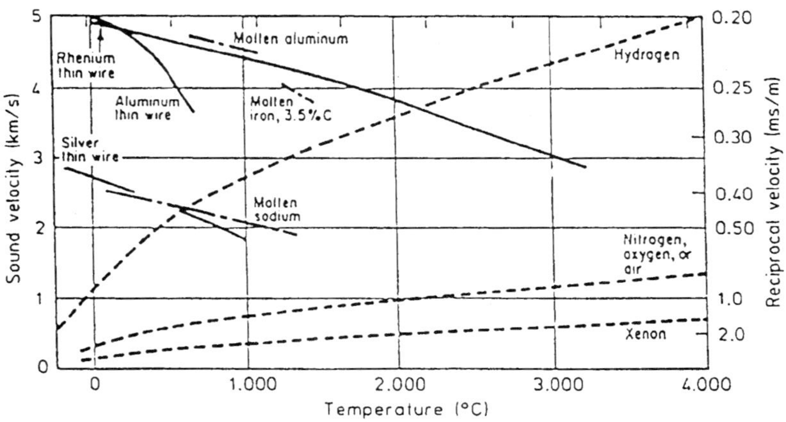

FIG. 75. Sound speed c as a function of temperature in gases, liquids, and solids.

1. *Average Temperature*

In contrast to when one measures with thermocouples, resistance temperature detectors, integrated circuit temperature sensors, and SAW or other small devices that sense temperature essentially at one point, when one uses acoustic or ultrasonic waves one can measure the speed of sound over an extended path to obtain an average temperature reading. At sufficiently low carrier frequency (57 Hz) and with a sufficiently intense (221 dB) coded source, this has been done over a distance of 18,000 km, using the ocean's 1- to 2-km thick *SOFAR* channel, achieving a range that reached halfway around the world (Baggeroer and Munk, 1992). (This low-frequency measurement, insofar as *range* is concerned, dwarfs Laenen's (1984) *ultrasonic* measurement of flow across a river) (Figure 35), which for eight years stood as one of the best examples of a long-path acoustic/ultrasonic measurement. If the scope of flow or temperature applications is strictly limited to ultrasonic frequencies, Laenen's example perhaps still holds the world distance record.) On a somewhat smaller scale, as in combustors where gas temperature may be of the order of 1000°C, the path lengths may be 10 to 20 m, and the frequency may be 1500 Hz (Kleppe, 1989, 1995a, 1995b, 1996). In smokestacks at somewhat milder conditions of only 200°C and flow velocities up the stack of Mach 0.1, the frequency can be higher, reaching up to the lowest octave of ultrasound—20 to 40 kHz. A frequency as high as 100 kHz was used to measure sound speed and flow velocity across a 3-m path in a 150°C flue section leading to a smokestack (Matson and Davis, 1994). (See also Matson and Lynnworth, 1997.)

If a *wire waveguide* is used as the sensor for temperature (or elastic properties such as Young's modulus, shear modulus, and Poisson's ratio), the frequency used in the past has typically been around 100 kHz (Bell, 1957). Starting around 1960, various investigators in the United Kingdom proposed using a wire waveguide sensor as a replacement for a thermocouple (tc) for several reasons: to avoid tc errors attributed to insulator failure at very high temperature, to avoid tc diffusion errors, to avoid having to fit *two* different alloys and an insulator within a given sheath ID, and to achieve a more rugged design. Sometimes the sensor is a waveguide *added* to the system; sometimes an existing structural element or process boundary can be the sensor. In some nuclear fuel pin studies, some or all of these potential advantages were realized. However, at the time those studies were conducted, no practical way was found to extend the results to analogous and/or recurring industrial applications in an economical fashion.

Nowadays, however, the widespread use of microprocessors and precision ultrasonic intervalometers for other purposes (principally flow, thickness gaging, and liquid level) means that one could adapt, for example, contra-propagation flowmeters or digital thickness gages to wire waveguide single-zone or multizone temperature measurements. This has been done for single-zone, dual-mode *moduli* measurements. The simultaneous thin-wire determination of the two principal moduli requires simultaneous interrogation with extensional and torsional waves. As these waves propagate at different velocities (e.g., near 5000 and 3000 m/s, respectively) near room temperature in magnetostrictive materials like Ni, Remendur, or Permendur, a one-zone specimen generates echoes in different time zones, much like a multizone wire waveguide temperature sensor. See Collard and McLellan (1990) for high-temperature (>1000°C) moduli and other results obtained with a modified Panametrics Model 6468 ultrasonic flowmeter. In the 1990s, Thermosonics independently developed a wire waveguide thermometer. One of their initial furnace profile applications involved temperature around 1600°C (Fendrock and Varela, 1995).

2. *Temperature Profile along One Path*

Shortly after the wire waveguide was introduced into the United States (around 1965) as a candidate sensor to measure temperature in certain extreme environments, it was recognized that by intentionally creating a series of discontinuities along the waveguide, one could measure temperature profile. By 1970 this had been demonstrated in laboratory prototype sensors (Lynnworth and Patch, 1970). In other words, the "average" temperature measurement of Section II.C.1, utilizing the sound velocity–temperature data represented by the thin wire curves in Figure 75, can be applied to the smaller path lengths of a multizone sensor such as that in Figure 76. It is also possible

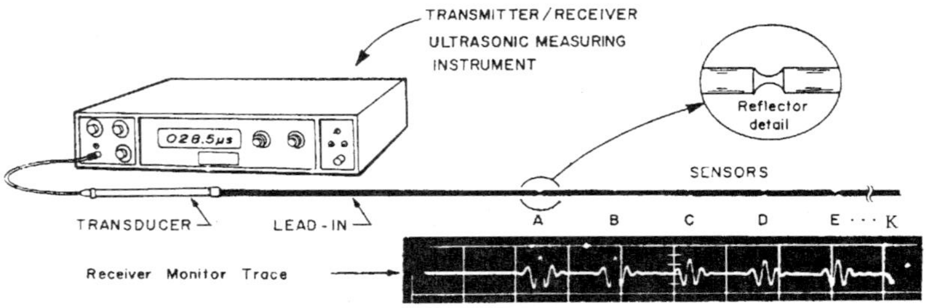

FIG. 76. An ultrasonic T profiler waveguide containing ten zones.

to "tap" a multizone wire waveguide at n points to extract profile information (Kim *et al.*, 1993), somewhat analogous to picking up signals at stations along Baggeroer and Munk's 18,000-km *SOFAR* path mentioned above.

Researchers at Euratom in Karlsruhe (Tasman, 1979; Tasman *et al.*, 1982), at Sandia National Laboratory in Albuquerque (Carlson *et al.*, 1977), and at the Idaho National Engineering Laboratory in Idaho Falls (Arave *et al.*, 1978) improved upon the refractory wire sensor or electronic (Panatherm®) ultrasonic thermometer instruments that were manufactured at the time by Panametrics. These workers and their colleagues collectively demonstrated the hysteresis-, drift-, and attenuation-resistant properties of doped or thoriated W, the reduction in minimum zone length from 50 down to 10 mm, the avoidance of "sticking" of sensor to its sheath, and improvement in time resolution from 50 down to 5 ns. The time resolution improvement was achieved by using either higher-frequency pulses or more sophisticated timing methods than were contained in the Panatherm Model 5010 of that era.

Note in Section II.B.2.b and Figure 43 the recent (ca. 1997) use of a *bundle* of thin rods as a buffer for high-temperature *flow* measurements (Liu *et al.*, 1998). These rods are interrogated from a *piezoelectric* source at 500 kHz in the examples cited. At 500 kHz, "Mexican hat" or other broadband pulses can be timed to ± 2 ns by a variety of techniques available since 1995, if the *SNR* (signal-to-noise ratio) is high (e.g., ≥ 10). The wire sensor was extended from measuring temperature to measuring the curing of epoxy by Papadakis (1974) at $f \sim 100$ kHz, and to a study of the aging of rubber by Doyle (1996) at $f = 300$ kHz. (See also Kim (1989); Kim and Bau (1989); Kim *et al.* (1991); Nagy and Nayfeh (1996); Li and Menon (1998).)

3. *Tomographic Reconstruction*

Audible sound and ultrasound offer potential advantages over competing technologies in applications that require long-path averages and in temperature profile determinations based on a multizone path or tomographic reconstruction. The "sensor" may be gas, liquid, or solid (Figure 75). A number of multizone waveguide applications are reviewed in Lynnworth (1989). Many of these used a waveguide similar to that shown in Figure 76.

Since the mid-1980s, CAT scanning of the hot gas flowing through an exit plane of large combustors and boilers has been proven practical. One of the early proposals to use multichord scanning for sensing temperature distributions in hot solid bodies appeared as a diagram (but without details) in 1970. One of the earliest papers (known to the present authors) on acoustic tomography with application to flow is due to Johnson *et al.* (1975, 1977). See also, Johnson, 1979. One of Johnson *et al.*'s diagrams is reproduced in Lynnworth (1979, p. 497). The first large-scale published working example, due to Green (1985), is shown in Figure 77. If two transducers are installed along each wall of a *rectangular* duct, such that each transducer communicates with all the other transducers except the one on its own wall (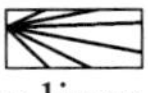 etc.), then there are 24 independent paths. More recent work along these lines is reported by Kleppe (1995a, 1995b, 1996), Basarab-Horwath and Dorozhevets (1994), and Rychagov and Ermert (1994, 1996). To overcome attenuation in the hot, sooty turbulent gas, Kleppe used high-intensity audible air blasts, sometimes ~1.5 kHz for paths of 10 to 20 m. Examples of Kleppe's work are reproduced in Figures 78–80.

D. Pressure

1. *Medium as Its Own Sensor*

In gases near ordinary atmospheric pressure, the pressure has very little influence on the speed of sound. However, fast adiabatic pressure variations are associated with temperature variations, making *pressure* pulsations accessible to ultrasonic measurement. At the previously described ultrasonic intake air mass meter (Section II.B.2.a), such measurements of fast pressure changes were performed successfully. Figure 81 shows the results of such a measurement of pulsating air flow in an ultrasonic air mass sensor compared to a conventional pressure sensor, while Figure 82(f) contains a plot of

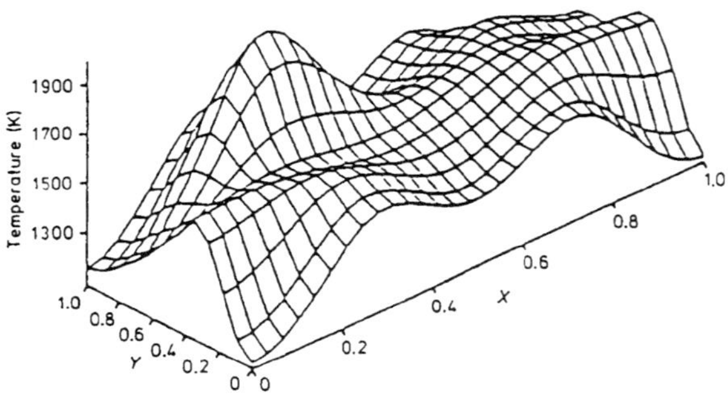

FIG. 77. Temperature distribution in the exit plane of a boiler, obtained by a tomographic reconstruction process. After Green (1985). © 1985 AIP and reproduced with permission.

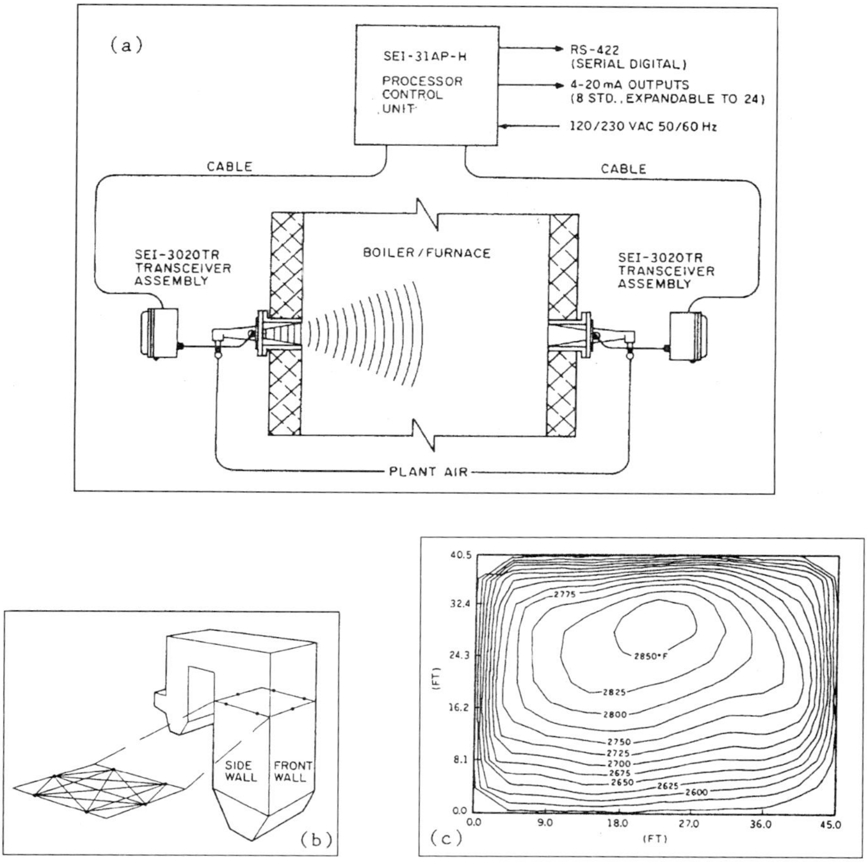

FIG. 78. (a) In BOILERWATCH® acoustic thermometry equipment manufactured by SEI for large boiler furnaces, the transmitted sound wave is generated by the air blast that also purges fly ash from the hollow waveguide. The broadband signal is in the midaudio band, < 10 kHz, low enough to be neither directional nor excessively attenuated. This means one transducer can communicate over long paths with several others. In addition, nondirectionality means these transducers, when applied to flow measurements in large flues, do not need to point at each other (Fig. 79). (b) Transducers placed around the furnace perimeter provide the multiplicity of paths needed for tomographic reconstruction. (c) Example of isotherms (°F) obtained for a 600-MW pulverized coal-fired boiler. The fireball of the combustion process is leaning toward or against one wall of the furnace. If not corrected, this condition can accelerate wall tube aging and can lead to tube failure. After Kleppe (1996).

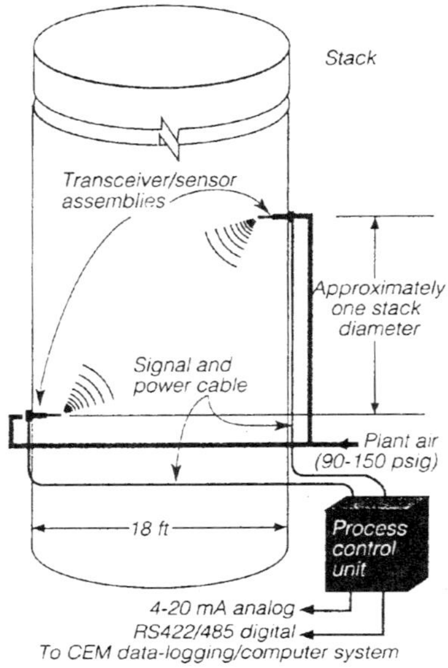

FIG. 79. SEI's STACKWATCH® flow monitor system uses a pair of air-blast sources spaced axially about one stack diameter, ~5.5 m in the example illustrated. The blast is generated by the momentary operation of a standard air valve. The start-to-stop transit times are obtained by cross-correlating noise signals picked up by transducers inside SS housings near each blaster. After Kleppe (1995a,b).

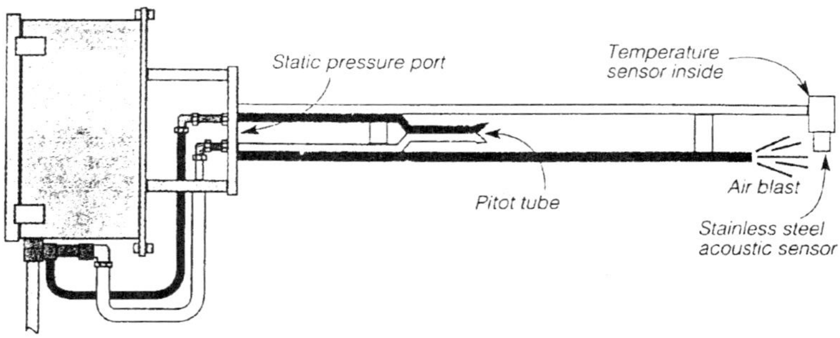

FIG. 80. Detail view of SEI's cem transducer probe, courtesy of its inventor John A. Kleppe, reveals *T* and *P* sensors, and pitot sensors that measure flow at two points not right at the wall but in a bit, so that combined with the theoretical zero velocity at the wall and the acoustic path average flow velocity, a better measure of flow profile and meter factor *K* can be derived. By measuring a reference *T* using thermocouples embedded in each transducer housing, the across-the-stack sound speed *c* can be related to average temperature independent of gas molecular weight. Also, if *T* is known, *c* can be related to average molecular weight to the extent that γ, the ratio of specific heats, is known. After Kleppe (1995).

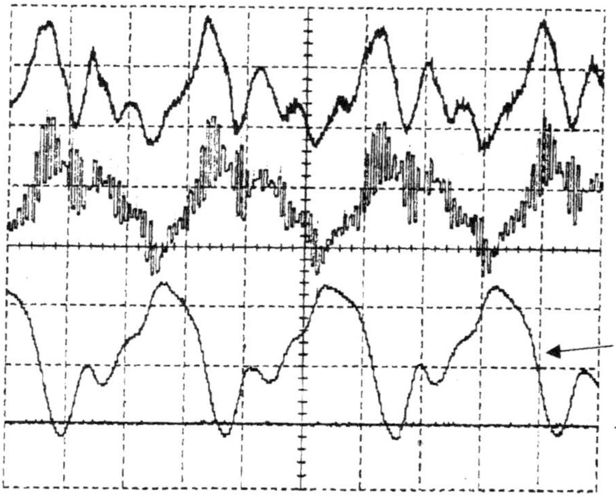

FIG. 81. Ultrasonic measurement of high-speed pressure variations compared to a conventional pressure sensor measurement. Measurement setup as in Figure 40.

amplitude of the received signal vs *steady* air pressure, obtained in laboratory tests under no-flow conditions.

In gases at *high* pressure (hundreds of bars), *c* increases by some 50% or more (Carey *et al.*, 1969). If composition and temperature were well known, this might provide a way to estimate or measure gas pressure based on *c*. The T-mode flowmeter vee-path geometry, already shown to be usable in some liquid level sensing applications, might be used on cylinders of compressed gas as a clamp-on pressure sensor. (A pulse-echo arrangement appears in Lynnworth, 1989, p. 181, Figure 3–43.) In the case where one seeks a clamp-on way to *estimate* gas pressure, the *amplitude* of the received signal, rather than transit time, might be the more useful parameter to measure. This is because the gas attenuation coefficient α and the acoustic impedance Z change so much more, on a fractional basis, than does the sound speed. The transmission coefficients into and out of the gas, and the attenuation along the path, must respond to gas pressure P. The amplitude jitter may also provide an indication of gas pressure, not in the sense of measuring pulsating flow (see above) but responsive to turbulence which in some experiments appears to depend in part on gas pressure. If P were in the range of a few tenths of a bar up to ten bar, and possibly to much higher pressure, and if flow velocity were low enough so turbulence would not be a significant source of excess attenuation, then it might be practical to determine or estimate P from

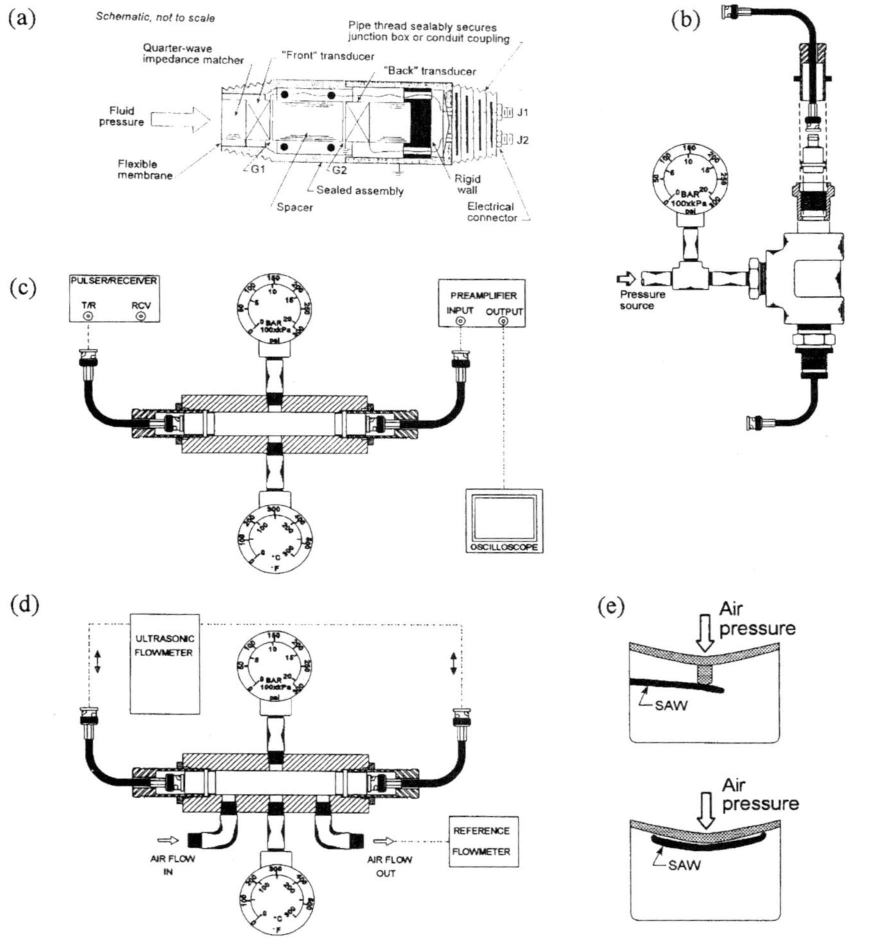

FIG. 82. Laboratory investigations of ultrasonic measurement of gas pressure using standard or modified flowmeter transducers or *air-coupled* transducers, mainly based on changes in the sound pressure transmission coefficient. (a) Concept based on pressure coupling data such as that in Crecraft (1964), proposed by Lynnworth (1996). (b) Test of T7 air transducers shown in Fig. 15, with each transducer mounted in a one-inch pipe plug. (c) Test similar to (b) except: one-inch pipe plugs are eliminated; T7 transducers are mounted directly to each end of the air (or gas) path; path is increased from ~ 50 mm to 100 mm; gas temperature T is measured near middle of gas path. (Cell design by C. D. Smart.) (d) Conversion of pressure cell (c) into a flowcell to investigate effect of flow on amplitude measurement and to find out if a simple offset flowcell with only two transducers can yield the flow velocity V of a known gas, gas temperature T from sound speed c, and gas pressure P from the amplitude A of the received signal. (e) Experiment using SAW sensor to measure air pressure in a tire. Top: pressure bends the cantilevered SAW sensor. Bottom: SAW sensor is fitted to the membrane. Membrane and SAW curvature exaggerated. Diagrams after Pohl *et al.* (1997).

(f)

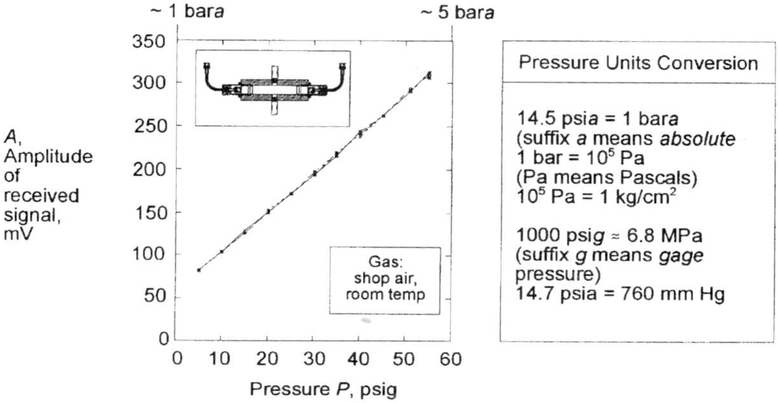

FIG. 82. (*continued*) (f) Laboratory test data at $f = 100$ kHz, above atmospheric pressure but without flow. Tests used the cell shown in (c).

the amplitude of the received signal. Alternatively, one might be able to use an echo off a clean reference reflector placed not in the freestream. Pressure coupling, studied long ago by several investigators (e.g., Crecraft, 1964), also might be used as a basis for constructing a pressure sensor (Figure 82(a)). But these ideas have not yet been reduced to a practical device. Perhaps they will be of interest to some readers as subjects for research. For precise measurements of pressure, the only known commercially available ultrasonic solutions at present use intrusive sensors, e.g., the products discussed in the next section.

2. *Intrusive Sensors*

Practical devices for sensing P so far have taken the form of a quartz tuning fork resonator, a quartz crystal resonator, or a SAW device where the path is loaded or stressed. The load or stress imparted to the device is converted to a resonant frequency, or a change in transit time, for a sensitive and wide-ranging means of sensing pressure. Examples of devices in this category include products made by Quartztronics (illustrated in EerNisse *et al.*, 1988 and in Lynnworth, 1996, p. 535) and by Paroscientific (illustrated in Figure 83).

Paroscientific's Digiquartz® quartz crystal resonators vary in resonant frequency as a function of pressure. A second quartz resonator provides a temperature compensation signal. Pressures range up to 40,000 psi, or ~270 MPa. Background technical information is available in Paros and Wearn (1988) and Busse (1987).

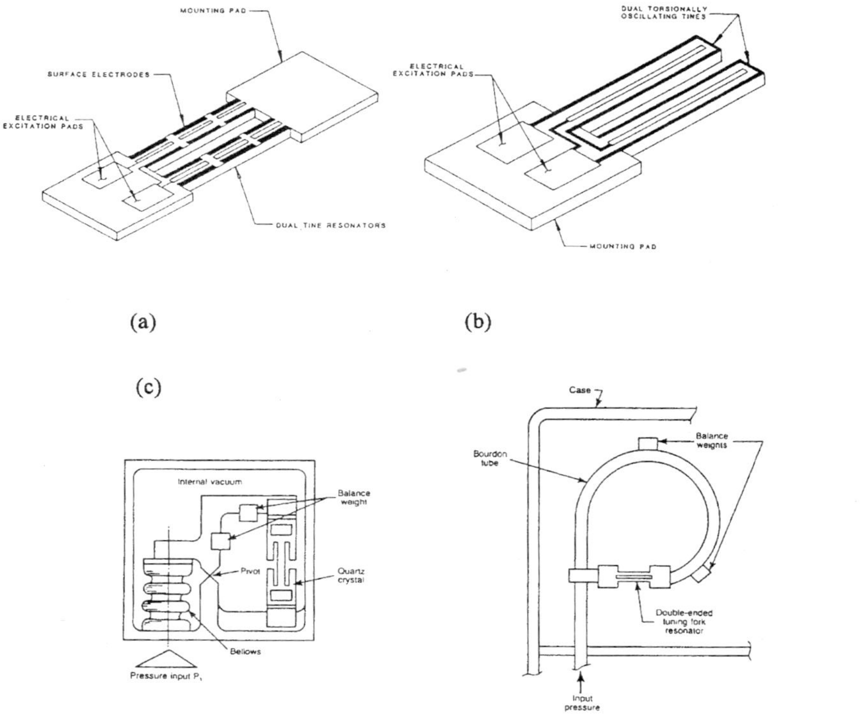

FIG. 83. Quartz-based pressure sensors from Paroscientific. (a) Double-ended tuning fork resonator. (b) Temperature-sensitive quartz resonator. (c) Bellows or Bourdon tubes convert input pressure to an axial force applied to the crystal resonator.

A wireless-linked SAW pressure sensor is described in Buff *et al.* (1997), and a *nonintrusive* SAW pressure sensor prototype that attaches to the valve of a car's tire is reported by Pohl *et al.* (1997). This prototype (experimental device) comes close to being a clamp-on pressure sensor, as far as the valve is concerned, but it is not a clamp-on in the usual sense of being attachable without changing the pressure boundary.

III. Analyzer Applications

Analyzer applications of sensors for moisture, viscosity, color, opacity, etc. are obviously in use for on-line closed-loop process control, not just for off-line analysis. In this section the emphasis is on measurands other than temperature, pressure, flow, and level. Applications are limited to *c*-based (sound-speed-based) equipment for gases and, separately, for liquids. The analyzed parameter, e.g., a density-related term, may enhance a process control measurement, as will be illustrated with the flare gas flowmeter.

A. Concentration Measurements in Gases

At a given temperature and pressure, the speed of sound c in an ideal gas equals $(\gamma RT/MW)^{1/2}$ where γ is the ratio of specific heats, R is the gas constant, T is the absolute temperature, and MW is the molecular weight. That is, c is inversely proportional to the square root of molecular weight. In a mixture of two ideal gases, the sound speed "averages" the two molecular weights according to their relative concentrations (Valdes and Cadet, 1991; Stagg *et al.*, 1992). This is the basis for binary gas analyzers. The method also applies to pseudobinary gas mixtures, where one of the gases is a known mixture of two or more gases, e.g., air. If pressure is not too high and is known in a gas, it is usually an easy matter to go from average molecular weight to gas density. As a numerical example for ambient air, $MW \approx 29$ and density ≈ 1.3 grams/liter $= 1.3\ \mathrm{kg/m^3}$ (Figure 23).

1. *Binary Gas Mixtures*

Ultrasonic flowmeters, having a c output, can be adapted to measuring the concentration of binary gas mixtures, as in the work reported by Valdes and Cadet (1991). The ratio of specific heats, γ, also plays a role in determining c. In fact, in some early studies (ca. 1934) c was measured to determine the specific heat or γ. More often, however, binary gas analyzers are not adaptations of ultrasonic flowmeters but are designed expressly for analyzer purposes, as for example in the products of Tracor or Thomas Swan & Co. Ltd. A cell manufactured by Tracor is illustrated in Lynnworth (1989, p. 577). An up-to-date example of a binary gas analyzer available from Thomas Swan is shown in Figure 84. By operating the cell at a frequency f at or near 1 MHz, the cell can be quite compact. This is a very desirable feature in many semiconductor applications, especially where flows are low and fast response is desired.

Among the dilemmas facing the designer of such cells is the need for the cell to operate reliably despite the gas pressure being at $\frac{1}{2}$ bar or even $\frac{1}{4}$ bar (in some applications), which suggests lower frequencies, coupled with the need for small volume and high resolution, which suggests higher frequencies. The U-shaped binary gas analyzer cell shown in Lynnworth *et al.* (1997a, p. 1095)

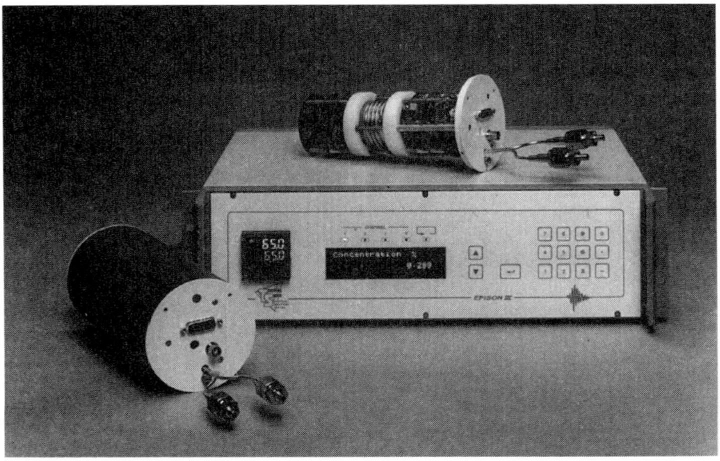

FIG. 84. Epison gas phase reagent concentration monitor, showing control unit and measurement cell (with and without cover removed). Illustration courtesy of Thomas Swan & Co. Ltd.

was used at $f = 100$ kHz by Reinoso (1996). The electronics in Reinoso's application was a four-channel flowmeter adapted to measuring sound speed c in up to four cells.

2. *Flare Gas as Example of a Multicomponent Gas*

According to simple theory one would *not* expect c to bear a useful and unique relationship to *MW* when there are many components present. This negative prediction is related to the apparently large fractional uncertainty in γ. For example, referring to Figure 23, if $\gamma = 1.333 \pm 0.333$, then $(\Delta\gamma)/\gamma = \pm 25\%$. However, in petrochemical flare lines, the hydrocarbon gases are often related in such a way that the ratio of specific heats, γ, is *not* independent of *MW*. This means one can find a unique relationship between the average *MW* and c. Accordingly, accuracies in *MW* on the order of 2% are attainable for *MW* from 2 to 58. This is ten times better than the pessimistic prediction. Complications arise, however, if the flare is purged with an *unknown* amount of nitrogen, because nitrogen's $\gamma = 1.4$, unlike that of the flare gases. If the nitrogen concentration can be estimated, or if the nitrogen flowrate into the flare line can be measured, then the algorithm can be adjusted to take the nitrogen concentration or the nitrogen estimate into account.

The measurement of sound speed as well as flow velocity was introduced very early in the prototype flare gas flowmeters developed between about 1982 and 1984 by Panametrics under a project initiated by Exxon and reported by Smalling *et al.* (1984). Smalling extracted samples of flare gas and compared sound speed with the average molecular weight of the sampled gas, after compensating for temperature. The relationship is graphed in Figure 85(c) (Smalling *et al.*, 1988). The empirical relationship, corrected for temperature *and* pressure, is used to determine flare gas *mass flow rate* to $\pm 2\%$ for MW from 2 to 58. These results are typically obtained using a gas path of 275 mm and a frequency f of 100 kHz. The mass flow rate (M_F) output is useful to plant operators for checking energy and material balances and for controlling the amount of steam sent to the flarestack tip to inspire air for complete and smokefree combustion. In the flare gas example, the flow velocity (V) output or a volumetric (Q) output generally does not suffice. V and Q do *not* meet the needs of the customer, who is the flare gas operator. The need is for *mass* flow rate. The MW determination by itself is also useful, for it indicates where in the refinery or chemical plant the major sources of flare gases are coming from, e.g., the possibility of a hydrogen valve stuck open, if $MW = 2$. (In instances where this hypothetical example actually occurred, the flare gas flowmeter paid for itself in one day—much more quickly than the six month payback period normally estimated for this product.)

Flare gas flowmeter applications generally require the transducers to be installed by a hot tapping procedure, although spoolpieces were sometimes provided. Examples of flare gas flowmeter equipment appear in Figure 85. Over one thousand such systems were installed in the period 1984–1997. In many ways this flare gas equipment, including improvements in the late 1980s, became the basis for subsequent gas flowmeters manufactured by Panametrics, e.g., the CEM68 introduced in the early 1990s for continuous emissions monitoring of gases exiting up smokestacks, the GP68 general purpose gas flowmeter, and natural gas and steam flowmeters that are currently under development. Gas conditions, the dimensions of the conduit, and other constraints dictate the operating frequencies, transducer details, signal processing algorithms, and other design parameters.

B. Concentration Measurements in Liquids

Graphs showing the dependence of sound speed c on concentration or density of a mixture of two liquids appear in many texts (e.g., Babikov, 1960; Bhatia,

(a)

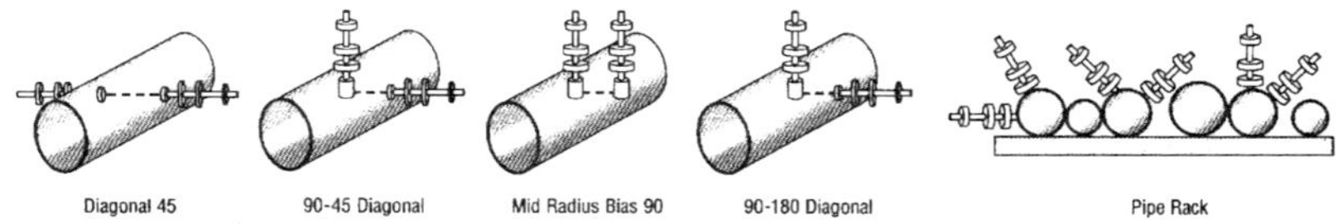

(b)

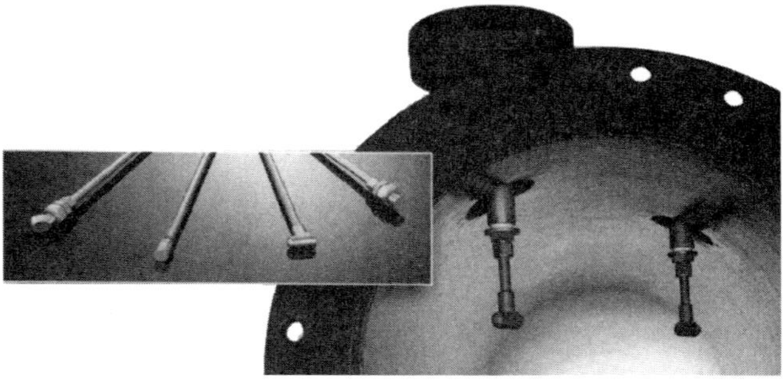

(c)

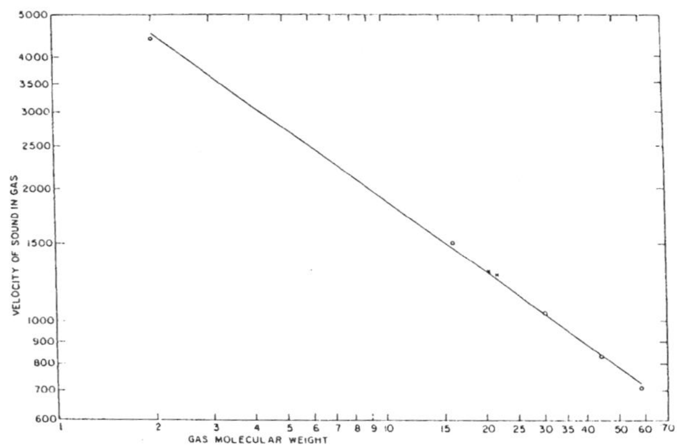

FIG. 85. Flare gas equipment from Panametrics. (a) Standard transducer mounting configurations. (b) Close-up view of transducers that launch 100-kHz pulses typically at 0°, 45°, or 90° to the axis of the long tubular portion, combined with a cutaway view showing "bias-90" transducers communicating over a gas path of ~350-mm within a spoolpiece. the bias-90 configuration is often installed via hot-tapped ports. (c) Empirically determined relationship between flare gas sound speed c (units: ft per sec) and average MW (molecular weight), at one temperature T. This relationship underlies the ultrasonic determination of average molecular weight to about 2% of reading, for $MW = 2$ to 58. This assists in locating the source of leaks into a flare system, energy balancing, steam control to inspire air and mix at the flare tip to reduce black smoke (a sign of incomplete combustion). Details in Smalling *et al.* (1984, 1986, 1988) or Lynnworth (1989).

1967). In Lynnworth (1989, p. 236), data are reproduced for various electrolytic solutions at 25°C. A graph of *c* vs density ρ for water appears on p. 424 therein. Examples for aqueous solutions, mixtures of hydrocarbons, and other combinations may be found in the literature. As mentioned previously, sometimes attenuation can sort out the concentration, especially in two-phase mixtures; in some cases the attenuation can resolve ambiguities when a ternary mixture is the problem at hand (Babikov, 1960).

1. Mole Fraction Analysis of Heavy Water

The nuclear industry is required to measure the D_2O/H_2O ratio, or mole fraction of heavy water in the field, in factories or in warehouses inside sealed containers. A clamp-on (nonintrusive, noninvasive) pulse-echo technique was developed during a collaboration between M. S. Zucker of Brookhaven National Laboratory and Panametrics, wherein the speed of sound was found to be a reliable indicator of the sought parameter. In other words, when light and heavy water are mixed, *c* depends on their relative concentrations. In tests comparing the speed of sound determination of the mole fraction with the value determined from a mass spectrometer, agreement was found to be within ±0.002 mole fraction units. The portable ultrasonic instrument was designated the Model 5246 Mole Fraction Gage. The operator's obligation, besides coupling the one transducer to the container, is to enter the container diameter (ID) and the temperature. The instrument then displays the mole fraction of deuterium oxide.

2. Pipeline Interface Detector

The concentration analyzer pioneered by Zacharias of NuSonics is obviously similar in principle to that company's pipeline interface detector (Figure 86a–c). Details differ in part according to the installation of such *c*-based analyzers being in a pipeline, in a laboratory, or in a chemical plant environment.

Some twenty-eight years ago, Zacharias (1970) found that different gasoline products had sufficiently different sound speeds, so that they could be detected as they crossed a measuring sensor installed in a pipeline. The sensor could stick into the fluid stream, or it could reside in a recess (nozzle) so that the assembly, including its reflector, would not be in the way when the line is pigged. [The recess may be purged, as shown in a Mapco design in Lynnworth (1979, p. 432, Figure 13(g)).] Zacharias' high-precision *T*-compensated intervalometer led to a pipeline interface detector having applications in many locations, where different grades pass through the same point at different

(a)

(b)

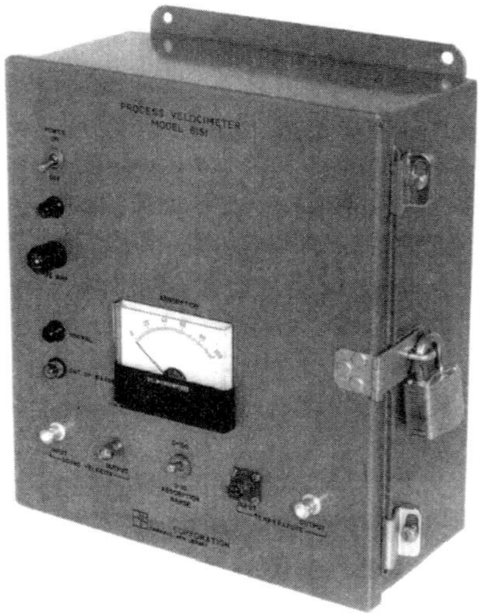

FIG. 86. Pipeline interface detector sensor (a) and early concentration analyzer electronics (b), courtesy Ellis M. Zacharias, Jr. The probe includes an RTD for accurate temperature measurement and a reflector for precise control of the path used for *c* measurement. Application examples are given in the papers by Zacharias and colleagues at NuSonics in the 1970s. (A NuSonics corner reflector is shown in Lynnworth, 1989, p. 48). Some twenty-eight years later the pipeline interface detector takes the form illustrated in (c) for Model 86 PID, manufactured by NuSonics Division of Mesa Laboratories. This instrument detects gasoline–gasoline interfaces, and other liquid hydrocarbons including crude oils, at several hundred sites throughout the world.

(c)

FIG. 86. *(Continued)*

times and need to be steered to appropriate routes for further storage or processing. Equipment of this type has been available at various times from Mapco, Mesa Labs, or NuSonics. A graph of *c* vs density for various petroleum products, due to Zacharias and Ord (1981), appears in Lynnworth (1989, p. 425). The sound speed ranges from about 750 to 1400 m/s, with propane being at the low-*c* end and fuel at the high-*c* end. Gasolines fall a bit above the middle, with *c* around 1150 m/s.

IV. Contactless (Wireless) Ultrasonic Sensors Including Remote SAW Sensors

Contactless sensing can be a special case of noninvasive sensing. The objective is to infer the characteristics of a remote object or remote region, without a physical connection to that object or region (Mágori, 1993a and 1993b). Two kinds of solutions are presented. One is where the link is acoustic/ultrasonic, i.e., using airborne or waterborne ultrasound. (See also Section II.A.) The other employs a wireless electromagnetic link to a remote sensor. In some applications at sea, the system includes both kinds of contactless solutions, e.g., an electromagnetic link to a satellite and an acoustic link to a robotic device on the sea bed (Figure 97). (Compare with NDT immersion testing, e.g., Figure 4(a), or Chapter 3 by Papadakis in this volume.)

The wireless electromagnetically linked applications to be described are presented mainly with reference to SAW sensors. (Noncontact emat [electromagnetic acoustic transducers] are described in Igarashi *et al.* (1997).) Some wireless-linked non-SAW sensors are: triaxial seismic sensor (Figure 95), downward-looking sonar (Figure 96), a combination of long-distance wireless and acoustic links (Figure 97), and a laser used to probe a surface to "see" the ultrasonic wave (Figure 98).

A. Readout with Airborne Ultrasound

An example where the link is airborne *acoustic* is represented by the measurement of paper web speed (Jarrti and Luukkala, 1977), reproduced in Lynnworth, 1979, p. 442). In that work a Lamb or plate wave was launched in the paper web by air-coupled ultrasound, and the observed Doppler shift was attributed to reradiation from a moving source. A more recent example of an air-coupled measurement to a moving source is the remote measurement of vibration (e.g., Bou Matar *et al.*, 1996, 1997). Here the explanation requires that a nonlinear parametric effect be included. Contactless sensing of the level of a liquid or solids by downward-looking ultrasound could also be put into this category. Note that if the level of interest lies within a closed tank, it is usually necessary to utilize an existing manhole or equivalent opening at the top (or to make such an opening) to accommodate the noncontact transducer. This might or might not be considered noninvasive.

According to an article in *Sensors* (Rosa, 1991), ultrasonics appears to be in wide use in woodworking and sawmill operations. Board sizing, in this 1991 example, is controlled with an ultrasonic measurement and control system manufactured by Massa, reducing waste and yielding the butcher block of the desired size, according to an "Opti-Sizer" design concept by Taylor Manufacturing (Figure 19(e)). (Compare with Figure 11, and with Figure 16 and the associated text on the parking garage sensor.) A bowling alley pinsetter system based on Massa's determination of which pins remain standing is described in Lynnworth (1989, pp. 609–611), based on an earlier Massa publication in *Sensors*.

Recent issues of *Sensors* (e.g., the July 1997 issue) and their annual *Buyer's Guide* contain numerous advertisements, or offers for product literature, for air-coupled sensors. Applications mentioned therein for air-coupled sensors include proximity, sort/select, wind/unwind, motion control, measuring the diameter of rolls of sheet material, and web control. For technical articles on air transducers, the reader is referred to *IEEE Trans. UFFC*, *Proc. Ultras.*

Symp. (e.g., Hayward, 1997; Ladabaum *et al.*, 1997), *Proc. UI '95*, *Proc. UI '97*, *Ultrasonics*, *J. Phys. E*, or *J. Acoust. Soc. Am.* as starting points. Four examples of recent air-coupled transducers currently at the laboratory stage (reported in the July 1997 issue of *IEEE Trans. UFFC*) are the miniature hollow sphere transducers (BBs) of Alkoy *et al.* (1997), the cymbal transducer of Tressler and Newnham (1997), ferroelectric nylon materials (Brown *et al.*, 1997), and the thick-film composite of De Cicco *et al.* (1997). Evidently, there is much sensing research activity for the development and use of such transducers. The Polaroid air transducer (included among the several Polaroid transducers comprising Figure 19(i)) is an example of a device in use in the millions, not only in its original target market as an echo ranger for an autofocusing camera, but also for air-ranging the dimensions of rooms and numerous other inexpensive applications involving time of flight in air.

Most R&D on air transducers understandably concentrates on *low* acoustic impedance sources. (Exceptions: Sections II.A.2 and II.A.3.) The possibility of using a *high* acoustic impedance device for air-coupled ultrasound is investigated in Lynnworth *et al.* (1997a, 1998b). This device, Figure 15, is derived in part from applications in cem (continuous emissions monitoring), which involve high temperatures and corrosive gases, and binary gas applications where the pressures might suddenly increase to 30 bar (3 MPa). It turns out that the *high* acoustic impedance source, tracing its ancestry to high-temperature and high-pressure devices, offers some advantages with respect to ruggedness and the ability to be clamped or coupled to the outside of plastic pipe or plastic windows in wind tunnels. This means a clamp-on ("air") transducer can be considered practical in particular circumstances, e.g., monitoring the vortex shedding frequency or studying the wake structure in a wind tunnel. (Manufacturers of *ultrasonic vortex shedding flowmeters* include J-Tec and Yokogawa. The principles involved may be found in Miller, 1996, chapter 14, p. 14.14–14.18, or in Lynnworth, 1989, pp. 326–334.)

When scanning large areas such as boards in motion, usually one would like to obtain a representative reading, some type of average over the area. In other cases, one seeks detailed information from edge to edge. For either objective an array of many transxucers might be required. Because of the short time available in which to make a determination of product dimensions or internal qualities and properties, and the cost associated with each electronic channel of measurement, there is motivation to sense along many paths as simultaneously as practical, with the least number of electrical channels. One solution to this kind of problem is to employ a sheet or film transducer, e.g., PVDF or perhaps an evolving piezoelectric material like ferroelectric nylon (Brown *et al.*, 1997). By depositing electrodes over different regions, a multielement transducer is created.

Another possible solution is to differentially space pairs of discrete transducers, so that even if N are excited simultaneously, the received signals will be time-separated. This method is illustrated for a multigap liquid presence detector due to Van Valkenburg and Sansom (1959), reproduced in Lynnworth (1989, p. 493). An NDE version of this, with ten parallel measurements of delaminations, appears in Figure 4(a), adapted from Krautkrämer and Krautkrämer (1977, p. 415). In NDE applications, broadband transducers have been developed with damping behind the piezoelement, so that the assembly rings down in a cycle or two. However, if the transducer must withstand high pressure and be corrosion-resistant, the surrounding materials and backing materials often introduce ringing. At high pressures, then, a potential limitation of this form of N-banger method is that if the transducers ring too long, the space (path) differential must be large. A partial remedy is to space the transducers in the frequency domain too. If the transducers were narrowband, with different pairs having different resonant frequencies, then a chirp or frequency-hopping driver would excite different pairs, one pair at a time. On the other hand, if the transducers were sufficiently broadband, they could all be identical and be driven with different codes. This could satisfy simultaneity but would require one channel per coded path. This gives the designer three ways to separate signals launched over different paths: time, frequency, code. The problem presented here, for rapidly scanning a large area to obtain an area average or perhaps to obtain detailed information on the distribution of characteristics, resembles analogous problems in flow or in acoustic tomography for determining temperature in dynamic systems such as large combustors (Kleppe, 1989, 1995a, 1995b, 1996). There too the object is to obtain the correct area average. If multipath measurements are utilized, experience apparently teaches that the various paths should be interrogated as simultaneously as practical.

Air-coupled resonant ultrasonic inspection of artificial defects in a fiber-reinforced thermoplastic composite plate is discussed in Schindel *et al.* (1996). Air-coupled transmission through aluminum is discussed in Ladabaum *et al.* (1997). For a recent general perspective on constraints and solutions for industrial implementation, see Hayward (1997).

One U.S. manufacturer of ultrasonic air-coupled dimensional gaging equipment is Ultrasonic Arrays. One of their systems, the BMS-1000 bond measurement system (Figure 19), is designed to monitor the internal integrity of composite products. Its specification includes detection of a defect 50-mm wide by 50- to 200-mm long at product speeds of 0.5 to 2 m/s.

Air-coupled NDE systems and results are found in the ultrasonic literature dating back at least twenty-five years, e.g., aircraft tire inspection (Van Valkenburg, 1973). A recent application under development at the Dow Chemical Company involves an air-coupled determination of the compressive strength of cellular polymers. This is mentioned in Lynnworth *et al.* (1997a), and is based on an air-coupled measurement of sound speed through the board that is in motion. The principle of measurement is similar to the substitution method for plastics submersible in water (Zacharias *et al.*, 1974). Among the transducers being evaluated in this application is the T7 O-ring isolated design shown in Figure 15. This type of transducer has been used in some laboratory studies of the pressure dependence of sound transmission (Figure 82), the data therein (Figure 82(f)) having been obtained after hydrostatically testing the transducers to 1000 psig (70 bar). Referring again to Figure 19, however, and applications in air at ordinary pressure, that figure shows many examples of air transducers, e.g., Delta Control Corporations' downward-looking air sonar for measuring liquid level at a Parshall, V-notch, Cipoletti, or Palmer-Bowlus weir or flume. See also the literature of Badger Meter, Endress + Hauser or other suppliers of such equipment listed in the April 1997 or April 1998 issue of *Measurements & Control.* Polaroid's air transducers include models such as their "K" series, $f = 40$, 120, and 200 kHz; "L" series at 40 kHz; water-resistant 9000 series at 45 kHz, and electrostatics at 40 to 100 kHz.

B. Surface Acoustic Wave (SAW) Sensors

Surface acoustic waves (SAWs) can be excited by electric signals on the surface of piezoelectric materials (quartz or lithium niobate ($LiNbO_3$) or on solids covered by a thin piezoelectric film. One employs thin (0.1 μm) metallic electrode stripes that can be deposited on the surface by evaporation. The patterning of this metal layer is performed by a photolithographic process, e.g., by the "lift-off" process. Typical electrode structures that serve to transform electric signals into surface waves or vice versa are the so-called interdigital transducers, which consist of two systems of interconnected electrode stripes like the stretched fingers of two hands penetrating into each other without touching. Figure 87 shows a scanning electron microscope picture of two surface wave packets on the surface of a crystal.

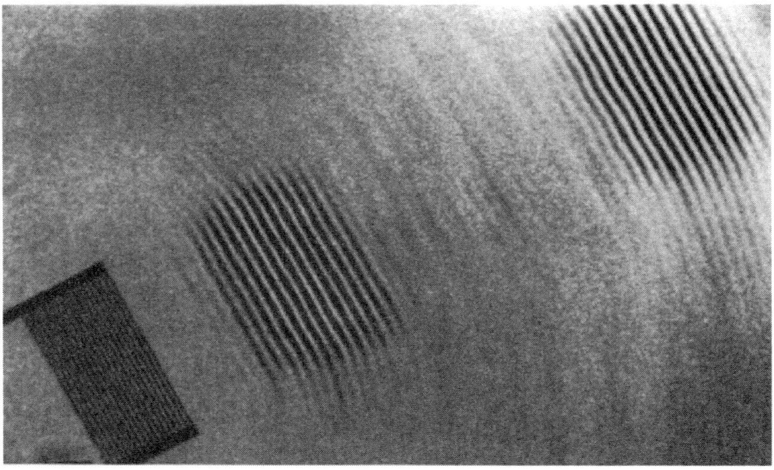

FIG. 87. Emission of SAW bursts on a piezoelectric crystal surface. Scanning electron microscopy picture courtesy of Siemens.

The propagation velocity of approximately 3500 m/s and the frequency range of 10 MHz to a few GHz lead to microscopic wavelengths and penetration depths of 35 μm, for example, at 100 MHz. The frequency range is limited at the high end by the technologically achievable resolution in fabricating the electrode structure. A large number of different high-quality components are made possible thereby with tailor-made long-term transmission characteristics (Ruppel *et al.*, 1993). The photolithographic process permits the deposition of many different electrode structures at the surface with high precision. The selection of the crystal and its orientation as well as the form of the guiding structure have a bearing on the wave's characteristics. These determine in a highly reproducible way the transmission characteristics of the electrical signals that are then transferred to the SAW. In addition to these, filters, resonators, correlators, and so on are produced in huge quantities nowadays, and new SAW applications continue to be introduced in the field of sensors and accepted in the sensor market (Bulst and Ruppel, 1994), as cited previously in the March 1987 issue of *IEEE Trans. UFFC.* (See also Ballantine *et al.*, 1996 and the Sensors and Actuators website of the IEEE UFFC Sensors and Actuators section, currently hosted by the Sensor Technology Laboratory of the University of Maine (R. M. Lec., 1997, priv. comm.).)

The various sensoric principles that are usable for SAW sensors are (Mágori, 1993b):

- Alterations of the bulk parameters by temperature and the introduction of different forms of mechanical stress to the crystal (e.g., temperature or acceleration sensors).
- Alterations of the surface parameters at the wave-bearing surface portions by deposition of specific layers whose mass distribution, elastic properties, or complex electrical conductivity are changed by the measurand (e.g., moisture or chemosensors).
- Connection of electrodes to sensor elements, whose impedance depends on the sensoric influence (e.g., photosensor by connecting a photoresistor to an electrode).

SAW sensors possess an extremely high sensitivity to the slightest change in mass on their surface, which strongly increases with increasing operational frequency. Thus, sensitive SAW chemical sensors can be made by the deposition of thin films, e.g., on SAW resonators that change in a specific way under the influence of chemicals. These films can be tailor-made to adsorb selectively certain gas molecules according to their concentration and thereby change the mass on the surface of the resonator. This manifests itself as a frequency change of an oscillator circuit, which utilizes the SAW resonator as its frequency-determining element, compared to an identical reference oscillator using a reference SAW resonator, which is mounted on the same substrate but which does not have the evaporated adsorbing film (Dickert *et al.*, 1990). By mixing both oscillators' output frequencies, a low frequency beat signal results whose frequency is proportional to the measurand's influence without being dependent on temperature (Venema *et al.*, 1987). Similarly, oscillator circuits have been built incorporating SAW resonators that become detuned under other external influences such as temperature, mechanical stress (Figure 88), and so on. Thus the oscillator utilizes this frequency difference to become a sensitive sensor—for example, a torque sensor through the proper application of a shear force acting on the two resonators in opposite directions. The sensor operates on the rotating shaft without slip-rings (Baldauf and Schrüfer, 1992). A comprehensive overview of SAW sensors and related efforts is given by Fischerauer *et al.* (1995). See also Chapter 3 by F. Hickernell in Volume B of this book.

Nevertheless, in many cases it is still hard to find good reasons for replacing existing proven sensors with SAW sensors. A very strong argument,

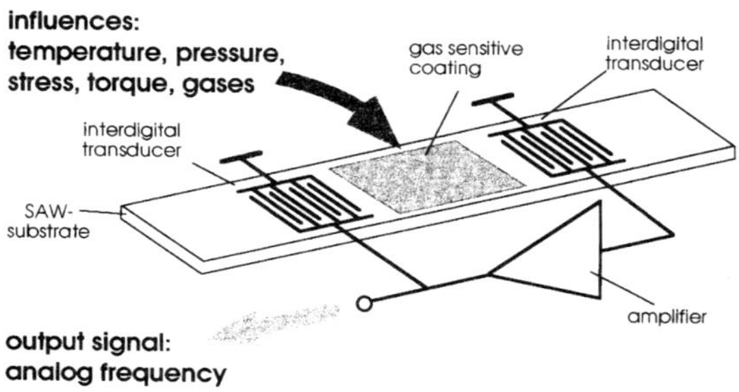

FIG. 88. Principle of a frequency analog SAW sensor.

however, is the possibility of designing a SAW sensor whose information can be read out from a distance by radio waves without a wire connection or any other material link to the passive sensor element and without a dedicated power supply (battery) on the sensor. This possibility is discussed next.

C. WIRELESS IDENTIFICATION OF REMOTE FAST-MOVING OBJECTS

SAW components can be used in sensor systems as passive identification tags, so called *SAW ID tags*, for remote identification. Such systems are already in use at turnpike toll stations in Norway and the subway railway system in Munich for the automatic identification of vehicles. The cars that carry SAW ID tags are recognized at a distance of many meters by an interrogator transmitting UHF radio pulses, which receives and evaluates echoes coming from the SAW components (Mágori, 1993a, 1993b).

Figure 89 shows schematically such a SAW ID-tag. The radar pulses are picked up by a simple antenna attached to the tag and transformed into surface acoustic waves. The electrodes that serve as reflectors are positioned in a characteristic sequence (similar to a bar code) and reflect these waves. The reflected waves are transformed back into radar pulses and evaluated by the receiver. The specific arrangement of the reflectors is manifested in the return signal as a characteristic pattern of partial pulses that allow, for example, 4×10^9 different combinations with 32 bits. Figure 90(a) shows the typical response signal of an ID tag, a chain of partial bursts with different mutual spacing.

Instead of this multiple tap reflective delay line, a bank of SAW resonators on the ID tag, whose existing resonant frequencies are used as a code, could be connected to the antenna. After the excitation by an interrogator signal, the

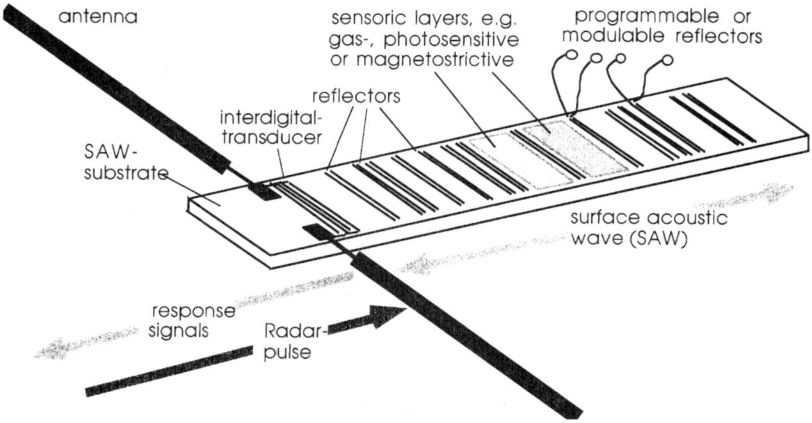

FIG. 89. Schematic of a remote readout SAW ID tag SAW sensor system.

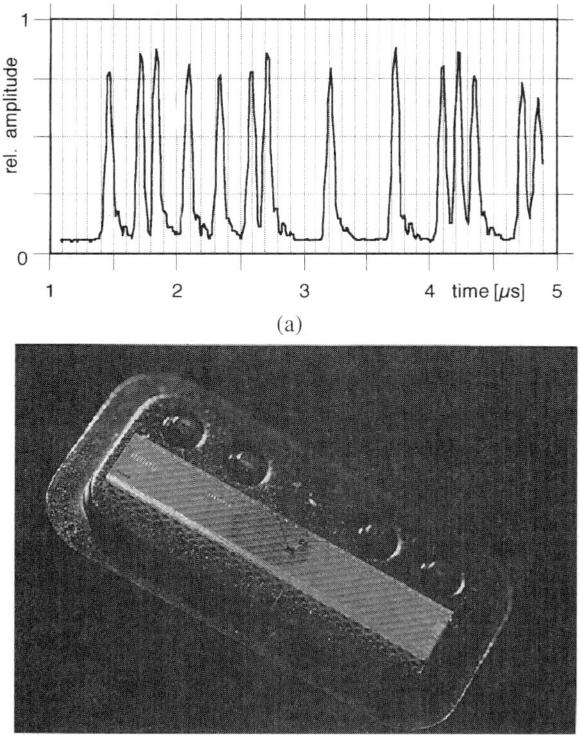

(a)

(b)

FIG. 90. (a) Typical response signal of a SAW ID tag. (b) SAW ID tag for 2.45 GHz operation. Illustrations courtesy of Siemens.

amplitude of the resonators will decay and an associated signal, containing the resonance frequencies of the resonators, will be sent back to the interrogator. A disadvantage, however, is that the narrowband partial signals will be sensitive to interferences by multipath propagation of the interrogator and response signal.

A photograph of a SAW ID tag is shown in Figure 90(b). Designed for 2.45 GHz operation, this device is made on a lithium niobate crystal, which is mounted in a standard metal housing (cover removed). A SAW transducer, contacted by bond wires, can be seen in the middle of the crystal. At both sides of this transducer, which is intended for bidirectional SAW radiation, reflectors are positioned on the surface.

In addition to the mere recognition of the existence of partial signals, their mutual position can be measured with high accuracy. By the influences of various physical parameters, as shown in Section IV.B, the mutual positions of the partial signals are changed, and the changes can be recognized from a distance by the interrogator without a material link. Since the mutual positions become evaluated, the propagation of the radio wave between interrogator and the remote readout sensor element is eliminated ("reference on chip"). Figure 91 shows the architecture of this type of remote readout SAW sensor system.

By this method, compared to the sensor possibilities as described in Section IV.B, two additional sensor principles are utilized:

- Remote recognition of patterns, given by the distribution of the electrodes on the surface.
- Evaluation of the propagation parameters of the electromagnetic wave

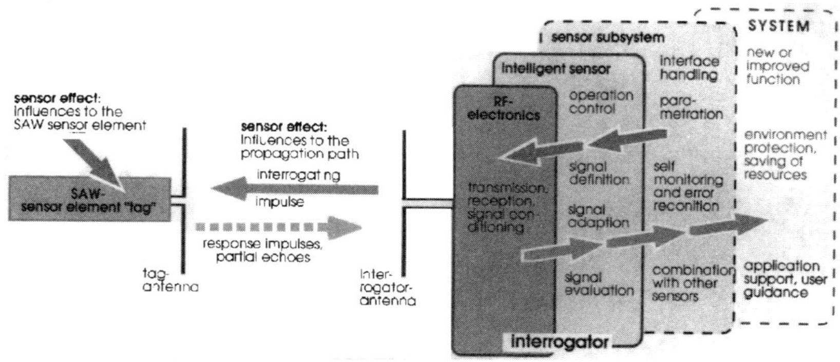

FIG. 91. Remote readout SAW sensor system architecture.

between the interrogator signal emission and the echoes from the interrogated SAW devices (transit time, Doppler-shift, attenuation, polarization, etc.)

The advantages of these remote readout ID tags are:

- Remote interrogation through windows and walls at inaccessible locations, e.g., dangerous rooms or other hazardous areas
- Passive operation without electrical power supply on the ID tag (no battery required, saves costs, saves maintenance, and is safe)
- Small, light-weight, easy to mount
- Defined range of interrogation; interference by echoes from other objects are eliminated by the short transit time
- High reliability, uses stable crystals

Because of these advantages, SAW ID tags open up a number of technically feasible and economically attractive applications. Interesting applications in the field of process control would be:

- Identification of rail or road vehicles that deliver raw materials or products
- Identification of containers, parts, and workpieces, and their progress in the production process
- Checking the license to access of persons, recording the access of persons into dangerous zones
- Road markings and orientation aids for driverless, autonomically navigating vehicles, speed measurement, determination of position, distance measurement
- Selective reflex barriers, danger recognition, room security
- Indication of malfunctioning by passive means by setting a switchable reflector (read-me-flag), e.g., in case of damage by impact or excessive temperature, to prevent more damage from occurring

In some cases, severe application problems arise from the immediate response of the SAW sensor elements to the interrogation pulses, causing response signals to interfere with each other if more than one device is in the detection range of the interrogator at the same time. Thus, it must be assured that only one device can be "seen" or techniques must be developed to differentiate different devices that are in the interrogation range at the same time.

For the interrogation of SAW ID tags, coherent methods offers special advantages (Figure 92). The emitted pulses are derived from a reference oscillator in phase synchronism and the received signals are rectified in phase synchronism as well. As opposed to envelope demodulation, the phase information contained in the complex receiver signal is fully preserved in this case. After digitizing the received signal and reading it into a digital signal processor (DSP), there are no more obstacles to its evaluation with algorithms specific to the sensor. This is an enormous advantage: For a generic hardware the decisive DSP software defines what kind of sensor quantities are measured.

Through the coherent integration of a large number of successively received signals, the signal-to-noise ratio is improved and the range of the interrogator is increased accordingly. Further extension of the interrogation distance and a high immunity to disturbances has been achieved by spread spectrum techniques, e.g., pulse expansion/compression using dedicated low-cost dispersive SAW filters in the interrogator (Ruppel *et al.*, 1994). As an alternative to a pulsed system like that shown in Figure 92, methods using frequency modulated continuous waves (FMCW) or pseudo-noise signals etc. could be applied for interrogators. Furthermore, one can determine, with the aid of the Doppler effect in case of relative motion between interrogator and tag, the relative velocity and incremental displacement of the interrogator and tag, and with this the relative position. The coding of the tag is accomplished during the manufacture of the tag. The goal is to develop a tag programmable by the client. Prototypes already exist that can be programmed by the opening and closing of electric contacts (Reindl and Ruile, 1993).

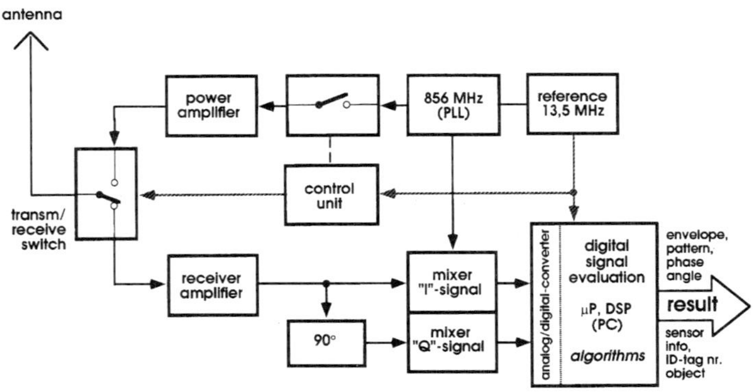

FIG. 92. Schematic of a coherent interrogator (as a homodyne system). I-Q output signals.

D. Radio Communication to Remote SAW Sensors for Temperature or Other Measurands

A further potential for innovative sensors is provided by the combination of the sensing properties of SAW devices and the interrogation by radio of ID tags. Such sensors can be positioned very flexibly and then be interrogated by radio waves. They can be mounted at practically any location, even inaccessible and dangerous places, and installed without any major effort (e.g., using a stick-on magnet or tape). The quantity of interest can then be determined from a safe distance, even through walls and windows, with (for example) a portable interrogator and transmitted to another location. An ID tag mounted on lithium niobate is already a sensitive interrogatable sensor because of its high temperature sensitivity (approximately 90 ppm/Kelvin) (Bao *et al.*, 1987). By evaluating transit-time differences of the various "bits" in the received signal, the temperature of the tag can be determined from a distance. Figure 93 shows the measured difference of transit times as a function of temperature; note the high linearity. Such a sensor was used to measure the temperature of brakes of a railway car during its slowing down and stopping in the station (Figure 94) (Pohl and Seifert (1996)). By the design of a dispersive transducer at the sensor element, a significant enhancement of the sensor sensitivity was achieved (Reindl and Mágori, 1995). Another advan-

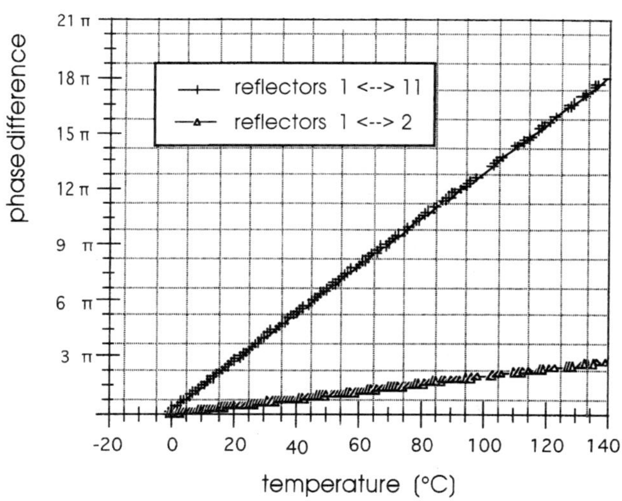

FIG. 93. Remote readout SAW temperature sensor showing sensitivity at different reflector spacing. Courtesy of Siemens.

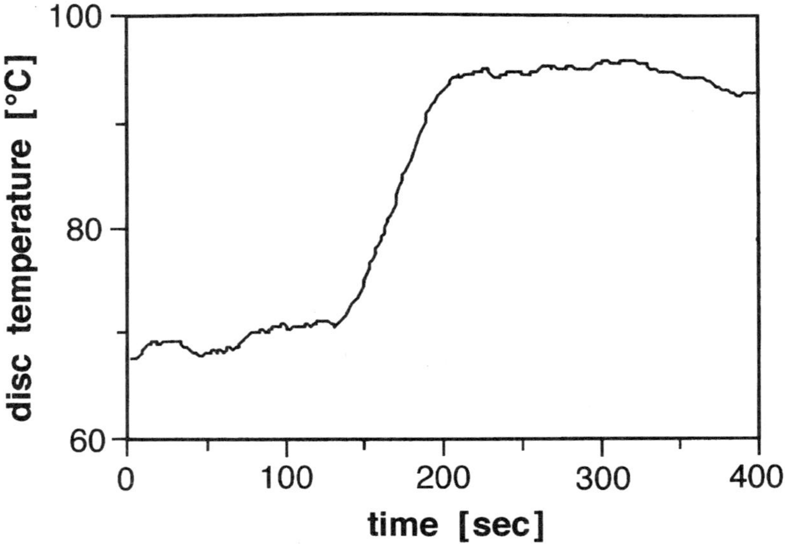

FIG. 94. Brake temperature of a railway car during braking, as measured with a SAW temperature sensor, after Pohl and Seifert (1996).

FIG. 95. Passive (triaxial seismic) sensor system includes a wireless link to satellite. Seismometer bandwidth: 0.01 to 50 Hz, achieved using separate boreholes. For details see Herrington (1991).

TABLE 6.
REMOTE READOUT SAW SENSOR OVERVIEW

Measured Quantity	Physical Effect	Remarks
Identity	Evaluation of signal patterns (e.g., bar-codes)	e.g., 32 bit
Temperature, force (acceleration, pressure, torque, strain)	Delay time changes of partial response signals due to temperature, mech. deformation, etc.; simultaneous evaluation of signals of individual reflectors ("on-chip" reference)	Resolution: 1 K accuracy: better than $\pm 1\%$ ($-200°C$ to $+350°C$)
Chemical quantities, humidity, recognition of gases	Alterations of the propagation velocity and attenuation of the SAW due to selective gas sensitive layers	Sensitivity in the ppm range, status: predevelopment
Motion (velocity, direction), relative position	Doppler signal evaluation, integration of Doppler phase increments	Resolution: better than ± 2 cm
Distance	Direct delay time measurement	Resolution: better than ± 20 cm
Rotation angle	Antenna polarization (relative)	Also circular polarization
Light, radiation, magnetic field, current, position, rotation angle, level limit, status monitoring; customer spec. programming; etc.	Modulation of individual "bits" of the response signal by connection of reflectors to a photoresistor, field sensitive resistor, potentiometer, limit level switch, etc.	20-dB dynamic range, differential setups preferable

FIG. 96. (Left): Active (liquid level, pitch-catch) sensor. The plastic housing contains the battery power supply and a wireless link to a remote data logger receiving station, remote being ≤ 1.6 km. Courtesy of Arichell Technologies. (Right): Flowline Model LU-30 measures air distances from ~ 150 mm to about five times that distance, with relay output; further details in *m&c* (April 1996), pp. 24 and 189. Illustration courtesy of Flowline.

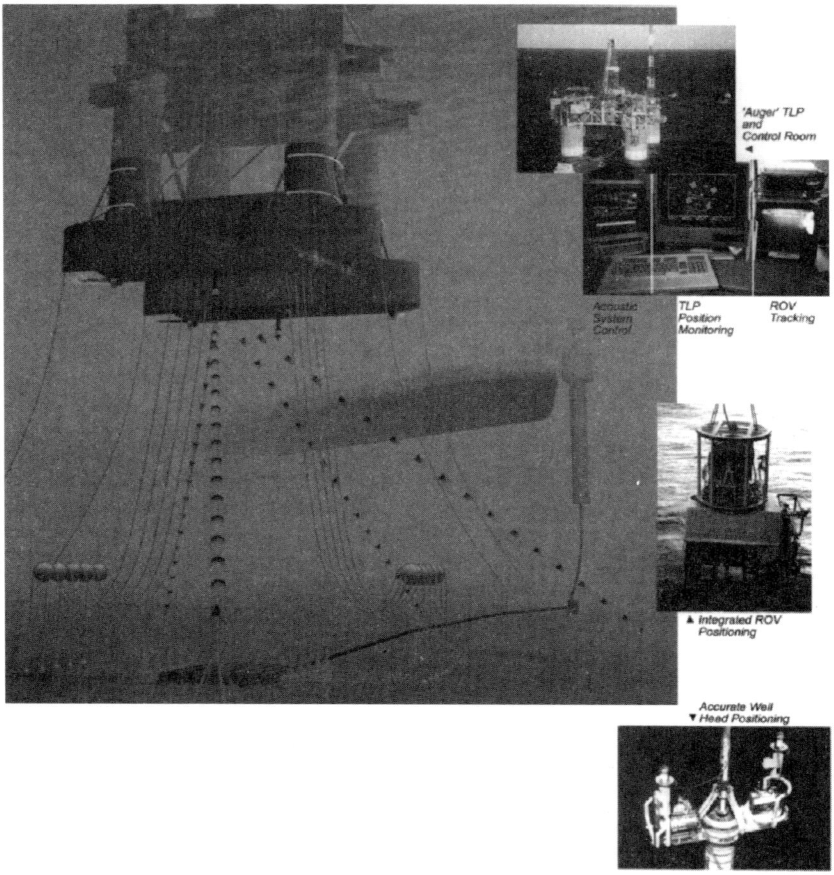

FIG. 97. Electromagnetic link to satellite for global position information, acoustic link to seabed reference targets, and for exploration and characterization of seabed. Acoustic positioning systems integrate ranges and bearings from hull-mounted transceivers to seabed transponder arrays. This enables the system to calculate the rig's position relative to seabed targets. Courtesy Sonardyne.

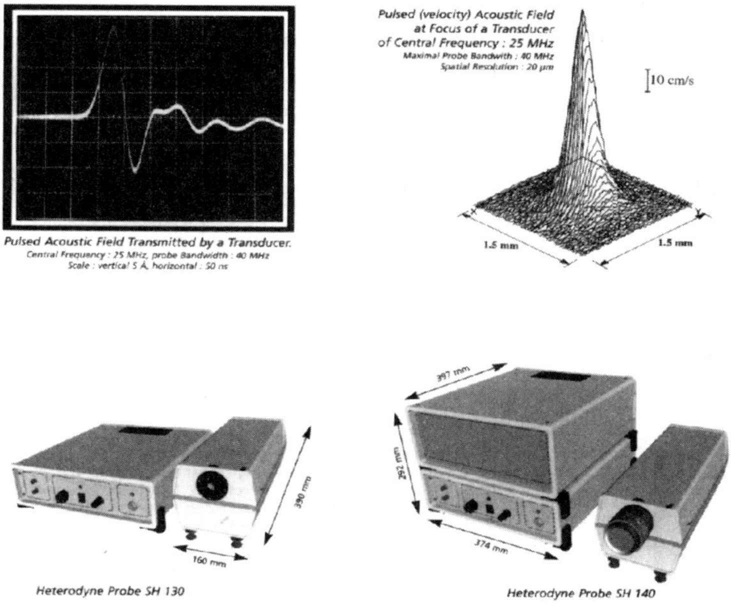

FIG. 98. Remote monitoring of a surface using a laser. Illustration courtesy B. M. Industries.

tage of the method is that the transit-time difference is independent of the distance between interrogator and tag: The reference is on the substrate itself. With this technology, mechanical quantities such as torque can be measured very conveniently from moving parts, e.g., from a motor shaft during rotation (Wolff *et al.*, 1996). Table 6 shows further application possibilities for passive radio interrogatable SAW sensors.

ACKNOWLEDGMENTS

The authors acknowledge important contributions from their colleagues at Panametrics and Siemens, and the permission of these organizations to reproduce herein illustrations, tables, or passages copyrighted by them, especially from UR-226 (Panametrics) and a 1994 R&D special issue (Siemens). The help of Alex von Jena (Siemens) and Petra Heggenberger (Universität der Bundeswehr München) was especially important in the final stages. The authors also acknowledge the contributions from the firms whose products are illustrated. Illustrations of equipment manufactured in the 1960s and 1970s were kindly supplied from the respective inventor's files of Robert L. Rod and Ellis M. Zacharias, Jr. Thanks also to Roberta M. Lynnworth and Tracey A. Russell who prepared the manuscript. Permission from the various publishers to reproduce illustrations that appeared in their journals or books is gratefully acknowledged. These publishers include: AIP, IEEE, PenWell Publ. Co. (for I&CS), and Helmers Publ. Inc. (for Sensors), respectively.

REFERENCES

Adamowski, J. C., Buiochi, F., and Sigelmann, R. A. (1998). Ultrasonic measurement of density of liquids flowing in tubes. *IEEE Trans. UFFC* **45**(1), 48–56.

Ageeva, N. S. (1960). Ultrasonic method for measuring the height of the fluid level in a vessel by means of flexural oscillation of a thin elastic strip. *Sov. Phys., Acoustics* **6**(1), 116–117.

Alkoy, S., Dogan, A., Hladky, A.-C., Langlet, P., Cochran, J. K., and Newnham, R. E. (1997). Miniature piezoelectric hollow sphere transducers (BBs). *IEEE Trans. UFFC* **44**(5), 1067–1076.

Allen, C. S. (1995). "In-Flow Acoustic Sensor." U.S. Patent No. 5,477,506.

Andreev, V. G., Dmitriev, V. N., Pishchal'nikov, Yu. A., Rudenko, O. V., Sapozhnikov, O. A., and Sarvazyan, A. P. (1997). Observation of shear waves excited by focused ultrasound in a rubber-like medium. *Acoust. Phys* **43**(2), 123–128. See also Sarvazyan, A. P. (1997).

Anon. (October 1990). Ultrasonic flow meter. *Pipeline & Gas J.*

Anon. (1996). "Measurement of fluid flow in closed conduits—methods using transit time ultrasonic flowmeters." ISO/TC 30/WG 20 N 106 E, Reference number: ISO/TR 12765.

Anon. (1997). New technique debuts for multiphase flow. *InTech* **44**(9) ISA, 17–18.

Anon. (June 1998). "Measurement of gas by multipath ultrasonic flowmeters." Transmission Measurement Committee Rpt. No. 9, AGA (American Gas Association), Cat. No. XQ9801.

Arave, A. E., Fickas, E., and Shurtliff, W. (1978). Instrumentation in the aerospace industry. *Proc. 24th ISA Int. Instrumentation Symp.* **24**, 609–620.

Asher, R. C. (1997). "Ultrasonic Sensors for Chemical Process Plant." Institute of Physics Publishing, London.

Babb, M. (November 1996). New ultrasonic level switch works from outside the tank. *Contr. Eng.*, 40–41.

Babikov, O. I. (1960). "Ultrasonics and Its Industrial Applications." Consultants Bureau, Plenum Press, New York.

Baggeroer, A., and Munk, W. (1992). "The Heard Island feasibility test." *Phy. Today* **45**(9), 22–30.

Baker, R. C., and Thompson, E. J. (March 14, 1978). "Measurement of Fluid Flow." U.S. Patent No. 4,078,428.

Baldauf, W., and Schrüfer, E. (1992). "Dehnungs- und Drehmomentmessung mit Oberflächenwellen-Resonatoren." Lecture at Second DFG-Workshop, Sensorsysteme für die Fertigungstechnik, Aachen.

Ballantine, Jr., D. S., White, R. M., Martin, S. J., Ricco, A. J., Frye, G. C., Zellers, E. T., and Wohltjen, H. (1996). "Acoustic Wave Sensors: Theory, Design, & Physico-Chemical Applications." Academic Press, Boston.

Bao, X. Q., Burghard, W., Varadan, V. V., and Varadan, K. V. (1987). SAW temperature sensor and remote reading system. *Proc. IEEE Ultras. Symp.*, 583–585.

Basarab-Horwath, I., and Dorozhevets, M. M. (1994). Measurement of the temperature distribution in fluids using ultrasonic tomography. *Proc. IEEE Ultras. Symp.*, 1891–1894.

Baumoel, J. (September 26, 1995). "Method and Apparatus for Leak Detection and Pipeline Temperature Modeling Method and Apparatus." U.S. Patent No. 5,453,944.

Bell, J. F. W. (1957). The velocity of sound in metals at high temperatures. *Phil. Mag.* **2**, 1113–1120.

Bergmann, L. (1954). "Der Ultraschall und seine Anwendungen in Wissenschaft und Technik." 6th Edition, Hirzel-Verlag, Stuttgart.

Bernst, T., Löschberger, J., and Mágori, V. (1986). Moderne Verfahren der Signalverarbeitung bei Ultraschall-Robotik-Sensoren. *NTG-Fachberichte* **93**, 253–263.

Bhatia, A. B. (1967). "Ultrasonic Absorption." Clarendon Press, Oxford.

Blickley, G. J. (1996). Flowmeter selection isn't easy, but tools are here. *Contr. Eng.* **43**(15), 43–50.

Bonnefous, O. (1992). Time domain color flow imaging: Methods and benefits compared to Doppler. *Acoust. Imag.* **19**, 301–309.

Bonnefous, O. (1994). Stenoses dynamics with ultrasonic wall motion images. *Proc. IEEE Ultras. Symp.*, pp. 1709–1712. See also Edler & Hertz (1954), especially their figure reproduced on p. 191 in Lynnworth (1989), showing how they photographically recorded distance to the moving heart wall.

Bonnefous, O., and Pesqué, P. (1986). Time domain formulation of pulse-Doppler ultrasound and blood velocity estimation by cross-correlation. *Ultras. Imag.* **8**, 73–85.

Bou Matar, O., Roncin, A., Remenieras, J. P., and Patat, F. (1996). Noncontact measurement of vibration using airborne ultrasound. *Proc. 1996 IEEE Int. Ultras. Symp.*, 697–700.

Bou Matar, O., Remenieras, J. P., Pizarro, L., Roncin, A., and Patat, F. (1997). Performances of the parametric acoustic vibrometer for vibration sensing. *Proc. 1997 IEEE Int. Ultras. Symp.*, 605–608.

Brandestini, M. (1978). Topoflow—a digital full range Doppler velocity meter. *IEEE Trans. Son. & Ultrason.* **SU-25**(5), 287–293.

Brandestini, M. A., and Forster, F. K. (1978). Blood flow imaging using a discrete-time frequency meter. *Proc. IEEE Ultras. Symp.*, 348–352.

Brown, L. F., Mason, J. L., Kinkenborg, M. L., Scheinbeim, J. I., and Newman, B. A. (1997). Ferroelectric nylon materials and their feasibility for ultrasound transducers. *IEEE Trans. UFFC* **44**(5), 1049–1059.

Bruner, R. F. (1977). Theoretical and experimental assessment of uncertainties in non-intrusive, ultrasonic flow measurement. *Proc. Symp. on Flow in Open Channels and Closed Conduits.* National Bureau of Standards Special Publication 484, pp. 277–292.

Buff, W., Rusko, M., Goroll, M., Ehrenpfordt, J., and Vandahl, T. (1997). Universal pressure and temperature SAW sensor for wireless applications. *Proc. 1997 IEEE Ultras. Symp.*, 359–362.

Bulst, W., and Ruppel, C. (1994). Developments to watch in surface acoustic wave technology. *Siemens Review*, R&D-Special, Spring issue.

Busse, D. W. (1987). Quartz transducers for precision under pressure. *Mech. Eng.* **109**(5), 52–56.

Carey, C. A., Carnevale, E. H., Uva, S., and Marshall, T. (1969). "Experimental determination of gas properties at high temperatures and/or pressures." AEDC-TR-69-78 (March).

Carlson, G. A., Sullivan, W. H., and Plein, H. G. (1977). Application of ultrasonic thermometry in LMFBR safety research. *Proc. 1977 IEEE Ultra. Symp.*, 24–28.

Chernyshev, V. A. (1994). Ultrasonic flowmeter probe for large pipes. *Dev. and Contr. Sys.* **9**, 3–5 (in Russian).

Chilowski, C., and Langevin, P. (Oct. 23, 1923). "The Production of Submarine Signals and the Location of Submarine Objects." U.S. Patent No. 1,471,547.

Collard, S. M., and McLellan, R. B. (1990). Modulus defect in the noble metals. *Scripta Metallurgica et Materialia* **24**, 623–627.

Coulthard, J. (1973). Ultrasonic cross-correlation flowmeters. *Ultrasonics* **11**(2), 83–88.

Craster, R. V. (1996). A canonical problem for fluid-solid interfacial wave coupling. *Proc. Roy. Soc. Lond. A*(452), 1695–1711.

Craster, R. V. (1997). The light fluid loading limit for fluid/solid interactions. *J. Appl. Math.* **7**, 485–506.

Crecraft, D. I. (1964). Launching ultrasonic shear waves into solids at normal incidence by pressure coupling. *J. Sound Vibr.* **1**(4), 381–387.

Dabirikhah, H., and Turner, C. W. (1994). Leaky plate wave airborne ultrasonic transducer. *Elect. Lett.* **30**(18), 1549–1550.

De Cicco, G., Morten, B., and Prudenziati, M. (1997). A new ultrasonic composite transducer implemented with thick film technology. *IEEE Trans. UFFC* **44**(5), 992–996.

Del Grosso, V. A., and Spurlock, E. M. (1957). The feasibility of using wholly external ultrasonics to measure fluid flow within thick-walled metal pipes. *NLR 4967*, AD149409.

Dever, D. (July 1998). New kid on the block. *FC (Flow Control)* 29–32.

Dickert, F. L., Bertlein, G., Mages, G., and Bulst, W. E. (1990). Triphenyl methane dyes as sensor materials for solvent detection with surface-acoustic-wave devices. *Adv. Mater.* **2**(9), 420–422.

Dogan, A., Yoshikawa, S., Uchino, K., and Newnham, R. E. (1994). The effect of geometry on the characteristics of the moonie transducer and reliability issue. *Proc. 1994 IEEE Ultras. Symp.*, 935–939.

Doyle, T. E. (1996). Using acoustic waveguides to probe materials ultrasonically. *NASA Tech. Briefs*, p. 66. (September.)

Drenthen, J. G. (May 23, 1989). "Device for Determining the Flow Velocity of a Medium in a Cylindrical Conduit." U.S. Patent No. 4,831,884. (Other midradius U.S. patents include: Baker, R. C., and Thompson (March 14, 1978); Lynnworth, L. C. (August 1, 1978). See also Lynnworth (1989), pp. 285, 293, 294.)

Drost, C. J. (October 14, 1980). Volume flow measurement system. U.S. Patent No. 4,227,407.

Eccardt, P.-C., Landes, H., and Lerch, R. (1996). Finite element simulation of acoustic wave propagation within flowing media. *Proc. 1996 IEEE Ultras. Symp.*, 991–994.

EerNisse, E. P., Ward, R. W., and Wiggins, R. B. (1988). Survey of quartz bulk resonator sensor technologies. *IEEE Trans. UFFC* **35**(3), 323–330.

Eichfeld, H., Künemund, T., and Klimke, M. (1993). A 8-bit coprocessor for fuzzy control. *Proc. ISSCC 1993* **286**, 180–181.

Embree, P. M., and O'Brien Jr., W. D. (1985). The accurate ultrasonic measurement of the volume flow of blood by time domain correlation. *Proc. 1985 IEEE Ultras. Symp.*, 963–966.

Evans, J. (April 1997). Ultrasonics comes of age for solids and liquids. *InTech*, 38–42.

Fendrock, C., and Varela, D. W. (1995). Ultrasonic thermometry for industrial furnace temperature measurement. *Sensors* **12**(1), 23–25, 27, and 37.

Fink, M. (1992). Time reversal of ultrasonic fields—Part I: Basic principles. *IEEE Trans. UFFC* **39**(5), 555–566.

Fischer, B., Mágori, V., and von Jena, A. (1995). "Ultraschall-Dichtemesser zum Messen der spezifischen Dichte eines Fluids." European Patent Nos. 0 482 326 B1 and 0 483 491 B1.

Fischerauer, G., Mauder, A., and Müller, R. (1995). Acoustic wave devices. *Sensors* **8**, VCH Verlagsanstalt, pp. 136–157.

Frederick, J. R. (1947). A study of the elastic properties of various solids by means of ultrasonic pulse techniques. Ph.D. Thesis, Univ. of Michigan. See also Frederick, J. R. (1948). *J. Acoust. Soc. Am.* **20**, 586A.

Frederick, J. R. (1965). "Ultrasonic Engineering." John Wiley & Sons, New York, p. 213.

Gazis, D. C., Kane, W. F., and von Gutfeld, R. J. (August 11, 1998). "Ultrasonic Liquid Level Gauge for Tanks Subject to Movement and Vibration." U.S. Patent No. 5, 793, 705.

Green, S. F. (1985). An acoustic technique for rapid temperature distribution measurement. *J. Acoust. Soc. Am.* **77**(2), 759–763.

Gutterman, J. (December 3, 1985). "Reflective Acoustic Fluid Flow Meter." U.S. Patent No. 4,555,951.

Hamidullin, V. K. (1989). "Ultrasonic control instrumentation systems" (in Russian). Leningrad (St. Petersburg) University. See also *Proc. 1993 IEEE Ultras. Symp.*, 395–399.

Hayward, G. (1997). Air-coupled NDE—constraints and solutions for industrial implementation. *Proc. 1997 Int. Ultras. Symp.*, pp. 665–673.

Heimisch, W. (1996). Blutflußmessung in der Herz- und Gefäßchirurgie: Methodik–Technik–Praxis. *In* Gefäßchirurgie. Springer-Verlag, Heidelberg/Triltsch, Würzburg.

Helmholtz, H. von (1860). Vorlesungen über die mathematischen Prinzipien der Akustik. Leipzig, 1898. See also, Pierce, 1981, pp. 195–203.

Henning, B., and Hauptmann, P. (1997). Miniaturized ultrasonic sensor for analysis of liquids. *Proc. Sensor '97 Conf.*, 8th International Fair and Congress for Sensors, Transducers & Systems.

Herrington, P. (1991). Seismic verifier listens for bombs bursting in air. *Sensors* **8**(6), 8.

Hertz, T. G. (1993). Applications of acoustic streaming. Ph.D. Thesis, Lund Institute of Technology.

Hertz, T. G., Dymling, S. O., Lindstöm, K., and Persson, H. W. (1991). Viscosity measurement of an enclosed liquid using ultrasound. *Rev. Sci. Instr.* **62**, 457–462. See also Hertz, T. G. (1993).

Hill, J. E., and Ruoff, A. L. (1965). Velocity of sound measurements in liquid metals. *Rev. Sci. Instr.* **36**(10), 1465–1472.

Hill, W. (February 4, 1997). "Method and System for Analyzing a Two Phase Flow." U.S. Patent No. 5,600,073.

Holing, D. (1994). The sound and the fury. *Amicus J.* **16**(3), 19–23.

Horne, R. (1997 and June 19, 1998). Private communications.

Hutchins, D. A., and Schindel, D. W. (1994). Advances in non-contact and air-coupled transducers. *Proc. 1994 IEEE Ultras. Symp.*, 1245–1254.

Igarashi, B., Alers, G. A., and Purtscher, P. T. (1997). An ultrasonic measurement of magnetostriction. *Proc. 1997 IEEE Ultras. Symp.*, 709–712.

Jacobson, S. A. (October 24, 1995). "Flow Measurement System Including Ultrasonic Transducers." U.S. Patent No. 5,460,047.

Jacobson, S. A., Denbigh, P. N., and Naudé, D. E. H. (1985). A new method for the demodulation of ultrasonic signals for cross-correlation flowmeters. *Ultrasonics* **23**(3), 128–132.

Jacobson, S. A., Korba, J. M., Lynnworth, L. C., Nguyen, T. H., Orton, G. F., and Orazietti, A. J. (1987). Low-gravity sensing of liquid/vapor interface and transient liquid flow. *IEEE Trans. UFFC* **34**(2), 212–224.

Jacobson, S. A., Korba, J. M., Lynnworth, L. C., McGrath, W. F., and Nguyen, T. H. (1992). Panametrics Final Report. DOE SBIR Contract DE-AC02-88ER80587.

Jacobson, S. A., Lynnworth, L. C., and Korba, J. M. (November 29, 1988). "Differential Correlation Analyzer." U.S. Patent No. 4,787,252.

Jarrti, P., and Luukkala, M. (1977). Ultrasonic method for web speed measurement. *Tappi* **60**(11), 167.

Jen, C.-K., and Legoux, J. G. (1996). High performance clad metallic buffer rods. *Proc. 1996 IEEE Ultras. Symp.*, 771–776.

Jen, C.-K., Chen, J.-Y., Hoa, S. V., Nguyen, K. T., Legoux, J.-G., and Hébert, H. (1997). Clad buffer rods for in-situ process monitoring. *Proc. 1997 IEEE Ultras. Symp.*, 801–806.

Jena, A. von, and Mágori, V. (1992). Die Ansaugströmung von Kfz-Motoren mit Ultraschallmessen. *Feinwerktechnik und Meßtechnik* **3**, 95–98.

Jena, A. von, Mágori, V., and Rußwurm, W. (1993). Ultrasound gas flow meter for household application. *Sensors & Actuators* **37–38**, 135–140.

Jensen, J. A. (1996). "Estimation of Blood Velocities Using Ultrasound: A Signal Processing Approach." Cambridge University Press.

Jepson, W. P., Lee, A. H., and Sun, J. Y. (1993). A study of flow regime transitions of oil-water-gas mixtures in horizontal pipelines. *Proc. Int. Soc. of Offshore & Polar Engineers* (ISOPE), Paper No. JSC-052 (Singapore, June).

Johnson, S. A. (July 31, 1979). "Measurement and Reconstruction of Three-Dimensional Fluid Flow." U.S. Patent No. 4,162,630.

Johnson, S. A., Greenleaf, J. F., Samayoa, W. A., Duck, F. A., and Sjöstrand, J. (1975). Reconstruction of three-dimensional velocity fields and other parameters by acoustic ray tracing. *Proc. 1975 IEEE Ultras. Symp.*, 46–51.

Johnson, S. A., Greenleaf, J. F., Tanaka, M., and Flandro, G. (1977). Theoretical and experimental assessment of uncertainties in non-intrusive, ultrasonic flow measurement. *Proc. Symp. on Flow in Open Channels and Closed Conduits*. National Bureau of Standards Special Publication 484, pp. 335–359.

Každys, R.-J. (1986). "Ultrasonic Information Measuring Systems." Mokslas, Vilnius (in Russian).

Kelly, S. P., Farlow, R., and Hayward, G. (July 1996). Applications of through-air ultrasound for rapid NDE scanning in the aerospace industry. *IEEE Trans. UFFC* **43**(4), 581–591.

Kim, J. O. (1989). The interaction between stress waves transmitted in solid waveguides and adjacent media. Ph.D. Dissertation, Univ. of Pennsylvania, Philadelphia.

Kim, J. O., and Bau, H. H. (1989). Instrument for simultaneous measurement of density and viscosity. *Rev. Sci. Instr.* **690**(6), 1111–1115.

Kim, J. O., Bau, H. H., Liu, Y., Lynnworth, L. C., Lynnworth, S. A., Hall, K. A., Jacobson, S. A., Korba, J. M., Murphy, R. J., Strauch, M. A., and King, K. G. (1993). Torsional sensor applications in two-phase fluids. *IEEE Trans. UFFC* **40**(5), 563–576.

Kim, J. O., Wang, Y., and Bau, H. H. (1991). The effect of an adjacent viscous fluid on the transmission of torsional stress waves in a submerged waveguide. *J. Acoust. Soc. Am.* **89**(3), 1414–1422.

Kim, R. K. (1992). A new gas flowmeter based on cavity oscillations. *Flow Meas. Instr.* **3**(2), 103–111.

Kim, R. K. (1998). Private communication.

Kleinschmidt, P., and Mágori, V. (1981). Ultrasonic remote sensors for noncontact object detection. *Siemens Forschungs- und Entwicklungsberichte* **10**, 110–118.

Kleppe, J. A. (1989). "Engineering Applications of Acoustics." Artech House, Boston.

Kleppe, J. A. (1995a). Acoustic gas flow measurement in large ducts and stacks. *Sensors* **12**(5), 18–21; 85–87.

Kleppe, J. A. (1995b). Adapt acoustic pyrometer to measure flue-gas flow. *Power* **139**(8), 46–47.

Kleppe, J. A. (1996). High-temp gas measurement using acoustic pyrometry. *Sensors* **13**(1), 17–22.

Knapp Boetticher, C. (1958). Geschwindigkeits- und Mengenmessung strömender Flüssigkeiten mittels Ultraschalls. Promotionsarbeit, (Prom.-Nr. 2795), ETH Zürich. See also Knapp (Boetticher) (1964).

Knapp (Boetticher), C. (1964). Geschwindigkeits- und Mengenmessung strömender Flüssigkeiten mittels Ultraschall. *VDI-Ber.* **86**, 65–71.

Knoll, A. C. (1991). Ultrasonic holography techniques for localizing and imaging solid objects. *IEEE Trans. Robot. & Automation* **7**(4).

Krautkrämer, J., and Krautkrämer, H. (1977). "Ultrasonic Testing of Materials, 2nd Edition." See esp. p. 415. Springer-Verlag, Berlin. (First Edition, 1966, *Materialprüfung mit Ultraschall*, in German.)

Kroemer, N. (1993). Ultraschall-Distanzsensor mit Fuzzy-Auswertung. *Proc. Sensor '93 Conf.* **II**, 113–121.

Lach, M., and Ermert, H. (1992). Object recognition using an ultrasonic sensor system. *Acoust. Imag.* **19**, 957–962.

Ladabaum, I., Jin, X. C., and Khuri-Yakub, B. T. (1997). Air coupled through transmission of aluminum and other recent results using MUTs. *Proc. IEEE Int. Ultras. Symp.*, 983–986.

Laenen, A. (1984). The use of acoustic velocity meters in the measurement of streamflow. *Proc. 30th ISA Int. Instr. Symp.*, 201–208.

Lake, D. B. (1962). "Flow Meters." U.S. Patent No. 3,050,997.
Lange, S. T. (September 1994). Are your water rates accurate? *Public Works Mag.*
Lec, R. M. (1997). Private communication.
Lester, R. A., and Thompson, J. H. (1965). "Acoustic Flow Meters." U.S. Patent No. 3,209,591.
Li, X., Wang, Z., and Cheeke, J. D. N. (1997). Determination of the dispersion curves of circumferential waves in a cylindrical shell using a chirp IDT. *Proc. 1997 IEEE Int. Symp.*, 705–708.
Li, Y., and Menon, S. (1998). Ultrasonic sensing of composite materials during the heat-cure cycle. *Sensors* **15**(2), 14–19.
Liu, Y., and Lynnworth, L. C. (1993). Flexural wave sidewall sensor for noninvasive measurement of discrete liquid levels on large storage tanks. *Proc. 1993 IEEE Ultras. Symp.*, 385–390.
Liu, Y., and Lynnworth, L. C. (October 10, 1995). Elastic wave sensing system. U.S. Patent No. 5,456,114.
Liu, Y., Lynnworth, L. C., and Zimmerman, M. A. (1998). Buffer waveguides for flow measurement in hot fluids. *Ultrasonics* **36**(1–5), 305–315.
Lockyer, R. (1996). A sonic anemometer for general meteorology. *Sensors* **13**(5), 12–14, 17, 18.
Löschberger, J., and Mágori, V. (1988). Ultrasonic robotic sensors with lateral resolution. *Proc. 1988 Ultras. Symp.*, 547–551.
Lynnworth, L. C. (1967). Ultrasonic probes using shear wave crystals—principles. *Mater. Eval.* **25**(12), 265–277.
Lynnworth, L. C. (1975). Industrial applications of ultrasound—a review. II. Measurements, tests and process control using low-intensity ultrasound. *IEEE Trans. Son. & Ultrason.* **SU-22**(2), 71–101.
Lynnworth, L. C. (August 1, 1978). "Ultrasonic Measuring System for Differing Flow Conditions." U.S. Patent No. 4,103,551.
Lynnworth, L. C. (1979). Ultrasonic Flowmeters. *In* "Physical Acoustics." Vol. 14 (W. P. Mason and R. N. Thurston, eds.), Academic Press, Boston, pp. 407–525.
Lynnworth, L. C. (1988). Proposed downhole ultrasonic quadrant flow tool: technology background and design considerations. *Proc. Tech. Rev. on Adv. in Geothermal Reservoir Techology-Research in Progress*, pp. 78–83. Lawrence Berkeley Laboratory Report LBL-25635 (September).
Lynnworth, L. C. (1989). "Ultrasonic Measurements for Process Control." Academic Press, Boston.
Lynnworth, L. C. (1995). Ultrasonic gas flowmeters. *Measurements & Control* **29**(5), 92–101; (6), 187.
Lynnworth, L. C. (1996). "Sensors, Acoustic." *In* Encyclopedia of Applied Physics, Vol. 17 (G. L. Trigg, ed.), VCH/AIP, New York, pp. 521–541.
Lynnworth, L. C. (1998a). Acoustic/ultrasonic sensors for process measurands, Panametrics Ultras. Rpt. UR-212.
Lynnworth, L. C. (1998b). Air transducer with high acoustic impedance. *J. Acoust. Soc. Am.* **103**(5), Pt. 2, 2833(A); *Proc. 16th Intl. Cong. Acoust. & 135th Meeting Acoust. Soc. Am.* **I-IV**, 741–742.
Lynnworth, L. C., and Carnevale, E. H. (1972). Ultrasonic thermometry using pulse techniques. *In* "Temperature—Its Measurement and Control in Science and Industry." Vol. 4 (H. H. Plumb, ed.-in-chief), ISA.
Lynnworth, L. C., and Lynnworth, A. M. (1985). Calculated turbulent flow meter factors for nondiametral paths used in ultrasonic flowmeters. *ASME J. Fluids Eng.* **107**, 44–48.
Lynnworth, L. C., and Patch, D. R. (1970). New sensors for ultrasound: measuring temperature profiles. *MTRSA (Materials Research and Standards)* **10**(8), cover, 6–11, 40.
Lynnworth, L. C., and Pedersen, N. E. (1972). Ultrasonic mass flowmeter. *Proc. 1972 IEEE Ultras. Symp.*, 87–90.
Lynnworth, L. C., Jacobson, S. A., Korba, J. M., McGrath, W. F., Murphy, R. J., Nguyen, T. H., Strauch, M. A., and True, J. A. (1993). "Downhole quadrant flow tool (QFT) for geothermal applications to 350 degrees celsius and 10,000 psi (68 MPa)." Final Report, Contract DE-AC02-88ER80587, Phase II.

Lynnworth, L. C., Jossinet, G., and Chérifi, E. (1996). 300°C clamp-on ultrasonic transducers for measuring water flow and level. *Proc. 1996 IEEE Ultra. Symp.*, 407–412.

Lynnworth, L. C., Matson, J. E., Nguyen, T. H., and Powers, W. T. (1992). Small-inertia clamp-on cryogenic flowmeter transducer. *1992 Conf. Adv. Earth-to-Orbit Propulsion Techn.*, **Vol. I**, 207–216.

Lynnworth, L. C., Nguyen, T. H., Smart, C. S., and Khrakovsky, O. K. (1997a). Acoustically isolated paired air transducers for 50-, 100-, 200-, or 500-kHz applications. *IEEE Trans. UFFC* **44**(5), 1087–1100.

Lynnworth, L. C., Papadakis, E. P., Patch, D. R., Fowler, K. A., and Shepard, R. L. (1971). Nuclear reactor applications of new ultrasonic transducers. *IEEE Trans. Nuc. Sci.* **NS-18**(1), 351–362.

Lynnworth, L. C., Seger, J. L., and Bradshaw, J. E. (1982). "Ultrasonic System for Measuring Fluid Impedance or Liquid Level." U.S. Patent No. 4,320,659.

Lynnworth, L. C., Steingard, L., Khrakovsky, O. K., Machado, E. J., Smart, C. D., and Nguyen, T. H. (1997b). Improved shear wave hockey stick transducer measures liquid flow and liquid level. *Proc. 1997 IEEE Ultras. Symp.*, 865–870.

Macedo, P., and Litovitz, T. A. (1965). Ultrasonic viscous relaxation in molten boron trioxide. *Phys. Chem. Glasses* **6**(3), 69–80.

Mágori, V. (1980). "Ultraschallwandler." German Patent Nos. 2537788DE and 2541492DE.

Mágori, V. (1985). A novel ultrasonic flow sensor based on lambda locked loop principle with interdigital transducers. *Proc. IEEE Ultras. Symp.*, 525–530.

Mágori, V. (1989). Signal procesing for smart ultrasonic sensors. *Proc. 3rd Ann. Europ. Computer Conf.*, 3-21–3-26.

Mágori, V. (1993a). How sensors can improve the efficiency of transportation systems. *Siemens Review*, R&D Special, Spring issue, pp. 26–30.

Mágori, V. (1993b). Neue Meßverfahren für klassische und neue Meßgrößen. *In* "Beitrag zu Sensorik in der Prozeßleittechnik." VDE-Verlag, Frankfurt, pp. 37–59.

Mágori, V. (1994). Ultrasonic sensors in air. *Proc. 1994 Ultras. Symp.*, 471–481.

Mágori, V., and Walker, H. (1987). Ultrasonic presence sensors with wide range and high local resolution. *IEEE Trans. UFFC* **34**(2), 202–211.

Manthey, W., Kroemer, N., and Mágori, V. (1992). Ultrasonic transducers and transducer arrays for applications in air. *Meas. Sci. Techn.* **3**, 249–261.

Matson, J., and Davis, R. (February 1994). Ultrasonic flowmeters offer reliable solution to stack monitoring. *I&CS* **67**(2), 67–69.

Matson, J., and Lynnworth, L. C. (1997). Ultrasonic flowmeters. *Measurements & Control* **31**(5), 238–240.

Mayranen, T., Koukkari, S., Hiismaki, P., Knuuttila, M., Kall, L., and Tiitta, A. (May 7, 1996). "Method and Device for Measurement of the Flow Velocities of Gases and/or of Quantities That Can Be Derived from the Same." U.S. Patent No. 5,513,535.

McGunigle, R. D. (June 25, 1974). "Tracking System For Time-Displaced Signals." U.S. Patent No. 3,819,919.

Menz, B., and Dittes, M. (1997). Vortex-Durchflußmesser mit akustischer Wirbeldetektion in Clamp-on-Technik. *Technisches Messen* **64**, 54–59.

Michaelides, E. E., and Zissis, K. L. (1983). Velocity of sound in two-phase mixtures. *Int. J. Heat & Fluid Flow* **4**(2), 79–84.

Miller, R. W. (1996). "Flow Measurement Engineering Handbook." 3rd Edition, McGraw-Hill, New York.

Mohr, G. (1995). Technology overview: ultrasonic detection. *P/PM TECHNOLOGY* **8**(4), 56– 61. See also: Anon. (1998). Looking Ahead, *NASA Tech Briefs* **22**(6), 30.

Müller, R. (1996). Industrielle Grenzstandmessung von außen mit Ultraschall. *VDI-Berichte 1255*, 247–252.

Nagy, P. B., and Nayfeh, A. H. (1996). Viscosity-induced attenuation of longitudinal guided waves in fluid-loaded rods. *J. Acoust. Soc. Am.* **100**(3), 1501–1508.

Nygaard, O. G. H., and Mylvaganam, K. S. (1993). Ultrasonic time-domain reflectometry for level measurement in molten metals. *Technisches Messen* **60**(1), 4–14.

Papadakis, E. P. (1974). Monitoring the moduli of polymers with ultrasound. *J. Appl. Physics* **45**(3), 1218–1222.

Paros, J. M., and Wearn, R. B. (May 2–5, 1988). Measurements of dead weight tester performance using high resolution quartz crystal pressure transducers. Presented at Instrument Society of America 34th International Instrumentation Symposium, Albuquerque, New Mexico.

Pfau, B. (1970). Erweiterung des Meßbereiches eines Ultraschalldurchflußmessers unter Ausnutzung des Doppler-Effektes. *Chemie-Ing.-techn.* **42**(17), 1103–1109.

Pierce, A. D. (1981). "Acoustics." McGraw-Hill, New York.

Pohl, A., and Seifert, F. (1996). Wireless interrogable SAW sensors for vehicular applications. *Proc, IEEE Instr. and Meas. Conf.*, 1465–1468.

Pohl, A., Ostermayer, G., Reindl, L., and Seifert, F. (1997). Monitoring the tire pressure at cars using passive SAW sensors. *Proc. 1997 IEEE Ultras. Symp.*, 471–474.

Povey, M. J. W. (1997). "Ultrasonic Techniques for Fluids Characterization." Academic Press, New York.

Puttmer, A., Hauptmann, P., Henning, P., and Dierks, K. (1996). Ultraschallsensor zur Messung der akustischen Impedanz von Flüssigkeiten. *VDI-Berichte 1255*, 241–245.

Rao, M. R. (1940). A relation between velocity of sound in liquids and molecular volume. *Indian J. Phys.* **14**, 109–116.

Rayleigh, J. W. S. (1896). "The Theory of Sound." 2nd Edition, Dover, New York, pp. 145–146.

Reindl, L., and Mágori, V. (1995). Funksensorik mit passiven Oberflächenwellen-Komponenten (OFW). *Fortschr. Ber. VDI Reihe* **8**(Nr. 515), 62–79.

Reindl, L., and Ruile, W. (1993). Programmable reflectors for SAW-ID tags. *Proc. IEEE Ultras. Symp.* **1**, 125–130.

Reinoso, J. J. (1996). Metallorganic chemical vapor deposition of cadmium zinc telluride. Ph.D. Thesis, Carnegie Mellon University, Pittsburgh.

Remenieras, J. P., Kouamé, D., Gregoire, J. M., Jaraczewski, P., and Patat, F. (1996). Multiphase pipe flow velocity profile measurements by Doppler ultrasound containing a high level of colored noise. *Proc. 1996 IEEE Ultras. Symp.*, 629–634.

Rencken, W. D., Leuthaeuser, I., Bauer, R., Feiten, W., Lawitzki, G., and Moeller, M. (1993). Autonomous low-cost mobile robots for complex non-production environments. *1st IFAC Int. Workshop on Intelligent Autonomous Vehicles*, Univ. of Southampton, U.K., 31–36.

Rod, R. L. (1962). Propellant gaging and control. *Instr. & Contr. Sys.* **35**(10), 119–121.

Rod, R. L., and Massa, J. A. (November 1962). A digital propellant utilization system. *Aerospace Eng.*, 27–32.

Rosa, D. (Ed.) (1991). An ultrasonic echo-ranging sensor for board inspection and selection. *Sensors* **8**(6), 12–13.

Rountree, S. P., and Berjaoui, S. W. (October 22, 1996). "Methods and Apparatus for Use in Ultrasonic Ranging." U.S. Patent No. 5,568,449.

Roux, P., and Fink, M. (1996). Vorticity measurements with an acoustic time-reversal mirror. *Proc. 1996 IEEE Ultras. Symp.*, 1249–1254.

Royer, D., and Dieulesaint, E. (1996). "Ondes Élastique dans les Solides, **1**. Propagation Libre et Guidée." Masson, Paris. (English translation in press, Springer (1998).) **2**. Generation, Acousto-Optic Interaction and Applications. In Press (1998), in French; English translation in progress.

Royer, D., Dieulesaint, E., and Legras, O. (1992). Capteurs à ondes élastiques guidées. *J. Phys. III France* **2**, 145–168.

Rschevkin, S. N. (1963). "A Course of Lectures on the Theory of Sound." Pergamon Press, Oxford, and MacMillan, New York.

Rudenko, O. V., Sarvazyan, A. P., and Emelianov, S. Y. (1996). Acoustic radiation force and streaming induced by focused nonlinear ultrasound in a dissipative medium. *J. Acoust. Soc. Am.* **99**(5), 2791–2798.

Ruppel, C. C. W., Dill, R., Fischerauer, A., Fischerauer, G., Gawlik, W., Machui, J., Müller, F., Reindl, L., Ruile, W., Scholl, G., Schropp, I., and Wagner, K. C. (1993). SAW devices for consumer communication applications. *IEEE Trans. UFFC* **40**(5), 438–452.

Ruppel, C. W., Reindl, L., and Wagner, K. C. (1994). Optimum design of low time bandwidth product SAW filters. *Proc. IEEE Ultras. Symp.* **1**, 61–65.

Ruser, H., and Mágori, V. (1997). Sweep linearization of a microwave FMCW Doppler sensor by an ultrasonic reference. *Proc. 1997 IEEE International Frequency Control Symposium*, 201– 206.

Rychagov, M. N., and Ermert, H. (1994). Applicability of wave tomography methods for 2-D flow imaging. *Proc. 1994 IEEE Ultras. Symp.*, 1731–1735.

Rychagov, M. N., and Ermert, H. (1996). Cross-flow visualization by acoustic CT measurements. *Ultrasonics* **34**, 517–522.

Sachs, D. C., Meagher, T. F. V., and Peckham, V. D. (1977). Measurement of boundary layer transition using acoustic techniques. *Proc. Symp. on Flow in Open Channels and Closed Conduits*. Natl. Bur. Stand. (U.S.), Spec. Publ. **484**, 687–704.

Sarvazyan, A. P. (March 4, 1997). "Method and Device for Shear Wave Elasticity Imaging." U.S. Patent No. 5,606,971.

Sather, A. (1968). Ultrasonic buffer-rod technique for the high temperature measurement of the elastic moduli of short specimens. *J. Acoust. Soc. Am.* **43**(6), 1291–1294.

Scelzo, M., and Jacobson, S. (1994). Transit-time ultrasonic flowmeter with patented signal-coding. *Measurements & Control* **28**(2), Issue 164, pp. 84–87.

Scelzo, M. J., and Munk, W. D. (1987). Field test of an ultrasonic flowmeter for natural gas pipelines. In E. J. Seiders (Ed.), *Pipeline Engineering Symposium, 1987: PD-Vol. 6* (pp. 111–114). ASME.

Schaaffs, W. (1963). "Molekularakustik." Springer-Verlag, Berlin, esp. pp. 174–176, 278–280.

Schindel, D. W., Hutchins, D. A., and Grandia, W. A. (1996). Capacitive and piezoelectric air-coupled transducers for resonant ultrasonic inspection. *Ultrasonics* **34**, 621–627.

Schmidt, D. W. (1975). Acoustical method for fast detection and measurement of vortices in wind tunnels. *ICIASF '75 Record*, 216–228.

Schmidt, F., Sczesny, O., Ruppel, C., and Mágori, V. (1996). Wireless interrogator system for SAW identification marks and SAW sensor components. *Proc. IEEE Freq. Contr. Symp.*, 208–215.

Schoenwald, J. S., Martin, J. A., and Ahlberg, L. A. (1982). Acoustic scanning for robotic range sensing and object pattern recognition. *Proc. IEEE '82 Ultras. Symp.*, 945–949.

Sheen, S.-H., Chien, H.-T., and Raptis, A. C. (1996). Measurement of shear impedances of viscoelastic fluids. *Proc. 1996 IEEE Ultras. Symp.*, 453–457.

Shen, J. (May 26, 1992). "Measurement of Fluid Properties of Two-Phase Fluids Using an Ultrasonic Meter." U.S. Patent No. 5,115,670.

Smalling, J. W., Braswell, L. D., and Lynnworth, L. C. (June 24, 1986). "Apparatus and Methods for Measuring Fluid Flow Parameters." U.S. Patent No. 4,596,133.

Smalling, J. W., Braswell, L. D., and Lynnworth, L. C. (July 5, 1988). Apparatus and Methods for Measuring Fluid Flow Parameters." U.S. Patent No. 4,754,650.

Smalling, J. W., Braswell, L. D., Lynnworth, L. C., and Wallace, D. R. (1984). Flare gas ultrasonic flow meter. *Proc. 39th ISA Ann. Symp. on Instr. for the Process Industries*, 27–38.

Smith, T. L., Koman, L. A., and Mosberg, A. T. (1994). Cardiovascular physiology and methods for toxicology. *In* "Principles and Methods of Toxicology." 3rd Edition (A. W. Hayes, ed.), Raven Press, New York, pp. 917–935.

Smith Jr., R. H. (1994). The direct measurement of circulation in free surface vorticies. MS Thesis, Worcester Polytechnic Institute, Worcester, Massachusetts.

Smith Jr., R., Durgin, W., and Johari, H. (1995). The direct measurement of circulation in free surface vorticies. *Proc. AIAA Conf.*, Paper 95-0104.

Stagg, J. P., Christer, J., Thrush, E. J., and Crawley, J. (1992). Measurement and control of reagent concentrations in MOCVD reactor using ultrasonics. *J. Crystal Growth* **120**, 98–102.

Stein, P. J., Fitzgerald, M. G., and Euerle, S. E. (1996). "Non-invasive sensors for shipboard sewage systems." SSI (Scientific Solutions, Inc.) Final Report, Navy SBIR Contract N00167-95-C-0014.

Sylvia, D., Liu, Y., Flannery, P. N., and Lynnworth, L. C. (1994). External sidewall sensor for noninvasive measurement of high and high-high liquid levels in storage tanks, including external verification. *Proc. EPA AST (Aboveground Storage Tank) '94 Conf.*, 595–622. NTIS/U.S. Dept. of Commerce Report PB95-251591.

Takeda, Y. (1995). Velocity profile measurement by ultrasonic Doppler method. *Exper. Thermal and Fluid Sci.* **10**, 444–453; sequel work (1996) with collaborators summarized as extended abstracts in *Abstr. of the First Int. Symp. Ultras. Doppler Methods for Fluid Mech. and Fluid Eng.*, Paul Scherrer Inst., Villigen Switzerland.

Tallon, S., and Davies, C. E. (1996). Velocity and attenuation of sound: application to flow rate measurement in pneumatic conveying. *Power Hand. & Proc.* **8**(4), 329–335.

Talmadge, R., and Appley, K. E. (1991). Magnetostrictive ultrasonic transducer measures displacement and velocity. *Sensors* **8**(6), 18–19.

Tanter, M., Thomas, J.-L., and Fink, M. (1997). Comparison between time reversal focusing in absorbing medium and inverse filtering. *Proc. 1997 IEEE Ultras. Symp.*, 1741–1745.

Tasman, H. A. (1979). Nuclear applications of ultrasonic thermometry. *Proc. 1979 IEEE Ultras. Symp.*, 380–383.

Tasman, H. A., Campana, M., Pel, D., and Richter, J. (1982). Ultrasonic thin-wire thermometry for nuclear applications. *In* "Temperature—Its Measurement and Control in Science and Industry." AIP, New York, pp. 1191–1196.

Tomioka, G. (July 30, 1968). "Apparatus for Measuring Liquid Level." U.S. Patent No. 3,394,589.

Tornberg, J. (1986). "Correlation Methods for Ultrasonic Flow Measurements in Paper Pulp Suspensions." Acta Universitatis Ouluensis, Series C, Technica No. 35, Electronica No. 8.

Tornberg, J. S., Karras, E., Härkönen, M., and Hirsimäki, O. (1983). Analysis of ultrasonic correlation flowmeters for pulp suspension. *IEEE Trans. Son. & Ultrason.* **SU-30**(4), 264–269.

Trahey, G. E., Allison, J. W., Hubbard, S. M., and von Ramm, O. T. (1987). Measurement of local speckle pattern displacement to track blood flow in two dimensions. *Proc. 1987 Ultras. Symp.*, 957–961.

Transonic Systems Inc. (1996). "Clinical & Research Flowmeters Catalog." Ithaca, New York.

Tressler, J. F., and Newnham, R. E. (1997). Doubly resonant cymbal-type transducers. *IEEE Trans. UFFC* **44**(5), 1175–1177.

Urata, Y., Freeke, J., Matson, J. E., and Lynnworth, L. C. (1998). Steam flowmeters, Panametrics Rpt. UR-245, *Proc. 1998 IEEE Ultras. Symp.*

Valdes, J. L., and Cadet, G. (1991). Ultrasonic time-of-flight method for on-line quantitation of in situ generated arsine. Analy. Chem. **63**(4), 366–369.

Van Deventer, J., and Delsing, J. (1997). An ultrasonic density probe. *Proc. 1997 IEEE Ultras. Symp.*, 871–875.

Van Valkenburg, H. E. (1973). Air-coupled ultrasound as a production inspection technique for aircraft tires. *Proc. 1973 Symp. Nondestr. Testing of Tires*, 55–60. NTIAC 74-1 (P. E. J. Vogel, ed.) AMMRC, Watertown, Massachusetts.

Van Valkenburg, H. E., and Sansom, R. E. (April 28, 1959). "Liquid Level Indicator." U.S. Patent 2,883,861.

Venema, A., Nieuwkoop, E., Vellekoop, M. J., Ghijsen, W. J., Barendsz, A. W., and Nieuwenhuizen, M. S. (1987). NO_2 gas-concentration measurement with a SAW chemosensor. *IEEE Trans UFFC* **34**(2), 148–155.

Viktorov, I. A. (1967). "Rayleigh and Lamb Waves." Plenum Press, New York.

Vontz, T., and Mágori, V. (1996). An ultrasonic flowmeter for industrial applications using a helical sound path. *Proc. 1996 IEEE Ultra. Symp.*, 1047–1050.

Vossiek, M., Mágori, V., and Ermert, H. (1994). An ultrasonic multielement sensor system for position invariant object identification. *Proc. 1994 IEEE Ultras. Symp.* **2**, 1293–1247.

Wenzel, S. W., Costello, B. J., and White, R. M. (1994). Flexural-plate waves for sensors and micropumps. *Sensors* **11**(12), 47–49.

Werlink, R., Kelley, A., and Margasahayam, R. (1996). "Space shuttle hypergol load determination using nonintrusive ultrasonic flowmeters." NASA TM-111884.

Wilhjelm, J. E., and Pedersen, P. C. (1993). Target velocity estimation with FM and PW echo ranging Doppler systems—Part I: signal analysis. *IEEE Trans. Ultrason., Ferroelec., Freq. Contr.* **40**, 366–372. Part II: system analysis, ibid., 373–380. See also Wilhjelm, J. E. (1991).

Wilhjelm, J. E. (1991). Ultrasound FM Doppler systems for flow profiling with comparison to PW Doppler systems. Ph.D. Thesis, Worcester Polytechnic Institute, Worcester, Massachusetts.

Wolff, U., Schmidt, F., Scholl, G., and Mágori, V. (1996). Radio accessible SAW sensors for non-contact measurement of torque and temperature. *Proc. 1996 IEEE Ultras. Symp.*, 359–362.

Yeh, T. T., and Mattingly, G. E. (June 22–26, 1997). Computer simulations of ultrasonic flow meter performance in ideal and non-ideal pipeflows. *1997 ASME Fluids Eng. Div. Summer Meeting*, FEDSM97-3012.

Yon, M. C. (1991). Densitometers in oil and gas flow metering applications. *Sensors* **8**(6), 28–33.

Zacharias Jr., E. M. (1970). The sonic interface detector meets field tests in pipelining. *Oil & Gas J.* **68**(27), 96–103.

Zacharias Jr., E. M. (1972). Sonic detectors see gasoline interfaces. *Oil & Gas J.* **70**(34), 79–81.

Zacharias Jr., E. M., and Ord, R. (November 1981). *Oil & Gas J.*, 80–89.

Zacharias Jr., E. M., Zacharias, W., and Parnell Jr., R. A. (1974). Measuring sound velocity and bulk modulus of plastics. *Mod. Plast.* **51**(5), 88–90.

Zenzinger, A., Mathiae, S., Ruser, P., Reindl, L., and Weigel, R. (1997). MSK direct sequence spread spectrum system incorporating SAW tapped delay line and SAW shaping filter. *Proc. 1997 IEEE Ultras. Symp.*, 313–316.

Zipser, L. (1997). Acoustic gas sensor for extreme process conditions. *Proc. 1997 IEEE Ultras. Symp.*, 445–448.

Index

B

C

D

E

J

K

L

M

Q

R

U

V

W

X

Z

ISBN 0-12-477923-9